Lecture Notes in Computer Science 14436

Founding Editors

Gerhard Goos
Juris Hartmanis

Editorial Board Members

Elisa Bertino, *Purdue University, West Lafayette, IN, USA*
Wen Gao, *Peking University, Beijing, China*
Bernhard Steffen, *TU Dortmund University, Dortmund, Germany*
Moti Yung, *Columbia University, New York, NY, USA*

The series Lecture Notes in Computer Science (LNCS), including its subseries Lecture Notes in Artificial Intelligence (LNAI) and Lecture Notes in Bioinformatics (LNBI), has established itself as a medium for the publication of new developments in computer science and information technology research, teaching, and education.

LNCS enjoys close cooperation with the computer science R & D community, the series counts many renowned academics among its volume editors and paper authors, and collaborates with prestigious societies. Its mission is to serve this international community by providing an invaluable service, mainly focused on the publication of conference and workshop proceedings and postproceedings. LNCS commenced publication in 1973.

Qingshan Liu · Hanzi Wang · Zhanyu Ma · Weishi Zheng · Hongbin Zha · Xilin Chen · Liang Wang · Rongrong Ji
Editors

Pattern Recognition and Computer Vision

6th Chinese Conference, PRCV 2023 Xiamen, China, October 13–15, 2023 Proceedings, Part XII

Editors
Qingshan Liu
Nanjing University of Information Science and Technology
Nanjing, China

Zhanyu Ma
Beijing University of Posts and Telecommunications
Beijing, China

Hongbin Zha
Peking University
Beijing, China

Liang Wang
Chinese Academy of Sciences
Beijing, China

Hanzi Wang
Xiamen University
Xiamen, China

Weishi Zheng
Sun Yat-sen University
Guangzhou, China

Xilin Chen
Chinese Academy of Sciences
Beijing, China

Rongrong Ji
Xiamen University
Xiamen, China

ISSN 0302-9743 ISSN 1611-3349 (electronic)
Lecture Notes in Computer Science
ISBN 978-981-99-8554-8 ISBN 978-981-99-8555-5 (eBook)
https://doi.org/10.1007/978-981-99-8555-5

© The Editor(s) (if applicable) and The Author(s), under exclusive license
to Springer Nature Singapore Pte Ltd. 2024

This work is subject to copyright. All rights are reserved by the Publisher, whether the whole or part of the material is concerned, specifically the rights of translation, reprinting, reuse of illustrations, recitation, broadcasting, reproduction on microfilms or in any other physical way, and transmission or information storage and retrieval, electronic adaptation, computer software, or by similar or dissimilar methodology now known or hereafter developed.
The use of general descriptive names, registered names, trademarks, service marks, etc. in this publication does not imply, even in the absence of a specific statement, that such names are exempt from the relevant protective laws and regulations and therefore free for general use.
The publisher, the authors, and the editors are safe to assume that the advice and information in this book are believed to be true and accurate at the date of publication. Neither the publisher nor the authors or the editors give a warranty, expressed or implied, with respect to the material contained herein or for any errors or omissions that may have been made. The publisher remains neutral with regard to jurisdictional claims in published maps and institutional affiliations.

This Springer imprint is published by the registered company Springer Nature Singapore Pte Ltd.
The registered company address is: 152 Beach Road, #21-01/04 Gateway East, Singapore 189721, Singapore

Paper in this product is recyclable.

Preface

Welcome to the proceedings of the Sixth Chinese Conference on Pattern Recognition and Computer Vision (PRCV 2023), held in Xiamen, China.

PRCV is formed from the combination of two distinguished conferences: CCPR (Chinese Conference on Pattern Recognition) and CCCV (Chinese Conference on Computer Vision). Both have consistently been the top-tier conference in the fields of pattern recognition and computer vision within China's academic field. Recognizing the intertwined nature of these disciplines and their overlapping communities, the union into PRCV aims to reinforce the prominence of the Chinese academic sector in these foundational areas of artificial intelligence and enhance academic exchanges. Accordingly, PRCV is jointly sponsored by China's leading academic institutions: the Chinese Association for Artificial Intelligence (CAAI), the China Computer Federation (CCF), the Chinese Association of Automation (CAA), and the China Society of Image and Graphics (CSIG).

PRCV's mission is to serve as a comprehensive platform for dialogues among researchers from both academia and industry. While its primary focus is to encourage academic exchange, it also places emphasis on fostering ties between academia and industry. With the objective of keeping abreast of leading academic innovations and showcasing the most recent research breakthroughs, pioneering thoughts, and advanced techniques in pattern recognition and computer vision, esteemed international and domestic experts have been invited to present keynote speeches, introducing the most recent developments in these fields.

PRCV 2023 was hosted by Xiamen University. From our call for papers, we received 1420 full submissions. Each paper underwent rigorous reviews by at least three experts, either from our dedicated Program Committee or from other qualified researchers in the field. After thorough evaluations, 522 papers were selected for the conference, comprising 32 oral presentations and 490 posters, giving an acceptance rate of 37.46%. The proceedings of PRCV 2023 are proudly published by Springer.

Our heartfelt gratitude goes out to our keynote speakers: Zongben Xu from Xi'an Jiaotong University, Yanning Zhang of Northwestern Polytechnical University, Shutao Li of Hunan University, Shi-Min Hu of Tsinghua University, and Tiejun Huang from Peking University.

We give sincere appreciation to all the authors of submitted papers, the members of the Program Committee, the reviewers, and the Organizing Committee. Their combined efforts have been instrumental in the success of this conference. A special acknowledgment goes to our sponsors and the organizers of various special forums; their support made the conference a success. We also express our thanks to Springer for taking on the publication and to the staff of Springer Asia for their meticulous coordination efforts.

We hope these proceedings will be both enlightening and enjoyable for all readers.

October 2023

Qingshan Liu
Hanzi Wang
Zhanyu Ma
Weishi Zheng
Hongbin Zha
Xilin Chen
Liang Wang
Rongrong Ji

Organization

General Chairs

Hongbin Zha	Peking University, China
Xilin Chen	Institute of Computing Technology, Chinese Academy of Sciences, China
Liang Wang	Institute of Automation, Chinese Academy of Sciences, China
Rongrong Ji	Xiamen University, China

Program Chairs

Qingshan Liu	Nanjing University of Information Science and Technology, China
Hanzi Wang	Xiamen University, China
Zhanyu Ma	Beijing University of Posts and Telecommunications, China
Weishi Zheng	Sun Yat-sen University, China

Organizing Committee Chairs

Mingming Cheng	Nankai University, China
Cheng Wang	Xiamen University, China
Yue Gao	Tsinghua University, China
Mingliang Xu	Zhengzhou University, China
Liujuan Cao	Xiamen University, China

Publicity Chairs

Yanyun Qu	Xiamen University, China
Wei Jia	Hefei University of Technology, China

Local Arrangement Chairs

Xiaoshuai Sun	Xiamen University, China
Yan Yan	Xiamen University, China
Longbiao Chen	Xiamen University, China

International Liaison Chairs

Jingyi Yu	ShanghaiTech University, China
Jiwen Lu	Tsinghua University, China

Tutorial Chairs

Xi Li	Zhejiang University, China
Wangmeng Zuo	Harbin Institute of Technology, China
Jie Chen	Peking University, China

Thematic Forum Chairs

Xiaopeng Hong	Harbin Institute of Technology, China
Zhaoxiang Zhang	Institute of Automation, Chinese Academy of Sciences, China
Xinghao Ding	Xiamen University, China

Doctoral Forum Chairs

Shengping Zhang	Harbin Institute of Technology, China
Zhou Zhao	Zhejiang University, China

Publication Chair

Chenglu Wen	Xiamen University, China

Sponsorship Chair

Yiyi Zhou	Xiamen University, China

Exhibition Chairs

Bineng Zhong	Guangxi Normal University, China
Rushi Lan	Guilin University of Electronic Technology, China
Zhiming Luo	Xiamen University, China

Program Committee

Baiying Lei	Shenzhen University, China
Changxin Gao	Huazhong University of Science and Technology, China
Chen Gong	Nanjing University of Science and Technology, China
Chuanxian Ren	Sun Yat-Sen University, China
Dong Liu	University of Science and Technology of China, China
Dong Wang	Dalian University of Technology, China
Haimiao Hu	Beihang University, China
Hang Su	Tsinghua University, China
Hui Yuan	School of Control Science and Engineering, Shandong University, China
Jie Qin	Nanjing University of Aeronautics and Astronautics, China
Jufeng Yang	Nankai University, China
Lifang Wu	Beijing University of Technology, China
Linlin Shen	Shenzhen University, China
Nannan Wang	Xidian University, China
Qianqian Xu	Key Laboratory of Intelligent Information Processing, Institute of Computing Technology, Chinese Academy of Sciences, China
Quan Zhou	Nanjing University of Posts and Telecommunications, China
Si Liu	Beihang University, China
Xi Li	Zhejiang University, China
Xiaojun Wu	Jiangnan University, China
Zhenyu He	Harbin Institute of Technology (Shenzhen), China
Zhonghong Ou	Beijing University of Posts and Telecommunications, China

Contents – Part XII

Object Detection, Tracking and Identification

OKGR: Occluded Keypoint Generation and Refinement for 3D Object Detection

Mingqian Ji, Jian Yang, and Shanshan Zhang(✉)

PCA Lab, Key Lab of Intelligent Perception and Systems for High-Dimensional Information of Ministry of Education, and Jiangsu Key Lab of Image and Video Understanding for Social Security, School of Computer Science and Engineering, Nanjing University of Science and Technology, Nanjing, China
{mingqianji,csjyang,shanshan.zhang}@njust.edu.cn

Abstract. Lidar-based 3D object detectors utilize point clouds to detect objects in autonomous driving. However, the point clouds are sparse and incomplete, which affects the detectors' learning of shape knowledge and limits the 3D detection performance. Previous works improve performance through completing object shape at the point level or representation level, such as voxel. The former increases computational burden, while the latter has poor generalization ability to point-based detectors. In this paper, we present an approach, namely Occluded Keypoint Generation and Refinement (OKGR), which is effective to improve 3D detection performance by completing object features at the keypoint level. Specifically, Occluded Keypoint Generation (OKG) generates occluded keypoints to densify raw keypoints and learns the offsets between the generated keypoints and prototypes, while retaining the raw keypoints unchanged. Occluded Keypoint Refinement (OKR) assigns weights to the generated keypoints and conducts these weights to features to obtain high-quality complete features for detection. We apply our approach to two representative detectors, PV-RCNN++ and PDV, and evaluate the detectors on KITTI and Waymo Open Dataset. The experiments show significant performance improvement. Particularly, our OKGR applied on PV-RCNN++ achieves improvements of Pedestrian and Cyclist of +3.19%, +2.53% AP on average difficulty levels on KITTI, and +2.18%, +2.29% mAPH on Waymo Open Dataset. For more information, the supplementary material and code are available at https://github.com/Mingqj/OKGR.

Keywords: 3D Object Detection · Point Clouds · Object Shape Completion

1 Introduction

Lidar-based 3D object detection plays an important role in autonomous driving systems. A large number of research have been conducted focusing on improving 3D detection performance. In early years, some Lidar-based 3D object detectors

© The Author(s), under exclusive license to Springer Nature Singapore Pte Ltd. 2024
Q. Liu et al. (Eds.): PRCV 2023, LNCS 14436, pp. 3–15, 2024.
https://doi.org/10.1007/978-981-99-8555-5_1

[1–7] are proposed mainly about designing various structures and extracting features using different representations such as voxels and keypoints, etc. These detectors show impressive speed and accuracy in real-world scenarios. Yet, these works disregard the effect of the properties of point clouds on detection performance. More recently, some works [8,9] analyze the influence of the sparsity and incompleteness of raw point clouds on detection performance, and address this issue from the perspective of completing raw point clouds or voxels. SIENet [8] uses a completion network (PCN) [10] to complete the raw point clouds within candidate boxes. However, the limitation of completing the entire raw point clouds is the increased computational burden in scenarios with numerous objects. Instead, SPG [9] designs a supervision scheme to complete object voxels. However, the limitation of completing the voxels is that it only fits voxel-based 3D object detectors and can not be applied to point-based 3D object detectors, because the virtual voxels can hardly be converted back to the real points. In summary, the above issues motivate us to design a completion method with less computational burden and it can be applied to both voxel-based and point-based detectors.

In this paper, we propose a completion method at the keypoint level. Different from completing the entire raw point clouds, completing keypoints as a small part of the raw point clouds is efficient. Also different from completing the voxels, keypoint completion can be directly used in point-based detectors and used in voxel-based detectors through voxelization. Specifically, our approach consists of two parts, namely Occluded Keypoint Generation (OKG) and Refinement (OKR). In OKG, we design a Keypoint Densification (KD) module to generate occluded keypoints and a Shape Learning (SL) module to learn shape knowledge. In this way, we can obtain the dense complete keypoints. In OKR, we design a Density-and-Distance-based Weight Assignment (DWA) module to assign a weight to each generated keypoint, and the weights will be conducted to the corresponding features. In this way, the features of high confidence will be enhanced, and these of low confidence will be weakened. Thus, we can get high-quality features to improve detection performance. We apply our approach to state-of-the-art detectors and achieve significant performance improvement. Figure 1 shows the boxes and features of baseline and OKGR. We can see that our approach can enhance the features of the occluded regions of the objects. Thus, it can improve the detection ability for objects with severe occlusion. Our contributions are summarized as follows:

1) We design an Occluded Keypoint Generation (OKG) method to generate occluded keypoints via the Keypoint Densification (KD) module and learn the shape knowledge via the Shape Learning (SL) module.
2) We present an Occluded Keypoint Refinement (OKR) method, and it combines the priors of density and distance to refine the generated occluded keypoints via the Density-and-Distance-based Weight Assignment (DWA) module.
3) Our approach is applied to two state-of-the-art detectors, PV-RCNN++ [11] and PDV [12], and is evaluated on two benchmark datasets, KITTI [13] and Waymo Open Dataset [14]. The results show the significant performance improvement.

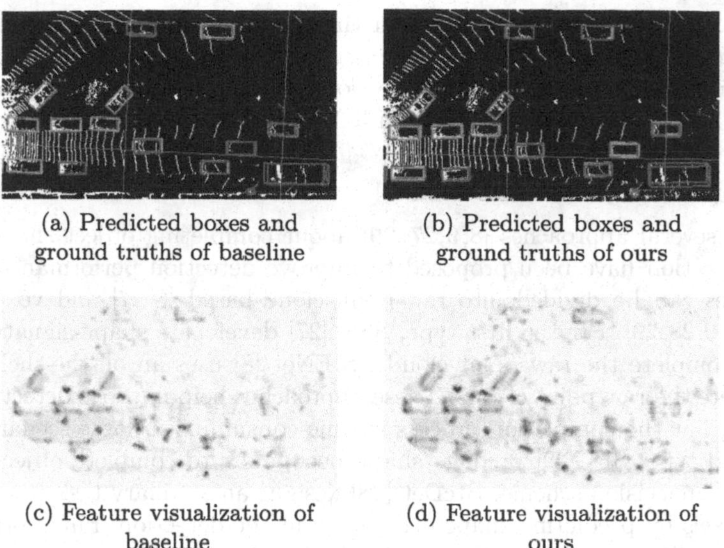

(a) Predicted boxes and ground truths of baseline

(b) Predicted boxes and ground truths of ours

(c) Feature visualization of baseline

(d) Feature visualization of ours

Fig. 1. The box and feature visualization of baseline and ours. (a) and (b) are the predicted boxes of baseline and ours. Blue boxes are the ground truths and green boxes are the predicted boxes. (c) and (d) are the feature visualization of the baseline and ours in BEV. The darker the color, the higher the feature quality. (Color figure online)

2 Related Works

Since we use LiDAR-based detectors as the base detectors and design completion modules on this basis, we review the related works on LiDAR-based 3D object detection and object shape completion.

2.1 LiDAR-Based 3D Object Detection

Currently, research on LiDAR-based 3D object detection has been comprehensively developed. The 3D object detectors are generally divided into different categories according to the data representation [15], including point-based, grid-based, and point-voxel-based methods. Point-based methods [1–4,16] detect 3D objects from raw point clouds by using point-based backbones. For instance, PointRCNN [1] uses PointNet++ [17] to extract points features and designs a two-stage network to gradually get bounding boxes. Grid-based methods [5–7,18–21] divide point clouds into grids by means of voxels or pillars to extract features. Typically, SECOND [6] uses 3D sparse convolution [22] to extract voxel features and designs a direction classifier to assist in predicting bounding boxes. Point-voxel-based methods [11,12,23–25] combine the features of points and voxels using a set abstraction module for detection. For instance, PV-RCNN [23] first uses 3D sparse convolution [22] and farthest point sampling [26] to extract the features of voxels and points, respectively. Then, it designs a set abstraction and a RoI feature abstraction to finish the processes of voxel-to-keypoint and keypoint-to-grid. These detectors achieve good performance, but they overlook the sparsity and incompleteness of the point clouds. Instead, we consider this issue and make improvements based on 3D object detectors.

2.2 Object Shape Completion

Recently, several approaches [8,9,27–29] about completing object shape in 3D object detection have been proposed to improve detection performance. These approaches can be divided into raw-point-cloud-based [8,27] and voxel-based methods [9,28,29]. For the first type, SSN [27] develops a shape signature network to complete the raw point clouds. SIENet [8] uses an off-the-shelf model to complete the raw point clouds. These approaches help improve detection performance, but the completion process is time-consuming in large scenarios. For the second type, SPG [9] predicts shape occupancy to complete object voxels under its supervision scheme. BtcDet [28] designs an auxiliary task to complete object voxels by predicting shape occupancy in the detection framework. Similarly, Sparse2Dense [29] completes object voxels in the constructed distillation framework. These works complete the object voxels successfully by predicting the shape occupancy, but this method can hardly transfer to point-based detection. Based on these previous works, we present an effective approach to complete the object shape at the keypoint level, which completes the object keypoints with little computational cost. Furthermore, our approach can be inserted on both voxel-based and point-based detectors. Substantial experiments demonstrate the effectiveness of our approach.

3 Methodology

Based on the above motivation and related work, we propose to complete object shape at the keypoint level via Occluded Keypoint Generation and Refinement (OKGR) to improve detection performance.

3.1 Overview

As illustrated in Fig. 2, our approach based on a typical two-stage detection framework mainly consists of Occluded Keypoint Generation (OKG) and Refinement (OKR). The workflow of pipeline is summarized as follows:

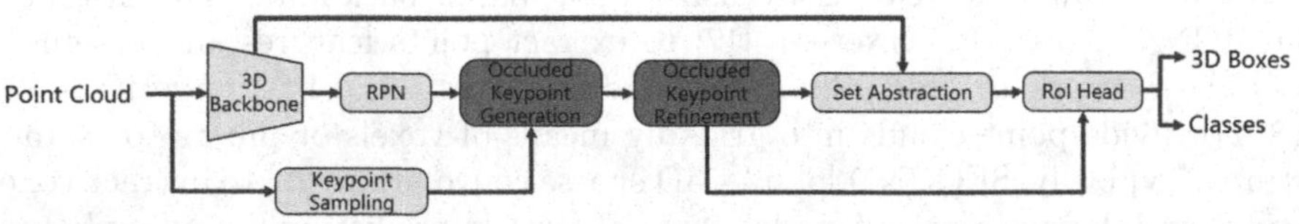

Fig. 2. The overview of our method. Our approach mainly consists of two modules, namely Occluded Keypoint Generation (OKG) and Occluded Keypoint Refinement (OKR).

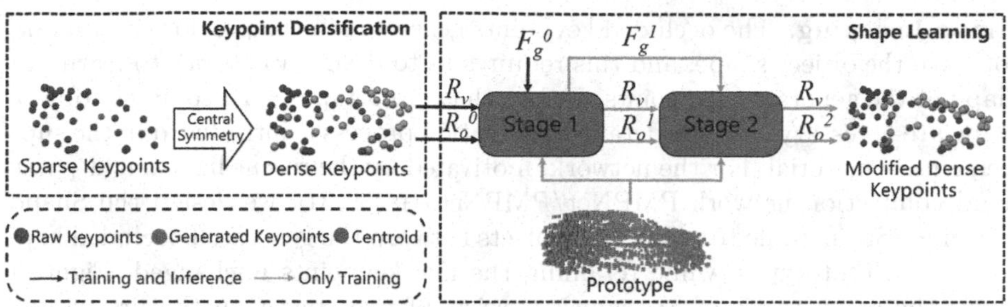

Fig. 3. Occluded Keypoint Generation. It consists of two modules, namely Keypoint Densification (KD) and Shape Learning (SL) module.

First of all, the detector utilizes a 3D backbone to extract features from raw point clouds, while the raw keypoints are generated in the Keypoint Sampling module. Then, the Region Proposal Network (RPN) generates proposals based on these features. Next, the raw keypoints and proposals are sent to OKG to generate the occluded keypoints, and then the generated keypoints are sent to OKR for refinement. After that, the whole keypoints aggregate the features from 3D Backbone to enhance the features in Set Abstraction. Finally, the detector predicts the 3D boxes and classes in RoI Head based on the enhanced features. In detail, OKG is described in Sect. 3.2 and OKR is described in Sect. 3.3.

3.2 Occluded Keypoint Generation

As shown in Fig. 3, the Occluded Keypoint Generation consists of the Keypoint Densification (KD) module and the Shape Learning (SL) module. Specifically, in KD, we first densify keypoints with the aim of filling the keypoints in the occluded regions with less computational complexity. Then, in SL, a recurrent neural network based on predicting offsets is designed to learn the shape of the generated keypoints while retaining the raw keypoints unchanged.

Keypoint Densification. Inspired by using a mirror to get complete object shapes offline as a supervision in BtcDet [28] and Sparse2Dense [29], we also use a method of symmetry to generate occluded keypoints. Unlike them, we use the central symmetry to densify keypoints during detection, because it is difficult to predict the object direction accurately in RPN, while the centers are easy to predict. In detail, as shown in Fig. 3, the occluded keypoints R_o can be generated by central symmetry, and they are defined as:

$$R_o = L(R_v, C_a), \tag{1}$$

where $L(\cdot)$ is a linear function used to map coordinates through the ball query. We define the keypoints as R_v and the centers of objects as C_a. In summary, we can obtain dense keypoints via the KD module.

Shape Learning. The occluded keypoints generated in the previous section do not have the object shape, and this requires us to design a network to learn the shape of the generated keypoints. Before that, prototypes with complete object shape are necessary, and the detailed generation process is introduced in the supplementary material. For the network, motivated by the recent lightweight point cloud completion network PMPNet/PMPNet++ [30,31], we design the Shape Learning (SL) module to predict the offsets between the generated occluded keypoints and prototypes, while retaining the raw keypoints unchanged. Figure 3 shows the architecture of SL, which consists of two stages. In the first stage, the raw keypoints R_v and generated keypoints R_o are separately fed with path searching information F_g^0 [30] obeying a random normal distribution. The outputs are the modified generated keypoints R_o^1, retained raw keypoints R_v and previous path searching information F_g^1. Similarly, they are fed as inputs to the second stage for further modification. Finally, we can get the modified dense keypoints $\{R_v, R_o^2\}$ with complete object shape. For the specific network of each stage, as shown in Fig. 4, the SA modules are used to extract features and coordinates of raw keypoints and generated keypoints. The FP modules are used to propagate the global features from SA3 and local features of generated keypoints $\{F_{o1}, F_{o2}\}$ to the local coordinates $\{P_{o0}, P_{o1}, P_{o2}\}$ of generated keypoints (marked in red in Fig. 4). The RPA modules can manipulate the path searching information $\{F_{g0}^s, F_{g1}^s, F_{g1}^s\}$ of the generated keypoints. It should be noted that, to balance sufficient training and fast inference, we train both two stages but infer only the first stage. The experimental demonstration is in ablation studies in Sect. 4.6.

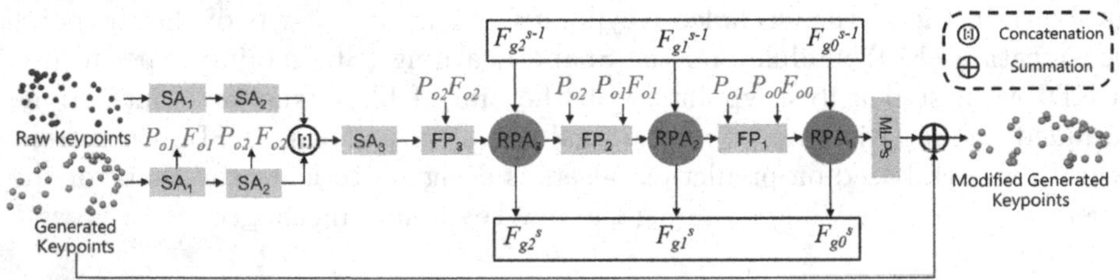

Fig. 4. One stage of the Shape Learning module. It consists of the Set Abstraction (SA) module [17], the Features Propagation (FP) module [17] and the Recurrent Path Aggregation (PRA) module [30,31].

3.3 Occluded Keypoint Refinement

Generating new keypoints brings the noise inevitably. Thus, it is necessary to evaluate the generated keypoints. We design the Density-and-Distance-based Weight Assignment (DWA) module to assign the confident weights to generated keypoints. This module consists of density calculation, distance calculation and weight calculation.

Density Calculation. The density information has been successfully used in 3D object detection, like PDV [12]. As prior knowledge, it can describe the distribution of point clouds. KDE [32] is a typical method to estimate the probability distribution through a kernel function. In the Density-and-Distance-based Weight Assignment (DWA) module, we first describe the density p_w of the generated keypoints by KDE, and it is defined as:

$$p_w = \frac{1}{N'w} \sum_{i=1}^{N'} K(\frac{r'_o - r'_{o_i}}{w}), \tag{2}$$

where $K(\cdot)$ is the kernel function [33]; w is the bandwidth and it is set to 0.2; N' is the number of r'_o; i is for counting from 1 to N'. For fine filtering, we use the cylinders to cover the average proposals for effective computation instead of the ball. Specifically, we choose the vertical cylinders to address the category of pedestrian and cyclist, and the horizontal cylinders to address the category of car. Thus, the filtered generated keypoints R'_o is defined as:

$$R'_o = \left\{ r'_{o_i} \middle| ||r'_{o_i} - l_i|| < d_m \right\}, \tag{3}$$

$$d_m = \begin{cases} \sqrt{dx_m^2 + dy_m^2}/2, dz_m > d_{max} \\ \sqrt{dz_m^2 + {d_{min}}^2}/2, dz_m \leq d_{max} \end{cases}, \tag{4}$$

where $||\cdot||$ is the 3D distance; we define one element of R_o as r_o, and r'_o is r_o under the condition in Eq. (3); l_i will be described in Eq. (5); d_m is the diameter of the cylinder; (dx_m, dy_m, dz_m) are the length, width and height of the mth proposal; d_{min} and d_{max} are the smaller and the larger between dx_m and dy_m, respectively. Therefore, we can calculate the density value of each generated keypoint.

Distance Calculation. As illustrated in Fig. 5, there is a phenomenon that the density of keypoints inside the object is similar to those of the keypoints outside the object near the proposal box. Therefore, we introduce the distance to address this issue. The distance l between the keypoints and the axis of the cylinder is defined as:

$$l = \begin{cases} E(r'_o, C_a), & dz_m > d_{max} \\ E(r'_o, C_a) \cdot |sin(\theta - \varphi)|, dz_m \leq d_{max} \end{cases}, \tag{5}$$

where $E(\cdot)$ is the Euclidean distance of the generated keypoints r'_o and center C_a in bird's eye view; θ is tilt angle of the average proposals; φ is the angle between the line of connecting the keypoints and the center, and the Y-axis. Then, the distance can be calculated by constructing the above geometric relation.

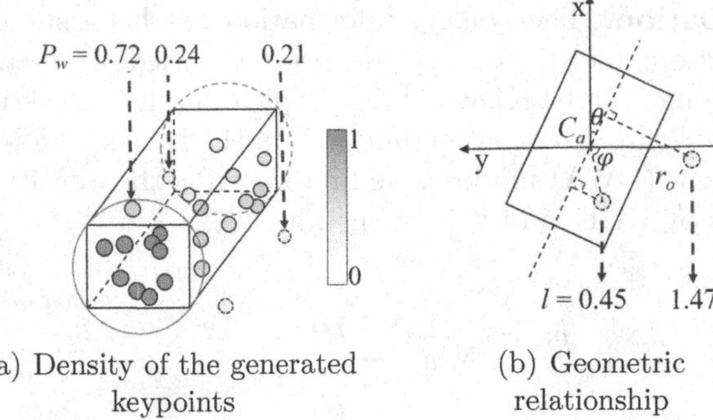

(a) Density of the generated keypoints

(b) Geometric relationship

Fig. 5. Density distribution and geometric relation. (a) The density distribution of generated keypoints. The darker the color, the greater the density value. (b) The geometry between distance and angle. The distances of different points with similar density are different.

Weight Calculation. We introduce the above density p_w and distance l to construct a linear weight assignment. It is defined as:

$$W = S_p \cdot (\mu \cdot p_w + (1 - \mu) \cdot l^{-1}), \tag{6}$$

where S_p is the scores of proposals; μ is the weighting factor to balance the density and distance and it is set to 0.7. We set a threshold T_{thr} to filter the generated keypoints with extremely low confidence, and the value is set to 0.2.

3.4 Loss Function

The detector with OKGR is trained with the loss of region proposal network L_{rpn}, shape learning L_{sl} and R-CNN L_{rcnn}. The whole loss of detector is the summation of L_{rpn}, L_{sl} and L_{rcnn}, and each loss have the equal weight. It is defined as:

$$\mathcal{L} = L_{rpn} + L_{sl} + L_{rcnn}, \tag{7}$$

where L_{rpn} and L_{rcnn} are the same as baseline. L_{sl} is defined as:

$$L_{sl} = L_{cd} + \delta L_{emd}, \tag{8}$$

where L_{cd} is the loss of Chamfer distance [30,31] for minimizing the moving path; L_{emd} is the loss of Earth Mover's distance [30,31] for minimizing the distance between generated keypoints and prototypes, and the weighting factor δ is set to 0.001.

4 Experiments

In this section, we will first introduce the datasets and evaluation metrics. Then, we will introduce our implementation details of experiments. After that, we will show the results of experiments on the above datasets. In the end, we will conduct the ablation studies and analyze each component of our methods.

4.1 Datasets and Evaluation Metrics

We evaluate our methods on KITTI [13] and Waymo Open Dataset [14]. The evaluation metric for KITTI is average precision (AP), and the evaluation metric for Waymo Open Dataset is mean average precision (mAP) and 3D mAP weighted by Heading (mAPH). The detailed introduction and division of both datasets are in the supplementary material.

4.2 Implementation Details

For KITTI, we apply an ADAM optimizer with an initial learning rate of 0.01 to train our network on 2 NVIDIA 2080Ti GPUs for 80 epochs with batch size 4, and the training time takes around 13 h on PV-RCNN++ with OKGR and 14 h on PDV with OKGR. For Waymo Open Dataset, we apply an ADAM optimizer with an initial learning rate of 0.01 to train our network on 4 NVIDIA 2080Ti GPUs for 30 epochs with batch size 8, and the training time takes around 34 h on PV-RCNN++ with OKGR and 36 h on PDV with OKGR. The learning rate strategy and proposal refinement strategy are the same as PV-RCNN [23]. During training, data augmentation strategies of 3D object detection are utilized as the same as the baseline [11,12].

Table 1. Comparisons on KITTI validation set.

Methods	Car 3D AP R40			Ped. 3D AP R40			Cyc. 3D AP R40			Average		
	Easy	Mod.	Hard	Easy	Mod.	Hard	Easy	Mod.	Hard	Easy	Mod.	Hard
PV-RCNN++	91.83	84.32	82.25	63.46	56.73	52.27	91.80	73.22	68.62	82.36	71.42	67.71
with ours	**92.33**	**84.75**	**82.66**	**67.45**	**59.67**	**54.91**	**94.89**	**75.51**	**70.96**	**85.89**	**73.31**	**69.51**
Improvement	+0.50	0.43	+0.41	+3.99	+2.94	+2.64	+3.09	+2.29	+2.34	+3.53	+1.89	+1.80
PDV	91.84	84.86	82.70	64.66	57.76	52.62	92.90	75.79	71.17	83.13	72.80	68.83
with ours	**92.26**	**85.24**	**82.96**	**66.92**	**58.76**	**54.11**	**94.37**	**75.82**	**72.00**	**84.52**	**73.27**	**69.69**
Improvement	+0.42	+0.38	+0.26	+2.26	+1.00	+1.49	+1.47	+0.03	+0.83	+1.39	+0.47	+0.86

4.3 Evaluation on KITTI Dataset

As shown in Table 1, it reports the comparison results on KITTI validation set. We re-implement PV-RCNN++ [11] and PDV [12] based on open-source codes and train our approach based on both detectors. The results show that our approach improves the performance of both baseline detectors on all categories and difficulty levels. Specifically, for PV-RCNN++, our approach improves 3D AP with 40 recall positions of average categories of +3.53%, +1.89%, and +1.80% AP on easy, moderate, and hard difficulty levels. For PDV, our approach improves 3D AP with 40 recall positions of average categories of +1.39%, +0.47%, and +0.86% AP on easy, moderate, and hard difficulty levels. Besides, for categories of pedestrian and cyclist, our approach applied on PV-RCNN++ achieves stable improvements of +3.19%, +2.53% AP on average difficulty levels and +1.58%, +0.78% AP based on PDV. The results show that our approach can improve the 3D detection performance of the baseline on KITTI.

4.4 Evaluation on Waymo Open Dataset

As shown in Table 2, the results show that our approach improves the performance of both baseline detectors on all categories and both LEVEL-1 and LEVEL-2 metrics. Specifically, compared to PV-RCNN++, our approach increases the performance of +1.44% LEVEL-1 mAP, +1.52% LEVEL-1 mAPH, +1.48% LEVEL-2 mAP and +1.59% LEVEL-2 mAPH on average categories, respectively. Compared to PDV, our approach also increases the performance of +0.93% LEVEL-1 mAP, +0.76% LEVEL-1 mAPH, +1.04% LEVEL-2 mAP and +0.94% LEVEL-2 mAPH on average categories, respectively. Particularly, for categories of pedestrian and cyclist, our approach applied on PV-RCNN++ achieves significant improvements of +2.18%, +2.29% mAPH on average difficulty levels. The results further demonstrate that our approach is effective.

4.5 Model Efficiency

Table 3 shows the efficiency of OKGR on KITTI validation set. The evaluation is based on a 2080Ti GPU with a batch size of 1. OKGR contains 1.7 million parameters and adds 4.1 million fewer parameters than PCN [10] used in SIENet [8]. Besides, we use OKGR to complete raw point clouds. OKGR costs about extra 110 milliseconds on baselines and adds about 80 milliseconds less than PCN. Compared to completing raw point clouds using PCN, OKGR completing representation only costs extra 28.7 milliseconds on PV-RCNN++ and 30.3 milliseconds on PDV. These indicate that OKGR is highly efficient for practical application.

Table 2. Comparisons on the validation set of Waymo Open Dataset.

Methods	Difficulty	Vehicle		Pedestrian		Cyclist		Average	
		mAP	mAPH	mAP	mAPH	mAP	mAPH	mAP	mAPH
PV-RCNN++ with	LEVEL-1	77.82	77.32	77.99	71.36	71.80	70.71	75.87	73.13
OKGR Improvement		**78.14**	**77.45**	**79.66**	**73.48**	**74.13**	**73.03**	**77.31**	**74.65**
		+0.32	+0.13	+1.67	+2.21	+2.33	+2.32	+1.44	+1.52
PV-RCNN++ with	LEVEL-2	69.07	68.62	69.92	63.74	69.31	68.26	68.84	66.33
OKGR Improvement		**69.46**	**69.02**	**71.70**	**65.88**	**71.57**	**70.51**	**70.32**	**67.92**
		+0.38	+0.40	+1.78	+2.14	+2.26	+2.25	+1.48	+1.59
PDV with OKGR	LEVEL-1	76.85	76.33	74.19	65.96	68.71	67.55	73.25	69.95
Improvement		**76.97**	**76.42**	**75.67**	**67.49**	**69.95**	**68.22**	**74.18**	**70.71**
		+0.12	+0.09	+1.48	+1.53	+1.24	+0.67	+0.93	+0.76
PDV with OKGR	LEVEL-2	69.30	68.81	65.85	58.28	66.49	65.36	67.21	64.15
Improvement		**69.55**	**69.30**	**67.49**	**59.95**	**67.72**	**66.03**	**68.25**	**65.09**
		+0.25	+0.49	+1.64	+1.67	+1.23	+0.67	+1.04	+0.94

Table 3. Inference time and model parameters on KITTI validation set.

Methods	Completing Point Clouds	Completing Representation	Inference Time (ms)	Parameters (M)
PV-RCNN++	–	–	273.4	13.0
with PCN [10]	✓	–	463.9(+190.5)	18.8(+5.8)
with OKGR	✓	–	385.2(+111.8)	14.7(+1.7)
with OKGR	–	✓	302.1(+28.7)	14.7(+1.7)
PDV	–	–	281.4	12.9
with PCN [10]	✓	–	479.2(+197.8)	18.7(+5.8)
with OKGR	✓	–	394.6(+113.2)	14.7(+1.8)
with OKGR	–	✓	311.7(+30.3)	14.7(+1.8)

4.6 Ablation Studies

In ablation studies, we conduct experiments based on PV-RCNN++ to evaluate the components of OKGR and verify the design of the SL module. The results are shown in Table 4 and Table 5, respectively. In Table 4, by adopting the KD module, the raw keypoints are densified by generating occluded keypoints without the shape knowledge. The KD module improves the 3D detection performance of +0.51% AP on moderate levels. The SL module is adjacent to the KD module and designed to learn shape knowledge, which improves the performance of +0.79% AP on moderate levels. Through the refinement via the DWA module, we can get high-quality features for detection, and this module can continue to improve the performance of +0.59% AP on moderate levels. In summary, our approach can boost the baseline performance of +1.89% AP on moderate levels, which demonstrates the effectiveness of our KD, SL and DWA modules. In Table 5, we conduct ablation studies on the SL module to verify the rationality of SL module design. We can see that training and inference on both the first stage and the second stage can get the best AP (value with bold) on the moderate level, but it requires much inference time. Instead, we train both the first stage and the second stage, and only evaluate the first stage, which gets competitive AP and requires less inference time (values with underline) than the former. Besides, we only train and evaluate the first stage disregarding the previous path searching information, which shows the bad performance. In summary, the design of the SL module is reasonable, and it can achieve the balance between detection accuracy and inference speed.

Table 4. Ablation studies of OK GR.

Methods	Car 3D AP Mod.	Ped. 3D AP Mod.	Cyc. 3D AP Mod.	Average
Baseline	84.32	56.73	73.22	71.42
+ KD	84.59	57.39	73.82	71.93
+ SL	84.71	58.91	74.54	72.72
+ DWA	**84.75**	**59.67**	**75.51**	**73.31**

Table 5. Ablation studies of SL.

Methods	Training		Inference		Average AP Mod.	Inference Time (ms)
	Stage 1	Stage 2	Stage 1	Stage 2		
Baseline	–	–	–	–	71.42	**273.4**
+ OKGR	✓	✓	✓	✓	**73.47**	347.7
+ OKGR	✓	✓	✓	–	<u>73.31</u>	<u>302.1</u>
+ OKGR	✓	–	✓	–	72.13	303.7

5 Conclusion

In this paper, we propose to complete object features for 3D object detection via Occluded Keypoint Generation and Refinement. Concretely, we design the Occluded Keypoint Generation to generate points in object occluded region and learn the shape knowledge at the keypoint level. Besides, Occluded Keypoint Refinement is proposed to assign weights to the generated keypoints, and the weights conduct to features to obtain high-quality features for detection. Extensive experiments have demonstrated the effectiveness of our approach.

Acknowledgement. This work is supported by the National Natural Science Foundation of China (Grant No. 62322602, Grant No. 62172225), CAAI-Huawei MindSpore Open Fund.

References

1. Shi, S., Wang, X., Li, H.: PointRCNN: 3D object proposal generation and detection from point cloud. In: CVPR (2019)
2. Yang, Z., Sun, Y., Liu, S., Shen, X., Jia, J.: IPOD: intensive point-based object detector for point cloud. arXiv:1812.05276 (2018)
3. Yang, Z., Sun, Y., Liu, S., Shen, X., Jia, J.: STD: sparse-to-dense 3D object detector for point cloud. In: ICCV (2019)
4. Ngiam, J., et al.: StarNet: targeted computation for object detection in point clouds. arXiv:1908.11069 (2019)
5. Zhou, Y., Tuzel, O.: VoxelNet: end-to-end learning for point cloud based 3D object detection. In: CVPR (2018)
6. Yan, Y., Mao, Y., Li, B.: Second: sparsely embedded convolutional detection. Sensors (2018)
7. Lang, A.H., Vora, S., Caesar, H., Zhou, L., Yang, J., Beijbom, O.: PointPillars: fast encoders for object detection from point clouds. In: CVPR (2019)
8. Li, Z., Yao, Y., Quan, Z., Xie, J., Yang, W.: Spatial information enhancement network for 3D object detection from point cloud. PR (2022)
9. Xu, Q., Zhou, Y., Wang, W., Qi, C.R., Anguelov, D.: SPG: unsupervised domain adaptation for 3D object detection via semantic point generation. In: ICCV (2021)
10. Yuan, W., Khot, T., Held, D., Mertz, C., Hebert, M.: PCN: point completion network. In: 3DV (2018)
11. Shi, S., et al.: PV-RCNN++: point-voxel feature set abstraction with local vector representation for 3D object detection. IJCV (2022)
12. Hu, J.S., Kuai, T., Waslander, S.L.: Point density-aware voxels for lidar 3D object detection. In: CVPR (2022)
13. Geiger, A., Lenz, P., Urtasun, R.: Are we ready for autonomous driving? The KITTI vision benchmark suite. In: CVPR (2012)
14. Sun, P., et al.: Scalability in perception for autonomous driving: waymo open dataset. In: CVPR (2020)
15. Mao, J., Shi, S., Wang, X., Li, H.: 3D object detection for autonomous driving: a review and new outlooks. arXiv:2206.09474 (2022)
16. Shi, W., Rajkumar, R.: Point-GNN: graph neural network for 3D object detection in a point cloud. In: CVPR (2020)

17. Qi, C.R., Yi, L., Su, H., Guibas, L.J.: PointNet++: deep hierarchical feature learning on point sets in a metric space. In: NeurIPS (2017)
18. Shi, S., Wang, Z., Shi, J., Wang, X., Li, H.: From points to parts: 3D object detection from point cloud with part-aware and part-aggregation network. TPAMI (2020)
19. Yin, T., Zhou, X., Krahenbuhl, P.: Center-based 3D object detection. In: CVPR (2021)
20. Mao, J., et al.: Voxel transformer for 3D object detection. In: ICCV (2021)
21. Deng, J., Shi, S., Li, P., Zhou, W., Zhang, Y., Li, H.: Voxel R-CNN: towards high performance voxel-based 3D object detection. In: AAAI (2021)
22. Graham, B., Engelcke, M., Van Der Maaten, L.: 3D semantic segmentation with submanifold sparse convolutional networks. In: CVPR (2018)
23. Shi, S., et al.: PV-RCNN: point-voxel feature set abstraction for 3D object detection. In: CVPR (2020)
24. Li, Z., Wang, F., Wang, N.: Lidar R-CNN: an efficient and universal 3D object detector. In: CVPR (2021)
25. Mao, J., Niu, M., Bai, H., Liang, X., Xu, H., Xu, C.: Pyramid R-CNN: towards better performance and adaptability for 3D object detection. In: ICCV (2021)
26. Eldar, Y., Lindenbaum, M., Porat, M., Zeevi, Y.Y.: The farthest point strategy for progressive image sampling. TIP (1997)
27. Zhu, X., Ma, Y., Wang, T., Xu, Y., Shi, J., Lin, D.: SSN: shape signature networks for multi-class object detection from point clouds. In: Vedaldi, A., Bischof, H., Brox, T., Frahm, J.-M. (eds.) ECCV 2020. LNCS, vol. 12370, pp. 581–597. Springer, Cham (2020). https://doi.org/10.1007/978-3-030-58595-2_35
28. Xu, Q., Zhong, Y., Neumann, U.: Behind the curtain: learning occluded shapes for 3D object detection. In: AAAI (2022)
29. Wang, T., Hu, X., Liu, Z., Fu, C.W.: Sparse2dense: learning to densify 3D features for 3D object detection. In: NeurIPS (2022)
30. Wen, X., et al.: PMP-Net: point cloud completion by learning multi-step point moving paths. In: CVPR (2021)
31. Wen, X., et al.: PMP-Net++: point cloud completion by transformer-enhanced multi-step point moving paths. TPAMI (2022)
32. Rosenblatt, M.: Remarks on some nonparametric estimates of a density function. Ann. Math. Stat. (1956)
33. Baudat, G., Anouar, F.: Generalized discriminant analysis using a kernel approach. Neural Comput. (2000)

Camouflaged Object Segmentation Based on Fractional Edge Perception

Xia Yuan[1(✉)], Junjie Cui[1], Zhengyu Liu[1], Shuting Yang[2], and Xuejian Zhang[2]

[1] School of Computer Science and Engineering, Nanjing University of Science and Technology, Nanjing 210094, China
yuanxia@njust.edu.cn

[2] Institute of Agricultural Economy and Information Technology, Ningxia Academy of Agriculture and Forestry Sciences, Yinchuan 750002, China

Abstract. Camouflaged object detection is a challenging task because of intrinsic similarity between background and foreground. In the existing models, many confusing edge features eventually lead to bad predictions. In this paper, we propose an Interactive Task Learning Network, in which an improved fractional-order differential operator is used to calculate the gradient intensity of the image. By calculating average gradient and spatial frequency of image, it can adaptively learn the fractional-order v and extract features from different directions. Experiments on three camouflaged datasets indicate that the proposed method is effective and progressiveness.

Keywords: Camouflaged object detection · Fractional edge · Interactive task learning

1 Introduction

Camouflage is a survival skill of creatures learned in the long-term evolution process to hide themselves from recognition by changing colors, approximating contrast, mimicking shapes, and so on. Broadly speaking, camouflaged objects also refer to the objects that are extremely small, highly similar to the background, or heavily obscured. Camouflage object detection (COD) is not only beneficial to scientific value, but also important for many basic tasks, including medical diagnosis (such as polyp segmentation [1] and lung infection analysis [2]), agriculture (such as pest detection), industry (such as defective parts detection), search and rescue [3].

COD is a very challenging task due to the nature of camouflage, that is, the high intrinsic similarities between candidate objects and chaotic background, which make it difficult to spot camouflaged objects for humans. To tackle this issue, numerous deep learning-based methods have been broadly divided into three types. First is to design more complex network for detecting camouflaged objects, such as C2FNet [4], UGTR [5]. Second is to incorporate auxiliary tasks into joint learning, including edge extraction, salient object detection, fixation graph and camouflaged object ranking. Third is to mimic bio-inspired method, such as SINet [6].

© The Author(s), under exclusive license to Springer Nature Singapore Pte Ltd. 2024
Q. Liu et al. (Eds.): PRCV 2023, LNCS 14436, pp. 16–28, 2024.
https://doi.org/10.1007/978-981-99-8555-5_2

Object detection models based on edge guidance has achieved better results than ordinary semantic segmentation models. But for the camouflaged objects with fuzzy edges and occlusion, existing models are usually difficult to identify the complete structure or details of objects, resulting in unsatisfactory prediction results. This paper uses both integral and fractional edge to highlight the weak gradient between foreground and background. Factional-order derivative can produce a coarser edge transition compared with integral-order derivative, which is conducive to strengthening high-frequency edge and texture details while retaining part of low-frequency contour information. The workflow of the interactive task learning network (ITLNet) proposed in this paper is shown in the Fig. 1. It is composed of two sub-task modules: camouflaged object edge detection module (CEDM) and camouflaged object detection module (CODM). Through the mutual feedback and interactive learning of the two sub tasks, ITLNet can obtain rich semantic guidance information and edge details, meanwhile, it maximizes the mining of different levels of image features to capture camouflaged objects with accurate edges. To sum up, our main contributions are as follows:

- A plug and play lightweight fractional edge perception module is proposed, which improves the performance of camouflage object segmentation model.
- Experiments show that the proposed module can be used as an independent lightweight end-to-end edge detection network.

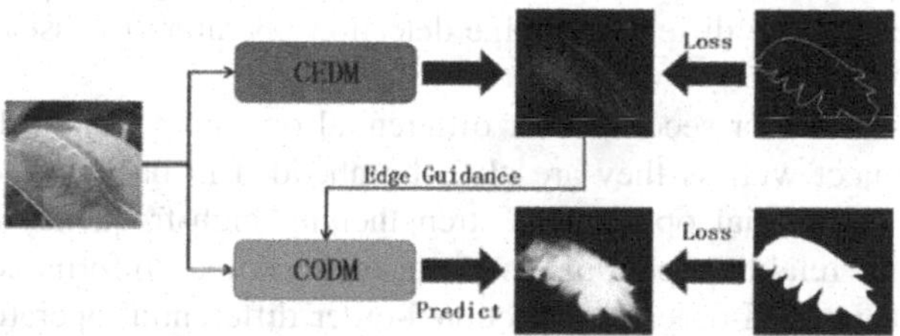

Fig. 1. Interactive camouflaged object detection model. CEDM refers to camouflaged object edge detection, CODM refers to camouflaged object detection. Edge map obtained through CEDM can enhance the edge of the prediction map when CODM works.

2 Related Work

In salient object detection task, Chen et al. [7] proposed a new Contour Loss, which uses the object contour to guide model to learn more information within the boundary and enhance local salient prediction while retaining accurate target boundaries. BPFINet proposed by Zhou et al. [8] aggregated high-level semantic features, low-level edge information and global features step by step through a U-shaped network model, and proposed a new loss function to highlight pixels near the boundary to solve the imbalance between background and objects. AFNet proposed by Feng et al. [9] refined the rough prediction map step by step by constructing the attention feedback module, used additional boundary supervision in the last two decoding stages. They extracted the boundary details through average pooling, and calculated loss function using ground truth as a part of whole loss function to guide the network training.

In the Camouflaged object detection task, Xu et al. [10] put forward an edge guidance module to build connections between the regional branch and the boundary branch of each side output. ERRNet proposed by Ji et al. [11] first integrated the low-level features to obtain edge prior. And then crossed models them with high-level semantic information and compares potential camouflage regions and their complementary regions by considering neighbor prior, global prior, edge prior, and semantic prior. MGL proposed by Zhao et al. [12] designed two interactive modules named region induced graph reasoning and edge constrained graph reasoning to guide training process mutually. BGNet proposed by Sun et al. [13] used edge prior to help recover the object structure step by step in the decoding module to improve the performance of camouflaged object detection.

3 Interactive Task Learning Network

The texture of camouflaged object is very close to the background. The human visual system often first discovers the edge position with the greatest difference from the background when searching for camouflaged objects. Therefore, edges are important clues for detecting camouflaged objects. This paper design a lightweight CNN module CEDM to generate edge prediction map, and insert it into CODM to train end-to-end. In order to overcome the weak feature extraction ability of the lightweight CNN module, CEDM uses interpretable differential edge detection operators to assist itself to learn edge feature.

Common first-order or second-order differential operators cannot detect the edge of camouflage object well as they are almost embedded in background. We employ fractional-order differential operator to strengthening high-frequency edge and texture details, while retaining part of low-frequency contour information. Therefore, CEDM uses both integral-order and fractional-order differential operators to improve the accuracy of edge prediction.

3.1 Integral and Fractional Edge

Integral Edge: Sobel [14] is a simple and efficient operator which combines the first-order derivative and gaussian smooth function. We generate horizontal convolution kernel G_x and vertical convolution kernel G_y based on a 3 × 3 convolution kernel, as shown in Eq. (1).

$$G_x = \begin{bmatrix} -1 & 0 & 1 \\ -2 & 0 & 2 \\ -1 & 0 & 1 \end{bmatrix}, \ G_y = \begin{bmatrix} -1 & 2 & -1 \\ 0 & 0 & 0 \\ 1 & 2 & 1 \end{bmatrix} \tag{1}$$

After each point in the image is combined with convolution, we can calculate the gradient as Eq. (2), where A is the image.

$$G = \sqrt{(G_x * A)^2 + (G_y * A)^2} \tag{2}$$

To simplify the operation, we change it to Eq. (3).

$$G = |G_x * A| + |G_y * A| \tag{3}$$

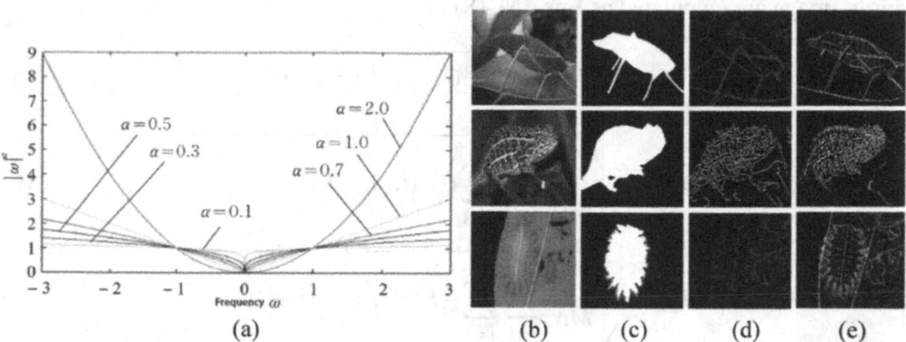

Fig. 2. Fractional edge. (a) Amplitude-frequency curve. (b) original image. (c) object mask. (d) Integral edge (Sobel). (d) Fractional edge (Tiansi).

Fractional Edge: Figure 2(a) shows the amplitude-frequency curve of the second derivative, the first derivative and [0, 1] fractional-order derivative. We can see that different orders in differential operation have different degrees of weakening effects on low frequency signals, and their lifting effects in high frequency signals also show different degrees of nonlinear growth. The fractional-order derivative between [0, 1] enhances the intermediate frequency and high frequency signals while retaining the nonlinear information of the low frequency signals.

$$\begin{bmatrix} \frac{v^2-v}{2} & 0 & \frac{v^2-v}{2} & 0 & \frac{v^2-v}{2} \\ 0 & -v & -v & -v & 0 \\ \frac{v^2-v}{2} & -v & 8 & -v & \frac{v^2-v}{2} \\ 0 & -v & -v & -v & 0 \\ \frac{v^2-v}{2} & 0 & \frac{v^2-v}{2} & 0 & \frac{v^2-v}{2} \end{bmatrix} \tag{4}$$

Equation (4) is the fractional-order differential Tiansi [22]. It calculates separately in 8 directions while they have the same rotation direction. However, Tiansi operator can't achieve good results in the edge transition area and smooth area. To solve this problem, we propose a method to adaptively calculating the fractional order between [0, 1] based on image content.

The fractional-order v relates to the gradient and spatial frequency of the image, and the larger the average gradient and spatial frequency of a certain area in the image, the greater the likelihood of it being an edge, which needs to be enhanced. On the contrary, when the average gradient and spatial frequency of a certain region are smaller, it indicates that the probability of the region being a smooth region is higher, and some detailed information of the region needs to be preserved in the calculation process.

Therefore, an adaptive [0, 1] order fractional operator is proposed by calculating the average gradient and spatial frequency of the image, and finding the relationship between them and the order v to construct the corresponding mapping function.

The input image is divided into image blocks with a width of M and a height of N, respectively. The spatial frequency SF(i, j) and the average gradient $\overline{G}(i, j)$ at pixel (i, j) in input image are calculated as Eqs. (5)–(8):

$$SF(i.j) = \sqrt{CF^2(i,j) + RF^2(i,j)} \tag{5}$$

$$CF(i.j) = \sqrt{\frac{1}{MN}\sum_{i=1}^{M}\sum_{j=1}^{N}((F(i,j) - F(i,j+1)))^2} \tag{6}$$

$$RF(i.j) = \sqrt{\frac{1}{MN}\sum_{i=1}^{M}\sum_{j=1}^{N}((F(i,j) = F(i+1,j)))^2} \tag{7}$$

$$\overline{G}(i.j) = \frac{1}{(M-1)(N-1)}\sum_{i=1}^{M-1}\sum_{j=1}^{N-1}\sqrt{\frac{(F(i,j) - F(i,j+1))^2 + (F(i,j) - F(i+1,j)^2}{2}} \tag{8}$$

CF(i, j) and RF(i, j) are column frequency and row frequency of pixel (i, j) respectively. After calculating the spatial frequency SF(i, j) and average gradient $\overline{G}(i, j)$, the order v will be obtained by constructing a mapping function. The mapping function is as Eqs. (9) and (10):

$$Y = \frac{1}{\pi}\left(\tan(SF)\_\tan\left(\overline{G}\right)\right) \tag{9}$$

$$v = (alpha - beta) * \frac{\tanh(Y) - \min(Y)}{\max(Y) - \min(Y)} + beta \tag{10}$$

where we set alpha to 0.7 and beta to 0.5 in Eq. (10). They are two parameters we set to limit the range of fractional-order parameter v. Compared with the integer order operator shown in Fig. 2, the fractional operator can detect the weak contrast edge between the camouflage object and the background better, which can be used as the prior supervision information of the camouflage target detection model to improve its performance.

3.2 Camouflaged Edge Detection Module

The proposed CEDM consists of two branches: horizontal branch and vertical branch. First, the input images are sent into convolutional layer with kernel size of 1×7 and 7×1. The low-level features of image are extracted along the horizontal and vertical directions respectively. Then, *ResBlock* will reduce the number of channels. Considering that the top-level feature map X_4 generated in ResNet-50 has rich semantic information, which is used to assist to get location of camouflage objects. To this end, we design selection-enhancement module to integrate features and improve the representation ability. Sobel operator will yield information with abundant edge details, while improved fractional-order Tiansi operator can retain low frequency texture details better. Finally, through channel attention, the predicted edge map can guide CODM detect objects better. The workflow in Fig. 4 is named a interactive task learning network (ITLNet) in this paper.

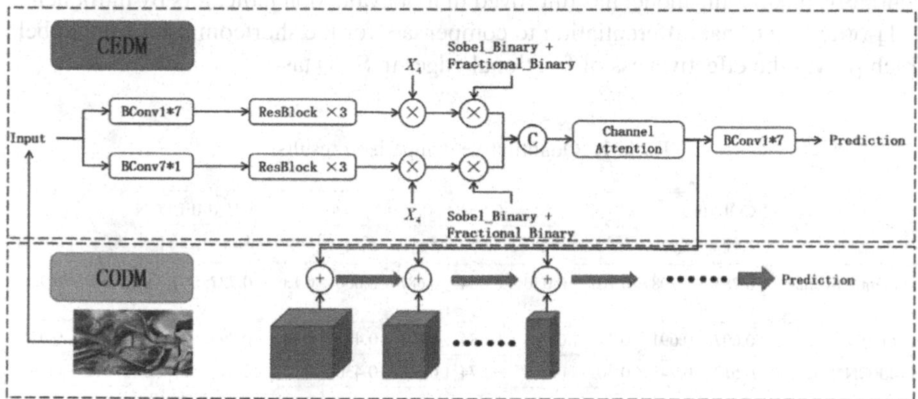

Fig. 4. The proposed interactive task learning network (ITLNet).

4 Performance Evaluation

4.1 Datasets and Experiment Settings

We carry out our experiments on three camouflaged datasets: CAMO, CHAMELEON and COD10K. The evaluation criteria contains: MAE (mean absolute error), S_α (structure measure), E_ϕ (Enhanced-alignment measure), F_β^ω (weighted F-measure).

All experiments are run on a TITAN RTX. In order to verify the effectiveness of the proposed CEDM, we select three camouflaged object detection models as CODM module which are SINet, PFNet and C2FNet. At the same time, the predicted edge maps generated by CEDM are processed using the sigmoid function as weights to guide the training of CODM (the weight matrix is represented by r_{edge} in the following text). Through interactive guidance, a predicted image with precise edges is generated.

4.2 Quantitative Evaluation

By analyzing the quantitative comparison data of multiple models presented in Table 1 on three datasets for four evaluation indicators. *+I_CEDM and *+IF_CEDM in Table 1 represent the use of integer edges and integer+fractional edges, respectively. It can be concluded that the detection performance of *+I_CEDM and *+IF_CEDM models proposed in this paper are significantly superior to the SOD and GOD models (the 2–6 rows in the Table 1). Compared to the COD benchmark model selected as an end-to-end COD network, the multi task interaction network proposed in this paper performs better. For example, C2FNet+IF_CEDM showed an average improvement of 0.58%, 1.22%, 1.01%, and 6.64% compared to C2FNet in four evaluation indicators; On the COD10K dataset, the four evaluation indicators of SINet+IF_CEDM increased by 0.52%, 3.23%, 8.89%, and 15.7% compared to SINet, respectively. According to S_{α}, it can be seen that the method proposed in this article detects the edges of camouflaged targets more accurately. Finally, the model has improved in four evaluation indicators by introducing [0, 1]-order fractional differentiation to compensate for the shortcomings of the Sobel, which proves the effectiveness of fractional edges in SOD task.

Table 1. Quantitative comparison results.

	COD10K				CAMO				CHAMELEON			
	$S\alpha\uparrow$	$E_{\phi}\uparrow$	$F_{\beta}^{\omega}\uparrow$	$M\downarrow$	$S\alpha\uparrow$	$E_{\phi}\uparrow$	$F_{\beta}^{\omega}\uparrow$	$M\downarrow$	$S\alpha\uparrow$	$E_{\phi}\uparrow$	$F_{\beta}^{\omega}\uparrow$	$M\downarrow$
Segment Anything [23]	0.783	0.798	0.701	0.050	0.684	0.687	0.606	0.132	0.727	0.734	0.639	0.081
FPN [19]	0.697	0.691	0.411	0.075	0.684	0.677	0.483	0.131	0.794	0.783	0.590	0.075
MaskRCNN [15]	0.613	0.748	0.402	0.080	0.574	0.715	0.430	0.151	0.643	0.778	0.518	0.099
Unet++ [20]	0.623	0.672	0.350	0.086	0.599	0.653	0.392	0.149	0.695	0.762	0.501	0.094
MSRCNN [21]	0.641	0.706	0.419	0.073	0.617	0.669	0.454	0.133	0.637	0.686	0.443	0.091
PFANet [16]	0.636	0.618	0.286	0.128	0.659	0.622	0.391	0.172	0.679	0.648	0.378	0.144
EGNet [17]	0.737	0.779	0.509	0.056	0.732	0.768	0.583	0.104	0.848	0.870	0.702	0.050
SINet [6]	0.771	0.806	0.551	0.051	**0.751**	0.771	0.606	0.100	0.869	0.891	0.740	0.044
SINet+I_CEDM	0.773	0.827	0.587	0.044	0.749	**0.773**	0.610	**0.098**	0.875	0.909	0.792	0.037
SINet+F_CEDM	**0.775**	**0.832**	**0.600**	**0.043**	0.733	**0.773**	**0.612**	**0.098**	**0.877**	**0.916**	**0.797**	**0.035**
PFNet [18]	0.800	0.868	0.660	0.040	0.782	0.852	0.695	0.085	0.882	0.942	0.810	0.033
PFNet+I_CEDM	0.809	0.876	0.664	0.038	0.785	0.860	0.700	0.082	**0.886**	0.936	0.819	**0.032**
PFNet+IF_CEDM	**0.812**	**0.879**	**0.667**	**0.037**	**0.789**	**0.863**	**0.702**	**0.081**	**0.886**	**0.936**	**0.824**	**0.032**
C2FNet [4]	0.813	0.809	0.686	0.036	0.796	0.854	0.719	0.080	0.888	0.935	0.828	0.032
C2NFet+I_CEDM	0.815	0.896	0.690	0.034	**0.799**	0.859	0.722	0.077	0.893	0.944	0.832	0.030
C2FNet+IF_CEDM	**0.816**	**0.899**	**0.693**	**0.033**	**0.799**	**0.864**	**0.724**	**0.076**	**0.897**	**0.949**	**0.839**	**0.029**

4.3 Qualitative Evaluation

SINet: The first benchmark of end-to-end camouflage target detection model. It divides the detection process into two stages: target search and recognition, and its structural design is robust and universal. Considering that SINet will generate two camouflage maps: rough prediction map C_s and refined results C_i. We get $C_s = C_s * r_{edge} + C_s$. Supplement structural details to optimize the network as much as possible within the target during the recognition phase.

Figure 5 shows a visualization example. Overall, the ITLNet (Fig. 5 (d) and (e)) is significantly better than SINet. Specifically, whether it is images containing a single camouflage target or images with multiple camouflage targets, SINet is prone to incomplete target detection, blurred edges, and misjudgment of the detected target object internally. The ITLNet based on Sobel edge can focus more on the internal target for training and optimization under the interactive learning of rich edge information and Semantic information, which solves the above problems well. However, due to the limitations of the Sobel operator in grayscale smooth regions, some edge regions of the predicted image are blurry. Based on this, we try to compensate for this defect by introducing [0, 1]-order fractional differentiation, as shown in Fig. 5 (e). The improved model can effectively recognize complete edges in areas with high confusion in the front background, such as adjacent parts of two fishes in the second row of Fig. 5.

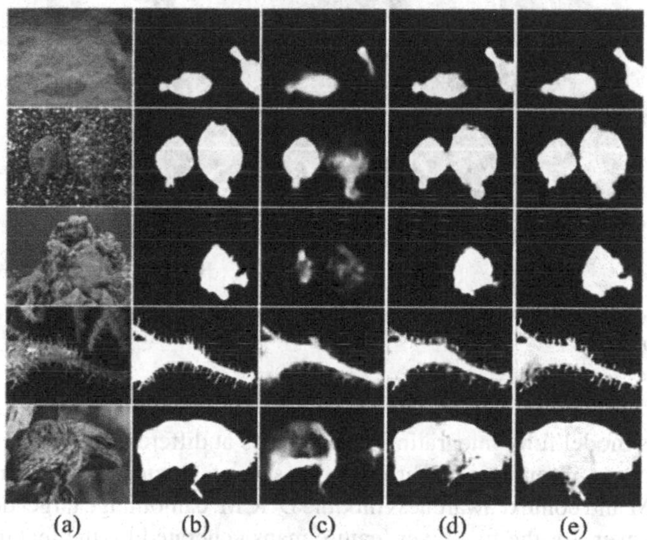

Fig. 5. Results of detection by adding differential operators. We choose SINet as our comparison network. From left to right are (a) original image, (b) ground truth, (c) SINet, (d) SINet+I CEDM, (e) SINet+IF_CEDM.

PFNet: PFNet divides the entire detection process into two stages: positioning and focusing. It adopts a U-shaped network architecture to achieve rough positioning and focus refinement of camouflaged targets step by step, while generating four side output prediction graphs cri (i = 1, 2, 3, 4). Similar to C2FNet as the reference network, in this experiment, r_{edge} and cri multiply and then add the elements, i.e., cri = cri * r_{edge} + cri (i = 1, 2, 3, 4,) to obtain a selected enhanced prediction map with more accurate edges. According to Fig. 6, the PFNet prediction results shown in the second, the fourth, and the fifth line in the figure are affected by background interference with highly similar colors and texture structures in the surrounding area. The camouflage object in the fifth line is missed in the overlapping area with the background, resulting in incomplete object. We construct an edge detection module by combining shallow convolutional networks with traditional hard coding operators, and retains as much contour information as possible to conduct interactive learning with PFNet to improve detection integrity and accuracy. The results are shown in Fig. 6 (d) and (e).

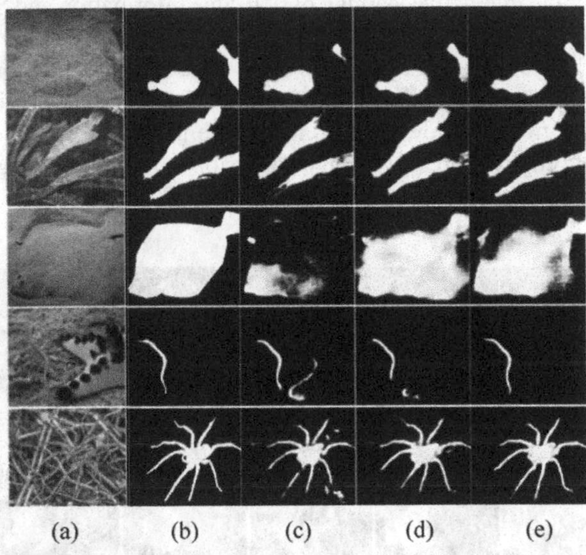

Fig. 6. Results of PFNet. (a) original image, (b) ground truth, (c) PFNet, (d) PFNet+I CEDM, (e) PFNet+IF_CEDM.

C2FNet: This model fully integrating feature maps at different levels and considering rich global context information. Through two sets of feature cross layer integration modules ACFM and context awareness module DGCM, camouflage target detection was achieved. However, for the five-layer feature maps generated by the feature extraction network, the network only integrates mid to high level semantic features, completely discarding the underlying features, which results in the loss of the rich structural details contained in them, leading to issues such as edge blur and edge errors in the predicted camouflage map. In this experiment, by utilizing r_{edge} processed feature map X_{43} and X_{432} for each DGCM is enhanced and optimized, i.e., $X_i = X_i * r_{edge} + X_i, i = 43, 432$.

Figure 7 is a partial visualization example image selected from the COD10K test set. From the image, it can be seen that the prediction image generated by C2FNet has poor detection performance in the front background gradient smooth area as mentioned earlier, and the prediction image part shows no obvious contour. The results of proposed ITLNet (Fig. 7 (d) and (e)) are superior to C2FNet in object's integrity and edge accuracy.

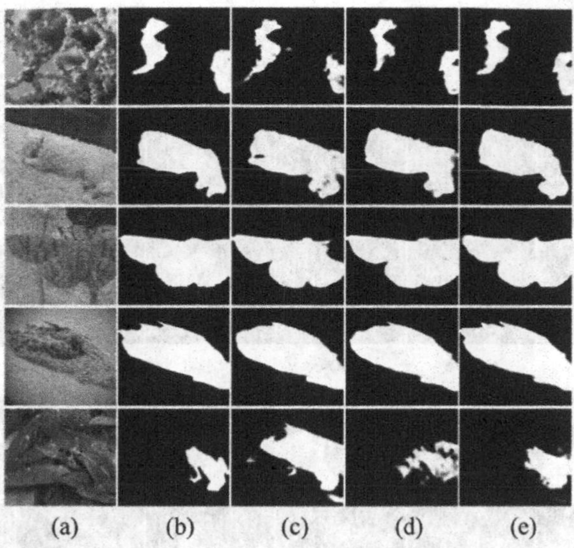

Fig. 7. Results of C2FNet. (a) original image, (b) ground truth, (c) C2FNet, (d) C2FNet+I CEDM, (e) C2FNet+IF_CEDM.

4.4 Generalization of Edge Detection

Many object detection tasks, including camouflaged object detection face the common problem of edge detection accuracy. In order to further analyze the effectiveness of the proposed CEDM, we separate it into an end-to-end edge detection network. We select some edge detection results from COD and SOD test dataset shown in Fig. 8. The CEDM achieves good results whether they are salient objects or camouflaged objects. So, it has more applications in the field of object segmentation and detection.

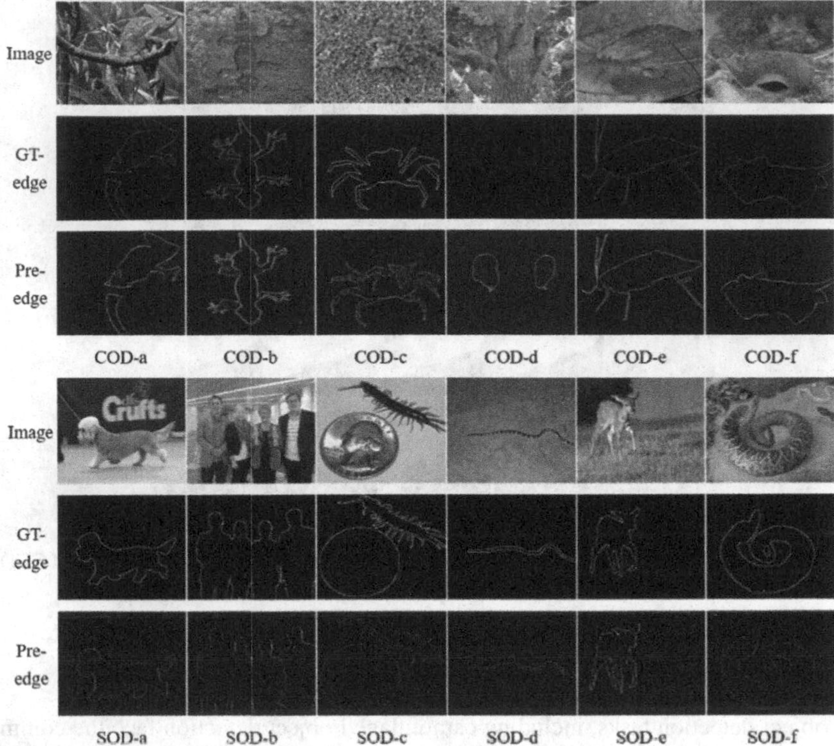

Fig. 8. Edge detection results of the proposed CEDM on COD and SOD dataset.

5 Conclusion

In this paper, the problem of blurring edges in the detection result image existing in camouflage target detection is deeply studied, and a camouflage target detection model ITLNet based on edge perception is proposed. The model captures rich semantic information and low-level details of the original image through two interactive guidance task modules, CEDM and CODM, to achieve high accuracy of the detection result map. We use Sobel operator to compensate for the loss of underlying structure information caused by the deepening of network layers, and use improved Tianci operator to detect gradient smooth areas and weak edge areas, and adaptively generate parameter v according to the gradient and spatial frequency of the image. After applying CEDM to several camouflage detection models, the experimental results verify the progressiveness nature of the proposed ITLNet model.

Acknowledgements. This work was supported by the project of "Research and Demonstration of Key Technologies for Smart Planting of Wine Grapes in Ningxia" under grant NKYG-23-02, and by the National Natural Science Foundation of China under grant 12071218.

References

1. Fan, D.-P., et al.: PraNet: parallel reverse attention network for polyp segmentation. In: Martel, A.L., et al. (eds.) MICCAI 2020. LNCS, vol. 12266, pp. 263–273. Springer, Cham (2020). https://doi.org/10.1007/978-3-030-59725-2_26
2. Shamim, S., Awan, M.J., Zain, A.M., Naseem, U., Mohammed, M.A., Begonya, G.Z.: Automatic Covid-19 lung infection segmentation through modified UNet model. J. Healthc. Eng. **2022**, Article ID 6566982 (2022)
3. Barbosa, A., et al.: Cuttlefish camouflage: the effects of substrate contrast and size in evoking uniform, mottle or disruptive body patterns. Vision. Res. **48**(10), 1242–1253 (2008)
4. Chen, G., Liu, S.J., Sun, Y.J., Ji, G.P., Wu, Y.F., Zhou, T.: Camouflaged object detection via context-aware cross-level fusion. IEEE Trans. Circuits Syst. Video Technol. **32**(10), 6981–6993 (2022)
5. Yang, F., et al.: Uncertainty-guided transformer reasoning for camouflaged object detection. In: Proceedings of the IEEE/CVF International Conference on Computer Vision, pp. 4146–4155. IEEE, Virtual (2021)
6. Fan, D.P., Ji, G.P., Sun, G.L., Cheng, M.M., Shen, J.B., Shao, L.: Camouflaged object detection. In: Proceedings of the IEEE/CVF Conference on Computer Vision and Pattern Recognition, pp. 2777–2787, Virtual (2020)
7. Chen, Z.X., Zhou, H.J., Lai, J.H., Yang, L.X., Xie, X.H.: Contour-aware loss: boundary-aware learning for salient object segmentation. IEEE Trans. Image Process. **30**, 431–443 (2021)
8. Huang, Z., Xiang, T.Z., Chen, H.X., Dai, H.: Scribble-based boundary-aware network for weakly supervised salient object detection in remote sensing images. arXiv preprint arXiv: 2202.03501 (2022)
9. Feng, M.Y., Lu, H.C., Ding, E.R.: Attentive feedback network for boundary-aware salient object detection. In: Proceedings of the IEEE/CVF Conference on Computer Vision and Pattern Recognition, pp. 1623–1632, IEEEE, Long Beach (2019)
10. Xu, X.Q., Zhu, M.Y., Yu, J.H., Chen, S.H., Hu, X.L., Yang, Y.Q.: Boundary guidance network for camouflage object detection. Image Vis. Comput. **114**, 104283 (2021)
11. Ji, G.P., Zhu, L., Zhuge, M.C., Fu, K.R.: Fast camouflaged object detection via edge-based reversible re-calibration network. Pattern Recogn. **123**, 108414 (2022)
12. Zhai, Q., Li, X., Yang, F., Chen, C.L.Z., Cheng, H., Fan, D.P.: Mutual graph learning for camouflaged object detection. In: Proceedings of the IEEE/CVF Conference on Computer Vision and Pattern Recognition, pp. 12997–13007. IEEE, Virtual (2021)
13. Chen, T.Y., Xiao, J., Hu, X.G., Zhang, G.F., Wang, S.J.: Boundary-guided network for camouflaged object detection. Knowl.-Based Syst. **248**, 108901 (2022)
14. Gao, W.S., Zhang, X.G., Yang, L., Liu, H.Z.: An improved Sobel edge detection. In: 3rd International Conference on Computer Science and Information Technology, pp. 67–71. IEEE, Chengdu (2010)
15. He, K.M., Gkioxari, G., Dollár, P., Girshick, R.: Mask R-CNN. In: Proceedings of the IEEE International Conference on Computer Vision, pp. 2961–2969. IEEE, Venice (2017)
16. Zhao, T., Wu, X.Q.: Pyramid feature attention network for saliency detection. In: Proceedings of the IEEE/CVF Conference on Computer Vision and Pattern Recognition, pp. 3085–3094. IEEEE, Long Beach (2019)
17. Zhao, J.X., Liu, J.J., Fan, D.P., Cao, Y., Yang, J.F., Cheng, M.M.: EGNet: edge guidance network for salient object detection. In: Proceedings of the IEEE/CVF international Conference on Computer Vision, pp. 8779–8788. IEEE, Seoul (2019)
18. Zhang, J.C., Shao, J.B., Chen, J.L., Yang, D.G., Liang, B.G., Liang, R.G.: PFNet: an unsupervised deep network for polarization image fusion. Opt. Lett. **4**(6), 1507–1510 (2020)

19. Lv, Y., et al.: Simultaneously localize, segment and rank the camouflaged objects. In: Proceedings of the IEEE Conference on Computer Vision and Pattern Recognition, pp. 11591–11601. IEEE, Virtual (2021)
20. Zhou, Z., Rahman Siddiquee, M.M., Tajbakhsh, N., Liang, J.: UNet++: a nested U-Net architecture for medical image segmentation. In: Stoyanov, D., et al. (eds.) DLMIA/ML-CDS-2018. LNCS, vol. 11045, pp. 3–11. Springer, Cham (2018). https://doi.org/10.1007/978-3-030-00889-5_1
21. Huang, Z., Huang, L., Gong, Y., Huang, C., Wang, X.: Mask scoring R-CNN. In: Proceedings of the IEEE Conference on Computer Vision and Pattern Recognition, pp. 6409–6418. IEEE, Long Beach (2019)
22. Yang, Z.Z., Zhou, J.L., Huang, M., Yan, Y.X.: Fractional differentiation for edge extraction. Comput. Eng. Appl. **43**(35), 15–18 (2007)
23. Kirillov, A., et al.: Segment anything. arXiv preprint, arXiv:2304.02643 (2023)

DecTrans: Person Re-identification with Multifaceted Part Features via Decomposed Transformer

Yan Zhang, Guangyu Gao[(✉)], Qianxiang Wang, and Jing Ge

School of Computer Science and Technology, Beijing Institute of Technology, Beijing, China

guangyugao@bit.edu.cn

Abstract. Utilizing part-level features provides a more detailed representation, leading to improved results in person re-identification (ReID). Yet existing works either use external tasks like pose estimation or struggle to define part features, which limit the model's learning capability. In this work, we propose the Decomposed Transformer (DecTrans), a transformer-based person ReID framework which exploits multifaceted part features. In particular, DecTrans extracts local features using the Vision Transformer (ViT) and then maps them into latent parts through a novel Token Decomposition (TD) layer. In the TD layer, soft clustering of ViT tokens forms clusters, and each token is decomposed into components based on its similarity to all cluster centroids. Token components referencing the same cluster are then regrouped to produce part features, thereby retaining more feature details. To ensure tokens from different pedestrians but referring to the same part are sufficiently clustered together, we propose to remove id information from tokens before clustering. Besides, we also propose a simple yet efficient data augmentation named Image Graying, which has been experimentally validated when used in conjunction with the TD layer. The DecTrans achieves remarkable performance, *e.g.*, mAP and Rank1 of 70.8%&87.1%, and 61.6%&67.7% on MSMT17 and Occluded-Duke, significantly outperforming state-of-the-arts.

Keywords: Person ReID · Vision Transformer

1 Introduction

Person re-identification (ReID) verifies pedestrians' identities captured by non-overlapping cameras. Some approaches adopt global features, characterizing the entire body [34,37]. However, this representation lacks fine-grained information and has limited discriminative power for similar-looking pedestrians with different IDs, especially in occluded scenarios. Alternatively, part-to-part matching-based ReID methods, such as [18,22], have shown superior performance.

© The Author(s), under exclusive license to Springer Nature Singapore Pte Ltd. 2024
Q. Liu et al. (Eds.): PRCV 2023, LNCS 14436, pp. 29–42, 2024.
https://doi.org/10.1007/978-981-99-8555-5_3

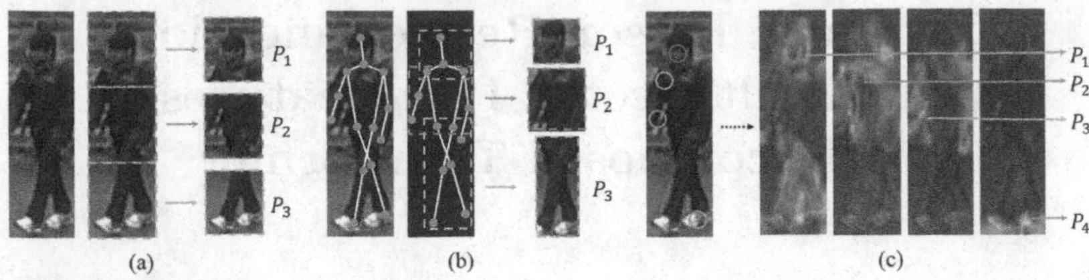

Fig. 1. Illustration of part feature responses. (a) Part features by uniform partitioning. (b) Part features based on pose estimation. (c) Latent part features by our DecTrans.

These methods aim to estimate critical semantic parts and extract well-aligned part features. Some [7,18,25] utilize body-part features from external tasks like pose estimation. However, the reliability of such tasks is uncertain, and the features might not adapt well to ReID due to cross-task differences. Others [22,26] extract part features directly on uniform grids of pedestrian images. Additionally, Sun [22] refines part pooling to assign outliers to the nearest part, enhancing within-part consistency. However, due to diverse poses and non-rigid body deformations, these models struggle with accurate part alignment.

Recently, transformers have made strides in person ReID [11], achieving notable results, which randomly rearrange and group patches within each pedestrian image. This haphazard grouping lacks a specific meaning for each patch group and doesn't ensure group alignment for matching purposes. Based on these insights, we propose applying clustering to guide transformer tokens into a series of *latent parts*, as depicted in Fig. 1. Specifically, we employ soft k-means clustering [8] on images within each training batch to ensure gradient backpropagation. To enhance cluster centroid stability, we integrate the dynamic routing mechanism [21]. Through soft clustering, our aim is for tokens corresponding to the same latent part across pedestrians to cluster around centroids with consistent semantics, despite their distinct IDs. Therefore, we refine each token by stripping its ID information, and creating an id-irrelevant token. This involves subtracting its projection in the direction of the ID vector.

We integrate id-irrelevant token refinement, soft clustering with dynamic routing, and cluster-based token decomposition into a unified Token Decomposition (TD) layer. This layer more explicitly guides tokens towards relevant latent parts with multifaceted representations. Finally, we construct the Decomposed Transformer (DecTrans) comprising a transformer token extractor, the TD layer, parts-based classification heads, and a global classification head featuring the common [class] token, as shown in Fig. 2.

Our main contributions can be summarized as follows: *i)* We introduce a plugin Token Decomposition (TD) layer rooted in soft clustering. This layer decomposes local features (tokens) into groups of feature components aligned with latent parts, facilitated by dynamic routing. *ii)* We introduce the notion of id-irrelevant tokens within the TD layer. This involves removing ID information or splitting each token into two vectors for clustering: an id vector and a purely latent part vector. *iii*) DecTrans achieve state-of-the-art performance on prevalent ReID datasets, by synergizing with the TD layer and a simple data augmentation technique called Image Graying.

2 Related Work

To tackle challenging issues in person ReID, *e.g.* background interference, partial occlusion, viewpoint change, pose change, previous works focus on extracting discriminative features. Some works capture contextual information from a global perspective [34,37], while some other works attain fine-grained features from local image parts [7,18,22,25,26]. For example, inspired by the spatial structure of the human body, some works [22,26] use horizontal partitioning to obtain fine-grained parts. However, a prerequisite assumption for them is that the body parts of pedestrians are relatively aligned; otherwise, there would be a problem of spatial semantic misalignment. To achieve part alignment, external cues are harnessed, *e.g.* pose estimation [7,18], human parsing [25], but this would significantly increase model complexity and harm efficiency.

The Transformer [24] was originally proposed for natural language processing, which can effectively process sequence data. Later, many works [9,14] have also proved its effectiveness in computer vision. The success of Vision Transformer (ViT) [5] well proves that a pure transformer-based architecture can effectively perform image classification. Hence a lot of research efforts have been made to explore transformer-based models for addressing various vision tasks [1,17,27,39]. In person re-identification, [11] utilizes a pure transformer with side-information embedding and a Jigsaw patch module to learn reliable feature representations; [38] adds the "partially marked" learnable vector to learn discriminative features, and integrates part arrangement into self-attention; [16] propose a novel end-to-end Part-Aware Transformer (PAT) to discover diverse discriminative human parts with a set of learnable part prototypes; [31] use Transformer to aggregate multiscale features of pedestrian images.

3 Methodology

3.1 Vision Transformer as Feature Extractor

Following TransReID [11], the image $I \in \mathbb{R}^{H \times W \times C}$ is divided into $N-1$ patches $\{I_p^i | i = 1, 2, \ldots, N-1\}$. These patches undergo flattening and linear projection, yielding N patch embeddings, together with an extra learnable [class] token that serves as the global pedestrian representation. The position and side information embeddings [11] are appended to the patch embeddings as N initial tokens, *i.e.*, the [class] token x_1^0 and $N-1$ patch tokens $\{x_i^0 | i = 2, 3, \ldots, N\}$. These initial tokens traverse $l-1$ transformer layers, resulting in encoded tokens $x^{l-1} = \{x_i^{l-1} | i = 1, 2, \ldots, N\}$, where x_1^{l-1} means the encoded [class] token.

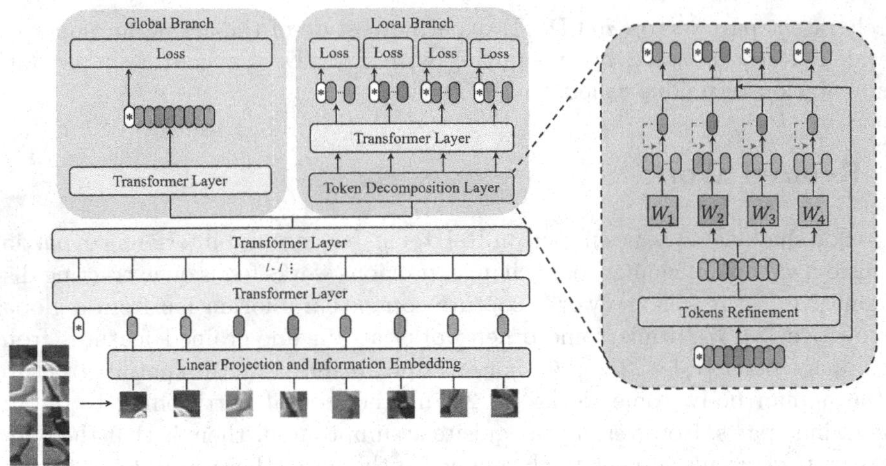

Fig. 2. Overall architecture of the proposed DecTrans. We first extract Vision Transformer (ViT) tokens as local features, and then design the Token Decomposition (TD) layer to generate part features via soft clustering with dynamic routing on these tokens.

3.2 Token Decomposition (TD) Layer

Soft Clustering with Dynamic Routing. Soft k-means clustering starts with an initial estimation of K centroids, which can be manually defined or randomly selected, and iteratively updated through similarity calculation. To stabilize the cluster centroids, we adopt dynamic routing [21], which facilitated by a trainable projection matrix, ensures consistent and stable feature mapping. For each encoded token $\boldsymbol{x}_i^{l-1}$ (abbr. $\boldsymbol{x}_i$), we project it onto K prediction vectors $\hat{\boldsymbol{x}}_{k|i}$ using the k_{th} cluster oriented trainable projection matrix $\boldsymbol{W}_k$. The centroid c_k is a weighted sum over $\hat{x}_{k|i}$ from all the input tokens as: $\boldsymbol{c}_k = \sum_i s_{k|i}\hat{\boldsymbol{x}}_{k|i}$, $\hat{\boldsymbol{x}}_{k|i} = \boldsymbol{W}_k\boldsymbol{x}_i$. Where $s_{k|i}$ is the scaled similarity to the k_{th} centroid, serving as the coupling coefficient. Coupling coefficients for x_i across clusters sum to 1 and are calculated using an iterative dynamic routing:

$$s_{k|i}^t = \text{softmax}\left(\boldsymbol{W}_k\boldsymbol{x}_i \cdot \frac{\boldsymbol{c}_k}{||\boldsymbol{c}_k||}\right), \boldsymbol{c}_k^{t+1} = \frac{\sum_{i=1}^N c_{k|i}^t \boldsymbol{x}_i}{\sum_{i=1}^N s_{k|i}^t} \tag{1}$$

where $\boldsymbol{c}_k^{t+1}$ represents the k_{th} centroid vector in the $(t+1)_{th}$ iteration. We propose a novel Token Decomposition (TD) layer to direct local features toward latent parts, as outlined in Algorithm 1.

Centroids start as the mean of token components: $\boldsymbol{c}_k^0 = \frac{1}{N-1}\sum_{i=2}^N \boldsymbol{W}_k\boldsymbol{x}_i$, and we iterative update centroids and coupling coefficients. This involves cosine similarity between centroids and token projections onto those centroids via the projection matrix. Rather than using cluster centroids as part features, we decompose each input token into K components according to its similarity to each centroid, and group the components referring to the same centroid together.

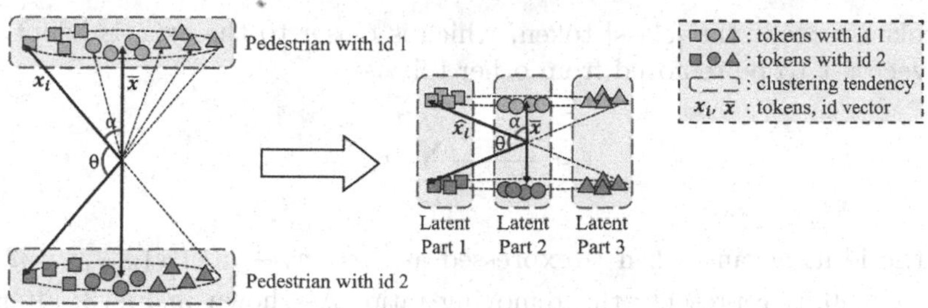

Fig. 3. Illustration of id-irrelevant tokens generation.

Finally, the [class] token and the decomposed components $\{\boldsymbol{r}_k(i)x_i\}_{i=2}^N$ are concatenated, *i.e.*, $\boldsymbol{Z}_k$, to go through the transformer to generate the part feature.

Id-irrelevant Tokens. The tokens of an image share the same id information, whereas we expect to cluster them into different latent parts each with a specific semantic. For different images (ids), we expect the cluster centroids with the same semantics to be close to each other. See an illustration in Fig. 3, where one color denotes one id while different shapes represent different semantics.

Algorithm 1. Token Decomposition Layer

Input: Image $\boldsymbol{X} = \{x_i\} \in \mathbb{R}^{N\times C}$, # of clusters K, projection matrix $\boldsymbol{W} \in \mathbb{R}^{K\times C\times C}$.
Output: Component groups $\boldsymbol{Z}_k \in \mathbb{R}^{N\times C}, k = 1, \ldots, K$.

for $i = 2$ to N **do**
  $\hat{x}_i = x_i - \mu(x_i^T \overline{x})\frac{\overline{x}}{||\overline{x}||}$, $\hat{x}_i = \text{normalize}(\hat{x}_i)$.
end for
for $k = 1$ to K **do**
  Centroids Initialization: $\boldsymbol{c}_k^0 = \frac{1}{N-1}\sum_{i=2}^N \boldsymbol{W}_k\hat{x}_i$
end for
for $t = 1$ to T **do**
  $\boldsymbol{r}_k^t = \text{softmax}\left(\boldsymbol{W}_k\hat{x}_i \cdot \frac{u_k^{t-1}}{||u_k^{t-1}||}\right)$
  $\boldsymbol{u}_k^t = \sum_{i=2}^N \boldsymbol{r}_k^t(i)\boldsymbol{W}_k\hat{x}_i / \sum_{i=2}^N \boldsymbol{r}_k^t(i)$
end for
for $k = 1$ to K **do**
  $\boldsymbol{Z}_k = [x_1, \{\boldsymbol{r}_k(i)x_i\}_{i=2}^N]$
end for
return $\boldsymbol{Z} = \{\boldsymbol{Z}_k\}$

We propose the concept of the id-irrelevant token, *i.e.*, removing id information from tokens, and then effectively clustering the refined feature vectors. In particular, the most intuitive choice for the id vector is to use the [class] token. However, the global [class] token focuses on representing id information for salient regions of the body rather than for all tokens. Therefore, the average of all tokens except the [class] token, which is closer to the axis, is adopted as the id vector $\overline{x}$ to be removed from other tokens:

$$\overline{x} = \frac{1}{N-1}\sum_{i=2}^{N} x_i. \tag{2}$$

Then, the id-irrelevant token is expressed as $\hat{x_i} = x_i - \mu(x_i^T\overline{x})\frac{\overline{x}}{||\overline{x}||}, i = 2...N$. Where $\mu \in [0,1]$ represents the removing ratio. As shown in Fig. 3, after the refinement by removing id information, tokens with the same semantics (shapes) are closer, to whatever id (color) they have.

3.3 Data Augmentation for TD Layer

We empirically find that most wrongly retrieved samples have a similar color as the query, as shown in Fig. 4(a). Thus, we propose a data augmentation method called Image Graying to alleviate the impact of color information. As shown in Fig. 4(b), the wrongly retrieved samples are sorted to lower ranks with the Image Graying. We use the probability of 0.1 to perform the Image Graying operation, which is calculated as $Gray = 0.299R + 0.587G + 0.114B$. In addition, we also use a data augmentation method called CutMix [30], which replaces some part of an image with the same part of another image, and the final label is weighted according to the cropping ratio, as shown in Fig. 4(c).

Fig. 4. Data augmentation operations. (a) Results before Image Graying. (b) Results after Image Graying. (c) Principle of CutMix operation.

3.4 Training and Inference

We optimize the whole network by constructing ID loss and triplet loss on both the global feature and latent part features. The ID loss L_{ID} is the cross entropy loss without label smoothing: $\mathcal{L}_{ID} = -\sum_{i=1}^{N} y_i \log p_i$.

$$\mathcal{L}_{ID} = -\sum_{i=1}^{N} y_i \log p_i \tag{3}$$

where y_i is the ground truth, p_i is the predicted probability, and N is the number of classes. For a set of triplets (a, p, n), the triplet loss L_T with soft margin is

$$\mathcal{L}_T = \log(1 + exp(\|f_a - f_p\|_2^2 - \|f_a - f_n\|_2^2)). \tag{4}$$

Finally, the DecTrans outputs both the global feature f_g and K local features referring to K latent parts as $\{f_l^1, f_l^2, ..., f_l^K\}$, and the overall loss is,

$$\mathcal{L} = \mathcal{L}_{ID}(f_g) + \mathcal{L}_T(f_g) + \frac{1}{K}\sum_{j=1}^{K}(\mathcal{L}_{ID}(f_l^j) + \mathcal{L}_T(f_l^j)) \tag{5}$$

In inference, we concatenate the global and part features from two images and calculate their cosine similarity for ranking.

4 Experiments

4.1 Datasets and Evaluation Metrics

We validate the performance of DecTrans on typical person ReID datasets, including Market-1501 [35], DukeMTMC-reID [20], MSMT17 [29] and Occluded-Duke [18]. Occluded-Duke is obtained by re-splitting the training, query, and gallery sets in DukeMTMC-reID such that all query images contain occlusion. For the evaluation metrics, we use the Cumulative Matching Characteristic (CMC) at Rank1 and the mean Average Precision (mAP) on all datasets.

4.2 Implementation Details

Pedestrian images are resized to 256×128. We use data augmentation methods including random horizontal flip, padding, random crop and random erase, the proposed Image Graying, as well as CutMix. For the input of the transformer, we split each image into patches of size 16, and the step size of the sliding window is 12. The batch size is set to 64, with four images per ID. In DecTrans, we use the same pre-trained model as TransReID for a fair comparison. The SGD optimizer is employed with a momentum of 0.9 and a weight decay of $1e-4$. The learning rate is initialized to 0.08 and cosine annealing is used as the scheduling of the optimizer. All the experiments are performed with two Nvidia 2080Ti GPUs using the PyTorch toolbox and APEX with FP16 training.

4.3 Comparisons to State-of-the-arts

Table 1. Comparison with the state-of-the-art methods, based on CNN (upper part) and ViT-B/16 (lower part) in the general Person ReID task.

Model	MSMT17		Market1501		DukeMTMC	
	mAP	Rank1	mAP	Rank1	mAP	Rank1
OSNet(ICCV19) [37]	52.9	78.7	84.9	94.8	73.5	88.6
ABDNet(ICCV19) [3]	60.8	82.3	88.3	95.6	78.6	89
PGFA(ICCV19) [18]	–	–	76.8	91.2	65.5	82.6
CBN(ECCV20) [40]	42.9	72.8	77.3	91.3	67.3	82.5
RGA-SC(CVPR20) [34]	57.5	80.3	88.4	96.1	–	–
SAN(AAAI20) [13]	55.7	79.2	88.0	96.1	75.7	87.9
SCSN(CVPR20) [4]	58.5	83.8	88.5	95.7	79	91
HOReID(CVPR20) [25]	–	–	84.9	94.2	75.6	86.9
PAT(ICCV21) [16]	–	–	88.0	95.4	78.2	88.8
ISE(CVPR2022) [33]	51.0	76.8	87.8	95.6	–	–
PNL+BDB(CVPR2022) [6]	53.4	79.0	88.4	95.4	79.0	89.2
TransReID(ICCV21) [11]	67.4	85.3	88.9	95.2	82	90.7
DecTrans	**70.8**	**87.1**	**90.9**	**96.0**	**83.5**	**90.9**

We compare the DecTrans with current state-of-the-art methods in Table 1. We can see the proposed DecTrans achieves the best performance on all these typical person ReID datasets. Especially, compared to TransReID, the previous best-performing transformer-based method, our DecTrans gains large performance improvements, *i.e.*, 3.4% in mAP, 1.8% in Rank1 on the large scale MSMT17 dataset, and 2.0%, 0.8% on Market1501 with a small scale. These results can also show the great potential of our method on large-scale datasets, as DecTrans involves the trainable projection matrices in dynamic routing, which would be more adequately trained over larger datasets. As shown in Table 2, for the occluded dataset of Occluded-Duke, the DecTrans has an improvement of 2.4% in mAP, and 1.3% in Rank1, demonstrating the robustness of DecTrans against the interference of pedestrian occlusion.

4.4 Ablation Study

Analysis of Components in DecTrans. We experiment on the largest dataset MSMT17 to verify the effects of TD layer and data augmentation methods, as shown in Table 3. For the baseline, we adopt a pre-trained ViT model

Table 2. Comparison with the state-of-the-art methods based on CNN (upper part) and ViT-B/16 (lower part) in the *Occluded-Duke* dataset.

Method	Rank1	Rank5	Rank10	mAP
HACNN(CVPR18) [15]	34.4	51.9	59.4	26.0
DSR(CVPR18) [10]	40.8	58.2	65.2	30.4
PCB(ECCV18) [22]	42.6	57.1	62.9	33.7
PGFA(ICCV19) [18]	51.4	68.6	74.9	37.3
RandErasing(AAAI20) [36]	40.5	59.6	66.8	30.0
HOReID(cvpr20) [25]	55.1	–	–	43.8
SORN(TCSVT20) [32]	57.6	73.7	79.0	46.3
SGSFA(ACML20) [19]	62.3	77.3	82.7	47.4
MoS(AAAI21) [12]	61.0	–	–	49.2
OAMN(ICCV21) [2]	62.6	77.5	–	46.1
MHSA(TNNLS22) [23]	59.7	74.3	79.5	44.8
BPBreID(CVPR23) [28]	66.7	–	–	54.1
PAT(CVPR21) [16]	64.5	–	–	53.6
TransReID(ICCV21) [11]	66.4	–	–	59.2
DecTrans	67.7	**81.7**	**86.4**	**61.6**

the same as TransReid, also using viewpoint and camera ID as side information, and use the [class] token as the final representation of the image.

Observing the results, CutMix leads to a 0.3% enhancement in mAP. Image Graying correlates with improvements: 0.7% in Rank1, 0.5% in Rank5, and 0.6% in Rank10. We then proceed to combine CutMix and Image Graying with the TD layer individually. Notably, both data augmentation techniques contribute positively to the extraction of intricate features facilitated by the TD layer. Lastly, a comparison is made between the TD layer and the JPM layer [11] within the context of data augmentation. The results reveal the TD layer's superiority, achieving a higher mAP at 70.8% compared to JPM's 70.0%.

Analysis of Token Decomposition (TD) Layer. The TD layer in DecTrans includes id-irrelevant token processing, soft k-means clustering with dynamic routing, and regrouped token components. We further verify the validity of the TD layer by *TD w/o id-irr.* (TD layer with originally encoded tokens rather than the id-irrelevant tokens), *TD w/o decomp.* (TD layer without token decomposition), and k-means clustering with different cluster numbers K.

As shown in Table 4, when tokens are not refined through the removal of id information, referred to as *TD w/o id-irr.*, there is a notable decrease in performance. This decline underscores the challenge of disentangling a significant number of highly correlated tokens. Similarly, utilizing the centroid vector alone

Table 3. Ablation study results of DecTrans.

Methods	Rank1	Rank5	Rank10	mAP
baseline	83.8	92.0	94.1	65.6
+CutMix	83.8	91.8	94.1	65.9
+Gray	84.5	92.5	94.7	65.6
+TD	85.4	92.4	94.2	67.3
+TD +CutMix	85.5	92.8	94.6	68.8
+TD +Gray	85.9	93.0	94.9	68.2
+JPM +Gray +CutMix	86.8	93.5	95.3	70.0
+TD +Gray +CutMix	**87.1**	**93.5**	**95.4**	**70.8**

as the representation of the latent part (*TD w/o decomp.*) also leads to performance degradation due to the reduced information within centroids, thereby affirming the efficacy of the decomposition process. Additionally, we explore various cluster numbers, detailed in the lower part in Table 4. The results show that the cluster number $K = 4$ yields the best performance.

Visualisation and Analysis. To clearly demonstrate the effects of the id-irrelevant tokens, Fig. 5 depicts the statistical distribution of the cosine similarity of tokens to the vector representing id information, *i.e.*, $\overline{x}$ (the average of all tokens), before refining (left part) and after refining (right part). The are two peaks in each sub-figure, which represent the centroids of background and foreground tokens respectively. The higher peak shows that most tokens are very similar before refining (around 0.8 on the left), and spread out (about 0.6 on the right) after the id-irrelevant refining. The "spread out" means, most of the

Table 4. Ablation study results of the TD layer.

Methods	Rank1	Rank5	Rank10	mAP
TD w/o id-irr	86.4	93.5	95.3	69.9
TD w/o decomp	86.5	93.3	95.2	70.0
TD	87.1	93.5	95.4	70.8
TD (#clusters=2)	86.4	93.5	95.3	69.5
TD (#clusters=3)	86.4	93.4	95.2	69.8
TD (#clusters=4)	**87.1**	**93.5**	**95.4**	**70.8**
TD (#clusters=5)	87.0	93.5	95.2	70.5
TD (#clusters=6)	86.7	93.6	95.3	70.4

refined tokens are far away from the vector of id information, and these refined tokens are more conducive to clustering.

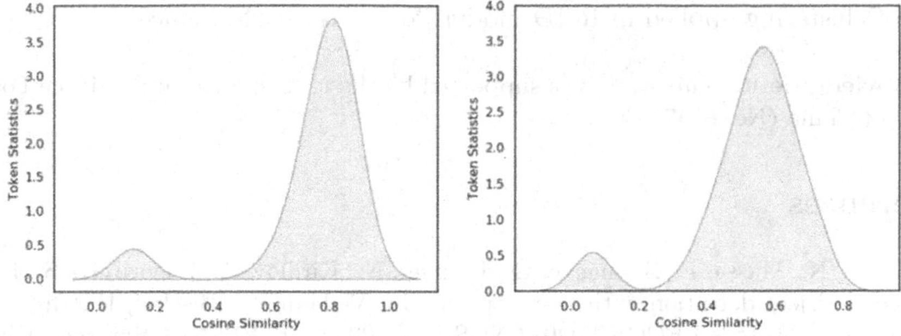

Fig. 5. Distribution of the cosine similarity of tokens to the id vector ($\overline{x}$). The left and right sub-figures are similar distributions before and after the id-irrelevant refinement.

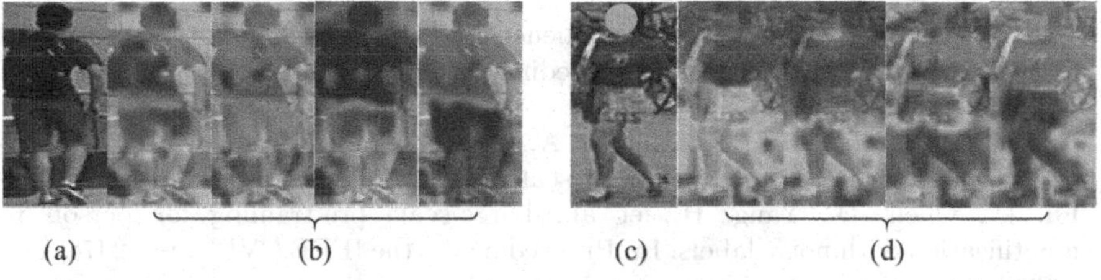

Fig. 6. Visualisation of attention maps about the latent part represented by groups of decomposed token components. (a) and (c) is the input pedestrian images, and (b) and (d) is the attention maps of latent parts on these images.

As shown in Fig. 6, the proposed DecTrans successfully attends to multiple semantically meaningful parts on pedestrians. Furthermore, the k_{th} cluster centroid for different input images attends to the part with similar semantics (Fig. 6(b) vs. (d)). Meanwhile, our proposed DecTrans can adapt to the input image and localize to flexible, irregular, and more semantic body parts compared to the uniform partition and pose estimation methods.

5 Conclusion

In this paper, we propose to apply soft clustering to force transformer tokens to be mapped into a series of latent parts for part-to-part matching based on person ReID. We develop the DecTrans that maps ViT tokens into latent parts each with the same semantic using a novel Token Decomposition (TD) layer. Besides, we also devise a data augmentation method called Image Graying, which has been experimentally proven effective when used with our TD layer. Extensive experiments validate the effectiveness and superiority of DecTrans over lots of state-of-the-art methods. In the future, we will continue to explore more effective ways of clustering applied in ReID models for better performance.

Acknowledgment. This work was supported by the National Natural Science Foundation of China (No. 61972036).

References

1. Carion, N., Massa, F., Synnaeve, G., Usunier, N., Kirillov, A., Zagoruyko, S.: End-to-end object detection with transformers. In: Vedaldi, A., Bischof, H., Brox, T., Frahm, J.-M. (eds.) ECCV 2020. LNCS, vol. 12346, pp. 213–229. Springer, Cham (2020). https://doi.org/10.1007/978-3-030-58452-8_13
2. Chen, P., Liu, W., Dai, P., et al.: Occlude them all: occlusion-aware attention network for occluded person re-id. In: Proceedings of the IEEE International Conference on Computer Vision, pp. 11833–11842 (2021)
3. Chen, T., Ding, S., Xie, J., et al.: ABD-Net: attentive but diverse person re-identification. In: Proceedings of the IEEE International Conference on Computer Vision, pp. 8351–8361 (2019)
4. Chen, X., Fu, C., Zhao, Y., et al.: Salience-guided cascaded suppression network for person re-identification. In: Proceedings of the IEEE CVPR, pp. 3300–3310 (2020)
5. Dosovitskiy, A., Beyer, L., Kolesnikov, A., et al.: An image is worth 16×16 words: transformers for image recognition at scale (2021)
6. Fu, D., Chen, D., Yang, H., et al.: Large-scale pre-training for person re-identification with noisy labels. In: Proceedings of the IEEE CVPR, pp. 2476–2486 (2022)
7. Gao, S., Wang, J., Lu, H., Liu, Z.: Pose-guided visible part matching for occluded person reid. In: Proceedings of the IEEE CVPR, pp. 11744–11752 (2020)
8. Ghosh, S., Dubey, S.K.: Comparative analysis of k-means and fuzzy c-means algorithms. Int. J. Adv. Comput. Sci. Appl. **4**(4) (2013)
9. Han, K., Wang, Y., Chen, H., et al.: A survey on vision transformer. arXiv e-prints, pp. arXiv-2012 (2020)
10. He, L., Liang, J., Li, H., Sun, Z.: Deep spatial feature reconstruction for partial person re-identification: alignment-free approach. In: Proceedings of the IEEE CVPR, pp. 7073–7082 (2018)
11. He, S., Luo, H., Wang, P., et al.: TransReID: transformer-based object re-identification. In: Proceedings of the IEEE International Conference on Computer Vision, pp. 15013–15022 (2021)
12. Jia, M., Cheng, X., Zhai, Y., et al.: Matching on sets: conquer occluded person re-identification without alignment. In: Proceedings of the AAAI, vol. 35, pp. 1673–1681 (2021)
13. Jin, X., Lan, C., Zeng, W., et al.: Semantics-aligned representation learning for person re-identification. In: Proceedings of the AAAI, vol. 34, pp. 11173–11180 (2020)
14. Khan, S., Naseer, M., Hayat, M., et al.: Transformers in vision: a survey. arXiv preprint arXiv:2101.01169 (2021)

15. Li, W., Zhu, X., Gong, S.: Harmonious attention network for person re-identification. In: Proceedings of the IEEE CVPR, pp. 2285–2294 (2018)
16. Li, Y., He, J., Zhang, T., et al.: Diverse part discovery: occluded person re-identification with part-aware transformer. In: Proceedings of the IEEE CVPR, pp. 2898–2907 (2021)
17. Liu, Z., Lin, Y., Cao, Y., et al.: Swin transformer: hierarchical vision transformer using shifted windows. arXiv:2103.14030 (2021)
18. Miao, J., Wu, Y., Liu, P., et al.: Pose-guided feature alignment for occluded person re-identification. In: Proceedings of the IEEE International Conference on Computer Vision, pp. 542–551 (2019)
19. Ren, X., Zhang, D., Bao, X.: Semantic-guided shared feature alignment for occluded person re-identification. In: Proceedings of the ACML, pp. 17–32 (2020)
20. Ristani, E., Solera, F., Zou, R., Cucchiara, R., Tomasi, C.: Performance measures and a data set for multi-target, multi-camera tracking. In: Hua, G., Jégou, H. (eds.) ECCV 2016. LNCS, vol. 9914, pp. 17–35. Springer, Cham (2016). https://doi.org/10.1007/978-3-319-48881-3_2
21. Sabour, S., Frosst, N., Hinton, G.E.: Dynamic routing between capsules. In: Proceedings of the Neural Information Processing Systems, pp. 3859–3869 (2017)
22. Sun, Y., Zheng, L., Yang, Y., et al.: Beyond part models: person retrieval with refined part pooling (and a strong convolutional baseline). In: Proceedings of the ECCV, pp. 480–496 (2018)
23. Tan, H., Liu, X., Yin, B., Li, X.: MHSA-Net: multihead self-attention network for occluded person re-identification (2022)
24. Vaswani, A., Shazeer, N., Parmar, N., et al.: Attention is all you need. In: Proceedings of the Neural Information Processing Systems, pp. 5998–6008 (2017)
25. Wang, G., Yang, S., Liu, H., et al.: High-order information matters: learning relation and topology for occluded person re-identification. In: Proceedings of the IEEE CVPR, pp. 6449–6458 (2020)
26. Wang, G., Yuan, Y., Chen, X., et al.: Learning discriminative features with multiple granularities for person re-identification. In: Proceedings of the ACM MM, pp. 274–282 (2018)
27. Wang, W., Xie, E., Li, X., et al.: Pyramid vision transformer: a versatile backbone for dense prediction without convolutions. In: Proceedings of the IEEE International Conference on Computer Vision, pp. 568–578 (2021)
28. Wang, Z., Zhu, F., Tang, S., et al.: Feature erasing and diffusion network for occluded person re-identification. In: Proceedings of the IEEE CVPR, pp. 4754–4763 (2022)
29. Wei, L., Zhang, S., Gao, W., Tian, Q.: Person transfer GAN to bridge domain gap for person re-identification. In: Proceedings of the IEEE CVPR, pp. 79–88 (2018)
30. Yun, S., Han, D., Oh, S.J., et al.: Cutmix: regularization strategy to train strong classifiers with localizable features. In: Proceedings of the IEEE International Conference on Computer Vision (2019)
31. Zhang, G., Zhang, P., Qi, J., Lu, H.: HAT: hierarchical aggregation transformers for person re-identification. In: Proceedings of the ACM MM, pp. 516–525 (2021)
32. Zhang, X., Yan, Y., Xue, J.H., et al.: Semantic-aware occlusion-robust network for occluded person re-identification. IEEE TCSVT **31**(7), 2764–2778 (2020)
33. Zhang, X., Li, D., Wang, Z., et al.: Implicit sample extension for unsupervised person re-identification. In: Proceedings of the IEEE CVPR, pp. 7369–7378 (2022)
34. Zhang, Z., Lan, C., et al.: Relation-aware global attention for person re-identification. In: Proceedings of the IEEE CVPR, pp. 3186–3195 (2020)

35. Zheng, L., Shen, L., Tian, L., et al.: Scalable person re-identification: a benchmark. In: Proceedings of the IEEE International Conference on Computer Vision, pp. 1116–1124 (2015)
36. Zhong, Z., Zheng, L., Kang, G., Li, S., Yang, Y.: Random erasing data augmentation. In: Proceedings of the AAAI, vol. 34, pp. 13001–13008 (2020)
37. Zhou, K., Yang, Y., Cavallaro, A., Xiang, T.: Omni-scale feature learning for person re-identification. In: Proceedings of the IEEE International Conference on Computer Vision, pp. 3702–3712 (2019)
38. Zhu, K., Guo, H., Zhang, S., et al.: AAformer: auto-aligned transformer for person re-identification. arXiv:2104.00921 (2021)
39. Zhu, X., Su, W., Lu, L., et al.: Deformable DETR: deformable transformers for end-to-end object detection (2021)
40. Zhuang, Z., et al.: Rethinking the distribution gap of person re-identification with camera-based batch normalization. In: Vedaldi, A., Bischof, H., Brox, T., Frahm, J.-M. (eds.) ECCV 2020. LNCS, vol. 12357, pp. 140–157. Springer, Cham (2020). https://doi.org/10.1007/978-3-030-58610-2_9

AHT: A Novel Aggregation Hyper-transformer for Few-Shot Object Detection

Lanqing Lai[1,2], Yale Yu[1,2], Wei Suo[1,2], and Peng Wang[1,2](✉)

[1] School of Computer Science, Northwestern Polytechnical University, Xi'an, China
peng.wang@nwpu.edu.cn

[2] Ningbo Institute, Northwestern Polytechnical University, Xi'an, China

Abstract. Few-shot object detection aims to detect novel objects with few annotated examples and this task has been extensively investigated by meta-learning-based paradigm. However, most of the previous approaches suffer from: 1) Most of the previous methods only perform two-branch interaction in the detection head which lacks the interaction of low-level semantic features. 2) Traditional method is difficult to capture the fine-grained differences between categories due to fixed weights. To alleviate these issues, we proposed a simple yet effective method, named Aggregation Hyper-Transformer (AHT) framework, which can generate corresponding weights into the primary network by hypernetworks mechanism. In particular, we design a novel Dynamic Aggregation Module and a Conditional Adaptation Hypernetworks, which apply the aggregated category vectors as conditions to dynamically generates class-specific parameters. Benefiting from the above two modules, our method significantly exceeds the previous meta-learning methods and provides new insights for my community.

Keywords: Few-shot object detection · Hypernetwork · Meta-learning

1 Introduction

Object detection is one of the important topics in the field of computer vision, which aims to predict the coordinates and corresponding categories of each instance in the image. There have been significant advancements in object detection techniques [21,22], with deep learning-based approaches achieving remarkable results. However, traditional object detection methods require large amounts of labeled data for training, which are expensive and time-consuming to collect. Therefore, Few-shot object detection(FSOD) task [9,26,31] is proposed and tries to address this difficulty by enabling the detector with very limited training data.

In particular, there are mainly two kinds of approaches to address such challenges: transfer-learning-based and meta-learning-based methods [26]. For transfer-learning-based framework [20,23,26,30], newly added detection head

© The Author(s), under exclusive license to Springer Nature Singapore Pte Ltd. 2024
Q. Liu et al. (Eds.): PRCV 2023, LNCS 14436, pp. 43–55, 2024.
https://doi.org/10.1007/978-981-99-8555-5_4

dimensions are directly fine-tuned on the novel class through the generalization knowledge derived from training on the base classes. It is extremely sensitive to the sample in the fine-tuning stage which requires carefully designing the frozen portion of the model [30]. More importantly, this paradigm suffers from significantly negative generalization effects, such as catastrophic forgetting for base classes [23,33]. On the contrary, the meta-learning-based methods [9,28,31,33] refer to simulating few-shot tasks through a two-branch structure formed by sampling support sets in base classes. With the process of learning to learn, the two-branch framework has a stronger generalization ability in the novel class, compared with the traditional transfer-learning-based framework.

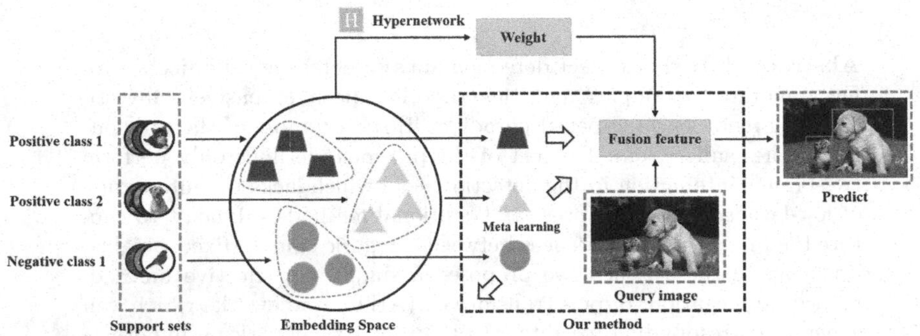

Fig. 1. Illustrations of hypernetwork for few shot object detection.

Although the aforementioned meta-learning-based methods have achieved pretty-well performance, it still restricted by the following limitations:

1) Most meta-learning-based methods [5,16,31] only perform two-branch interaction in the detection head, resulting in the misalignment of low-level semantic features, which significantly affects the final object detection performance.
2) [8] exploits the deep cross-branch interactions by fully Feature Pyramid Cross Transformer. Nevertheless, the updated method based on fixed weights is difficult to capture the fine-grained differences between categories during the interactions of query and support branch [2,14,19].

To solve the above problems, we propose a new Aggregation Hyper-Transformer (AHT) approach, which can effectively provide multi-level fine-grained two-branch interactions and dynamically generate class-specific primary network weights with a feed-forward meta-learning manner. Concretely, we first build a Dynamic Aggregation Module (DAM). By deriving the global dependency of each instance's local parts in the support sets, the inter-image prominent features are adaptively generated into aggregated weight. On the other hand, as shown in Fig. 1, we propose a novel Conditional Adaptation Hypernetworks (CAH) module for FSOD. Here, the hypernetworks refers to using another network to predict the weights of a primary network. Inspired by previous Hypernetwork-based methods [34], our CAH uses the aggregated weight vectors as the conditions of hypernetwork and dynamically generates class-specific parameters. Benefitting from the above two modules, our method provides fine-grained two-branch fusion and effectively improves performance.

In summary, the contributions of this work are as follows: 1) We propose a novel few-shot object detection method AHT that firstly incorporates corresponding weights into the meta-learning-based method by hypernetworks. Different from previous interactions in the detection head, the AHT can achieve deeper and fine-grained cross-branch fusion. 2) We design two new modules, including Dynamic Aggregation Module and Conditional Adaptation Hypernetworks. They are utilized to extract inter-image prominent features and generate class-specific parameters. 3) Extensive experiments show that, without bells and whistles, AHT exhibits competitive performance compared to state-of-the-art methods, revealing the effectiveness of our approach.

2 Related Work

2.1 Object Detection

Object Detection. Object detection is a critical task in computer vision that involves identifying and localizing objects. Early methods of object detection can be divided into two general approaches: one-stage and two-stage methods. One-stage detectors [15,21] directly predict the bounding boxes and the corresponding class labels on the extracted features. On the contrary, the two-stage detectors [22,24] need to generate region proposals through the Region Proposal Network (RPN) network, and then predict the detection results based on the proposals. However, these detectors are trained with sufficient labeled data and are not designed for data-scarce scenarios.

Few-Shot Object Detection. Few-shot object detection aims to recognize novel class objects using a few labeled examples. Existing works mostly focus on two paradigms: transfer-learning and meta-learning paradigms. Specifically, the transfer-learning-based approaches [23,26,30] first train a basic detector with large base samples. Then, the detection head of the pre-trained models would be finetuned to transform the acquired knowledge to the target few-shot field. However, due to serious negative transfer and catastrophic forgetting, the meta-learning-based approaches [9,28,31] are proposed, which apply the learning-to-learn (*i.e.,* meta-learning) to effectively adapt to new categories and learn to distinguish similar classes. Inspired by previous works, in this paper, we propose a novel Aggregation Hyper-Transformer (AHT) approach, which can improve the generalization ability of meta-learning-based method.

2.2 Hypernetworks

Hypernetworks [6] are neural networks that are utilized to generate the parameters of another neural network (called a main network). Hypernetworks can not only ensure the flexibility of the network but also reduce the amount of parameters of the network. For example, to improve the generalization ability, Ha et al. [6] use hypernetwork to replace the convolution layers for image recognition tasks. Hyperseg [18] uses hypernetwork to generate the parameters of the decoder in semantic segmentation. On the other hand, to improve the performance of few-shot tasks, TAFE-Net [27] generates task aware feature embeddings for few shot learning. Sylph [32] predict the parameters of classification head by hypernetworks. Hypertransformer [34] use transformer to generate a CNN model directly from few-shot samples. Inspired by these hypernetwork-based frameworks, we use support features as the conditions of hypernetworks to dynamically generate class-specific parameters which use for multi-level cross-branch fusion.

3 Method

3.1 Preliminaries

Following the previous works [5,16,20,26], the few-shot detection task involves a base set of classes, denoted as C_b, with a plentiful of annotated data, represented by D_b. Additionally, there is a novel set of classes, labeled as $C_n(C_n \cap C_b = \emptyset)$, with no overlap with C_b, and only N instances available for each category, denoted by D_n. The goal is to optimize a few-shot detection model that can detect objects r_q from both base and novel classes in testing. Here the r_q following the meta-learning paradigm [8,9,16]:

$$\underset{\theta}{argmax}\, P_\theta(r_q | I_q, \hat{I_s}), \tag{1}$$

where the probability $P_\theta(\cdot)$ is calculated by few-shot detection model parameters θ, using a query image I_q and N support instances (*N-shot*) of each category $\hat{I_s}$ as input data. Meanwhile, the whole learning procedure is organized into a two-stage paradigm (*i.e.*, the meta-knowledge is collected across D_b and finetunes it on D_n.).

3.2 Overview

Inspired by the feed-forward meta-learning manner [5,8,16], we propose a simple yet novel approach, called Aggregation Hyper-Transformer(AHT), to further enhance the capabilities of transformer-based detectors in few-shot scenarios. The fundamental insight driving AHT is that despite the traditional meta-learning paradigm having shown auspicious results, the interaction between the query and support branch only occurs within the detection head, resulting in the misalignment of low-dimensional features, which significantly affects the final object detection performance [8].

Recently, [8] exploits the deep cross-branch interactions by fully Feature Pyramid Cross Transformer. However, the method is fixed weights which has difficulty capturing fine-grained differences between categories [14,19], as well as the prominent features of a set of support images $\hat{I}_s$ cannot be effectively extracted due to coarse-grained averaging [8]. Therefore, as shown in Fig. 2, our model performs dynamic aggregation on the supported image $\hat{I}_s$ to enhance the aggregated weight and adaptively generates class-specific weights with hyper-network strategy.

Formally, features of the query and support branch would be fed into a feature pyramid cross transformer for instance-level cross-attention. In particular, we design the Dynamic Aggregation Module(DAM), it takes the set of support features as input for extracting inter-image prominent information when aggregating the support features into aggregated weight. On the other hand, we propose Conditional Adaptation Hypernetworks(CAH) which use the aggregated weight vectors as conditions to generate class-specific weights. By introducing the Dynamic Aggregation Module and Conditional Adaptation Hypernetworks, our model provides multi-level and higher-quality two-branch interactions. Finally, we use an RPN-based head and ROI Align to generate proposal features by contrastive training strategy [5], stimulating the detection head to accurately match the real category in the query image.

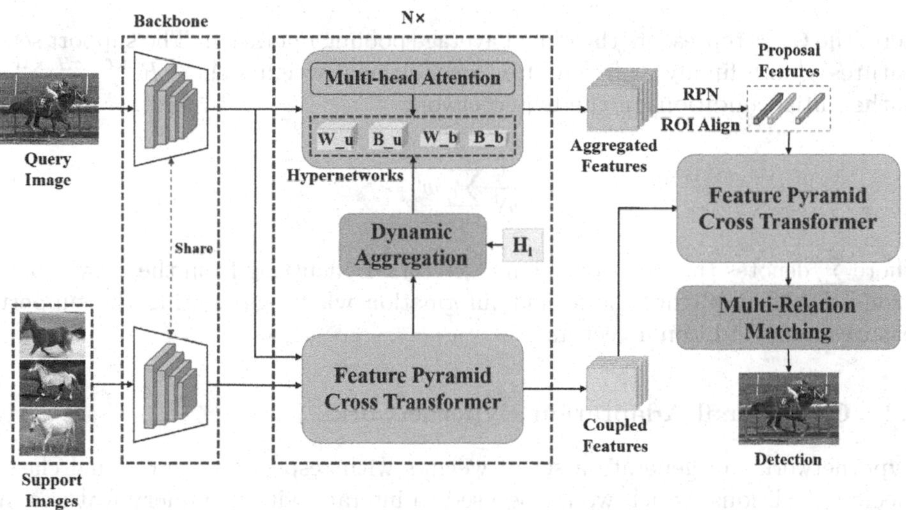

Fig. 2. The architecture of the Aggregation Hyper-Transformer (AHT) approach for few shot object detection.

3.3 Dynamic Aggregation Module

The Dynamic Aggregation Module (DAM) can generate aggregated weight vectors as the conditions of hypernetwork while preserving each category's notable information from N instances. It receives input from the support branch, denoted as $f_s \in R^{N \times C_i \times H_s \times W_s}$. Similar to a vanilla transformer, we map the input support sets f_s to K, V, while mapping a group of learned embeddings $H_i \in R^{C_i \times C_i}$ to Q. After the weighted combination of attention mechanism, features of each instance in the support branch are initially aggregated by:

$$\mathrm{f}_s^h = \mathrm{softmax}(\frac{QK^T}{\sqrt{d_k}})V, \tag{2}$$

where the d_k is scaling factor. It is worth noting that features $f_q \in R^{1 \times C_i \times H_q \times W_q}$ from the query branch in which the input sizes are different from the initial aggregated feature f_s^h. To ensure that the useless information in the support features is suppressed so that the two-branch interacts effectively, we execute the global representation of f_s^h and then derive the global dependency of each instance's local parts to acquire the aggregation weights $w_s^h \in R^{N \times C_i \times H_s \times W_s}$, which represent the proportion of the useful information carried in each support image. It is defined as:

$$w_s^h = \frac{GAP(f_s^h)\, f_s^h}{||GAP(f_s^h)||\ ||f_s^h||}, \tag{3}$$

Here, the GAP represents the global average pooling operation. The support sets features f_s^h are finally aggregated into conditional weights $A_c \in R^{1 \times C_i \times H_s \times W_s}$ as the initial conditions of the hypernetwork:

$$A_c = \frac{1}{N} \sum_{i=1}^{N} w_{s,i}^h \cdot f_{s,i}^h. \tag{4}$$

where $\sum$ denotes the sum operation of vectors. Benefitting from the above operations, DAM highlights the salient information when aggregating the support features into conditional vectors.

3.4 Conditional Adaptation Hypernetworks

Hypernetwork can generate a set of weights with respect to the original class-specific conditions, which would be used to interact with the query features in which after being encoded by the multi-head attention module.

$$\begin{cases} W\_u = dropout(A_c \cdot H_W, p_u^i), \\ B\_u = dropout(A_c \cdot H_B, p_b^i). \end{cases} \tag{5}$$

where p_u^i and p_b^i are ratio [11], $dim(H_W) = dim(H_B) \in R^{C_i \times C_i}$ are learned weights. Furthermore, W_b and B_b are directly generated by the linear projection of A_c. Based on this set of weights, a dense and fully-connected hypernetwork architecture is formed for the interaction with the query branch.

$$f_q^h = B\_u \cdot (W\_u \cdot f_q + W\_b) + B\_b. \tag{6}$$

As shown in Fig. 2, f_q^h and f_s will be used as the input of the joint feature extraction and interaction in the next Aggregate Hyper-Transformer block. After $N = 3$ stages of blocks, we gather Aggregated features that fully interact with the salient semantic features of the support branch. Additionally, we introduce query-perspective information to acquire Coupled features in the support branch.

3.5 The Classification-Regression Detection Head

After multi-level interactions between query and support branch, we follow [22] to adopt RPN for query and support branch respectively to generate more precise candidate proposals. Then, ROI Align is used to extract the initial RoI query features $f_q^r \in R^{P_q \times C_r \times H_q^r \times W_q^r}$ and support features $f_s^r \in R^{1 \times C_s \times H_s^r \times W_s^r}$ for each proposal. To achieve high-level semantic interaction, we directly send f_q^r and f_s^r to Feature Pyramid Cross Transformer [8] which conducts instance-level cross-attention. Then, we use Multi-Relation Matching [5] to enforce the final similarity learning which employs a contrastive training strategy. Finally, a binary classification and box regression layer are used for final detection.

Table 1. Experimental results on VOC dataset.

Method/ Shot	Novel Set 1					Novel Set 2					Novel Set 3				
	1	2	3	5	10	1	2	3	5	10	1	2	3	5	10
TFA w/fc [26]	36.8	29.1	43.6	55.7	57.0	18.2	29.0	33.4	35.5	39.0	27.7	33.6	42.5	48.7	50.2
TFA w/cos [26]	39.8	36.1	44.7	55.7	56.0	23.5	26.9	34.1	35.1	39.1	30.8	34.8	42.8	49.5	49.8
Meta-DETR [33]	40.6	51.4	58.0	59.2	63.6	37.0	36.6	43.7	49.1	54.6	41.6	45.9	52.7	58.9	60.6
FSOD-UP [29]	43.8	47.8	50.3	55.4	61.7	31.2	30.5	41.2	42.2	48.3	35.5	39.7	43.9	50.6	53.5
TTP-PRD [12]	54.6	–	56.5	–	61.4	37.9	–	44.4	–	47.6	42.8	–	46.6	–	51.2
DeFRCN [20]	53.6	57.5	61.5	64.1	60.8	30.1	38.1	47.0	53.3	47.9	48.4	50.9	52.3	54.9	57.4
CoCo-RCNN [17]	33.5	44.2	50.2	57.5	63.3	25.3	31.0	39.6	43.8	50.1	24.8	36.9	42.8	50.8	57.7
FCT [8]	49.9	57.1	57.9	63.2	67.1	27.6	34.5	43.7	49.2	51.2	39.5	54.7	52.3	57.0	58.7
LVC [10]	54.5	53.2	58.8	63.2	65.7	32.8	29.2	50.7	49.8	50.6	48.4	52.7	55.0	59.6	59.6
ours	53.9	**64.9**	**62.0**	**68.2**	**69.0**	27.1	33.9	41.5	45.7	**51.4**	40.9	50.4	**53.3**	**63.2**	**64.0**

4 Experiments

4.1 Experimental Setting

Dataset. We follow the previous works [9,26] and evaluate our approach on PASCAL VOC [4] and MS COCO [13] datasets, utilizing the same data splits provided by [26] for a fair comparison. For PASCAL VOC, our model is trained on the trainval sets of VOC 2007 and VOC 2012, and tested on VOC 2007 test set. We have three random split groups, where each group contains 15 base classes and 5 novel classes. Each novel class has K = 1, 2, 3, 5, 10 annotated instances for few-shot training. We report AP50 of novel classes (nAP50) on PASCAL VOC 07 test set. For MS COCO, we utilize a 5k subset of the COCO2014 validation set for evaluation and the remaining COCO2014 training and validation set for training. The 20 classes overlapped with PASCAL VOC are selected as novel classes with K = 10, 30 shots, the remaining 60 classes are selected as base classes. We report COCO-style mAP, mAP50, mAP75 on validation set.

Implementation Details. Our model based on Pyramid Vision Transformer (PVT) [25]. Our model is trained with AdamW optimizer over 4T A100 GPUs and the batch size is set to 8. The initial learning rate of 0.0002, weight decay of 0.0001. For PASCAL VOC dataset, the total number of training iterations is 10,000 and the learning rate divided by 10 after 7,500 and 10,000 iterations. For MS COCO dataset, the total number of training iterations is 110,000 and the learning rate divided by 10 after 85,000 and 100,000 iterations.

4.2 Comparison Results

PASCAL VOC. We present the results of our method on PASCAL VOC in Table 1. It can be observed that as the number of annotation instances increases, the performance of our model gradually improves. More importantly, the experimental results show that our method achieves significant improvement compared with the previous works. For example, for Novel Set 3, our approach surpasses the previously best method by 4.2% and 5.1% in 5-shot and 10-shot scenarios. In Novel Set 1, our approach surpasses the previously best method by 1.0% and 1.6% in 5-shot and 10-shot scenarios.

Table 2. Experimental results on MS COCO dataset.

Method/ Shot	10 shot			30 shot		
	AP	AP50	AP75	AP	AP50	AP75
MPSR [30]	9.8	17.9	9.7	14.1	25.4	14.2
TFA w/fc [26]	10.0	19.2	9.2	13.4	24.7	13.2
TFA w/cos [26]	10.0	19.1	9.3	13.7	24.9	13.4
FSOD-UP [29]	11.0	–	10.7	15.6	–	15.7
FSCE [23]	11.9	–	10.5	16.4	–	16.2
FADI [1]	12.2	22.7	11.9	16.1	29.1	15.8
Meta Faster R-CNN [7]	12.7	25.7	10.8	16.6	31.8	15.8
ours	**14.0**	24.7	**13.5**	**17.5**	29.3	**17.9**

MS COCO. We also evaluate our method on MS COCO dataset with the standard COCO metrics. As shown in Table 2, our method outperforms other methods in most evaluation metrics. Particularly, our model achieves 1.3% and 0.7% improvement in 10-shot and 30-shot AP respectively, which proves the effectiveness of our approach.

Table 3. Ablation study of different modules.

	Method		Shot					*Avg.*
	DAM	CAH	1	2	3	5	10	
(a)			53.9	60.3	59.8	68.3	65.7	61.6
(b)	✓		53.6	63.8	64.3	66.3	67.6	63.1
(c)		✓	55.9	63.2	62.7	66.1	66.4	62.9
(d)	✓	✓	54.1	64.9	62.0	68.2	69.0	63.6

4.3 Ablation Study

To verify the effectiveness of our model architecture, we conduct comprehensive ablation studies on the Novel Set 1 of Pascal VOC dataset. For fair comparison, the baseline detector based on FCT [8], which is the first vision transformer based detection model for few-shot object detection and also a robust meta-learning method that considers multi-level feature interactions. The architecture consists of a stack of cross transformer layers, which perform dual feature aggregation for both query and support branch. Ablation study results are shown in Table 3.

Impact of Dynamic Aggregation Module (DAM). The dynamic aggregation module highlights the salient information by aggregating the support features into conditions. As shown in Table 3(b), when only using the dynamic aggregation module, the average performance significantly improves by 1.5%. This indicates that the dynamic aggregation module is beneficial to highlight meaningful and salient information.

Effectiveness of Hypernetwork. Hypernetwork can generate class-specific parameters that will be used to interact with the query features. In Table 3(c), we find that the hypernetwork module gains 1.3% points on the average performance. Besides, The refined hyperparameters enable fine-grained interactions between the query and support branch, which improves few-shot detection performance under lower-shot settings. Specifically, the improvements are particularly significant under 1-shot (+2.0% mAP), 2-shot (+2.9% mAP), and 3-shot (+2.9% mAP) scenarios.

Table 4. Ablation study on each component in our model using different variants.

	variant	1 shot	2 shot	3 shot	5 shot	10 shot
(a)	linear	44.3	55.3	56.1	60.8	61.4
(b)	learned CLS	47.3	58.4	57.2	64.1	64.2
(c)	intra-support	53.2	60.6	59.3	65.2	66.1
(d)	w/o bernoulli	53.8	60.9	59.4	65.4	66.9
(e)	detection hpy	53.2	59.8	58.5	64.6	67.2
(f)	w/o cross	47.1	57.7	56.4	63.8	62.7
(g)	ours	54.1	64.9	62.0	68.2	69.0

Alternative Model Settings

1) Different aggregation manners. In Table 4, we explore several comparative experiments about the different model settings. In the rows of (a)-(b), we utilize the basic MLP method and prepend a learnable token [cls] (similarity to ViT [3]) to directly aggregate the features of support sets. The results show the effectiveness of our current hypernetwork scheme because more useless information being suppressed during the interactions of the two-branch.

2) Details of hypernetwork. In the DAM module, we use the average global pooling operation to highlight the salient information when aggregating the support features. Similarly, we explore prominent information of intra-image through simply global pooling (*i.e.,* Table 4(c)). We can observe that it has no obvious helpful effect on extracting the salient information of the support sets. Possibly, this is due to the fact that global pooling cannot directly eliminate the impact of intra-image useless information. In addition, Table 4(d) shows that smoothing hypernetwork weights distribution can slightly improve the performance of the model.

3) Location and input of hypernetwork. As shown in Table 4(e), hypernetwork weight generated based on coupled features have a negative impact on proposal features. Besides, as shown in Table 4(f), if the input of the hypernetwork does not introduce query-perceptual information, which means both the query and support branch adopt self-attention mechanism, the performance of the model will be significantly negatively affected. It effectively demonstrates the value of multi-level and high-quality interactions between the two-branch.

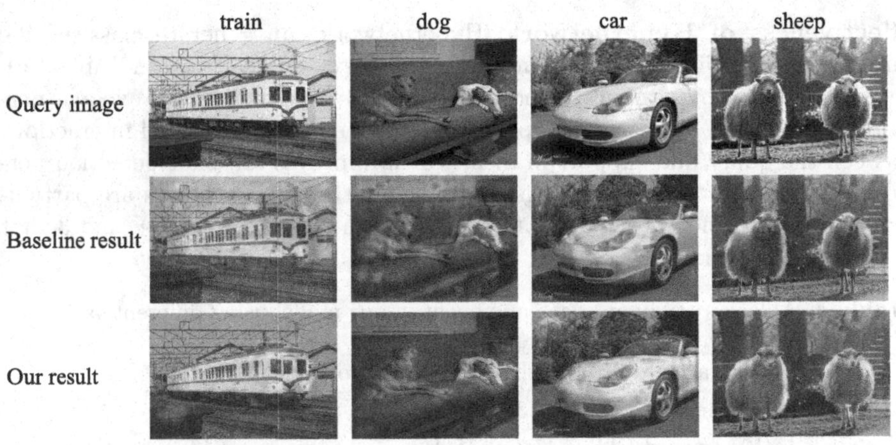

Fig. 3. Visualization results of the intermediate feature maps.

4.4 Visualization of Our Module

To analyze the behavior of our module, we visualize the feature maps output by AHT as a heatmap. As shown in Fig. 3, we find that our AHT enhances the aggregation of the query and support branch, highlighting information similar to class-specific support features in the query image.

5 Conclusion

We propose a novel and effective meta-learning framework AHT, which integrates hypernetwork to generate corresponding weights for the first time. We extensively experiment with different Hypernetwork strategies, in which Dynamic Aggregation Module is proposed to highlight the salient feature in support sets when generating aggregated weight. Furthermore, we build Conditional Adaptation Hypernetworks that dynamically generates class-specific weights interacting with the query branch. Despite its simplicity, Exhaustive experiments validate the effectiveness of our method which outperformed the previous meta-learning methods on the current benchmarks.

Acknowledgements. This work was supported by National Key R&D Program of China (No. 2020AAA0106900), the National Natural Science Foundation of China (No. U19B2037), Shaanxi Provincial Key R&D Program (No. 2021KWZ-03), and Natural Science Basic Research Program of Shaanxi (No. 2021JCW-03).

References

1. Cao, Y., et al.: Few-shot object detection via association and discrimination. In: Advances in Neural Information Processing Systems, vol. 34, pp. 16570–16581 (2021)
2. Chauhan, V.K., Zhou, J., Lu, P., Molaei, S., Clifton, D.A.: A brief review of hypernetworks in deep learning. arXiv preprint arXiv:2306.06955 (2023)
3. Dosovitskiy, A., et al.: An image is worth 16 × 16 words: transformers for image recognition at scale. arXiv preprint arXiv:2010.11929 (2020)
4. Everingham, M., Van Gool, L., Williams, C.K., Winn, J., Zisserman, A.: The pascal visual object classes (VOC) challenge. Int. J. Comput. Vis. **88**, 303–338 (2010)
5. Fan, Q., Zhuo, W., Tang, C.K., Tai, Y.W.: Few-shot object detection with attention-RPN and multi-relation detector. In: Proceedings of the IEEE/CVF Conference on Computer Vision and Pattern Recognition, pp. 4013–4022 (2020)
6. Ha, D., Dai, A., Le, Q.V.: Hypernetworks. arXiv preprint arXiv:1609.09106 (2016)
7. Han, G., Huang, S., Ma, J., He, Y., Chang, S.F.: Meta faster R-CNN: towards accurate few-shot object detection with attentive feature alignment. In: Proceedings of the AAAI Conference on Artificial Intelligence, vol. 36, pp. 780–789 (2022)
8. Han, G., Ma, J., Huang, S., Chen, L., Chang, S.F.: Few-shot object detection with fully cross-transformer. In: Proceedings of the IEEE/CVF Conference on Computer Vision and Pattern Recognition, pp. 5321–5330 (2022)
9. Kang, B., Liu, Z., Wang, X., Yu, F., Feng, J., Darrell, T.: Few-shot object detection via feature reweighting. In: Proceedings of the IEEE/CVF International Conference on Computer Vision, pp. 8420–8429 (2019)

10. Kaul, P., Xie, W., Zisserman, A.: Label, verify, correct: a simple few shot object detection method. In: Proceedings of the IEEE/CVF Conference on Computer Vision and Pattern Recognition, pp. 14237–14247 (2022)
11. Li, B., et al.: Dropkey for vision transformer. In: Proceedings of the IEEE/CVF Conference on Computer Vision and Pattern Recognition, pp. 22700–22709 (2023)
12. Lin, S., Zeng, X., Yan, S., Zhao, R.: Three-stage training pipeline with patch random drop for few-shot object detection. In: Proceedings of the Asian Conference on Computer Vision, pp. 1027–1043 (2022)
13. Lin, T.-Y., et al.: Microsoft COCO: common objects in context. In: Fleet, D., Pajdla, T., Schiele, B., Tuytelaars, T. (eds.) ECCV 2014. LNCS, vol. 8693, pp. 740–755. Springer, Cham (2014). https://doi.org/10.1007/978-3-319-10602-1_48
14. Littwin, G., Wolf, L.: Deep meta functionals for shape representation. In: Proceedings of the IEEE/CVF International Conference on Computer Vision, pp. 1824–1833 (2019)
15. Liu, W., et al.: SSD: single shot multibox detector. In: Leibe, B., Matas, J., Sebe, N., Welling, M. (eds.) ECCV 2016. LNCS, vol. 9905, pp. 21–37. Springer, Cham (2016). https://doi.org/10.1007/978-3-319-46448-0_2
16. Lu, X., et al.: Breaking immutable: Information-coupled prototype elaboration for few-shot object detection. arXiv preprint arXiv:2211.14782 (2022)
17. Ma, J., Han, G., Huang, S., Yang, Y., Chang, S.F.: Few-shot end-to-end object detection via constantly concentrated encoding across heads. In: Avidan, S., Brostow, G., Cissé, M., Farinella, G.M., Hassner, T. (eds.) ECCV 2022. LNCS, vol. 13686, pp. 57–73. Springer, Cham (2022). https://doi.org/10.1007/978-3-031-19809-0_4
18. Nirkin, Y., Wolf, L., Hassner, T.: HyperSeg: patch-wise hypernetwork for real-time semantic segmentation. In: Proceedings of the IEEE/CVF Conference on Computer Vision and Pattern Recognition, pp. 4061–4070 (2021)
19. Peng, H., Du, H., Yu, H., Li, Q., Liao, J., Fu, J.: Cream of the crop: distilling prioritized paths for one-shot neural architecture search. In: Advances in Neural Information Processing Systems, vol. 33, pp. 17955–17964 (2020)
20. Qiao, L., Zhao, Y., Li, Z., Qiu, X., Wu, J., Zhang, C.: DeFRCN: decoupled faster R-CNN for few-shot object detection. In: Proceedings of the IEEE/CVF International Conference on Computer Vision, pp. 8681–8690 (2021)
21. Redmon, J., Divvala, S., Girshick, R., Farhadi, A.: You only look once: unified, real-time object detection. In: Proceedings of the IEEE Conference on Computer Vision and Pattern Recognition, pp. 779–788 (2016)
22. Ren, S., He, K., Girshick, R., Sun, J.: Faster R-CNN: towards real-time object detection with region proposal networks. In: Advances in Neural Information Processing Systems, vol. 28 (2015)
23. Sun, B., Li, B., Cai, S., Yuan, Y., Zhang, C.: FSCE: few-shot object detection via contrastive proposal encoding. In: Proceedings of the IEEE/CVF Conference on Computer Vision and Pattern Recognition, pp. 7352–7362 (2021)
24. Wang, K., Zhang, L.: Reconcile prediction consistency for balanced object detection. In: Proceedings of the IEEE/CVF International Conference on Computer Vision, pp. 3631–3640 (2021)
25. Wang, W., et al.: PVT v2: improved baselines with pyramid vision transformer. Comput. Vis. Media **8**(3), 415–424 (2022)
26. Wang, X., Huang, T.E., Darrell, T., Gonzalez, J.E., Yu, F.: Frustratingly simple few-shot object detection. arXiv preprint arXiv:2003.06957 (2020)

27. Wang, X., Yu, F., Wang, R., Darrell, T., Gonzalez, J.E.: TAFE-Net: task-aware feature embeddings for low shot learning. In: Proceedings of the IEEE/CVF Conference on Computer Vision and Pattern Recognition, pp. 1831–1840 (2019)
28. Wang, Y.X., Ramanan, D., Hebert, M.: Meta-learning to detect rare objects. In: Proceedings of the IEEE/CVF International Conference on Computer Vision, pp. 9925–9934 (2019)
29. Wu, A., Han, Y., Zhu, L., Yang, Y.: Universal-prototype enhancing for few-shot object detection. In: Proceedings of the IEEE/CVF International Conference on Computer Vision, pp. 9567–9576 (2021)
30. Wu, J., Liu, S., Huang, D., Wang, Y.: Multi-scale positive sample refinement for few-shot object detection. In: Vedaldi, A., Bischof, H., Brox, T., Frahm, J.-M. (eds.) ECCV 2020. LNCS, vol. 12361, pp. 456–472. Springer, Cham (2020). https://doi.org/10.1007/978-3-030-58517-4_27
31. Yan, X., Chen, Z., Xu, A., Wang, X., Liang, X., Lin, L.: Meta R-CNN: towards general solver for instance-level low-shot learning. In: Proceedings of the IEEE/CVF International Conference on Computer Vision, pp. 9577–9586 (2019)
32. Yin, L., Perez-Rua, J.M., Liang, K.J.: Sylph: a hypernetwork framework for incremental few-shot object detection. In: Proceedings of the IEEE/CVF Conference on Computer Vision and Pattern Recognition, pp. 9035–9045 (2022)
33. Zhang, G., Luo, Z., Cui, K., Lu, S., Xing, E.P.: Meta-DETR: image-level few-shot detection with inter-class correlation exploitation. IEEE Trans. Pattern Anal. Mach. Intell. (2022)
34. Zhmoginov, A., Sandler, M., Vladymyrov, M.: HyperTransformer: model generation for supervised and semi-supervised few-shot learning. In: International Conference on Machine Learning, pp. 27075–27098. PMLR (2022)

Feature Refinement from Multiple Perspectives for High Performance Salient Object Detection

Xuan Li, Congao Wang, Ding Ma, and Xiangqian Wu(✉)

Faculty of Computing, Harbin Institute of Technology, Harbin, China
xuanli@stu.hit.edu.cn, wangcongao@gmail.com, martin3436@yeah.net, xqwu@hit.edu.cn

Abstract. Recently, deep-learning based salient object detection methods have gained great progress. However, there still exists some problems such as inefficient multi-level feature fusion, unstable multi-scale context-aware feature extraction, detail loss caused by upsampling and unbalanced distribution. To efficiently fuse multi-level features, we propose an attention-guided bi-directional feature refinement module (ABFRM) including top-down and bottom-up processes, which applies different attention-based feature fusion strategies for different directional processes. To obtain stable multi-scale contextual features, we design a serial atrous fusion module (SAFM), which uses serial atrous convolutional layers with small dilation rates. To reduce detail loss caused by upsampling with a large factor, we devise an upsampling feature refinement module (UFRM), which utilizes the combination of deconvolution and bilinear interpolation. To address unbalanced distribution from both foreground and background perspectives, we propose a novel hybrid loss, which contains Intersection-over-Union (IoU) and background boundary (BGB) losses. Comprehensive experiments on five benchmark datasets demonstrate that our proposed method outperforms 13 state-of-the-art approaches under four evaluation metrics. The code is available at https://github.com/xuanli01/PRCV210.

Keywords: Salient object detection · Attention-guided bi-directional feature refinement · Serial atrous fusion · Upsampling feature refinement · Background boundary loss

1 Introduction

Salient object detection (SOD), which aims to detect and segment the most attractive objects in an image, is an efficient pre-processing procedure of many computer vision tasks.

This work was supported in part by the National Key Research and Development Program of China under Grant 2020AAA0106502, in part by the Natural Science Foundation of China under Grant 62073105, in part by the Natural Science Foundation of Heilongjiang Province of China under Grant ZD2022F002, and in part by the Heilongjiang Touyan Innovation Team Program.

© The Author(s), under exclusive license to Springer Nature Singapore Pte Ltd. 2024
Q. Liu et al. (Eds.): PRCV 2023, LNCS 14436, pp. 56–67, 2024.
https://doi.org/10.1007/978-981-99-8555-5_5

Fig. 1. Challenges in SOD. (a) is the ground truth. (b) is our method. In (c), we replace the ABFRM in our method with a BiFPN layer [15]. In (d), we replace the SAFM in our method with ASPP module [2]. In (e), we replace the UFRM with bilinear interpolation. In (f), we replace our loss with the combination of BCE and IoU losses.

Although great progress has been made recently, there are still four big challenges in SOD. First, low-level features contain rich details but large amount of noises. High-level features contain accurate location information but lack enough details. To incorporate fine details and precise location at each level, researchers design many multi-level feature fusion algorithms [5,10,15,19,21] which can be divided into unidirectional and bi-directional methods. Compared with unidirectional methods [5,19], bi-directional methods [10,15,21], which contain both top-down and bottom-up processes, use both higher and lower level features to refine the features of each level. Top-down feature fusion is expected to be the process of utilizing accurate location information of high-level features to filter the low-level features. Bottom-up feature fusion is expected to be the process of utilizing rich detail information of low-level features to refine the high-level features. Most current bi-directional methods adopt the same fusion strategy in the top-down and bottom-up processes without fully considering the purposes of different directional processes, which may introduce noises and coarse boundaries (e.g., BiFPN [15] in Fig. 1(c)). Second, models are difficult to capture the visual context of scale-varying objects under the limited receptive field. To alleviate this issue, atrous spatial pyramid pooling (ASPP) module [2] or its variants are applied to capture multi-scale contextual information through parallel atrous convolutional layers with different dilation rates. However, a convolutional layer with a large dilation rate is hard to extract stable features (e.g., a failure case of ASPP module [2] to detect objects of different scales in Fig. 1(d)), which is caused by the weakness of association relationships among sampling points [26]. Third, many methods [17,18,25] directly apply interpolation algorithms to upsample the output of the final layer by a large factor to obtain the final saliency prediction maps, which is so rough for salient objects with elaborate structures that the details may be lost (see Fig. 1(e)). Fourth, unbalanced distribution in SOD refers to the phenomenon that images are usually dominated by the backgrounds. A model is biased to predict pixels to be background, if all pixels are treated equally by its loss function, e.g., Binary cross entropy (BCE) loss. To address the unbalanced distribution, some works [11,28] utilize Intersection-over-Union (IoU) loss or its variants to assist BCE loss by giving more focused on the foreground, which is incomplete without handling the background. In Fig. 1(f), some salient regions predicted by these methods may still have low confidence scores due to the bias of background.

For the first challenge, to fuse multi-level features efficiently, we propose an attention-guided bi-directional feature refinement module (ABFRM) which contains both top-down and bottom-up processes corresponding to top-down location refinement module (TLRM) and bottom-up detail refinement module (BDRM), respectively. For the purpose of the top-down process, TLRM utilizes high-level features to generate spatial attention to filter noises in the low-level features. To achieve the goal of the bottom-up process, BDRM uses the low-level features to generate channel attention to refine high-level features with the rich detail information. For the second challenge, we devise a serial atrous fusion module (SAFM). For the sake of stable multi-scale context-aware feature extraction, SAFM adopts multiple serial atrous convolutional layers with small dilation rates, which have almost the same receptive field as an atrous convolutional layer with a large dilation rate. Features extracted from atrous convolutional layers of different depths correspond to contextual information of different scales. For the third challenge, we propose an upsampling feature refinement module (UFRM), which aims to enhance the details lost during the process of upsampling with a large factor by an integration of deconvolution and bilinear interpolation. For the fourth challenge, we are inspired by the powerful unbalanced distribution handling capability of boundary loss [6] to design a background boundary (BGB) loss, which modifies boundary loss [6] to only regularize the background. Then, we combine BGB loss with IoU loss, forming a hybrid loss to settle the unbalanced distribution from both foreground and background perspectives.

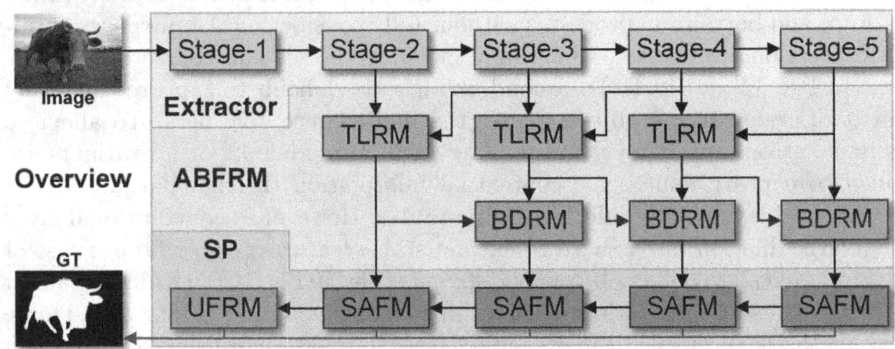

Fig. 2. Overall architecture of our proposed method. Red and black connectors indicate supervision and information flow, respectively (better viewed in color). (Color figure online)

2 Proposed Method

2.1 Overall Architecture

The overall architecture is shown in Fig. 2. Firstly, we utilize ResNet-50 [4] as the backbone to extract the multi-level features from the input images. For ease of statement, $i \in \{1, 2, 3, 4, 5\}$ indicates the i-th level from bottom to top.

We denote the backbone features as $\mathcal{F} = \{F_i\}_{i=1}^{5}$. To reduce computation, we discard F_1 from stage-1 due to its large spatial size. Then, we fuse multi-level backbone features by feeding them into the ABFRM. The output features are denoted by $\mathcal{R} = \{R_i\}_{i=2}^{5}$. After that, we feed $\mathcal{R}$ into the saliency predictor (SP) which is comprised of SAFM and UFRM to obtain and further refine multi-scale contextual features. We upsample the output of UFRM by a factor of 2 via a bilinear interpolation operation to get the final prediction S_1. Details of each component are discussed below.

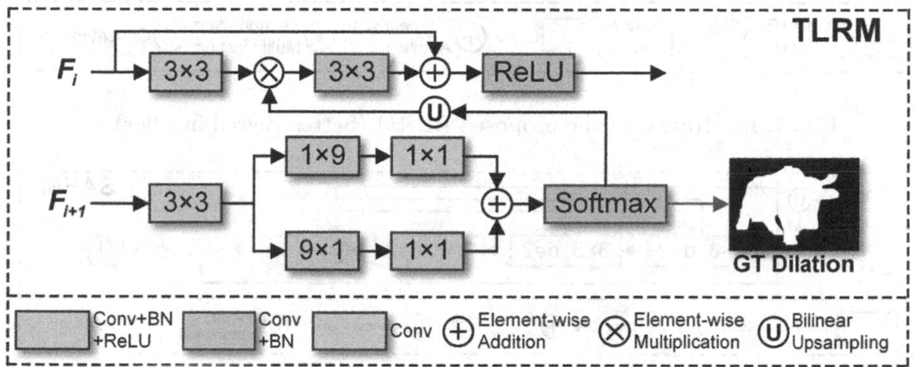

Fig. 3. Illustration of the proposed TLRM. Red and black connectors indicate supervision and information flow, respectively (better viewed in color). (Color figure online)

2.2 Attention-Guided Bi-directional Feature Refinement Module

The ABFRM, which is composed of TLRM and BDRM, aims to efficiently aggregate multi-level backbone features to obtain feature representation with fine detail and accurate location information. In practice, we first pass backbone features through the TLRM to obtain location refined features which are denoted by $\mathcal{B} = \{B_i\}_{i=2}^{4}$. Then, the location refined features are fed into the BDRM to get the detail refined features.

Top-Down Location Refinement Module. This module aims to introduce precise location information from high-level backbone features F_{i+1} to low-level backbone features F_i to highlight salient object regions and reduce the influence of complex background regions, which is shown in Fig. 3. Specifically, we first pass F_{i+1} through a 3×3 convolutional layer and two group convolutions with two convolutional layers in each group to increase receptive field. Then, we obtain spatial attention A_i via a softmax function. After that, we upsample the attention map by a factor of 2 and use it to weight the features generated by feeding F_i into a 3×3 convolutional layer. At last, a 3×3 convolutional layer and a residual connection are applied to acquire the final output of the TLRM.

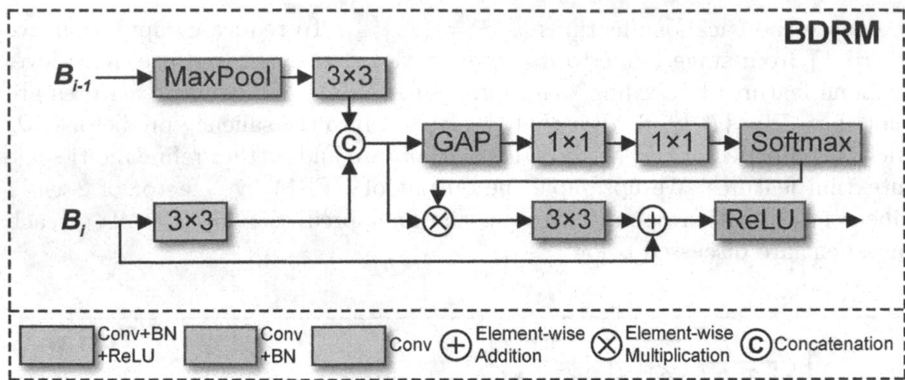

Fig. 4. Illustration of the proposed BDRM (better viewed in color).

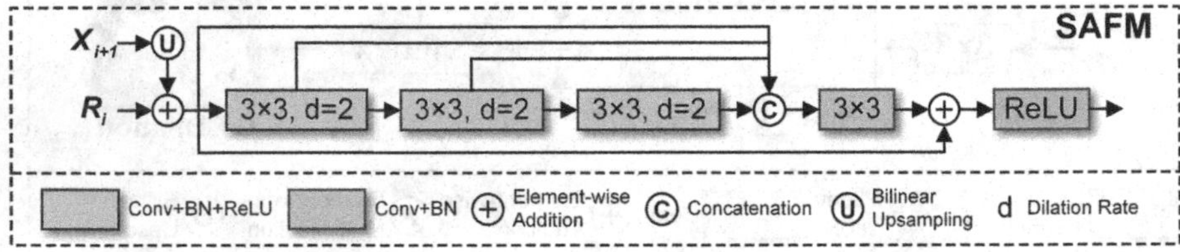

Fig. 5. Illustration of our proposed SAFM (better viewed in color).

Bottom-Up Detail Refinement Module. This module (see Fig. 4) aims to use rich detail information from low-level TLRM features B_{i-1} to refine the high-level TLRM features B_i. Specifically, we first dowsample B_{i-1} via a maxpooling operation with the kernel size of 2 and stride of 2. Two 3×3 convolutional layers are applied for transition of the B_i and dowsampled B_{i-1}, respectively. We utilize a cross-channel concatenation operation to merge these two transitioned features. Then, we adopt global average pooling (GAP), two 1×1 convolutional layers and a softmax function to get channel attention. After that, we weight the merged transitioned features with the channel attention via element-wise multiplication. At last, a 3×3 convolutional layer and a residual connection are applied to gain the final output of the BDRM.

2.3 Serial Atrous Fusion Module

Modeling multi-scale visual context is of vital importance to SOD, for the reason that salient objects have large variations in scale, shape and position. To achieve this, some methods apply ASPP-like modules, which use parallel atrous convolutional layers with different dilation rates to gather multi-scale context cues. Nevertheless, the information under a kernel with a large dilation rate is lack of steady relevance due to the sparsity of the kernel, leading to unstable feature extraction.

In this paper, we design a SAFM (see Fig. 5), which utilizes three serial atrous convolutional layers with a dilation rate of 2 to capture multi-scale context cues without large dilation rates. In SAFM, neurons can gain larger receptive field, as the serial atrous convolutional layers go deeper. Specifically, we first integrate the low-level ABFRM features R_i with the upsampled high-level SAFM features X_{i+1} through an element-wise addition operation to aggregate multi-level features. Then, we use three serial 3×3 atrous convolutional layers with the dilation rate of 2 to extract multi-scale contextual features. After that, we concatenate the input of the first atrous convolutional layer and outputs of three atrous convolutional layers along the channel dimension to fuse the multi-scale contextual features. At last, a 3×3 convolutional layer is adopted to enhance the fused features and a residual connection is applied to get the final output X_i of the SAFM.

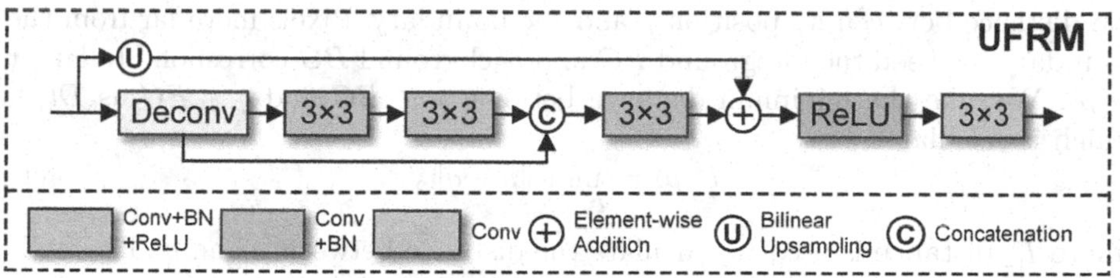

Fig. 6. Illustration of the proposed UFRM (better viewed in color).

2.4 Upsampling Feature Refinement Module

To acquire the final prediction, some methods [17,18,25] directly use interpolation algorithms to upsample the output of the final layer by a factor of 4. However, upsampling with a large factor via a simple interpolation algorithm (e.g., bilinear interpolation) may result in losing details.

To resolve this issue, we propose a UFRM (see Fig. 6) to enhance the details lost during the process of upsampling with a large factor. The UFRM contains two upsampling branches with different methods. For one branch, we apply bilinear interpolation to upsample the lowest-level SAFM features X_2 by a factor of 2. For the other branch, we first upsample X_2 via a 4×4 deconvolution with a stride of 2. Then, we apply two 3×3 convolutional layers to obtain large receptive field. After that, we combine features containing context cues of different scales via cross-channel concatenation and adopt a 3×3 convolutional layer to enhance the merged features. To better recover fine details, we combine the output of the two branches via an element-wise addition operation, which can attain complementation in methods. At last, a 3×3 convolutional layer is applied to get the final output of the UFRM. The ultimate prediction is obtained using bilinear interpolation to upsample the UFRM output by a factor of 2, not 4.

2.5 Objective Function

To avoid the impact of the unbalanced distribution in SOD, most existing methods apply IoU loss, which optimizes holistic regions rather than focusing on single pixel. The IoU loss pays more attention to the foreground. An intuitive thought is that the background also needs to be involved in dealing with the unbalanced distribution to improve the performance.

In this paper, we design a BGB loss, which only constrains the background by removing the foreground regularization of boundary loss [6]. Boundary loss [6] takes the form of a distance metric on the space of contours, instead of regions. To comprehensively address the unbalanced distribution, we combine IoU loss with BGB loss $\mathcal{L}_{bgb}^i$ to form a hybrid loss $\mathcal{L}_h^i$, which achieves complementation in the focuses (*i.e.* foreground and background) and regularization methods (*i.e.* regions and a distance metric on the space of contours). For BGB loss, concretely, each pixel in the background will be assigned with a weight $D(p)$ which measures the distance between its position p and the boundary. Pixels more far from the boundary between the foreground FG and background BG correspond to larger $D(p)$. We take the minimum distance between $p \in BG$ and $q \in FG$ as $D(p)$, which is calculated by

$$D(p) = \min_{q \in FG} \|p - q\|_2 \tag{1}$$

where L_2 distance is used to calculate the distance between p and q. The BGB loss is obtained by weighted average calculation, which is expressed as

$$\mathcal{L}_{bgb}^i = \frac{1}{\sum\limits_{p \in BG} 1} \sum_{p \in BG} S_i(p) \odot D(p) \tag{2}$$

where $\{S_i\}_{i=2}^5$ are obtained by passing $\{X_i\}_{i=2}^5$ through a 3×3 convolutional layer and a bilinear interpolation operation. $\odot$ is a Hadamard product. Finally, the hybrid loss is denoted by

$$\mathcal{L}_h^i = \mathcal{L}_{iou}(S_i, GT) + \mathcal{L}_{bgb}^i \tag{3}$$

where GT is the ground truth. The IoU loss $\mathcal{L}_{iou}$ and BCE loss $\mathcal{L}_{bce}$ have the same expressions as [28]. To gain better spatial attention to assist location in TLRM, we propose to explicitly guide the learning process of the spatial attention with pseudo-labels. In practice, we generate the pseudo-labels for the spatial attention by dilating GT, which can make the spatial attention more focused on the location of salient objects and less focused on details under the constraint of the pseudo-labels. BCE loss is applied for this constraint, which is

$$\mathcal{L}_{sa}^i = \mathcal{L}_{bce}(A_i, \mathrm{MaxPool}(GT)) \tag{4}$$

where the dilation operation is implemented by a maxpooling operation with the kernel size of 25 and stride of 1. The final loss $\mathcal{L}$ is the combination of $\mathcal{L}_h^i$ and $\mathcal{L}_{sa}^i$, which is

Table 1. Quantitative comparison. **Bold** and *italic* indicate the best and second-best performance, respectively.

Method	DUT-OMRON				DUTS-TE				ECSSD				HKU-IS				PASCAL-S			
	$F_\beta \uparrow$	$F_\omega \uparrow$	$M \downarrow$	$E_\gamma \uparrow$	$F_\beta \uparrow$	$F_\omega \uparrow$	$M \downarrow$	$E_\gamma \uparrow$	$F_\beta \uparrow$	$F_\omega \uparrow$	$M \downarrow$	$E_\gamma \uparrow$	$F_\beta \uparrow$	$F_\omega \uparrow$	$M \downarrow$	$E_\gamma \uparrow$	$F_\beta \uparrow$	$F_\omega \uparrow$	$M \downarrow$	$E_\gamma \uparrow$
F^3Net_{20}	.766	.747	.053	.876	.840	.835	.035	.918	.925	.912	.033	.946	.910	.900	.028	.958	.835	.816	.061	.895
$ITSD_{20}$	.756	.750	.061	.867	.804	.823	.041	.898	.899	.894	**.031**	.953	.895	.910	.034	.932	.785	.812	.066	.863
$MINet_{20}$	.755	.738	.056	.873	.828	.825	.037	.917	.924	.911	.033	.953	.909	.897	.029	.960	.829	.809	.064	.898
LDF_{20}	.773	.752	.052	.881	.855	.845	.034	.929	.930	.915	.034	.951	.914	.904	.028	.960	.843	.822	.060	.905
$GateNet_{20}$	.746	.729	.055	.868	.807	.809	.040	.903	.916	.894	.040	.943	.899	.880	.033	.953	.819	.797	.067	.884
$MSFNet_{21}$	.778	.757	*.050*	.876	.856	.841	.034	.931	.929	.916	.033	.954	.914	.903	.027	.959	.843	.822	.061	.901
$DCNet_{21}$	.774	.760	.051	*.885*	.843	.840	.035	.923	*.934*	*.920*	**.031**	**.957**	.914	.905	.027	.962	.837	.820	.062	.902
VST_{21}	.756	.755	.058	.872	.817	.828	.037	.916	.920	.910	.033	**.957**	.900	.897	.029	.960	.829	.816	.061	.902
$PFSNet_{21}$	.774	.756	.055	.883	.846	.842	.036	.922	.932	*.920*	**.031**	.953	.919	.910	.026	.962	.837	.819	.063	.895
$ICON_{22}$	.772	.761	.057	.879	.838	.836	.037	.919	.928	.918	*.032*	.954	.910	.902	.029	.958	.833	.818	.064	.893
$RCSBNet_{22}$	.778	.752	**.049**	.870	.856	.839	.035	.920	.927	.916	.034	.948	.923	.909	.027	.959	*.848*	.826	*.059*	*.907*
EDN_{22}	*.785*	*.770*	**.049**	*.885*	.851	.845	.035	.928	.932	.918	*.032*	.955	.919	.908	.026	.962	.847	.827	.062	.904
$DPNet_{22}$	.778	.767	**.049**	.882	*.861*	**.870**	**.028**	*.933*	.926	.918	**.031**	.947	*.924*	**.921**	**.023**	**.968**	.843	**.835**	**.054**	.898
Ours	**.798**	**.781**	**.049**	**.889**	**.872**	*.860*	*.032*	**.936**	**.940**	**.925**	**.031**	*.956*	**.929**	*.918*	*.024*	*.964*	**.855**	*.834*	.060	**.910**

$$\mathcal{L} = \sum_{i=1}^{5} \frac{1}{2^{i-1}} \mathcal{L}_h^i + \frac{1}{32} \sum_{i=2}^{4} \mathcal{L}_{sa}^i \tag{5}$$

3 Experiments

3.1 Experimental Setup

Datasets. The performance of our method is evaluated on five benchmark datasets: DUTS-TE [16], DUT-OMRON [23], ECSSD [22], HKU-IS [7] and PASCAL-S [8]. We choose DUTS-TR [16] as the training dataset like previous methods.

Evaluation Metrics. To quantitatively validate the proposed model, we adopt four evaluation metrics, including mean F-measure (F_β) [1], weighted F-measure (F_ω) [12], Mean Absolute Error (M) [14] and E-measure (E_γ) [3].

Implementation Details. DUTS-TR [16] is used to train our network. ResNet-50 [4], pre-trained on ImageNet, is used as the backbone network. Horizontal flipping, random cropping and multi-scale input images are applied to augment the images. We adopt the warm-up and linear decay strategies with a maximum learning rate of 0.005 for the backbone network and 0.05 for other parts. Our network is trained end-to-end using stochastic gradient descent (SGD) optimizer. Momentum and weight decay are set to 0.9 and 0.0005, respectively. We train the model for 48 epochs with a mini-batch size of 32. During testing, each image is resized to 352×352 to predict the saliency map without any post-processing.

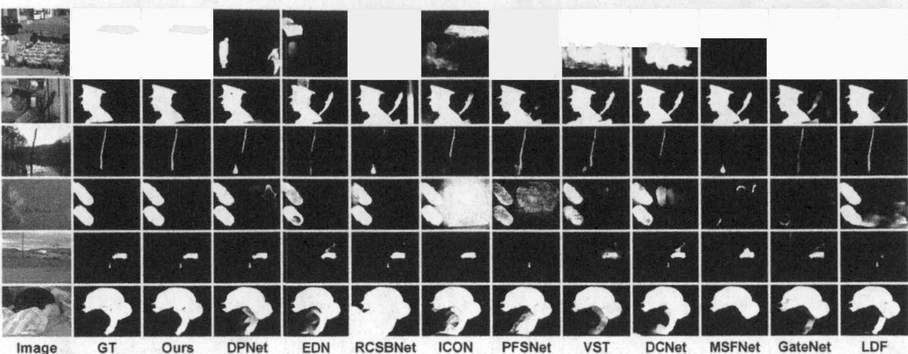

Fig. 7. Visual comparison of the proposed method with other state-of-the-art SOD models.

Table 2. Ablation study on different losses.

BCE	IoU	BGB	DUT-OMRON				DUTS-TE			
			$F_\beta \uparrow$	$F_\omega \uparrow$	$M \downarrow$	$E_\gamma \uparrow$	$F_\beta \uparrow$	$F_\omega \uparrow$	$M \downarrow$	$E_\gamma \uparrow$
✓			.733	.713	.059	.859	.799	.802	.041	.897
✓	✓		.752	.740	.059	.868	.824	.824	.039	.914
✓		✓	.748	.718	.057	.866	.820	.809	.040	.907
✓	✓	✓	.755	.738	.057	.870	.828	.824	.039	.915
	✓		.757	.740	.058	.871	.832	.823	.038	.920
	✓	✓	.764	.742	.056	.867	.842	.829	.038	.922

3.2 Comparison with State-of-the-Art Methods

We compare the proposed model with 13 state-of-the-art saliency detection methods, including F^{3}Net [17], ITSD [27], MINet [13], LDF [18], GateNet [26], MSFNet [25], DCNet [20], VST [9], PFSNet [11], ICON [28], RCSBNet [5], EDN [19] and DPNet [21]. VST [9] takes T2T-ViT [24] as the backbone and other methods take ResNet-50 [4] as the backbone. For fair comparisons, the saliency maps provided by these methods are evaluated by a unified code.

Quantitative Comparison. The comparison results of 13 methods under four evaluation metrics are shown in Table 1. It can be seen that our proposed method performs well on multiple datasets, especially DUT-OMRON and ECSSD. It is worth noting that we achieve the best F_β value for all datasets.

Visual Comparison. Figure 7 shows some saliency maps produced by our approach and other SOTA models. It can be observed that our method can deal with various challenging scenarios including cluttered backgrounds (1st row), large objects (2nd row), small objects (5th row), inverted reflection in water

Table 3. Ablation study on different architectures. #Params indicates the number of parameters of the model.

UFRM	SA	TLRM	BDRM	SAFM	DUT-OMRON				DUTS-TE				#Params (M)
					$F_\beta \uparrow$	$F_\omega \uparrow$	$M \downarrow$	$E_\gamma \uparrow$	$F_\beta \uparrow$	$F_\omega \uparrow$	$M \downarrow$	$E_\gamma \uparrow$	
					.764	.742	.056	.867	.842	.829	.038	.922	24.05
✓					.768	.751	.055	.873	.845	.834	.037	.924	24.20
✓	✓				.771	.755	.056	.878	.856	.845	.035	.930	24.59
✓		✓			.777	.757	.052	.879	.860	.846	.035	.930	24.59
✓		✓	✓		.789	.772	.052	.886	.863	.849	.035	.931	25.06
✓		✓	✓	✓	.798	.781	.049	.889	.872	.860	.032	.936	25.95

Table 4. Quantitative comparison of our ABFRM and other bi-directional multi-level feature fusion approaches [10,15] by replacing the ABFRM in our full model with a BiFPN layer [15] and the ASFF module [10]. #Params indicates the number of parameters of the model.

Model	DUT-OMRON				DUTS-TE				#Params (M)
	$F_\beta \uparrow$	$F_\omega \uparrow$	$M \downarrow$	$E_\gamma \uparrow$	$F_\beta \uparrow$	$F_\omega \uparrow$	$M \downarrow$	$E_\gamma \uparrow$	
ASFF	.784	.766	.051	.883	.865	.852	.033	.933	25.49
BiFPN	.788	.770	.051	.882	.867	.853	.033	.932	25.16
Ours	.798	.781	.049	.889	.872	.860	.032	.936	25.95

(3rd row), low-contrast (4th row) and multiple objects (4th, 5th and 6th rows). Compared with other competitors, the saliency maps generated by our approach are apparently clearer and more accurate.

3.3 Ablation Study

In this section, we investigate the importance of each component in our model on DUT-OMRON and DUTS-TE datasets. We remove the serial atrous convolutional layers, cross-channel concatenation and residual connection from SAFM to replace SAFM. The ABFRM is decomposed into TLRM and BDRM. The UFRM is replaced by bilinear interpolation.

Impact of Loss Functions. Table 2 shows the study on the impact of different loss functions, including BCE, IoU and BGB losses. It can be seen from the comparison results that BGB loss can improve the performance of the model regardless of how it is combined with BCE and IoU losses. The combination of IoU and BGB losses achieves the best result, which indicates the effectiveness of our hybrid loss.

Impact of Each Module. We progressively add UFRM, TLRM without the constraint of $\mathcal{L}_{sa}^i$ (SA), TLRM, BDRM and SAFM to evaluate their performance, which is shown in Table 3. The model is trained using IoU and BGB losses.

Table 5. Quantitative comparison of our SAFM and the ASPP module [2] by replacing the SAFM in our full model with the ASPP module [2]. #Params indicates the number of parameters of the model.

Model	DUT-OMRON				DUTS-TE				#Params (M)
	$F_\beta \uparrow$	$F_\omega \uparrow$	$M \downarrow$	$E_\gamma \uparrow$	$F_\beta \uparrow$	$F_\omega \uparrow$	$M \downarrow$	$E_\gamma \uparrow$	
ASPP	.791	.773	.049	.886	.869	.856	.032	.935	25.95
Ours	.798	.781	.049	.889	.872	.860	.032	.936	25.95

The model can be boosted by each module and obtain the best result with the combination of all components, which proves that all components are necessary for the proposed method. Moreover, we compare the ABFRM with other bidirectional multi-level feature fusion approaches (*i.e.* a BiFPN layer [15] and ASFF module [10]) in Table 4. The results show that our ABFRM performs better in the accuracy with little increasement of the number of parameters. In Table 5, we compare the SAFM with ASPP module [2] to verify the effectiveness of our multi-scale contextual feature extraction. It can be seen from the results that our SAFM performs better with the same number of parameters.

4 Conclusion

In this paper, our purpose is to address four SOD problems including inefficient multi-level feature integration, unstable multi-scale contextual feature acquisition, detail loss due to upsampling with a large factor and unbalanced distribution. First, we propose an ABFRM, which applies TLRM for top-down feature fusion and BDRM for bottom-up feature fusion, to efficiently aggregate multi-level features. Second, we design a SAFM to acquire stable multi-scale contextual features. Third, we devise a UFRM to supplement lost details caused by upsampling with a large factor. Fourth, we propose a new hybrid loss to solve unbalanced distribution from both foreground and background aspects. Comprehensive experiments on five public benchmark datasets demonstrate the superiority of our model over the state-of-the-art approaches.

References

1. Achanta, R., Hemami, S., Estrada, F., Susstrunk, S.: Frequency-tuned salient region detection. In: CVPR, pp. 1597–1604 (2009)
2. Chen, L.C., Papandreou, G., Kokkinos, I., Murphy, K., Yuille, A.L.: DeepLab: semantic image segmentation with deep convolutional nets, atrous convolution, and fully connected CRFs. IEEE TPAMI **40**(4), 834–848 (2017)
3. Fan, D., Gong, C., Cao, Y., Ren, B., Cheng, M., Borji, A.: Enhanced-alignment measure for binary foreground map evaluation. In: IJCAI, pp. 698–704 (2018)
4. He, K., Zhang, X., Ren, S., Sun, J.: Deep residual learning for image recognition. In: CVPR, pp. 770–778 (2016)

5. Ke, Y.Y., Tsubono, T.: Recursive contour-saliency blending network for accurate salient object detection. In: WACV, pp. 2940–2950 (2022)
6. Kervadec, H., Bouchtiba, J., Desrosiers, C., Granger, E., Dolz, J., Ayed, I.B.: Boundary loss for highly unbalanced segmentation. In: MIDL, pp. 285–296 (2019)
7. Li, G., Yu, Y.: Visual saliency based on multiscale deep features. In: CVPR, pp. 5455–5463 (2015)
8. Li, Y., Hou, X., Koch, C., Rehg, J.M., Yuille, A.L.: The secrets of salient object segmentation. In: CVPR, pp. 280–287 (2014)
9. Liu, N., Zhang, N., Wan, K., Shao, L., Han, J.: Visual saliency transformer. In: ICCV, pp. 4722–4732 (2021)
10. Liu, S., Huang, D., Wang, Y.: Learning spatial fusion for single-shot object detection. arXiv preprint arXiv:1911.09516 (2019)
11. Ma, M., Xia, C., Li, J.: Pyramidal feature shrinking for salient object detection. In: AAAI, vol. 35, pp. 2311–2318 (2021)
12. Margolin, R., Zelnik-Manor, L., Tal, A.: How to evaluate foreground maps? In: CVPR, pp. 248–255 (2014)
13. Pang, Y., Zhao, X., Zhang, L., Lu, H.: Multi-scale interactive network for salient object detection. In: CVPR, pp. 9413–9422 (2020)
14. Perazzi, F., Krähenbühl, P., Pritch, Y., Hornung, A.: Saliency filters: contrast based filtering for salient region detection. In: CVPR, pp. 733–740 (2012)
15. Tan, M., Pang, R., Le, Q.V.: EfficientDet: scalable and efficient object detection. In: CVPR, pp. 10781–10790 (2020)
16. Wang, L., et al.: Learning to detect salient objects with image-level supervision. In: CVPR, pp. 136–145 (2017)
17. Wei, J., Wang, S., Huang, Q.: F^3net: fusion, feedback and focus for salient object detection. In: AAAI, vol. 34, pp. 12321–12328 (2020)
18. Wei, J., Wang, S., Wu, Z., Su, C., Huang, Q., Tian, Q.: Label decoupling framework for salient object detection. In: CVPR, pp. 13025–13034 (2020)
19. Wu, Y.H., Liu, Y., Zhang, L., Cheng, M.M., Ren, B.: EDN: salient object detection via extremely-downsampled network. IEEE TIP **31**, 3125–3136 (2022)
20. Wu, Z., Su, L., Huang, Q.: Decomposition and completion network for salient object detection. IEEE TIP **30**, 6226–6239 (2021)
21. Wu, Z., Li, S., Chen, C., Qin, H., Hao, A.: Salient object detection via dynamic scale routing. IEEE TIP (2022)
22. Yan, Q., Xu, L., Shi, J., Jia, J.: Hierarchical saliency detection. In: CVPR, pp. 1155–1162 (2013)
23. Yang, C., Zhang, L., Lu, H., Ruan, X., Yang, M.H.: Saliency detection via graph-based manifold ranking. In: CVPR, pp. 3166–3173 (2013)
24. Yuan, L., et al.: Tokens-to-token ViT: training vision transformers from scratch on imagenet. In: ICCV, pp. 558–567 (2021)
25. Zhang, M., Liu, T., Piao, Y., Yao, S., Lu, H.: Auto-MSFNet: search multi-scale fusion network for salient object detection. In: ACMMM, pp. 667–676 (2021)
26. Zhao, X., Pang, Y., Zhang, L., Lu, H., Zhang, L.: Suppress and balance: a simple gated network for salient object detection. In: Vedaldi, A., Bischof, H., Brox, T., Frahm, J.-M. (eds.) ECCV 2020. LNCS, vol. 12347, pp. 35–51. Springer, Cham (2020). https://doi.org/10.1007/978-3-030-58536-5_3
27. Zhou, H., Xie, X., Lai, J.H., Chen, Z., Yang, L.: Interactive two-stream decoder for accurate and fast saliency detection. In: CVPR, pp. 9141–9150 (2020)
28. Zhuge, M., Fan, D.P., Liu, N., Zhang, D., Xu, D., Shao, L.: Salient object detection via integrity learning. IEEE TPAMI (2022)

Feature Disentanglement and Adaptive Fusion for Improving Multi-modal Tracking

Zheng Li[1,2,3], Weibo Cai[1,2,3], Junhao Dong[1,2,3], Jianhuang Lai[1,2,3], and Xiaohua Xie[1,2,3](✉)

[1] School of Computer Science and Engineering, Sun Yat-sen University, Guangzhou, China
{lizh525,caiwb6,dongjh8}@mail2.sysu.edu.cn, stsljh@mail.sysu.edu.cn,xiexiaoh6@mail.sysu.edu.cn

[2] Key Laboratory of Machine Intelligence and Advanced Computing, Ministry of Education, Guangzhou, China

[3] Guangdong Province Key Laboratory of Information Security Technology, Guangzhou, China

Abstract. Multi-modal tracking has increasingly gained attention due to its superior accuracy and robustness in complex scenarios. The primary challenges in this field lie in effectively extracting and fusing multi-modal data that inherently contain gaps. To address the above issues, we propose a novel regularized single-stream multi-modal tracking framework, drawing inspiration from the perspective of disentanglement. Specifically, taking into account the similarities and differences intrinsic in multi-modal data, we design a modality-specific weights sharing feature extraction module to extract well-disentangled multi-modal features. To emphasize feature-level specificity across different modal features, we propose a cross-modal deformable attention mechanism for the adaptive integration of multi-modal features with efficiency. Through extensive experiments on three multi-modal tracking benchmarks, including RGB+Thermal infrared and RGB+Depth, we demonstrate that our method significantly outperforms existing multi-modal tracking algorithms. Code is available at https://github.com/ccccwb/Multimodal-Detection-and-Tracking-UAV.

Keywords: Visual object tracking · Multi-modal tracking · Multi-modal Fusion · Cross-modal vision transformer

1 Introduction

Visual object tracking (VOT) [1,4,5,9,12,29] is appealing due to extensive applications like video surveillance, autonomous driving, and augmented reality. In

© The Author(s), under exclusive license to Springer Nature Singapore Pte Ltd. 2024
Q. Liu et al. (Eds.): PRCV 2023, LNCS 14436, pp. 68–80, 2024.
https://doi.org/10.1007/978-981-99-8555-5_6

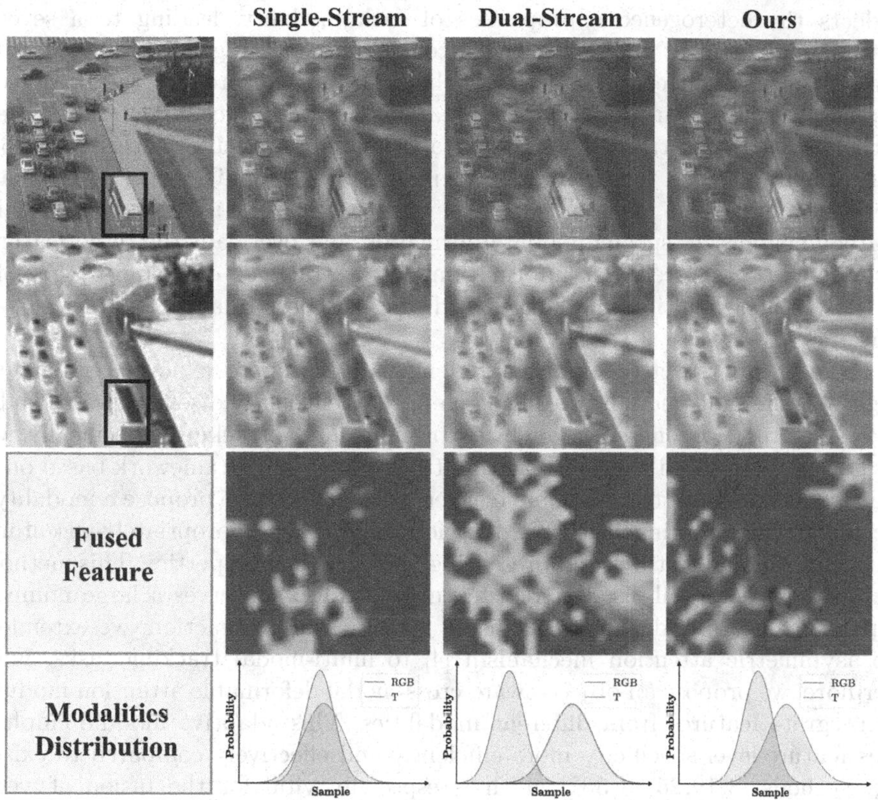

Fig. 1. Comparison of feature extraction strategies. The top three rows display the RGB, thermal infrared, and fused features, respectively. The final row illustrates the distribution relationships between modalities, with the x-axis representing sample values and the y-axis representing probabilities. The single-stream strategy suffers from a feature entanglement problem, whereas the dual-stream strategy exhibits an over-disentanglement phenomenon. In contrast, our method can disentangle multi-modal features well, providing more discriminative information for accurate localization.

recent years, numerous cutting-edge RGB trackers and datasets have been developed. However, the performance of RGB tracking tends to decrease in conditions such as rain, fog, and low light due to the limitations of visible sensors. Therefore, many recent works pay attention to the multi-modal tracking tasks, mainly including the RGB+Thermal infrared (RGB-T) and RGB+Depth (RGB-D) [22,30]. These approaches leverage either thermal infrared or depth images to provide additional temperature or depth information about the object, thereby enhancing the tracker's robustness and enabling all-weather tracking.

Previous trackers mainly use two feature extraction strategies. One is the single-stream strategy which simultaneously extracts multi-modal features [10,41], i.e., using identical network parameters. This data-friendly strategy neglects the heterogeneous properties of each modality, leading to a severe feature entanglement problem. The second column of Fig. 1 shows that this strategy focuses on common patterns but fails to capture the target object completely, yielding inadequate fused features for the target object. The other is the dual-stream strategy which extracts modality-specific features [28,36], which might overlook the collaborative cues between modalities and result in an over-disentanglement phenomenon. The third column of Fig. 1 shows that this strategy emphasizes specific modal information, but overloads the fused feature with much object-irrelevant information. Both strategies cannot fully exploit the potential of multiple modalities, reducing the performance of multi-modal trackers.

From the perspective of disentanglement, we propose a novel regularized single-stream multi-modal tracking framework to efficiently disentangle and fully fuse the features of different modalities for multi-modal tracking. Specifically, we design our regularized single-stream multi-modal tracking framework based on a one-stream one-stage transformers tracker [4,24]. Firstly, we propose a modality-specific weights sharing feature extraction module for appropriately disentangling multi-modal features and preserving homogeneous properties. This method not only extracts well-disentangled features but also conserves a large number of parameters. In addition, to facilitate faster feature extraction, we extended the asymmetric attention mechanism [4] to multi-modal tracking tasks. Furthermore, we propose an object-aware cross-modal deformable attention module to integrate features from different modalities. This adaptive module emphasizes feature-level specificity more efficiently and effectively compared to existing methods [8,19,25,34,36], which is especially vital for the fusion of well-disentangled features. As shown in the fourth column of Fig. 1, our method can appropriately disentangle multi-modal features and take full advantage of well-disentangled features, greatly improving the performance of the tracker.

Extensive experiments on several multi-modal tracking benchmarks prove that our proposed method sets a new state-of-the-art performance, achieving a real-time running speed of 36 FPS on a NVIDIA RTX3090 GPU. The main contributions are summarized as:

- From the perspective of disentanglement, we systematically analyze the importance of appropriately disentangling multi-modal features and propose a regularized single-stream multi-modal tracking framework.
- We propose a modality-specific weights sharing feature extraction module and an object-aware cross-modal deformable attention module to extract and fuse well-disentangled features.
- Extensive experiments demonstrate that our proposed method attains state-of-the-art performance across several multi-modal tracking benchmarks, including RGBT234, LasHeR, and DepthTrack.

2 Related Work

2.1 Multi-modal Tracking

Based on MDNet [21], Wang *et al.* [25] presented a novel tracking framework to diffuse instance patterns across RGB-T data in the spatial-temporal domain. Zhang *et al.* [36] designed a complementarity and distractor-aware RGB-T tracker based on the Siamese network [12]. Using DiMP [1] as a baseline, Gao *et al.* [8] proposed a dual-fused modality-aware tracker to learn informative and discriminative representations. Zhang *et al.* [35] jointly modeled appearance and motion cues using ECO tracker [5] and motion estimator. Different from the above methods, Zhao *et al.* [37] integrates RGB features with bird's-eye-view representations to better explore cross-modality 3D geometry for RGB-D tracking. From the perspective of disentanglement, we propose a regularized single-stream multi-modal tracking framework. A similar work is MANet++ [18], which proposed a multi-adapter network to jointly perform modality-shared, modality-specific, and instance-aware target representation learning. In contrast to the intricate network structure of MANet++, our method emphasizes the extraction and fusion of well-disentangled features in a simple but effective manner. This is achieved through a modality-specific weights sharing feature extraction module, augmented with a cross-modal deformable attention feature fusion module.

2.2 Transformers Tracking

Recently, several transformer-based VOT methods demonstrates excellent performance. Transformer trackers fall into two categories: CNN-Transformer-based trackers and fully-Transformer-based trackers [24]. The former category is mainly characterized by the use of two Siamese-like identical network pipelines. For instance, Yan *et al.* [29], building upon DETR [2] propose an encoder-decoder transformer tracking architecture. Chen *et al.* [3] present a Transformer tracking method based on siamese-like feature extraction and attention-based fusion mechanism. Feng *et al.* [7] designed a cross-modal model with shallow fusion and weight optimization based on [3]. The latter category exclusively employs transformers for feature extraction and integration. For example, Ye *et al.* [32] propose a one-stream one-stage tracking framework with a candidate early elimination module, while Cui *et al.* [4] design a similar synchronous modeling scheme with an asymmetric attention module. Drawing on [32], Zhu *et al.* [11] develop a visual prompt multi-modal tracking method, which learns the modal-relevant prompts to adapt the frozen pre-trained model to multi-modal tracking tasks. In this work, based on [4] we design a regularized single-stream multi-modal tracking framework to explore the importance of well-disentangled feature extraction in multi-modal tracking tasks.

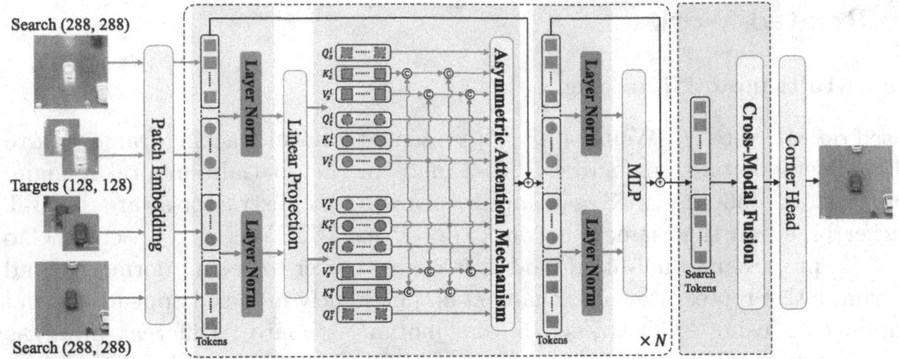

Fig. 2. Our proposed regularized single-stream multi-modal tracking framework. First, well-disentangled features are extracted using modality-specific weights sharing module with the asymmetric attention mechanism. Subsequently, multi-modal features are fused by applying a cross-modal fusion module. Lastly, the target object is localized employing a pyramidal corner head.

3 Methodology

3.1 Preliminary

Given a video with initial target bounding box B_0, VOT aims to estimate the location of the target in subsequent frames X_i. The tracking model can be expressed as $B_i = F(X_i; X_0, B_0)$, in the multi-modal tracking $X_i = (X_i^v, X_i^a)$, which symbolizes the RGB modality and another modality, respectively.

Given the simplicity and effectiveness of MixFormer [4], we adopt it as the foundation model for our proposed method. Specifically, the model input X_i^v contains the search region, template, and online template patch. First, the input will be mapped into non-overlapped patch embeddings using a convolutional layer with kernel size and stride of 16. Additionally, two different positional embeddings are added to the search and templates patch embeddings respectively. Subsequently, the concatenated search and templates tokens H_0^v are fed to an L-layer asymmetric encoder:

$$H_{l-1}^v = \text{AsymAttention}_l(\text{LN}^v(H_{l-1}^v)) + H_{l-1}^v, \qquad l = 1, 2, \ldots, L \qquad (1)$$

$$H_l^v = \text{MLP}_l(\text{LN}^v(H_{l-1}^v)) + H_{l-1}^v, \qquad l = 1, 2, \ldots, L \qquad (2)$$

where AsymAttention represents the asymmetric mixed attention between search and templates tokens. Finally, search tokens S_0^v are split from H_L^v, and the predicted result is obtained through a pyramidal corner head.

3.2 Our Approach

Overall Architecture. From the perspective of disentanglement, we propose a regularized single-stream multi-modal tracking framework, which is composed of two submodules. First, a modality-specific weights sharing module is introduced to extract well-disentangled features. Second, we propose an object-aware cross-modal feature fusion module, aimed to emphasize the feature-level specificity of different modal features. The overall architecture is illustrated in Fig. 2.

Modality-Specific Weights Sharing. Based on the foundation model in Sect. 3.1, we implement our regularized single-stream multi-modal tracking framework. Given the inputs $X_i = (X_i^v, X_i^a)$, tokens $H_0 = (H_0^v, H_0^a)$ are derived using an identical embedding layer and positional embeddings. Then the rest of forward propagation process can be formulated as:

$$H_{l-1} = \mathrm{CMAsymAttn}_l(\mathrm{LN}^v(H_{l-1}^v), \mathrm{LN}^a(H_{l-1}^a)), \quad l = 1, 2, \ldots, L \quad (3)$$

$$H_l = \mathrm{MLP}_l(\mathrm{LN}^v(H_{l-1}^v), \mathrm{LN}^a(H_{l-1}^a)), \quad l = 1, 2, \ldots, L \quad (4)$$

where CMAsymAttn denotes the cross-modal asymmetric attention mechanism. As shown in Fig. 2, this mechanism allows each search feature to query all template features and itself while each template feature only queries itself. In this module, only the layer normalization has a counterpart, while the attention mechanism and the multi-layer perceptron share the same parameters. This strategy greatly reduces the computation of the Vision Transformer (ViT) [6] backbone, while facilitating implicit cross-modal fusion for search frame features. An experimental analysis is conducted to examine the importance of modality-specific layer normalization in Sect. 4.3.

Object-Aware Cross-Modal Encoder. To emphasize feature-level specificity of well-disentangled features, we design an object-aware cross-modal fusion module. Current feature fusion methods [19,25,28,34,36] neglect the misalignment problem in multi-modal data. Consequently, they rely on implicitly learning alignment relations from a large volume of data. This impedes the full utilization of well-disentangled features, limiting further enhancement of performance. Drawing on the success of the deformable attention mechanism [27,40], we propose an object-aware cross-modal fusion module to address the misalignment problem in a highly efficient way.

Given the output search features $S_0 = (S_0^v, S_0^a)$ from the multi-modal backbone, we always use an adjusting layer (1×1 Conv and GroupNorm [26]) to change the channel numbers as necessary. The fusion process in the L'-Layer encoder can then be expressed as:

$$S_{l-1} = \mathrm{CMDAttn}_l(\mathrm{LN}(S_{l-1})), \quad l = 1, 2, \ldots, L' \quad (5)$$

$$S_l = \mathrm{MLP}_l(\mathrm{LN}(S_{l-1})), \quad l = 1, 2, \ldots, L' \quad (6)$$

where CMDAttn denotes the cross-modal deformable attention mechanism. In each block, we keep the modality-specific weights sharing strategy mentioned above. As a result, our fusion module can produce highly discriminative fused features, which aids in accurately localizing the target object. Finally, the fused features $S_{L'}$ are used to predict the target state via a pyramidal corner head.

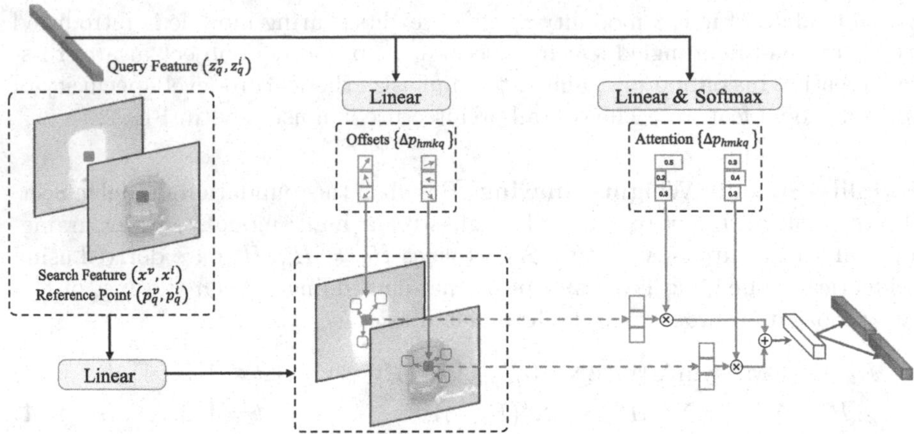

Fig. 3. Illustration of the proposed cross-modal deformable attention. We employ the query feature to compute the sampling offsets and attention weights for each reference point, followed by sampling and aggregating across all features. Finally, this aggregated feature is reintroduced to respective modal feature.

Cross-Modal Deformable Attention. In contrast to other fusion methods that focus on fixed spatial locations [19,25,34–36], our cross-modal deformable attention mechanism only attends to a small set of key sampling points around a reference point in each modality. This mechanism emphasizes the feature-level specificity of different modal features, resulting in more discriminatively fused features. The detailed design is depicted in Fig. 3.

Consider multi-modal features $\{x^m\}_{m=1}^M$, where $x^m \in \mathbb{R}^{C \times H \times W}$. For convenience, we let $x = \text{Concat}(\{x^m\}_{m=1}^M) \in \mathbb{R}^{MC \times H \times W}$ represents the concatenation of multi-modal features, $z_q \in \mathbb{R}^{MC}$ is a query element from x, and p_q is the corresponding reference point. The deformable attention module for each query z_q can then be expressed as:

$$\text{CMDAttn}_q(z_q, p_q, \{x^m\}) = \sum_{h=1}^{H} W_h \cdot \left[\sum_{m=1}^{M} \sum_{k=1}^{K} A_{hmkq} \cdot W_h' x^m \left(p_q + \Delta p_{hmkq}\right) \right], \tag{7}$$

where $W_h \in \mathbb{R}^{C' \times C}$ and $W_h' \in \mathbb{R}^{C \times C'}$ are learnable weights for h-th attention head. Δp_{hmkq} and A_{hmkq} are sampling offset and attention weight for k-th sampling point. $\Delta p_{hmkq} \in \mathbb{R}^2$, bilinear interpolation is used for features sampling. A_{hmkq} lies in the range $[0, 1]$, normalized by $\sum_{k=1}^{K} A_{hmkq} = 1$. Compared to standard attention fusion strategies [10], our method adaptively focuses on the region of interest in each modality, avoiding attention to potentially irrelevant global information. Additionally, our approach also greatly reduces computational load, by assigning only a small fixed number of keys to each query z_q.

3.3 Training and Inference

Training. In line with commonly used training configurations [4,29,32]. We train the model using AdamW [17] with weight decay 10^{-4}. For data augmentations, we use horizontal flip and brightness jittering. JET colormaps are applied to another modality [31]. We initialize the model with the pre-trained model from [4]. The base learning rate starts at 4×10^{-4} and decreases to 4×10^{-5}. The overall loss function combines ℓ_1 and CIoU [38] losses from predicted B_i and target $\hat{B}_i$ bounding boxes, as follows:

$$L_{loc} = \lambda_{\ell_1} L_1(B_i, \hat{B}_i) + \lambda_{ciou} L_{ciou}(B_i, \hat{B}_i), \tag{8}$$

where $\lambda_{\ell_1} = 5$ and $\lambda_{ciou} = 2$ are the respective weighting factors of two losses, determining the contribution of each loss component to the total loss.

Online template updates play an important role in VOT, capturing temporal information and handling object deformation. Similar to [4], we adopt a score prediction module to estimate the confidence score of current predicted bounding box. This module is initially pre-trained in the RGB modality and its loss is computed using standard cross-entropy loss.

Inference. During the inference phase, taking into account the computational load of multi-modal tracking, we utilize only a single online template. The online templates are updated only when the specified update interval is reached, at which point the sample with the highest confidence score is selected. In our work, we maintain a constant update interval of 200.

4 Experiments

4.1 Implementation Details

Basic Setting. Our trackers are implemented in Python using PyTorch, trained on 4 NVIDIA A100 GPUs. The regularized single-stream backbone employs the ViT-Base model [6] and the cross-modal fusion module utilizes a 2-layer encoder. Template images are 128×128 pixels, twice the target bounding box, while search images are 288×288 pixels, covering 4.5 times the target area. Each GPU holds 16 image pairs, resulting in a total batch size of 64. We conduct training for 150 and 100 epochs for RGB-T and RGB-D tracking, respectively, and train the score prediction module for 50 and 30 epochs.

Datasets and Metrics. In our work, we conduct comparative experiments on two RGBT datasets and one RGB-D dataset, namely LasHeR [14], RGBT234 [13], and DepthTrack [30]. LasHeR is a large-scale benchmark for RGB-T tracking, notable for its high alignment accuracy. It includes 979 training videos and 245 testing videos. In contrast, RGBT234 includes 234 videos with weakly aligned pairs of visible and thermal infrared images. For RGB-T tracking, we use

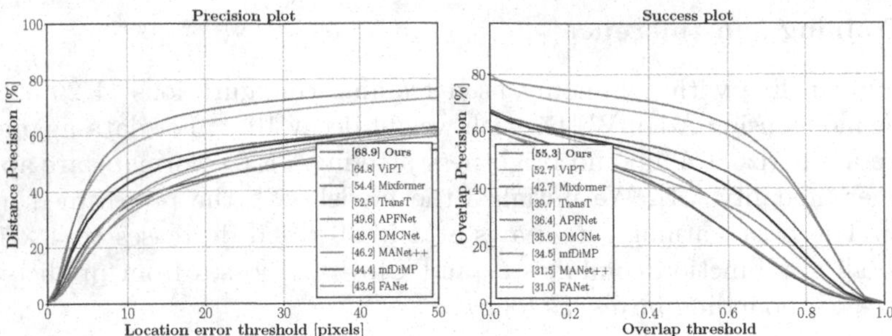

Fig. 4. Overall performance on LasHeR dataset.

Table 1. Overall performance on the RGBT234 dataset. The best three results are highlighted in red, blue, and green fonts.

	mfDiMP [34]	SiamCDA [36]	MIRNet [10]	DMCNet [19]	ProTrack [31]	Li et al. [15]	ViPT [11]	**Ours**
MPR(↑)	0.646	0.760	0.816	0.839	0.795	0.846	0.835	0.882
MSR(↑)	0.428	0.569	0.589	0.593	0.599	0.613	0.617	0.664

Table 2. Overall performance on the DepthTrack dataset. The best three results are highlighted in red, blue, and green fonts.

	CA3DMS [16]	DAL [23]	DeT [30]	SPT [39]	ProTrack [31]	ViPT [11]	DMTracker [8]	**Ours**
F-score(↑)	0.223	0.429	0.532	0.538	0.578	0.594	0.608	0.620
Re(↑)	0.228	0.369	0.506	0.549	0.573	0.596	0.597	0.615
Pr(↑)	0.218	0.512	0.560	0.527	0.583	0.592	0.619	0.625

the training split of LasHeR to train our network. The performance is evaluated using the maximum precision rate (MPR) and maximum success rate (MSR). DepthTrack is a large-scale RGB-D tracking benchmark, which comprises 150 training and 50 test videos. It employs precision (PR), recall (Re), and F-score [20] to assess the accuracy and robustness of target localization. For RGB-D tracking, we train our model using the training sets from DepthTrack and evaluate its performance on the test sets from the same benchmark.

4.2 Comparison with State-of-the-Arts Multi-modal Trackers

To demonstrate the effectiveness of our method, we conduct extensive experiments comparing it with state-of-the-art trackers on three multi-modal tracking benchmarks. The overall tracking performances of RGB-T benchmarks are shown in Fig. 4 and Table 1. Our method exhibits superior performance over other state-of-the-art trackers across all metrics on both RGB-T benchmarks. Specifically, on the LasHeR benchmark, our method achieves 55.3%/68.9% in MPR/MSR. On the RGBT234 benchmark, our method achieves 66.4%/88.1% in MPR/MSR. As presented in Table 2, our work has also surpassed all previous SOTA trackers on the RGB-D benchmark and obtained the highest F-score

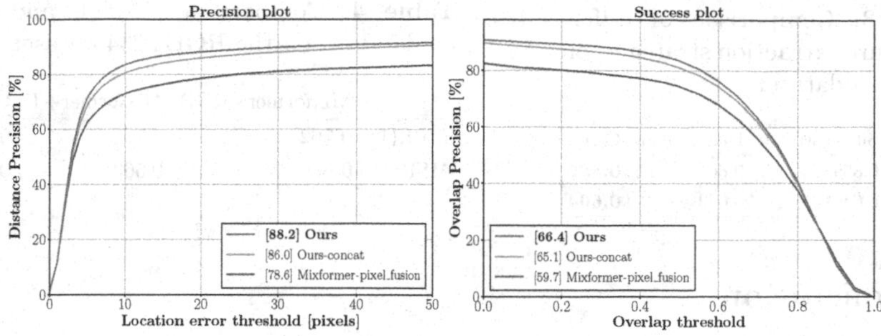

Fig. 5. Comparison of different features fusion methods on the RGBT234 dataset.

of 62.0%. The excited performance and significant promotion demonstrate the effectiveness of our proposed multi-modal tracking framework.

4.3 Ablation Study

To verify the effectiveness and give a thorough analysis of our proposed approach, we perform a detailed ablation study on the RGBT234 dataset.

Modality-Specific Layer Normalization. To demonstrate the effectiveness of our proposed feature extraction strategy, we performed comparative tests on the single-stream strategy, dual-stream strategy, and our method. The results are shown in Table 3, it demonstrates the importance of modality-specific layer normalization in multi-modal transformer tracking, and supports the claim of MSCLIP [33]: *transformers can support learning across multiple modalities and allow knowledge sharing.*

Different Feature Fusion Methods. To highlight the performance of our proposed feature fusion module, we juxtapose our method with the feature concatenation fusion method, which incorporates three layers of ConvBN. For more comprehensive comparision, we include a pixel-level fusion method [31] in our comparison. We plot the performance in Fig. 5, which demonstrates that our proposed method is vital to emphasize the specificity of different modalities.

Single/Dual-Modal Analysis. To underscore the benefits of multi-modal data in tracking, we devise two additional experiments using the large version of MixFormer: MixFormer+RGB and MixFormer+T. The former utilizes the pre-trained model from [4], while the latter is fine-tuned using the LasHeR dataset. As demonstrated in Table 4, our method can enhance multi-modal tracking performance without the need to specifically focus on the characteristics of thermal infrared or depth images.

Table 3. Comparison of different feature extraction strategies on RGBT234 dataset.

	Single-stream	Dual-stream	Ours
MPR(↑)	0.859	0.851	**0.882**
MSR(↑)	0.640	0.642	**0.664**

Table 4. Comparison of single and dual modal data on the RGBT234 dataset.

	Mixformer+RGB	Mixformer+T	Ours
MPR(↑)	0.762	0.765	**0.882**
MSR(↑)	0.580	0.563	**0.664**

5 Conclusion

We propose a regularized single-stream tracking framework for multi-modal tracking, from the perspective of disentanglement. With the aid of two submodules we proposed, our method is capable of extracting well-disentangled multi-modal features and subsequently fusing them by emphasizing feature-level specificity. Empirical evaluations demonstrate that our method offers notable improvement over existing trackers in the realm of multi-modal tracking.

Acknowledgments. This project is in part supported by the Key-Area Research and Development Program of Guangzhou (202206030003), and the National Natural Science Foundation of China (U22A2095, 62072482). We would like to thank Qi Chen and Jintang Bian for insight discussion.

References

1. Bhat, G., Danelljan, M., Van Gool, L., Timofte, R.: Learning discriminative model prediction for tracking. In: 2019 IEEE/CVF International Conference on Computer Vision (ICCV), pp. 6181–6190 (2019)
2. Carion, N., Massa, F., Synnaeve, G., Usunier, N., Kirillov, A., Zagoruyko, S.: End-to-End Object detection with transformers. In: Vedaldi, A., Bischof, H., Brox, T., Frahm, J.-M. (eds.) ECCV 2020. LNCS, vol. 12346, pp. 213–229. Springer, Cham (2020). https://doi.org/10.1007/978-3-030-58452-8_13
3. Chen, X., Yan, B., Zhu, J., Wang, D., Yang, X., Lu, H.: Transformer tracking. In: 2021 IEEE/CVF Conference on Computer Vision and Pattern Recognition (CVPR), pp. 8122–8131 (2021)
4. Cui, Y., Jiang, C., Wu, G., Wang, L.: Mixformer: end-to-end tracking with iterative mixed attention (2023)
5. Danelljan, M., Bhat, G., Khan, F.S., Felsberg, M.: Eco: efficient convolution operators for tracking. In: 2017 IEEE Conference on Computer Vision and Pattern Recognition (CVPR), pp. 6931–6939 (2017)
6. Dosovitskiy, A., et al.: An image is worth 16×16 words: transformers for image recognition at scale. In: International Conference on Learning Representations (2021)
7. Feng, M., Su, J.: Learning reliable modal weight with transformer for robust RGBT tracking. Knowl.-Based Syst. **249**, 108945 (2022)
8. Gao, S., Yang, J., Li, Z., Zheng, F., Leonardis, A., Song, J.: Learning dual-fused modality-aware representations for RGBD tracking. In: Computer Vision - ECCV 2022 Workshops, pp. 478–494. Springer, Cham (2023). https://doi.org/10.1007/978-3-031-25085-9_27

9. Henriques, J.F., Caseiro, R., Martins, P., Batista, J.: High-speed tracking with kernelized correlation filters. IEEE Trans. Pattern Anal. Mach. Intell. **37**, 583–596 (2015)
10. Hou, R., Ren, T., Wu, G.: Mirnet: a robust RGBT tracking jointly with multi-modal interaction and refinement. In: 2022 IEEE International Conference on Multimedia and Expo (ICME), pp. 1–6 (2022)
11. Jiawen, Z., Simiao, l., Xin, C., Wang, D., Lu, H.: Visual prompt multi-modal tracking. In: CVPR (2023)
12. Li, B., Wu, W., Wang, Q., Zhang, F., Xing, J., Yan, J.: Siamrpn++: evolution of siamese visual tracking with very deep networks. In: 2019 IEEE/CVF Conference on Computer Vision and Pattern Recognition (CVPR), pp. 4277–4286 (2019)
13. Li, C., Liang, X., Lu, Y., Zhao, N., Tang, J.: RGB-t object tracking: benchmark and baseline. Pattern Recogn. **96**, 106977 (2019)
14. Li, C.: Lasher: a large-scale high-diversity benchmark for RGBT tracking. IEEE Trans. Image Process. **31**, 392–404 (2022)
15. Li, F., Zha, Y., Zhang, L., Zhang, P., Chen, L.: Information lossless multi-modal image generation for RGB-T tracking. In: Yu, S., et al. (eds.) Pattern Recognition and Computer Vision, pp. 671–683. Springer, Cham (2022). https://doi.org/10.1007/978-3-031-18916-6_53
16. Liu, Y., Jing, X.Y., Nie, J., Gao, H., Liu, J., Jiang, G.P.: Context-aware three-dimensional mean-shift with occlusion handling for robust object tracking in RGB-D videos. IEEE Trans. Multimedia **21**, 664–677 (2019)
17. Loshchilov, I., Hutter, F.: Decoupled weight decay regularization. In: International Conference on Learning Representations (2019)
18. Lu, A., Li, C., Yan, Y., Tang, J., Luo, B.: RGBT tracking via multi-adapter network with hierarchical divergence loss. IEEE Trans. Image Process. **30**, 5613–5625 (2021)
19. Lu, A., Qian, C., Li, C., Tang, J., Wang, L.: Duality-gated mutual condition network for RGBT tracking. IEEE Trans. Neural Netw. Learn. Syst. (2022)
20. Lukežič, A., Zajc, L.V., Vojíř, T., Matas, J., Kristan, M.: Performance evaluation methodology for long-term single-object tracking. IEEE Trans. Cybern. **51**(12), 6305–6318 (2021)
21. Nam, H., Han, B.: Learning multi-domain convolutional neural networks for visual tracking. In: The IEEE Conference on Computer Vision and Pattern Recognition (CVPR) (2016)
22. Pengyu, Z., Zhao, J., Wang, D., Lu, H., Ruan, X.: Visible-thermal UAV tracking: a large-scale benchmark and new baseline. In: Proceedings of the IEEE Conference on Computer Vision and Pattern Recognition (2022)
23. Qian, Y., Yan, S., Lukežič, A., Kristan, M., Kämäräinen, J.K., Matas, J.: Dal: a deep depth-aware long-term tracker. In: 2020 25th International Conference on Pattern Recognition (ICPR), pp. 7825–7832 (2021)
24. Thangavel, J., Kokul, T., Ramanan, A., Fernando, S.: Transformers in single object tracking: an experimental survey. arXiv preprint arXiv:2302.11867 (2023)
25. Wang, C., et al.: Cross-modal pattern-propagation for RGB-T tracking. In: 2020 IEEE/CVF Conference on Computer Vision and Pattern Recognition (CVPR), pp. 7062–7071 (2020)
26. Wu, Y., He, K.: Group normalization. In: Proceedings of the European Conference on Computer Vision (ECCV) (2018)
27. Xia, Z., Pan, X., Song, S., Li, L.E., Huang, G.: Vision transformer with deformable attention. In: 2022 IEEE/CVF Conference on Computer Vision and Pattern Recognition (CVPR), pp. 4784–4793 (2022)

28. Xiao, Y., Yang, M., Li, C., Liu, L., Tang, J.: Attribute-based progressive fusion network for RGBT tracking. In: Proceedings of the AAAI Conference on Artificial Intelligence, vol. 36, no. 3, pp. 2831–2838 (2022)
29. Yan, B., Peng, H., Fu, J., Wang, D., Lu, H.: Learning spatio-temporal transformer for visual tracking. In: 2021 IEEE/CVF International Conference on Computer Vision (ICCV), pp. 10428–10437 (2021)
30. Yan, S., et al.: Depthtrack: unveiling the power of RGBD tracking. In: ICCV, pp. 10705–10713 (2021)
31. Yang, J., Li, Z., Zheng, F., Leonardis, A., Song, J.: Prompting for multi-modal tracking. In: Proceedings of the 30th ACM International Conference on Multimedia, MM 2022, p. 3492–3500. Association for Computing Machinery (2022)
32. Ye, B., Chang, H., Ma, B., Shan, S., Chen, X.: Joint feature learning and relation modeling for tracking: a one-stream framework. In: Brostow, G., Cisse, M., Farinella, G.M., Hassner, T. (eds.) Computer Vision - ECCV 2022, pp. 341–357. Springer, Cham (2022). https://doi.org/10.1007/978-3-031-20047-2_20
33. You, H., et al.: Learning visual representation from modality-shared contrastive language-image pre-training. In: Avidan, S., Brostow, G., Cisse, M., Farinella, G.M., Hassner, T. (eds.) Computer Vision - ECCV 2022, pp. 69–87. Springer, Heidelberg (2022). https://doi.org/10.1007/978-3-031-19812-0_5
34. Zhang, L., Danelljan, M., Gonzalez-Garcia, A., van de Weijer, J., Khan, F.S.: Multi-modal fusion for end-to-end RGB-T tracking. 2019 IEEE/CVF International Conference on Computer Vision Workshop (ICCVW), pp. 2252–2261 (2019)
35. Zhang, P., Zhao, J., Bo, C., Wang, D., Lu, H., Yang, X.: Jointly modeling motion and appearance cues for robust RGB-T tracking. IEEE Trans. Image Process. **30**, 3335–3347 (2021)
36. Zhang, T., Liu, X., Zhang, Q., Han, J.: Siamcda: complementarity- and distractor-aware RGB-T tracking based on siamese network. IEEE Trans. Circuits Syst. Video Technol. **32**(3), 1403–1417 (2022)
37. Zhao, H., Chen, J., Wang, L., Lu, H.: Arkittrack: a new diverse dataset for tracking using mobile RGB-D data. In: CVPR (2023)
38. Zheng, Z., Wang, P., Liu, W., Li, J., Ye, R., Ren, D.: Distance-iou loss: faster and better learning for bounding box regression. In: Proceedings of the AAAI Conference on Artificial Intelligence, vol. 34, no. 07, pp. 12993–13000 (2020)
39. Zheng, Z., Wang, P., Liu, W., Li, J., Ye, R., Ren, D.: Rgbd1k: a large-scale dataset and benchmark for RGB-D object tracking. In: Proceedings of the AAAI Conference on Artificial Intelligence (2023)
40. Zhu, X., Su, W., Lu, L., Li, B., Wang, X., Dai, J.: Deformable DETR: deformable transformers for end-to-end object detection. In: International Conference on Learning Representations (2021)
41. Zhu, Y., Li, C., Tang, J., Luo, B.: Quality-aware feature aggregation network for robust RGBT tracking. IEEE Trans. Intell. Veh. **6**(1), 121–130 (2021)

Modality Balancing Mechanism for RGB-Infrared Object Detection in Aerial Image

Weibo Cai[1,2,3], Zheng Li[1,2,3], Junhao Dong[1,2,3], Jianhuang Lai[1,2,3], and Xiaohua Xie[1,2,3](✉)

[1] School of Computer Science and Engineering, Sun Yat-sen University, Guangzhou, China
{caiwb6,lizh525,dongjh8}@mail2.sysu.edu.cn, {stsljh,xiexiaoh6}@mail.sysu.edu.cn

[2] Key Laboratory of Machine Intelligence and Advanced Computing, Ministry of Education, Guangzhou, China

[3] Guangdong Province Key Laboratory of Information Security Technology, Guangzhou, China

Abstract. RGB-Infrared object detection in aerial images has gained significant attention due to its effectiveness in mitigating the challenges posed by illumination restrictions. Existing methods often focus heavily on enhancing the fusion of two modalities while ignoring the optimization imbalance caused by inherent differences between modalities. In this work, we observe that there is an inconsistency between two modalities during joint training, and this hampers the model's performance. Inspired by these findings, we argue that the focus of RGB-Infrared detection should be shifted to the optimization of two modalities, and further propose a Modality Balancing Mechanism (MBM) method for training the detection model. To be specific, we initially introduce an auxiliary detection head to inspect the training process of both modalities. Subsequently, the learning rates of the two backbones are dynamically adjusted using the Scaled Gaussian Function (SGF). Furthermore, the Multi-modal Feature Hybrid Sampling Module (MHSM) is introduced to augment representation by combining complementary features extracted from both modalities. Benefiting from the design of the proposed mechanism, experimental results on DroneVehicle and LLVIP demonstrate that our approach achieves state-of-the-art performance. The code are available at (https://github.com/ccccwb/Multimodal-Detection-and-Tracking-UAV).

Keywords: RGB-Infrared object detection · Aerial image · Modality balancing mechanism · Multi-modal feature hybrid sampling

1 Introduction

Object detection in aerial images [4,6] plays a vital role in computer vision field with various applications, such as traffic control [13], video monitoring [14], and

© The Author(s), under exclusive license to Springer Nature Singapore Pte Ltd. 2024
Q. Liu et al. (Eds.): PRCV 2023, LNCS 14436, pp. 81–93, 2024.
https://doi.org/10.1007/978-981-99-8555-5_7

structural health monitoring [3]. Current studies for detecting objects in aerial images primarily use visible images, which perform well in favorable lighting and weather conditions. However, these methods are sensitive to poor lighting, limiting their effectiveness in low-light situations. To address this limitation, infrared images have gained popularity in recent years owing to their thermal sensing properties [12,20]. This advancement has boosted the development of RGB-IR detection through visible-infrared paired aerial images.

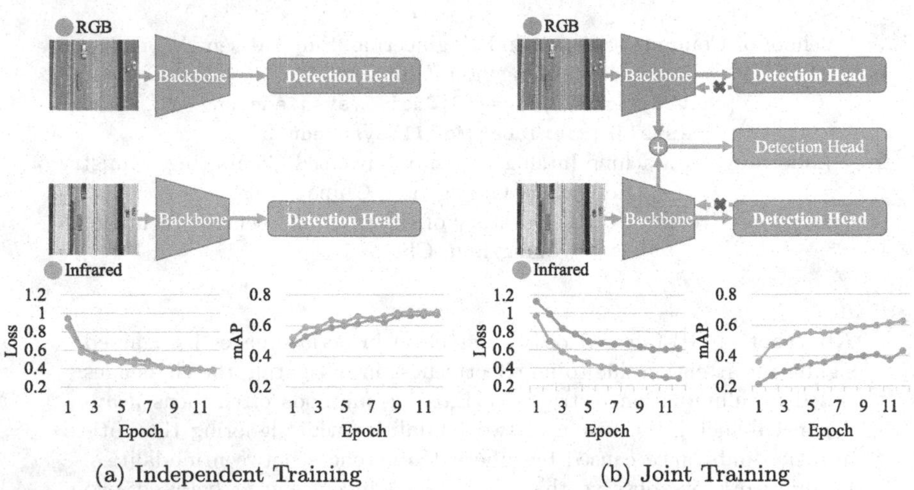

Fig. 1. Toy experiments illustrating the optimization imbalance. (a) Loss curve and performance of two independently trained detectors. (b) Loss curve and performance of two jointly trained detectors. In this case, the gradient from the detection heads of the two modalities is not propagated back to the backbone.

Current RGB-Infrared object detection methods [1,8,18,20,25,26] typically follow a two-step pipeline. Initially, separate backbones are employed to extract features from two modalities. These features are then combined using fusion modules, facilitating the classification and regression of bounding boxes. However, these methods encounter challenges in fusing the features from the two modalities due to disparities in data characteristics and variations in imaging positions. To address this issue, researchers have proposed methods such as MBNet [27] and TSFADet [24], which introduce carefully designed fusion modules to mitigate illumination imbalance and misalignment challenges in feature fusion. Nevertheless, it is noteworthy that these methods primarily emphasize feature fusion and do not comprehensively address the optimization imbalance arising from inherent modality differences [7,11,15,21].

In this paper, we conduct a few toy experiments to empirically show the optimization imbalance in multi-modal training. Figure 1(a) and Fig. 1(b) illustrate the model, loss curve, and performance during independent training and joint training, respectively. By comparing the loss curves under these two conditions, it can be observed that the optimization of visible modality is inhibited to a certain extent. Then comparing the performance of the model in the two cases, it is evident that the performance of the visible modality is significantly impacted. These findings demonstrate that the optimization imbalance in the multi-modal training process hampers the model's ability to effectively utilize and integrate information from both modalities.

To alleviate the optimization imbalance, we propose a simple yet effective strategy, Modality Balancing Mechanism (MBM). Specifically, we introduce an auxiliary detection head to assess the training process of both modalities. Then the learning rates of the two backbones are adaptively updated by the Scaled Gaussian Function (SGF). Furthermore, the Multi-modal Feature Hybrid Sampling Module (MHSM) is introduced to enhance feature representation by aggregating complementary features from both the visible and infrared modalities. It is worth emphasizing that the proposed method is a plug-and-play method, which can be seamlessly integrated into the existing detection framework. Extensive experimental results on DroneVehicle [20] and LLVIP [12] show that our method achieves new state-of-the-art performance.

The main contributions of the paper are as follow:

- We show the optimization imbalance between the two modalities during joint training and then propose a new plug-and-play method, which can be seamlessly integrated into the existing detection frameworks to alleviate the optimization imbalance.
- We propose a Modality Balancing Mechanism (MBM) to adaptively adjust the learning rate of two modalities and a Multispectral Feature Hybrid Sampling Module (MHSM) to effectively aggregate information from both modalities.
- We validate the effectiveness of our method on DroneVehicle and LLVIP, which outperforms other state-of-the-art models. Our method is also shown the capability to be combined with existing detectors to consistently improve performance at a low cost.

2 Related Work

2.1 Object Detection in Aerial Images

Current object detection techniques can be broadly classified into two categories: two-stage [19] and one-stage [16] methods. Two-stage methods first generate candidate regions, and then predict the category and refine the bounding box for each candidate region. In contrast, one-stage methods integrate recognition and detection into a straightforward deep neural network. Although these methods achieve promising performance in natural scenes, they can not be adept at effectively handling multi-angle objects in aerial imagery, like vehicles.

To address these issues, Oriented bounding boxes (OBB) based detection methods are emerging, which leverage rotating modules to predict the orientation of the object. By directly or indirectly introducing angles into the regional suggestion network, the two-stage detectors generate candidate boxes with orientation [5,23]. One-stage OBB detectors usually output the oriented bounding box with direction directly using feature refinement and step-wise regression [9]. However, the method solely relying on visible images is limited to adapting to light changes and diverse weather conditions. Therefore, recent researchers have incorporated infrared images to address this challenge.

2.2 RGB-Infrared Object Detection

The introduction of infrared images provides rich and diverse information for detection, compensating for the limitations of visible light in low-light and adverse weather conditions. Most of the RGB-Infrared object detection methods first utilize separate backbones to extract features from two modalities. Then fusion modules are designed to combine features for detection. According to the phrase of feature fusion, these methods can be classified into three categories: early fusion, mid fusion, and late fusion [28]. Mid fusion, occurring at the feature level, enables the integration of shared features between RGB and infrared modalities, making it the most extensively explored approach. [1,8,17,18,22]. CIAN [25] emphasizes the effectiveness of the cross-modal interactive attention mechanism in capturing complementary information and improving the accuracy of multispectral pedestrian detection. AR-CNN [26] effectively learns from weakly aligned multimodal data in the context of multispectral pedestrian detection. MBNet [27] proposed the difference fusion module and the illumination aware module to realize the adaptive alignment and fusion of features. UA-CMDet [20] incorporates uncertainty-aware learning to handle the challenges of cross-modal data captured by a drone. TSFDet [24] achieves feature alignment by learning geometric transformations on the network. While these methods improve the effectiveness of fusion, it is worth noting that existing joint training strategies fail to fully exploit the advantages offered by all modalities, leading to an imbalance in backbone optimization [7,15]. In this work, we aim to address this problem through adaptive control of optimization for each modality.

3 Method

In this section, we introduce the overview of the overall framework (§3.1), and elaborate on the proposed modality balancing mechanism (§3.2) and multi-modal feature hybrid sampling module (§3.3).

3.1 Overview

Given RGB-Infrared image pairs $\{(\mathcal{I}_i^{vis}, \mathcal{I}_i^{ir})\}_{i=1}^{N}$, our goal is to train a multi-modal detection model by exploring the consensus of RGB and IR modalities. Figure 2 gives the overall framework of the proposed method, which is seamlessly integrated into the existing detection methods. It consists of a modality balancing mechanism and a multi-modal feature hybrid sampling module. The features

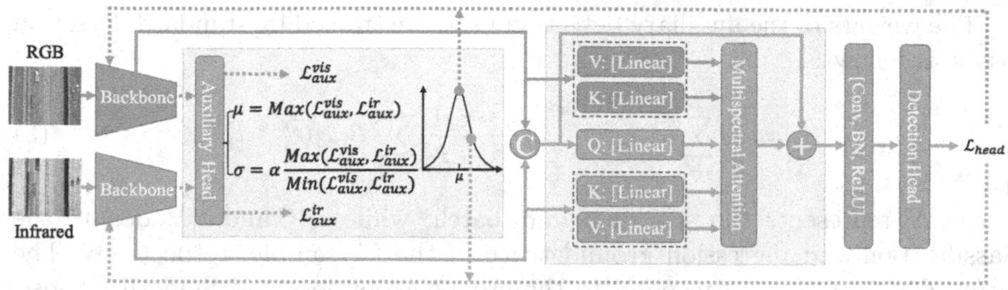

Fig. 2. Overview of the proposed method. The dotted lines in the figure connect operations that exist during training only.

of visible $f^{vis} \in \mathbb{R}^{C \times H \times W}$ and infrared $f^{ir} \in \mathbb{R}^{C \times H \times W}$ are first extracted by two independent feature encoders, where C denotes the number of channels, H and W denote the height and width, respectively. With these features, we inspect the training status by auxiliary detection head and dynamically adjust the learning rate of two modalities. Then, two types of features are adaptively aggregated by spatial location and modalities attention. Finally, following the previous detection methods [23,24], the total loss function is defined as follows:

$$\mathcal{L} = \mathcal{L}_{cls} + \mathcal{L}_{reg} + \mathcal{L}_{rpn} + \lambda \mathcal{L}_{aux}, \tag{1}$$

where $\mathcal{L}_{cls}$, $\mathcal{L}_{reg}$, and $\mathcal{L}_{rpn}$ is commonly used classification loss, regression, and rpn loss, respectively. $\mathcal{L}_{aux}$ serves as an extra loss to monitor the optimization of the two modalities. λ is a hyperparameter to balance different losses.

3.2 Modality Balancing Mechanism

Imbalanced optimization of backbones leads to the suppression of informative features. To alleviate the influence of imbalanced optimization, we propose a Modality Balancing Mechanism (MBM) to adaptively control the learning rate of two modalities. Firstly, we introduce an auxiliary detection head to inspect the optimization relationship between the two modalities. This is achieved by calculating the loss of the corresponding modality through the auxiliary head. Subsequently, utilizing these two losses, we employ a Scaled Gaussian Function (SGF) to compute the update ratio. Finally, the learning rate of the two modalities is dynamically adjusted throughout the training cycles.

Optimization Inspection. To obtain the optimization status of each modality in the joint training, an auxiliary detection head is introduced. The architecture of the auxiliary detection head is the same as the ordinary detector, which contains a projection linear layer F_{proj}, a classification branch F_{cls}, and a regression branch F_{reg}.

Taking the visible feature f^{vis} as an example, the classification logits c^{vis} and regression offsets t^{vis} of the auxiliary head can be expressed as:

$$\hat{c}^{vis} = F_{cls}(F_{proj}(f^{vis})), \quad \hat{t}^{vis} = F_{reg}(F_{proj}(f^{vis})). \tag{2}$$

The weights of the auxiliary detection head are trained by standard detection losses as follows:

$$\mathcal{L}_{aux}^{vis} = \frac{1}{N}\sum_{i=1}^{N}\mathcal{L}_{cls}(\hat{c}_i^{vis}, c_i^{vis}) + \frac{1}{N}\sum_{i=1}^{N}\mathcal{L}_{reg}(\hat{t}_i^{vis}, t_i^{vis}), \tag{3}$$

where N represents the size of a mini-batch, while c_i^{vis} and t_i^{vis} denote the classification and regression ground truth of the ith sample, respectively. The terms $\mathcal{L}_{cls}$ and $\mathcal{L}_{reg}$ correspond to the cross-entropy loss and Smooth L1 loss, respectively. The operation of the IR modal is the same as the RGB modal, and the total auxiliary loss is $\mathcal{L}_{aux} = \mathcal{L}_{aux}^{vis} + \mathcal{L}_{aux}^{ir}$. The auxiliary detection head is alternate optimized during the training phase. Its loss can reflect the optimization status of backbones in each training iteration.

Scaled Gaussian Function. To achieve balanced optimization between the two modalities, we employ an inhibitory coefficient that reduces the optimization rate of the more optimized modality. The optimization level of a given modality can be assessed by examining its associated auxiliary loss. A lower loss value signifies a more optimized backbone for the modality under consideration. As a result, the optimization disparity between two distinct modalities can be quantified by their corresponding auxiliary loss. To accomplish this, the Scaled Gaussian Function (SGF) is employed.

$$SGF(\mathcal{L}_{aux}) = A\frac{1}{\sigma\sqrt{2\pi}}e^{-\frac{(\mathcal{L}_{aux}-\mu)^2}{2\sigma^2}}, \tag{4}$$

where A serves as a scaling factor, which ensures that the sampling value at the mean μ is equal to the current learning rate. The mean value μ and standard deviation σ of the SGF are defined as follow:

$$\mu = Max(\mathcal{L}_{aux}^{vis}, \mathcal{L}_{aux}^{ir}), \quad \sigma = \alpha\frac{Max(\mathcal{L}_{aux}^{vis}, \mathcal{L}_{aux}^{ir})}{Min(\mathcal{L}_{aux}^{vis}, \mathcal{L}_{aux}^{ir})}, \tag{5}$$

where α is a hyperparameter to ensure the stability of training. Assigning the bigger auxiliary loss to μ ensures that the learning rate of the less optimized modality remains constant while suppressing the learning rate of the more optimized modality. Therefore, the learning rates modulated by SGF are obtained:

$$\eta^{vis} = SGF(\mathcal{L}_{aux}^{vis}), \quad \eta^{ir} = SGF(\mathcal{L}_{aux}^{ir}). \tag{6}$$

Parameter Updating. We further improve the Stochastic Gradient Descent (SGD) to balance the optimization of two backbones. With the modulated learning rate, the update of parameters is as follows:

$$\theta_{t+1}^{vis} = \theta_t^{vis} - \eta^{vis}\nabla_{\theta^{vis}}\mathcal{L}, \quad \theta_{t+1}^{ir} = \theta_t^{ir} - \eta^{ir}\nabla_{\theta^{ir}}\mathcal{L}, \tag{7}$$

where t signifies the iteration number, η denotes the modulated learning rate, and $\mathcal{L}(\cdot)$ represents the loss function. $\theta^{vis}, \theta^{ir}$ are the visible and infrared backbones parameter, respectively.

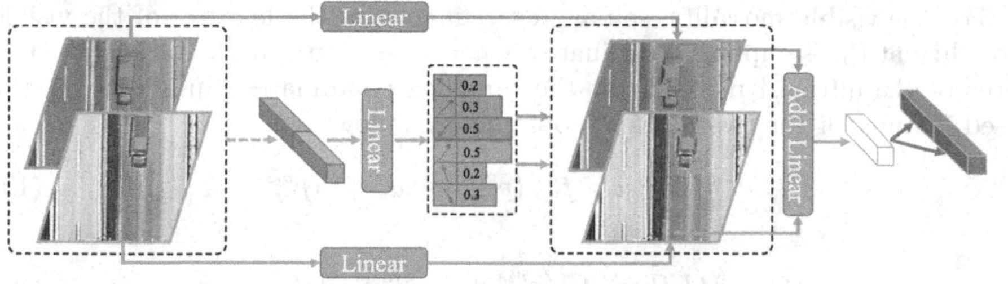

Fig. 3. Multi-modal feature hybrid sampling module.

3.3 Multimodal Feature Hybrid Sampling Module

To integrate and align the information from both features more effectively, we introduce a hybrid sampling method based on multi-modal characteristics to address the issue of weakly misaligned features.

Figure 3 presents a schematic representation of the proposed module. Given visible feature f^{vis} and infrared feature f^{ir}, the combined feature can be represented by $Q = (f^{vis} \bigoplus f^{ir}) \in \mathbb{R}^{2C\times H\times W}$, with concatenation along channel dimension. Firstly, for an arbitrary coordinate (x, y), we take its combined feature vector $q = Q(x, y) \in \mathbb{R}^{2C\times 1\times 1}$ as the query to generate sampling bias coordinates for the features of visible and infrared modalities by two distinct fully connected layers:

$$\{(\Delta x, \Delta y)\}^u = F^u_{offset}(q), \tag{8}$$

where $\{(\Delta x, \Delta y)\} \in \mathbb{R}^M$ represents the M offsets relative to query point (x, y). F_{offset} denotes the fully connected layers. Note that this operation is employed on both RGB and IR modalities. For simplicity, we denote u as the two modalities. With the offsets, the sampled coordinates can be obtained as follows:

$$p^u = \begin{cases} x^u = & x^u + \Delta x \\ y^u = & y^u + \Delta y \end{cases}. \tag{9}$$

The sampling weight is also determined linearly by the query q.

$$w^u = F^u_{weight}(q). \tag{10}$$

Secondly, we aggregate these features on sampling coordinates to produce fused features. The aggregation includes self-aggregation and cross-aggregation. Taking the visible modality as an example, the offset $\{(\Delta x, \Delta y)\}^{vis}$ is averagely divided into two groups, and the sampling coordinates of self-aggregation and cross-aggregation are calculated, as follows:

$$\{p^{vis}_{self},\ p^{vis}_{cross}\} \in p^{vis}. \tag{11}$$

For the visible modality, self-aggregation collects the features of the visible modality at the sampling coordinates, while cross-aggregation collects the features of the infrared modality at the sampling coordinates. Subsequently, the fused features of the two parts are obtained as follows:

$$\tilde{f}^{vis} = MLP(w^{vis} f^{vis}(p^{vis}_{self}) + w^{ir} f^{ir}(p^{vis}_{cross})), \tag{12}$$

$$\tilde{f}^{ir} = MLP(w^{ir} f^{ir}(p^{vis}_{self}) + w^{vis} f^{vis}(p^{ir}_{cross})). \tag{13}$$

The two features after sampling are combined to obtain a more comprehensive feature representation, enabling the model to better capture information from different perspectives of the modalities and improving the interaction between them. With those two features, the fused features can be represented by $f^{fused} = CBR(\tilde{f}^{vis} \bigoplus \tilde{f}^{ir}) \in \mathbb{R}^{C \times H \times W}$. Finally, the two features are reduced in dimension by CBR (Conv, BN, ReLu) and inputted into the detection head to get the final results, including scores and oriented bounding boxes.

4 Experiment

4.1 Settings

Datasets. To evaluate the efficacy of the proposed method, we evaluate two RGB-IR datasets captured from aerial photography perspectives: 1) The *DroneVehicle* dataset [20] is a comprehensive collection of drone-based data specifically designed for RGB-Infrared vehicle detection. It contains a total of 17,890 training samples and 1,469 validation samples. All images have a resolution of 640 × 512 pixels. It includes five categories: car, truck, bus, van, and freight car. 2) The *LLVIP* dataset [12] focuses on pedestrian detection and primarily consists of scenes captured in low-light environments. In comparison to the DroneVehicle dataset, the LLVIP dataset is more spatially aligned. It comprises 15,488 pairs of RGB-infrared images, with 12,025 pairs allocated for training and the remaining 3,463 pairs utilized for testing.

Metrics. The evaluation metric used is Mean Average Precision (mAP), which calculates the average value of Average Precision (AP) across different categories. In the DroneVehicle dataset, the IoU threshold is set at 0.5, while in the LLVIP dataset, the IoU threshold ranges from 0.50 to 0.95 with a step size of 0.05.

Implementation Details. Our method relies on the widely-used detection toolbox MMDetection [2]. During the training phase, we adopt the same hyperparameter settings as the original Oriented R-CNN [23] and utilize ResNet-50 [10] as the backbone network. To train our proposed method, we run 36 epochs with an initial learning rate of 0.005 and a batch size of 8. The weight decay and momentum values are set to 0.0001 and 0.9, respectively. The training process for both datasets is conducted on a single NVIDIA A100 GPU.

Table 1. Comparison of performances with state-of-the-art methods on DroneVehicle dataset.

Modality	Detectors	car	bus	truck	van	freight car	mAP
RGB	Faster R-CNN(OBB) [19]	67.88	66.98	38.59	23.20	26.31	44.59
	RetinaNet(OBB) [16]	79.51	70.54	35.77	26.89	35.34	47.61
	S2ANet [9]	90.05	89.00	52.69	45.90	55.36	66.60
	Oriented R-CNN [23]	88.25	89.00	61.47	47.16	49.19	67.01
IR	Faster R-CNN(OBB) [19]	88.58	67.86	37.20	33.70	33.80	53.25
	RetinaNet(OBB) [16]	90.03	85.09	40.64	37.11	36.29	57.83
	S2ANet [9]	90.11	89.62	59.09	48.29	52.90	68.00
	Oriented R-CNN [23]	90.13	89.65	60.00	49.89	53.96	68.72
RGB + IR	CIAN(OBB) [25]	89.98	88.90	62.47	59.59	60.22	70.23
	AR-CNN(OBB) [26]	90.08	89.38	64.82	51.51	62.12	71.58
	UA-CMDet [20]	87.51	87.08	60.70	37.95	46.80	64.01
	TSFADet [24]	90.01	89.70	69.15	55.19	65.45	73.90
	Ours	**90.38**	**90.17**	**69.53**	**63.04**	**66.46**	**75.92**

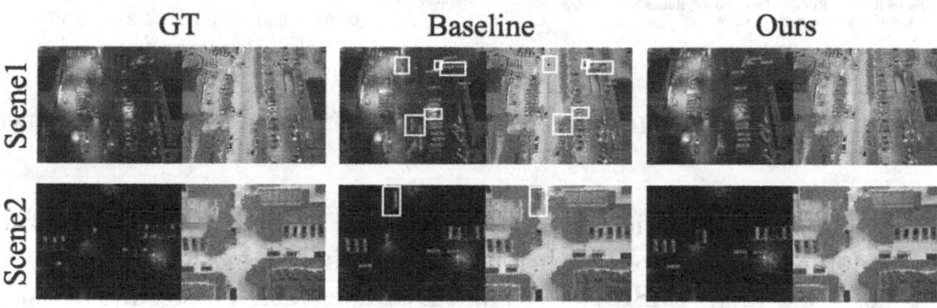

Fig. 4. Qualitative results of the proposed method. Scene 1 is a low-light scene, and Scene 2 is a dark scene. Red boxes are incorrect bounding boxes. (Color figure online)

4.2 Comparison with State-of-the-Art Methods

We compare our method with multiple state-of-the-art methods on the DroneVehicle and LLVIP. The results are shown in Table 1 and Table 2.

On the aerial imagery dataset DroneVehicle, our approach achieved state-of-the-art performance, with 75.92% mAP. Oriented R-CNN provides the best results among those single modality methods. However, compared with a single method, those multispectral methods generally achieve better performance. Among that multi-modal method, TSFADet obtains 73.90% mAP through fine characteristic fusion methods. Compared with TSFADet, the proposed method improves by 2.02% mAP. This improvement is primarily attributed to the fact that previous methods focused solely on feature fusion methods, neglecting the problem of unbalanced optimization between visible modality and infrared modality. Consequently, they failed to obtain a more effective feature representation from the backbone before fusion. As shown in Fig. 4 is the visualizations results of the proposed method. It can be seen that our method can provide more accurate detection in various scenarios compared to the baseline.

Table 2. Comparison of performances with state-of-the-art methods on LLVIP dataset.

Modality	Detectors	mAP
RGB+IR	UA-CMDet [20]	62.2
	CFT [17]	63.6
	AMSF-Net [1]	64.5
	DCMNet [22]	61.5
	Ours	**65.1**

Table 3. The ablation study of the proposed modules.

Methods	mAP
Baseline	71.74
Baseline + MBM	75.20
Baseline + MHSM	75.02
Baseline + MBM+ MHSM	**75.92**

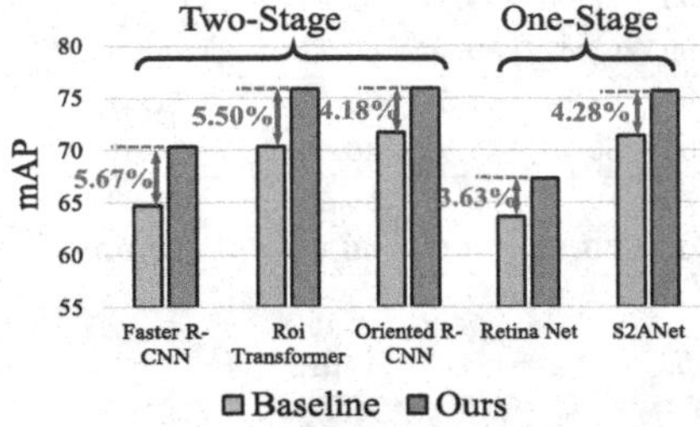

Fig. 5. The generalization experiment.

Fig. 6. Hyperparameter α experiment.

In the pedestrian detection dataset LLVIP, our approach also delivers competitive results. We compared with four state-of-the-art methods, namely UA-CMDet [20], CFT [17], AMSF-Net [1], and DCMNet [22]. The results are shown in the Table 2. Our method achieves the result of 65.1% mAP, better than other multi-modal methods. Unlike the DroneVehicle dataset, the pixel-level discrepancy in the visible-infrared images of the LLVIP dataset is smaller, resulting in a reduced modality imbalance phenomenon. Nevertheless, our method still achieves competitive performance.

4.3 Ablation Study

We conducted comprehensive ablation experiments to demonstrate the effectiveness of our method.

Baseline Comparison. We conduct comprehensive ablation experiments on the two proposed modules, and the corresponding experimental results are presented in Table 3. To ensure a fair comparison, Oriented R-CNN is used as the baseline. The baseline achieved 71.74% mAP. On this basis, we added the MBM and MHSM modules, resulting in an increase of 3.46% and 3.18% in mAP, respectively. These results demonstrate that the proposed method enables the model to capture more effective information and improve its performance.

Generality of the Proposed Method. We integrate our method into different popular detectors: Faster R-CNN [19], Roi-Transformer [5], Oriented R-CNN [23], Retina Net [16] and S2ANet [9]. The results are provided in Fig. 5. Compared with the baseline, integrating our method into Faster R-CNN, Roi-Transformer, Oriented R-CNN, Retina Net and S2ANet obtains 5.67%,5.50%, 4.18%, 3.63% and 4.28% mAP improvements, respectively. The experimental results validate the effectiveness and generality of our method.

Impacts of the Value of α in MBM. The α in Eq. 5 is used to adjust the standard deviation of SGF. We explore the impacts of the different settings of α. The results are shown in Fig. 6. Finally, we select $\alpha = 0.1$.

5 Conclusion

In this work, we present the optimization imbalance between two modalities during joint training. Inspired by these findings, we propose a Modality Balancing Mechanism (MBM) to inspect the training process and adaptively updated the learning rate of both modalities. Furthermore, an effective fusion module is introduced to enhance feature representation by aggregating complementary features from both modalities. Experimental results on DroneVehicle and LLVIP demonstrate that our approach achieves state-of-the-art performance.

Acknowledgments. This project is in part supported by the Key-Area Research and Development Program of Guangzhou (202206030003), and the National Natural Science Foundation of China (U22A2095, 62072482). We would like to thank Qi Chen for insight discussion.

References

1. Bao, W., Huang, M., Hu, J., Xiang, X.: Attention-guided multi-modal and multi-scale fusion for multispectral pedestrian detection. In: Pattern Recognition and Computer Vision: 5th Chinese Conference, PRCV 2022, Shenzhen, China, 4–7 November 2022, Proceedings, Part I, pp. 382–393. Springer, Heidelberg (2022). https://doi.org/10.1007/978-3-031-18907-4_30
2. Chen, K., et al.: Mmdetection: open mmlab detection toolbox and benchmark. arXiv preprint arXiv:1906.07155 (2019)
3. Chen, Q., Huang, Y., Sun, H., Huang, W.: Pavement crack detection using hessian structure propagation. Adv. Eng. Inf. **49**, 101303 (2021)
4. Cheng, G., Yuan, X., Yao, X., Yan, K., Zeng, Q., Han, J.: Towards large-scale small object detection: survey and benchmarks. arXiv preprint arXiv:2207.14096 (2022)
5. Ding, J., Xue, N., Long, Y., Xia, G.S., Lu, Q.: Learning ROI transformer for oriented object detection in aerial images. In: Proceedings of the IEEE/CVF Conference on Computer Vision and Pattern Recognition, pp. 2849–2858 (2019)
6. Ding, J.: Object detection in aerial images: a large-scale benchmark and challenges. IEEE Trans. Pattern Anal. Mach. Intell. **44**(11), 7778–7796 (2021)

7. Du, C., et al.: On uni-modal feature learning in supervised multi-modal learning. arXiv preprint arXiv:2305.01233 (2023)
8. Fu, H., et al.: LRAF-Net: long-range attention fusion network for visible-infrared object detection. IEEE Trans. Neural Netw. Learn. Syst. (2023)
9. Han, J., Ding, J., Li, J., Xia, G.S.: Align deep features for oriented object detection. IEEE Trans. Geosci. Remote Sens. **60**, 1–11 (2021)
10. He, K., Zhang, X., Ren, S., Sun, J.: Deep residual learning for image recognition. In: Proceedings of the IEEE Conference on Computer Vision and Pattern Recognition, pp. 770–778 (2016)
11. Huang, Y., Lin, J., Zhou, C., Yang, H., Huang, L.: Modality competition: what makes joint training of multi-modal network fail in deep learning?(provably). In: International Conference on Machine Learning, pp. 9226–9259. PMLR (2022)
12. Jia, X., Zhu, C., Li, M., Tang, W., Zhou, W.: LLVIP: a visible-infrared paired dataset for low-light vision. In: Proceedings of the IEEE/CVF International Conference on Computer Vision, pp. 3496–3504 (2021)
13. Kim, K., Kim, S., Shchur, D.: A UAS-based work zone safety monitoring system by integrating internal traffic control plan (ITCP) and automated object detection in game engine environment. Autom. Constr. **128**, 103736 (2021)
14. Li, S., Liu, Y., Zhao, Q., Feng, Z.: Learning residue-aware correlation filters and refining scale for real-time UAV tracking. Pattern Recogn. **127**, 108614 (2022)
15. Liang, P.P., Zadeh, A., Morency, L.P.: Foundations and recent trends in multimodal machine learning: principles, challenges, and open questions. arXiv preprint arXiv:2209.03430 (2022)
16. Lin, T.Y., Goyal, P., Girshick, R., He, K., Dollár, P.: Focal loss for dense object detection. In: Proceedings of the IEEE International Conference on Computer Vision, pp. 2980–2988 (2017)
17. Qingyun, F., Dapeng, H., Zhaokui, W.: Cross-modality fusion transformer for multispectral object detection. arXiv preprint arXiv:2111.00273 (2021)
18. Qingyun, F., Zhaokui, W.: Cross-modality attentive feature fusion for object detection in multispectral remote sensing imagery. Pattern Recogn. **130**, 108786 (2022)
19. Ren, S., He, K., Girshick, R., Sun, J.: Faster r-cnn: towards real-time object detection with region proposal networks. Adv. Neural Inf. Process. Syst. **28** (2015)
20. Sun, Y., Cao, B., Zhu, P., Hu, Q.: Drone-based RGB-infrared cross-modality vehicle detection via uncertainty-aware learning. IEEE Trans. Circuits Syst. Video Technol. **32**(10), 6700–6713 (2022)
21. Wu, J., Liang, Y., Akbari, H., Wang, Z., Yu, C., et al.: Scaling multimodal pretraining via cross-modality gradient harmonization. Adv. Neural. Inf. Process. Syst. **35**, 36161–36173 (2022)
22. Xie, J., et al.: Learning a dynamic cross-modal network for multispectral pedestrian detection. In: Proceedings of the 30th ACM International Conference on Multimedia, pp. 4043–4052 (2022)
23. Xie, X., Cheng, G., Wang, J., Yao, X., Han, J.: Oriented R-CNN for object detection. In: Proceedings of the IEEE/CVF International Conference on Computer Vision, pp. 3520–3529 (2021)
24. Yuan, M., Wang, Y., Wei, X.: Translation, scale and rotation: cross-modal alignment meets RGB-infrared vehicle detection. In: Computer Vision-ECCV 2022: 17th European Conference, Tel Aviv, Israel, 23–27 October 2022, Proceedings, Part IX, pp. 509–525. Springer, Heidelberg (2022). https://doi.org/10.1007/978-3-031-20077-9_30
25. Zhang, L., et al.: Cross-modality interactive attention network for multispectral pedestrian detection. Inf. Fusion **50**, 20–29 (2019)

26. Zhang, L., Zhu, X., Chen, X., Yang, X., Lei, Z., Liu, Z.: Weakly aligned cross-modal learning for multispectral pedestrian detection. In: Proceedings of the IEEE/CVF International Conference on Computer Vision, pp. 5127–5137 (2019)
27. Zhou, K., Chen, L., Cao, X.: Improving multispectral pedestrian detection by addressing modality imbalance problems. In: Vedaldi, A., Bischof, H., Brox, T., Frahm, J.-M. (eds.) ECCV 2020. LNCS, vol. 12363, pp. 787–803. Springer, Cham (2020). https://doi.org/10.1007/978-3-030-58523-5_46
28. Zhou, T., Fan, D.P., Cheng, M.M., Shen, J., Shao, L.: RGB-D salient object detection: a survey. Comput. Visual Media **7**, 37–69 (2021)

Pacific Oyster Gonad Identification and Grayscale Calculation Based on Unapparent Object Detection

Yifei Chen[1], Jun Yue[1](✉), Zhenbo Li[2], Jianmin Yang[1], and Weijun Wang[1]

[1] Ludong University, Yantai 264025, China
yuejuncn@sohu.com
[2] China Agricultural University, Beijing 100091, China

Abstract. The plumpness of the Pacific oyster gonad, the reproductive organs of both male and female oysters which are buried within the flesh of the oyster in the shell, has important implications for the quality and breeding of subsequent parents. At present, only the conventional method of breaking their shells allows for the observation and study of the interior tissues of Pacific oysters. In this paper, the gonad of Pacific oyster was observed by small animal Magnetic Resonance Imaging (MRI), and a multi-effective feature fusion network algorithm R-SINet was proposed for the detection of unapparent target, in Nuclear Magnetic Resonance (NMR) images, which can effectively solve the problem that the gonads of Pacific oysters are difficult to identify from the background images. In addition, the gray histogram of the segmented gonad region was calculated, and it was found that the female and male had differences in gray value. The sex of oyster was nondestructively detecting by this task. Firstly, established the Oyster gonad datasets; secondly, a compact pyramid refinement module that combines with high-level semantic features and low-level semantic features was proposed, designed a lightweight decoder to improve the accuracy of feature fusion; thirdly, a switchable excitation model capable of adaptive recalibration is proposed to obtain an attention map. Experimental results on the Oyster gonad datasets demonstrate the effectiveness of the method. Comparing R-SINet's experimental findings to those of popular algorithm models, such as the benchmark algorithm SINet_v2, revealed promising results.

Keywords: Pacific oyster gonad · Unapparent object detection · Gray value calculation · R-SINet

1 Introduction

Pacific oysters are very popular in aquaculture industry because of their large size, short breeding cycle, and high efficiency. The selection of Pacific oysters with mature and plump gonads for parental breeding is the key to quality and yield improvement. In addition, distinguishing between male and female individuals based on grayscale differences is of great significance for subsequent selection of specific gender oyster

© The Author(s), under exclusive license to Springer Nature Singapore Pte Ltd. 2024
Q. Liu et al. (Eds.): PRCV 2023, LNCS 14436, pp. 94–106, 2024.
https://doi.org/10.1007/978-981-99-8555-5_8

individuals for breeding. With the rapid development of small animal imaging technology and convolutional neural network, we can now use a small animal imaging system to obtain Pacific oyster Magnetic Resonance Imaging (MRI) images which can clearly and intuitively observe the gonad part of oysters without the harm to live Pacific oysters; it can solve the problem of high similarity between organs and tissues and insignificant color differences in MRI images when segmenting the gonads by detecting unapparent objects in MRI images with relatively complex backgrounds, which is important for improving the integrity and accuracy of segmenting the gonads of Pacific oysters.

Small animal MRI is a branch of magnetic resonance imaging, which is becoming an important tool for studying the internal structure of small animals [1]. In 2001, the first clinical study of canine intervertebral disc disease was conducted using MRI technology in China. In 2019, Zhang et al. [2] applied small animal MRI systems to study Alzheimer's disease (AD), providing multimodal imaging techniques to help diagnose early AD. In 2022, Hang et al. [3] used 7.0T small animal MRI equipment to noninvasively observe brain injury in a rat model of classic heat stroke. Small animal MRI techniques started earlier abroad. In 1990, Button et al. used 0.35T small animal MRI to observe the growth and morphology of tumors in mice. B Webster [4] wrote a practical small animal MRI manual in 2010. In 2021, S Gilchrist et al. [5] designed a gating unit for synchronized control of the small animal heart and respiration using small animal MRI. While in 2022, Liu et al. [6] conducted a multimodal animal MRI study for memory generalization in mice. The above-mentioned studies show that specialized small animal MRI techniques provide technical support for the study of small animal organ tissues. However, nowadays, small animal MRI techniques mainly focus on the detection and study of terrestrial animals, there are few studies in the field of marine organisms, especially in shellfish organisms.

In recent years, with the iterations and advances in computer vision, unapparent object detection identification techniques have developed rapidly. In 2020, Fan et al. [7] first proposed a camouflaged object detection technique, designed the SINet network architecture aimed at identifying small target objects in complex backgrounds. In 2021, Lv et al. [8] proposed a hierarchical localization of target regions, introducing reverse attention [9] to capture more details of the spatial structure and designed the LSR algorithm model; Zhai et al. [10] built an edge-shrinkage graph inference module to guide the learning of feature representations of camouflaged objects; Fan et al. [11] proposed the SINet_v2 network architecture with optimization improvements based on SINet. In 2022, Jia et al. [12] performed both amplification and repetition operations for camouflage object segmentation to achieve accurate localization of camouflage objects by iteration and target object amplification and proposed the SegMaR algorithm model. However, when these existing detection and segmentation algorithms for unapparent object detection are used to train MRI grayscale images, the local feature extractions in all of these algorithms are coarse and the global feature fusion is ignored. This results in inadequate local feature extractions and loss of global information in the segmented grayscale images, which in turn cause incomplete segmentation targets and unclear boundaries, and the overall level of the evaluation index of segmentation decreases.

In summary, to address the lack of small animal MRI technology in shellfish aquatic applications and the inability of existing unapparent object detection algorithms to accurately segment grayscale images, our main contributions are five-fold:

- We propose a new network algorithm (R-SINet), which can effectively enhance the integrity of feature extraction in gray scale images of Pacific oysters.
- A Compact Pyramid Refinement Module (CPRM) is proposed to integrate adjacent semantic features, and a lightweight decoder is designed to improve the accuracy of feature fusion.
- A Switchable Excitation Model (SEM) is proposed, by automatically choosing and changing activation operators based on the channel demands of different network layers, the model can achieve multi-effect feature fusion for unapparent object detection and improve segmentation accuracy.
- Extensive experimental results on self-built Oyster gonad datasets demonstrate the effectiveness of our R-SINet over other state-of-the-art methods.
- The gray value difference of male and female gonads was obtained by gray histogram, which laid a foundation for subsequent nondestructive detection of Pacific oyster sex.

2 Method

The technology roadmap of this paper is shown in Fig. 1.

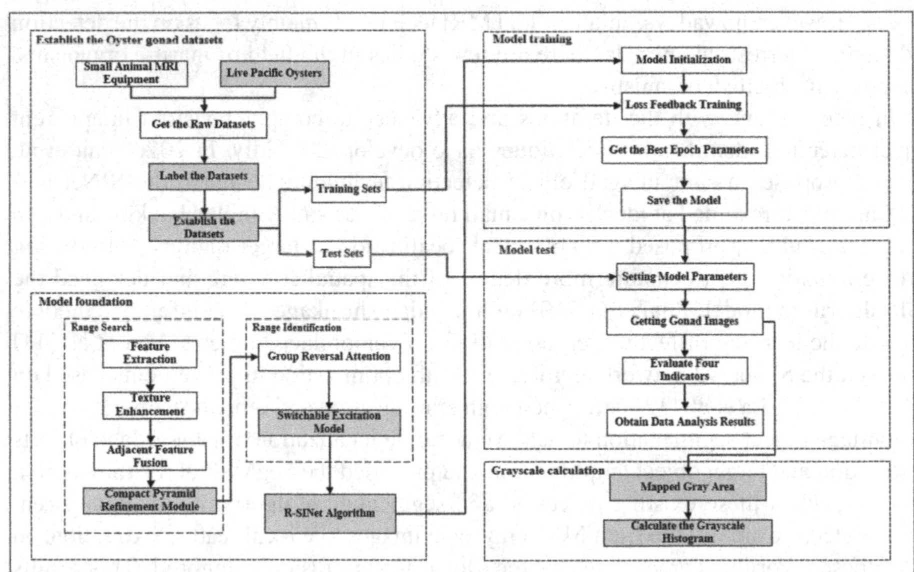

Fig. 1. Technology roadmap.

The network framework diagram of this paper is shown in Fig. 2, which consists of two phases.

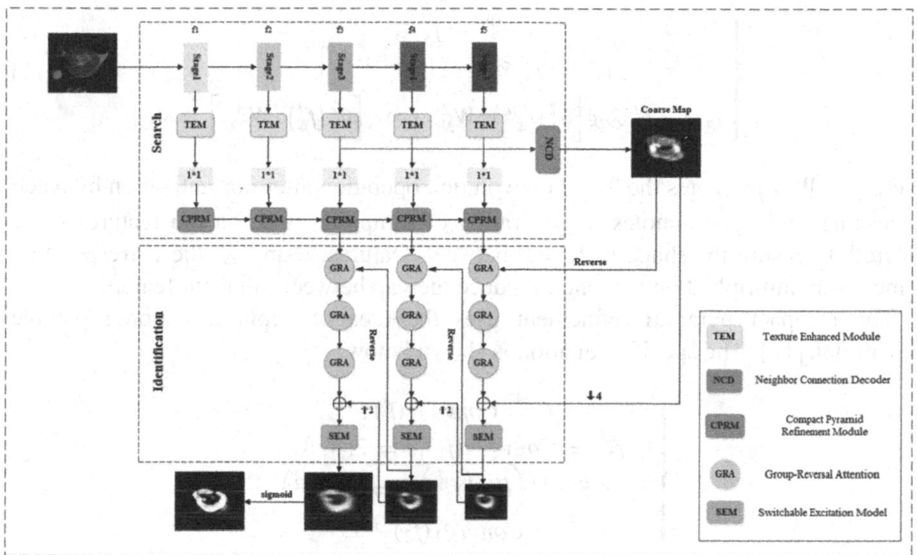

Fig. 2. R-SINet network architecture diagram.

The scope search stage mainly carries out feature extraction of grayscale images to obtain the position and scope of the target object. The scope recognition stage uses the group inversion attention module (GRA) [11] to obtain more detailed edge information and then switches the global adaptive channels through the switchable excitation model (SEM) to obtain more accurate global feature information, and finally completes the detection and segmentation of the target object in the grayscale images.

2.1 Compact Pyramid Refinement Module (CPRM)

The common pyramid models [13, 14] at this stage have problems such as large computation, large memory consumption, and slow inference speed. To solve such problems, this paper adds a lightweight feature pyramid module after effectively fusing neighboring features, deeply fuses high-level and low-level features, and proposes a Compact Pyramid Refinement Module (CPRM), which improves the efficiency while ensuring accuracy.

First, use Res2Net50 [15] network for image $I \in R^{W \times H \times 3}$ extracting features f_k ($k \in \{1, 2, 3, 4, 5\}$), obtaining 5 features with a resolution of $f_k = \frac{H}{2^k} + \frac{W}{2^k}$ of features and then expand the perceptual field by the texture enhancement module (TEM). The features to be selected are obtained from the TEM f_k'.. After that, the neighboring features are aggregated by using the Neighbor Connection Decoder (NCD) to keep the semantic information consistent within the same layer and semantically consistent across layers. Since low-level features have larger resolutions consume more computational resources and contribute less to the performance improvement, we only use f_3, f_4, and f_5 as the feature images $f_k^{nc} = F_{NC}\left(f_k'; W_{NC}^u\right), u \in \{1, 2, 3\}$, and each feature image is represented by the following equation:

$$\begin{cases} f_5^{nc} = f_5' \\ f_4^{nc} = f_4' \otimes g\left[\delta_\uparrow^2(f_5'); W_{NC}^1\right] \\ f_3^{nc} = f_3' \otimes g\left[\delta_\uparrow^2(f_4^{nc}); W_{NC}^2\right] \otimes g\left[\delta_\uparrow^2(f_4'); W_{NC}^3\right] \end{cases} \tag{1}$$

where $g\left(\cdot; W_{NC}^u\right)$ denotes the 3×3 convolution operation after normalization by batch processing, and $\delta_\uparrow^2(\cdot)$ denotes the operation of sampling twice on the features to be selected, to ensure the shape matching between features. Using $\otimes$, the corresponding elements are multiplied one by one to reduce the gap between adjacent features.

The compact pyramid refinement uses the idea of depth direction separable convolution [16]. The specific equation is shown below:

$$\begin{cases} f_1 = Conv_{1\times1}(f) \\ f_2^{d_i} = Conv_{3\times3}^{d_i}(f_1), i = 2, 4, 8 \\ f_2 = ReLU\left(BN\left(f_2^{d_i}\right), i = 2, 4, 8\right) \\ f_3 = Conv_{1\times1}(f_2) + f \end{cases} \tag{2}$$

where f denotes the input of the feature image, $\text{Conv}_{n\times n}$ denotes the input for the n $\times$ n the convolution operation, d_i denotes the dilation rates.

In this way, the lightweight decoder with feature pyramid refinement proposed in this paper can aggregate multi-level features from top to bottom and achieve efficient feature capture at all levels. The structure diagram of the compact pyramid refinement module is shown in Fig. 3.

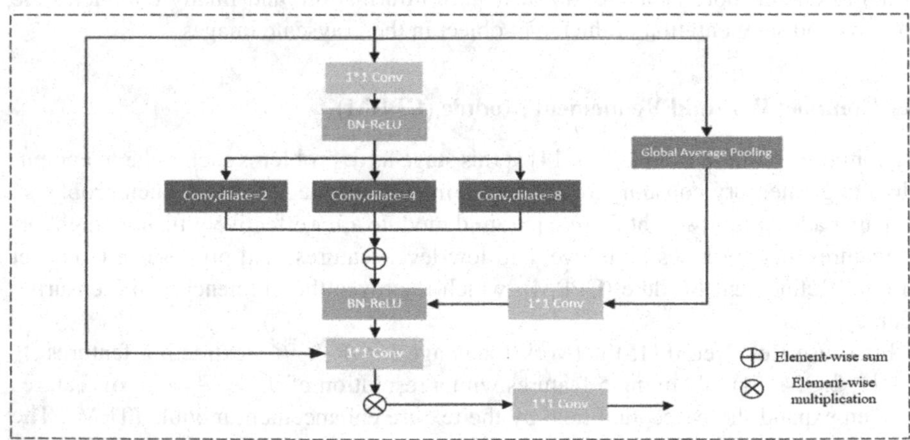

Fig. 3. Compact Pyramid Refinement Module.

2.2 Switchable Excitation Model (SEM)

In the scope recognition stage, the features between the levels are fused under the premise of ensuring the computational rate, and perform the image inversion operation, as shown in the following equation:

$$r_1^k = \begin{cases} \neg\left[\sigma\left(\delta_\downarrow^4(C_{k+1})\right), E\right], k = 5 \\ \neg\left[\sigma\left(\delta_\uparrow^2(C_{k+1})\right), E\right], k \in \{3, 4\} \end{cases} \quad (3)$$

where $\neg$ denotes the inverse operation, which is performed on the matrix E performs the inverse operation. The matrix E represents a matrix with all elements 1. $\delta_\downarrow^4$ and $\delta_\uparrow^2$ denote down-sampling 4 times and up-sampling 2 times, respectively.

By grouping inversion attention, more attention is paid to the local feature information of the target edges, but still lacks attention based on the global scope. In this paper, we propose a switchable excitation module that automatically decides to select and integrate attention operators to compute attention graphs. The Switchable Excitation Model (SEM) proposed in this paper is added before the iterative refinement operation, so that it trains the Sigmoid values of each channel and obtains the corresponding weights for each channel, and finally gives more attention to the channels with larger weights while suppressing the channels with smaller weight values. The structure of the switchable excitation model is shown in Fig. 4. SEM improves the excitation module of the attention model, which consists of two sub-modules, the decision module and the switching module. The feature map of the current network layer is defined as $x \in R^{C\times H\times W}$, and the procedure for x calculating the attention value of is as follows:

The $GAP(\cdot)$ the global average pooling extracts the global feature information from the squeeze module can be formulated as follows:

$$m = GAP(x) \quad (4)$$

where $m \in R^{C\times 1\times 1}$ represents the global information embedding, m as the input to the decision module and the switching module.

To use the information aggregated in the squeeze operation to determine the importance of different operations, this paper adds to Eq. $F(\cdot)$ that aims to fully capture the decision information from channel dependencies. This paper is designed to use a simple gating mechanism with Sigmoid activation:

$$w = \sigma(F(m)) = \sigma(W_d m) \quad (5)$$

where $\mathrm{F}(\cdot)$ denotes the decision function of the fully connected network, $\mathrm{W_d} \in \mathrm{R}^{\mathrm{N}\times\mathrm{C}}$ denotes the weight of the fully connected network.

Based on the decision vector w, in the switching module, define EO to denote a set of excitation operators and set its size to $N = 3$, , using the fully connected network (FC) [17], the convolutional neural network (CNN) [18] and the instance augmentation (IE) [19] as alternate excitation operators, and proposing the w the computational attentional feature map of $v \in R^{C\times 1\times 1}$ to adjust the proportion of each excitation operator in the switching module and combine the results of each operator in the form of dot product to obtain the final attentional feature map v which is formulated as follows:

$$v = \sigma\left(v_{fc} w_{fc}\right) \odot \sigma\left(v_{cnn} w_{cnn}\right) \odot \sigma\left(v_{ie} w_{ie}\right) \quad (6)$$

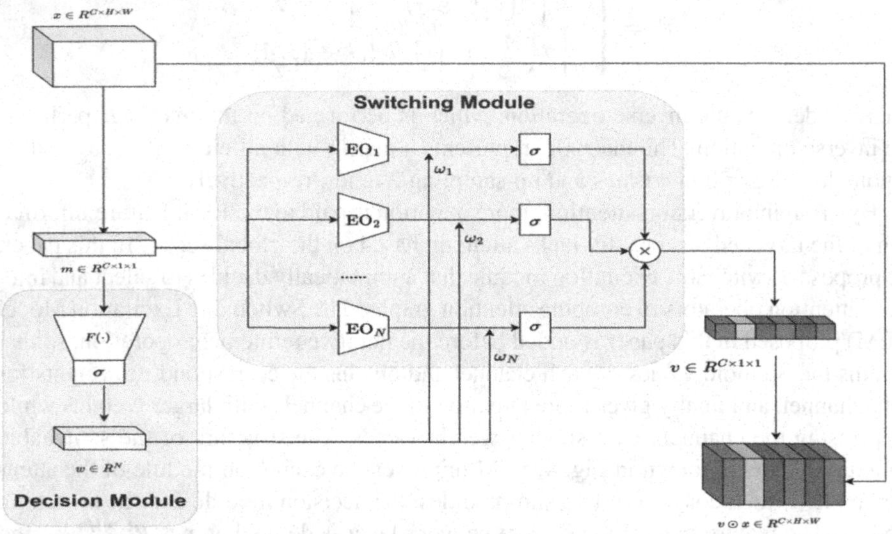

Fig. 4. Switchable Excitation Model.

3 Experiments and Analysis of Results

3.1 Establishment of the Datasets

To balance the variability of the growth of Pacific oysters in winter and summer seasons, a total of 300 Pacific oysters of similar shape and individual size were randomly selected in December 2021, June 2022 and February 2023, respectively, under the same culture environment. MRI images were acquired using 7.0T high field strength small animal magnetic resonance imaging system equipment, as shown in Fig. 5. The main technical specifications of the equipment: 7.0T magnet, aperture width of 20 cm, 660 mT/m gradient intensity, 7 groups of high-order uniform field coils, gradient power supply of 500 V/300 A, and the highest image pixel resolution of 10 μm. The main parameter settings: the longitudinal slice length was set to 2 mm, the number of slices per Pacific oyster was 20, and the echo time of transverse (T2) relaxation (TE) was set to 30 ms.

Fig. 5. Photographs taken with NMR equipment.

A total of 4000 Pacific oyster MRI images were obtained after screening, and the gonadal boundaries of the original images were labeled with labelme software to build the Oyster gonad datasets. The annotated images were randomly divided into training and test sets in the ratio of 7:1, of which 3500 were used for training the segmentation model and 500 were used for the tested model. An example map of partial Oyster gonad datasets annotation is shown in Fig. 6.

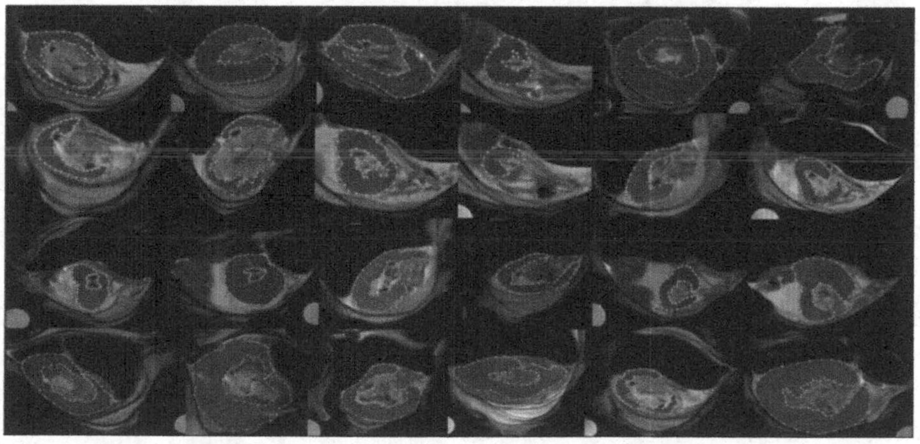

Fig. 6. Example of partial Oyster gonad datasets annotation.

3.2 Experimental Environment and Evaluation Index

The hardware and software parameters used in this study are configured as shown in Table 1.

Table 1. Software and hardware parameters configuration.

Software and hardware environment	Configuration
Small animal MRI system	Bruker BioSpec 70/20 USR
Processor	Inter(R) Core(TM) i9-9820X
Graphics Processor	NVIDIA Corporation GP100GL
Graphics processor computing platform	CUDA 10.2, cuDNN 7.4
Compile the program	Pycharm, Anaconda
Frame	Pytorch
Programming Languages	Python 3.6

The input image size of the experiments in this paper is 352 × 352, the epoch size is 50 during training, and the batch size is 8. The specific hyperparameter settings of the algorithm model in this paper are shown in Table 2. The training is performed using Adam optimizer [20], and the whole training process takes about 75 min.

Table 2. R-SINet algorithm hyperparameter settings.

Parameters	Numerical value
Input size	352 × 352
Learning rate	0.0001
Batch size	8
epoch	50
Number of iterations	3
Optimizer	Adam

The model evaluation metrics include S-measure (Sα) [21], enhanced-matching evaluation metrics E-measure (EΦ) [22], weighted F-measure (ωF) [23] and Mean absolute error (MAE) [24].

3.3 Ablation Experiments

The ablation experiments were tested in the Oyster gonad datasets using the same hyper-parameters, and the test results are shown in Table 3. The experimental results prove that the addition of the two models has a positive effect on the results of the algorithm, which performs well in all four evaluation indexes, and shows the best results after incorporating the two models into the overall framework at the same time, which verifies the effectiveness of the two models.

Table 3. Impact of two models proposed in this paper on the algorithm.

	SINet_v2	CPRM	SEM	Sα ↑	EΦ ↑	ωF ↑	MAE ↓
No. 1	√			0.866	0.910	0.865	0.037
No. 2	√	√		0.885	0.912	0.866	0.030
No. 3	√		√	0.867	0.923	0.880	0.028
No. 4	√	√	√	**0.889**	**0.929**	**0.882**	**0.020**

3.4 Comparative Experiments and Analysis of Results

To verify the performance level of the proposed R-SINet, it was tested with the SINet [7], LSR [8], SINet_v2 [11] and SegMaR [12] in the Oyster gonad datasets and the same running environment, respectively, and the quantitative evaluation results of the experimental comparison are shown in Table 4. The values of the proposed R-SINet algorithm are better than the results of the four unapparent object detection segmentation algorithms of the comparison experiments.

Table 4. Evaluation results of different algorithms in the four evaluation indexes.

Algorithm	Sα ↑	EΦ ↑	ωF ↑	MAE ↓
SINet (2020)	0.663	0.888	0.476	0.068
LSR (2021)	0.658	0.917	0.745	0.056
SINet_v2 (2021)	0.865	0.910	0.863	0.038
SegMaR (2022)	0.653	0.918	0.474	0.059
R-SINet (Ours)	**0.889**	**0.929**	**0.882**	**0.020**

3.5 Visualization Results

Some of the visualized segmentation results are shown in Fig. 7. This figure shows that the contour structure of the segmented gonads of Pacific oysters obtained by the method of this paper is closer to the true value map, and the gonad edge is clearer than the other four algorithms, which confirms the effectiveness of the method.

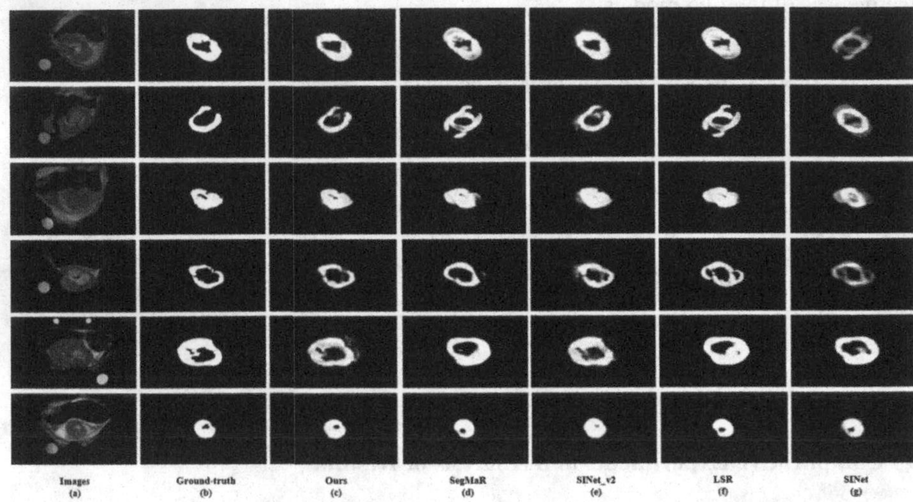

Fig. 7. Partial visualization of gonadal segmentation results of Pacific oyster compared.

3.6 Gray Value Calculation

Based on the method of Pacific Oyster gonad recognition proposed in this paper, the partitioned gonad part is mapped to the original image to obtain the grayscale image of the region. In the form of gray histogram, 50 of the 500 test images are randomly selected and the gray value is calculated. As can be seen from Fig. 8, there are two peaks in the gray histogram of the female and male of the oyster, and the distance between the peaks on the horizontal axis is large, which proves that there are obvious differences in the gray values of the gonads of the female and male oysters. The gray images of small animal NMR can be used to determine the sex of Pacific oysters.

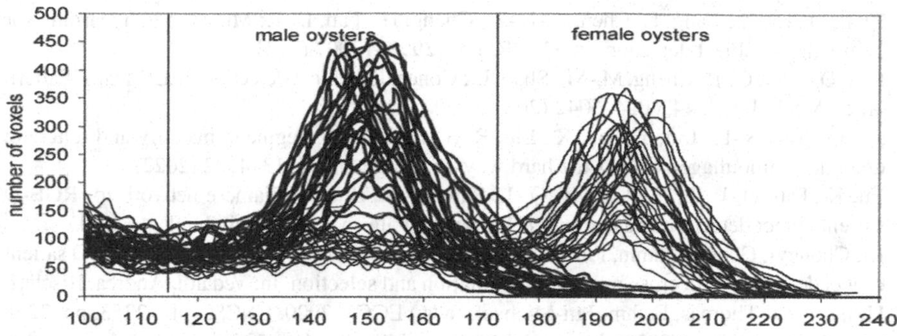

Fig. 8. Gray histogram curve of male and female oysters.

4 Conclusion

In this paper, based on the magnetic resonance imaging of the Pacific oyster, we identified and segmented the Pacific oyster gonads by the method of unapparent object detection, and proposed the R-SINet algorithm to improve the accuracy of oyster gonad segmentation. The model obtains the optimal test results in four evaluation metrics, which proves the effectiveness and robustness of the method. In addition, we intuitively found that there were large differences in gray values between the gonads of female and male oysters by using gray histogram, which provided a new technique and means for the subsequent selection of oyster sex.

References

1. Yang, L.: Introduction to the management of hospital Bruker BioSpec94/30 USR type small animal MRI research equipment. China Equip. Eng. 51–52 (2022)
2. Zhang, Z.-N., Zheng, Y., Wang, X.-M.: Application of 7.0T small animal MRI to study the progress of Alzheimer's disease. China Med. Imaging Technol. 930–933 (2019)
3. Hang, K.-B., Su, W.-W., Huang, J., Bao, G.-J., Liu, W.-H., Li, S.-P.: 7.0T small animal MR instrumentation to observe brain injury in a rat model of classic pyrexia. China Med. Imaging Technol. **38**, 481–485 (2022)
4. Webster, B.: Handbook of small animal MRI. Aust. Veterinary J. **88**, 407 (2010)
5. Gilchrist, S., et al.: A simple, open and extensible gating control unit for cardiac and respiratory synchronisation control in small animal MRI and demonstration of its robust performance in steady-state maintained CINE-MRI. Magn. Reson. Imaging **81**, 1–9 (2021)
6. Liu, W.-L., et al.: Enhanced medial prefrontal cortex and hippocampal activity improves memory generalization in APP/PS1 Mice: a multimodal animal MRI study. Front. Cell. Neurosci. **16**, 848967 (2022)
7. Fan, D.-P., Ji, G.-P., Sun, G.-L., Cheng, M.-M., Shen, J.-B., Shao, L.: Camouflaged object detection. In: CVPR, pp. 2774–2784 (2020)
8. Lv, Y.-Q., et al.: Simultaneously localize, segment and rank the camouflaged objects. In: CVPR, pp. 11591–11601 (2021)
9. Chen, S., Tan, X., Wang, B., Hu, X.: Reverse attention for salient object detection. In: Ferrari, V., Hebert, M., Sminchisescu, C., Weiss, Y. (eds.) ECCV 2018. LNCS, vol. 11213, pp. 236–252. Springer, Cham (2018). https://doi.org/10.1007/978-3-030-01240-3_15

10. Zhai, Q., Li, X., Yang, F., Chen, C.-L.-Z., Cheng, H., Fan, D.-P.: Mutual graph learning for camouflaged object detection. In: CVPR, pp. 12997–13007 (2021)
11. Fan, D.-P., Ji, G.-P., Cheng, M.-M., Shao, L.: Concealed object detection. IEEE Trans. Pattern Anal. Mach. Intell. **44**, 6024–6042 (2021)
12. Jia, Q., Yao, S.-L., Liu, Y., Fan, X., Liu, R.-S., Luo, Z.-X.: Segment, magnify and reiterate: detecting camouflaged objects the hard way. In: CVPR, pp. 4713–4722 (2022)
13. Fu, K., Fan, D.-P., Ji, G.-P., Zhao, Q.-J., Shen, J.-B., Zhu C.: Siamese network for RGB-D salient object detection and beyond. IEEE Trans. Pattern Anal. Mach. Intell. 1 (2021)
14. Li, Chongyi, Cong, Runmin, Piao, Yongri, Xu, Qianqian, Loy, Chen Change: RGB-D salient object detection with cross-modality modulation and selection. In: Vedaldi, Andrea, Bischof, Horst, Brox, Thomas, Frahm, Jan-Michael. (eds.) ECCV 2020. LNCS, vol. 12353, pp. 225–241. Springer, Cham (2020). https://doi.org/10.1007/978-3-030-58598-3_14
15. Gao, S.-H., Cheng, M.-M., Zhao, K., Zhang, X.-Y., Yang, M.-H., Torr, P.: Res2Net: a new multi-scale backbone architecture. IEEE Trans. Pattern Anal. Mach. Intell. **43**, 652–662 (2019)
16. Howard, A.-G., et al.: MobileNets: efficient convolutional neural networks for mobile vision applications. arXiv preprint arXiv:1704.04861 (2017)
17. Li, J.-H., Li, B., Xu, J.-Z., Xiong, R.-Q., Gao, W.: Fully connected network-based intra prediction for image coding. IEEE Trans. Image Process. **27**, 3236–3247 (2018)
18. Gu, J., Wang, Z., Kuen, J., Ma, L., Shahroudy, A., Shuai, B.: Recent advances in convolutional neural networks. Pattern Recogn. **77**, 354–377 (2018)
19. Liang, S., Huang, Z., Liang, M., Yang, H.: Instance enhancement batch normalization: an adaptive regulator of batch noise. In: Proceedings of the AAAI Conference on Artificial Intelligence, vol. 34, pp. 4819–4827 (2020)
20. Kingma, D.-P., Ba, J.: Adam: a method for stochastic optimization. In: International Conference on Learning Representation (2015)
21. Fan, D.-P., Cheng, M.-M., Liu, Y., Li, T., Borji, A.: Structure-measure: a new way to evaluate foreground maps. In: ICCV, pp. 4548–4557 (2017)
22. Fan, D.-P., Gong, C., Cao, Y., Ren, B., Cheng, M.-M., Borji, A.: Enhanced alignment measure for binary foreground map evaluation. In: IJCAI (2018)
23. Ran, M., Lihi, Z.-M., Ayellet, T.: How to evaluate foreground maps? In: IEEE CVPR, pp. 248–255 (2014)
24. Perazzi, F., Krähenbühl, P., Pritch, Y., Hornung, A.: Saliency filters: contrast based filtering for salient region detection. In: CVPR, pp. 733–740 (2012)

Multi-task Self-supervised Few-Shot Detection

Guangyong Zhang[1,2,3], Lijuan Duan[1,2,3](✉), Wenjian Wang[1,2,3], Zhi Gong[1,2,3], and Bian Ma[1,2,3]

[1] Faculty of Information Technology, Beijing University of Technology, Beijing 100124, China
{zhangguangyong,wangwj,gongzhi97,mabian}@emails.bjut.edu.cn
[2] Beijing Key Laboratory of Trusted Computing, Beijing 100124, China
[3] National Engineering Laboratory for Critical Technologies of Information Security Classified Protection, Beijing 100124, China
ljduan@bjut.edu.cn

Abstract. Few-shot object detection involves detecting novel objects with only a few training samples. But very few samples are difficult to cover the bias of the new class in the deep model. To address the issue, we use self-supervision to expand the coverage of samples to provide more observation angles for new classes. In this paper, we propose a multi-task approach that combines self-supervision with few-shot learning to exploit the complementarity of these two domains. Specifically, our self-supervision as an auxiliary task to improve the detection performance of the main task of few-shot learning. Moreover, in order to make self-supervision more suitable for few-shot object detection, we introduce the denoising module to expand the positive and negative samples and the team module for precise positioning. The denoising module expands the positive and negative samples and accelerate model convergence using contrastive denoising training methods. The team module utilizes location constraints for precise localization to improve the accuracy of object detection. Our experimental results demonstrate the effectiveness of our method on the Few-shot object detection task on the PASCAL VOC and COCO datasets, achieving promising results. Our results highlight the potential of combining self-supervision with few-shot learning to improve the performance of object detection models in scenarios where annotated data is limited.

Keywords: Few-shot object detection · Self-supervised learning · End-to-End Detector

1 Introduction

Object detection [8,15,19–22] is a fundamental task in computer vision that involves identifying and localizing objects within an image or video. However,

Supported in part by National Natural Science Foundation of China under grant 62176009, in part supported by Ant Group.

© The Author(s), under exclusive license to Springer Nature Singapore Pte Ltd. 2024
Q. Liu et al. (Eds.): PRCV 2023, LNCS 14436, pp. 107–119, 2024.
https://doi.org/10.1007/978-981-99-8555-5_9

traditional approaches to object detection rely heavily on large amounts of labeled data for training. This labeling process is both time-consuming and resource-intensive, which can be a significant challenge in scenarios where new objects need to be detected with limited training samples, such as rare cases and rare animals. To address this issue, few-shot object detection has emerged as a promising approach that aims to detect novel objects with only a few training examples. Few-shot object detection [7,11,17,23,24,29] has the potential to significantly reduce the amount of labeled data required for training, thereby easing the burden of the labeling process.

Despite significant advancements in few-shot object detection concerning the scarcity of labeled samples, effectively addressing the bias towards covering novel classes still poses a challenge in this field. At the same time, self-supervision [9,12] holds substantial promise in effectively mitigating the issue of limited samples within the realm of few-shot object detection, thereby furnishing it with a wider range of informative and varied data. By harnessing the capabilities of self-supervised learning, we can acquire more efficient object representations, consequently enhancing the accuracy and robustness of object detection.

In this paper, we propose a novel approach to few-shot object detection by combining self-supervision with DETR variants [2,16,25,33] in a multi-task manner. Our method leverages the self-supervised branch to predict embeddings of a separate self-supervised image encoder on object regions. Similar to the backbone, the self-supervised image encoder learns transformation-invariant embeddings, which are distilled into the detector's embeddings. While self-supervision is an auxiliary task, our primary goal is to improve the detection performance of few-shot object detection.

While self-supervised learning has shown the capability to extract diverse features by leveraging inter-sample correlations, enhancing the model's ability to generalize in low-data scenarios, further adaptations are needed to optimize self-supervision for few-shot object detection. To address this issue, we introduce a denoising module and a team module similar to Team-DETR [18] in the self-supervised branch. The denoising module employs a contrastive denoising training by adding both positive and negative samples, which helps the model avoid duplicate outputs of the same target samples with the same ground truth. The team module groups queries at the decoder side and provides guiding queries in terms of scale and spatial position. Effectively strengthening the constraint of the self-supervised branch on the model and improves detection performance.

The main contributions are summarized as follows:

- We combine self-supervised learning with few-shot object detection in a multi-task manner. Self-supervision as an auxiliary task mainly improves the performance of the main task of few-shot object detection.
- We only introduced a denoising module and a team module in the self-supervised branch. Improved the constraint ability of the self-supervised branch on the model while makes it more compatible with the main branch and accelerating its convergence speed.

- Thorough experimental results show that the proposed self-supervised auxiliary tasks are substantially effective for accurate few-shot object detection. Our method achieves significant performance improvements in both PASCAL VOC [6] and COCO [14] datasets.

2 Related Work

2.1 Self-supervised Learning

Self-supervised Learning (SSL) [4,27,31] has shown great potential in generating powerful representations, and has even outperformed supervised methods on challenging vision benchmarks. These learned representations have been shown to transfer well to object detection tasks. Recent research has proposed several methods that combine SSL with DETR, a popular object detection framework. One such method is UP-DETR [5], which pretrains DETR in a self-supervised manner by detecting and reconstructing random patches from the input image. Another method, DETReg [1], incorporates region priors from unsupervised region proposal algorithms to provide weak supervision for pretraining. Our research is focused on leveraging self-supervision as an auxiliary task to improve the detection performance of the main task. We aim to further advance the field of object detection by exploring the potential of self-supervised learning in conjunction with existing methods.

2.2 Few-Shot Object Detection

Recently, there has been extensive research focused on DETR and its variants. However, these methods often suffer from a significant performance drop when applied to few-shot detection scenarios. To address this limitation, Meta-DETR [32] was proposed, which incorporates meta-learning into the DETR framework. This allows for image-level detection by effectively leveraging the correlation among various support classes. Building upon this work, we introduce self-supervised learning to further improve the detection performance. By leveraging self-supervision as an auxiliary task, we aim to improve the performance of object detection, particularly in few-shot scenarios. Our research contributes to the ongoing effort to advance the field of object detection by exploring the potential of combining self-supervised learning and meta-learning with existing methods such as DETR.

3 Methodology

In this section, we first detail the setup for the FSOD problem. Then, we introduce multi-task self-supervised few-shot object detection method, as shown in Fig. 1. Our method takes Meta-DETR as the baseline. We combine self-supervision with DETR variants in a multi-task manner. Self-supervised branch aims to predict the embeddings of a separate self-supervised image encoder evaluated on object regions. In order to solve the problem of Meta-DETR convergence speed and improved the constraint ability of the self-supervised branch on the model. We only introduced a denoising module and a team module in the self-supervised branch.

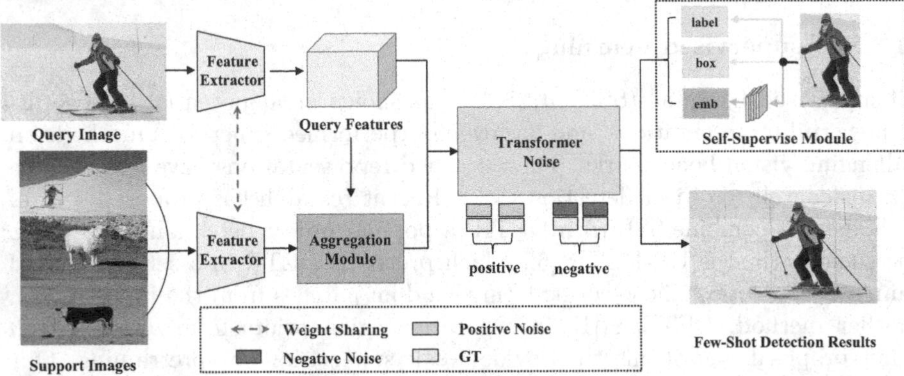

Fig. 1. Overview of our proposed method. Self-supervised auxiliary tasks are shared with the main task in feature extractor and transformer weights.

3.1 Problem Setting

As in previous work [7,24,34,35], we use the standard problem setting for FSOD in our paper. Specifically, the training data set consists of a base class set $D^{base} = \{x^{base}, y^{base}\}$ with a large number of samples and a novel class set $D^{novel} = \{x^{novel}, y^{novel}\}$ with only a few samples, where x and y represent training samples and labels, respectively. The number of samples for each class in the novel class set is K, thus constructing the k-shot problem. The classes in the base class set are C^{base}, the classes in the novel class set are C^{novel}, and the classes in the two sets are disjoint, that is, $C^{base} \bigcap C^{novel} = \emptyset$. We use an effective way to exploit the training set is to mimic the few-shot learning setting via episode based training. In each training iteration, an episode is formed by randomly selecting C classes from the training set with K labelled samples from each of the C classes to act as the sample set $S = \{(x_i, y_i)\}_{j=1}^{m})(m = K \times C)$, as well as a fraction of the remainder of those C classes samples to serve as the query set $Q = \{(x_j, y_j)\}_{j=1}^{n}$.This sample/query set split is designed to simulate the support/test set that will be encountered at test time. A model trained from sample/query set can be further fine-tuned using the support set, if desired. In this work we adopt such an episode-based training strategy.

3.2 Self-supervised Auxiliary Branch

Self-supervised task include their region localization and embedding components. Keeping the main task unchanged during pre-training, we only include self-supervised auxiliary tasks during fine-tuning. At a high level, we operates by predicting object localizations that match those from an unsupervised region proposal generator, while simultaneously aligning the corresponding feature embeddings with embeddings from backbone(RestNet-101)instead of a self-supervised image encoder like DETReg, see Fig. 1.

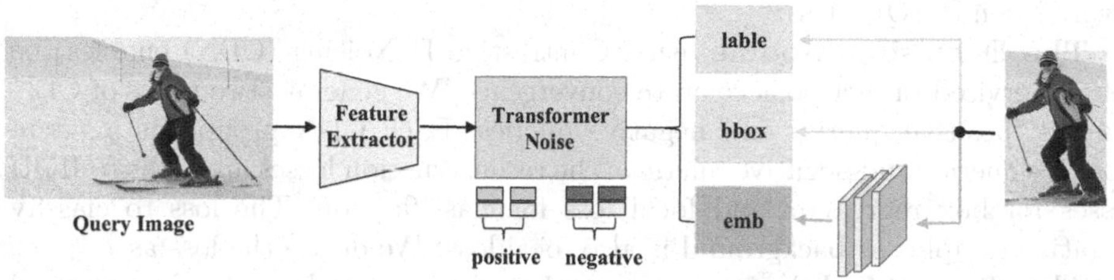

Fig. 2. Self-supervised Auxiliary Branch

Region localization takes a set of M boxes $b_1, ..., b_m$ output by an unsupervised region proposal method and optimizes a loss that minimizes the difference between the detector box predictions and these M boxes. The loss involves matching the predicted boxes and these M boxes. Selecting boxes policies is Top-K,similar to DETReg. Three prediction heads: f_{box} which outputs predicted bounding boxes, f_{cat} which predicts if the box is object or background, and f_{emb} which reconstructs the object embedding descriptor using backbone, These outputs as: $\hat{b}_i = f_{box}(v_i), \hat{z}_i = f_{emb}(v_i), \hat{p}_i = f_{cat}(v_i)$. We adopt a pairwise matching loss $L_{match}(y_i, \hat{y}_{\sigma(i)})$ to search for a bipartite matching between y and $\hat{y}$ with the lowest cost:

$$\hat{\sigma} = \underset{\sigma \in \Sigma_N}{\arg\min} \sum_{i}^{N} L_{match}(y_i, \hat{y}_{\sigma(i)}) \tag{1}$$

where $\hat{\sigma}(i)$ denotes the optimal assignment between predictions and targets. Since the matching also needs to consider both classification and localization, the matching loss is formalized as:

$$L_{match}(y_i, \hat{y}_{\sigma(i)}) = L_{cls}(c_i, \hat{c}_{\sigma(i)}) + L_{box}(b_i, \hat{b}_{\sigma(i)}) \tag{2}$$

We define the Self-supervised loss as:

$$L_{ssl}(y, \hat{y}) = \lambda_f L_{class}(c_i, \hat{P}_{\hat{\sigma}(i)}) + \lambda_b L_{box}(b_i, \hat{b}_{\hat{\sigma}(i)}) + \lambda_e L_{emb}(z_i, \hat{z}_{\hat{\sigma}(i)}) \tag{3}$$

Embedding components learn a strong object embedding, we encode each box region b_i via a separate encoder network and obtain embeddings z_i that are used as a target for the embeddings $\hat{z}_j$ (see the blue arrows in Fig. 2). The difference with DETReg is that we don't use SwAV [3] but backbone to ensure the invariance of image transformation embedding. We argue that swav gives the model additional guidance in addition to guaranteeing invariance. We introduce an additional MLP f_{emb} that predicts the object embedding $\hat{z}_j$ from the corresponding DETR query embedding $\hat{v}_j$. The loss is the L_1 loss between $\hat{z}_j$ and z_i .

$$L_{emb}(z_i, z_j) = \|z_i - \hat{z}_j\|_1 \tag{4}$$

where L_{class} is the class loss, that can be implemented via Cross Entropy Loss or Focal Loss, and L_{box} is based on the the L_1 loss and the Generalized Intersection Over Union (GIOU) loss.

The Denoising Module using Contrastive DeNoising (CDN) enables the self-supervised branch to accelerate convergence. We generate two types of CDN queries: positive queries and negative queries. Each CDN group has a set of positive queries and negative queries. The reconstruction losses are L_1 and GIOU losses for box regression and focal loss for classification. The loss to classify negative samples as background is also focal loss. We define the loss as L_{dn}.

The Team Module improves queries on the decoder side. The original queries are divided into groups,each responsible for predicting objects within a specific scale range. $\hat{B}_i$ is the prediction box of the i-th query q_i. When the distance between the center point of $\hat{B}_i$ and $\hat{A}_i$ exceeds the threshold η, a penalty is imposed on $\hat{B}_i$. σ is the number of boxes to be penalized. The loss function is expressed as

$$L_{pos} = \frac{1}{\sigma} \sum_{i=0}^{N-1} \mathbf{1}_{\{\|\hat{B}_i^{\{x,y\}} - A_i^{\{x,y\}}\|_2 > \eta\}} \|\hat{B}_i^{\{x,y\}} - A_i^{\{x,y\}}\|_2 \tag{5}$$

3.3 Multi-Task Learning

The Multi-Task Learning trains tasks together to overcome the shortage of annotated data. It provides each task with inductive bias to trigger regularization effect between one another. We combine Meta-DETR with a self-supervised branch in a multi-task manner.(see Fig. 1).

Here we briefly introduce the loss function of the Meta-DETR main task. Similar to Deformable DETR. $L_{meta}(y, \hat{y})$ is applied to every layer of the transformer decoder.

$$L_{meta}(y, \hat{y}) = \sum_{i=1}^{N} [L_{cls}(c_i, \hat{c}_{\sigma(i)}) + L_{box}(b_i, \hat{b}_{\sigma(i)})] \tag{6}$$

The total loss in training stage is :

$$L_{total} = L_{meta} + \lambda_{ssl}(L_{ssl} + L_{dn} + L_{pos}) \tag{7}$$

Experiments show that the best results are obtained when the value of λ_{ssl} is 0.17, which indicates that the self-supervised branch only plays an adjustment role for the main task.

4 Experiments

In this section, we first describe the details in the experiments, and then perform extensive experiments on the benchmarks using PASCAL VOC [6] and COCO [14] dataset. For fairness, we strictly adhere to the construction and evaluation protocol for FSOD data. Finally, we provide ablation analysis and visualizations.

4.1 Implementation Details

Our method is based on Meta-DETR [17], which uses Deformable DETR [22] with Resnet-101, and strictly keeps the parameters of Meta-DETR unchanged in all our experiments. We utilize the same data split as [24] to evaluate our method for fair comparison. When using the PASACL VOC and COCO datasets for evaluation, the latent knowledge is set to 64×256 and 128×256, namely N=64, m=256 and N=128, m=256, respectively. All our experiments are obtained on one 3090 GPU,using the AdamW optimizer with an initial learning rate of 2×10^{-4} and a weight decay of 1×10^{-4} and batch size is set to 32. In the base training stage, we train the model for 50 epochs for both Pascal VOC and MS COCO. Learning rate is decayed at the 45th epoch by 0.1. In the few-shot fine-tuning stage, the same settings are applied to fine-tune the model until convergence.

4.2 Few-Shot Object Detection Benchmarks

Results on PASCAL VOC. There are a total of 20 classes in the PASACL VOC dataset, which is divided into a base class set with 15 classes and a novel class set with 5 classes. We have three combinations of the base class set and the novel class set division, the same as the existing work. Each class in the base class set has a large number of samples, while each class in the novel class set has only k samples, in the experiment k=1, 2, 3, 5, 10, these k samples are in this class randomly selected from the sample. In the second stage, we extract k samples for each base class in the base class set, and fine-tune the model together with the novel class set. The experimental results are shown in Table 1.

As shown in Table 1, our experimental results show an improvement in accuracy when samples are scarce compared to existing works. This shows that self-supervision can make up for the sample defects of small sample object detection and improve the detection performance.

Results on COCO. The COCO datasetis a more challenging object detection dataset, which contains 80 classes including those 20 classes in Pascal VOC. We adopt the 20 shared classes as novel classes, and adopt the remaining 60 classes as base classes. Same as the baseline, we choose k=10, 30 for comparative experiments.We use train 2017 for training, and perform evaluations on val 2017. Standard evaluation metrics for MS COCO are adopted. Results are averaged over 5 randomly sampled support datasets. The experimental results are shown in Table 2.

Table 1. FSOD performance (novel AP50(%)) on three splits of PASCAL VOC dataset. *indicates the result of the baseline,which is the first model to use Deformable DETR for few-shot

Method/Shot	Class Split 1					Class Split 2					Class Split 3				
	1	2	3	5	10	1	2	3	5	10	1	2	3	5	10
Meta-YOLO [11]	14.8	15.5	26.7	33.9	47.2	15.7	15.2	22.7	30.1	40.5	21.3	25.6	28.4	42.8	45.9
metaDet [26]	18.9	20.6	30.2	36.8	49.6	21.8	23.1	27.8	31.7	43.0	20.6	23.9	29.4	43.9	44.1
Meta R-CNN [30]	19.9	25.5	35.0	45.7	51.5	10.4	19.4	29.6	34.8	45.4	14.3	18.2	27.5	41.2	48.1
TFA w/fc [24]	36.8	29.1	43.6	55.7	57.0	18.2	29.0	33.4	35.5	39.0	27.7	33.6	42.5	48.7	50.2
TFA w/cos [24]	39.8	36.1	44.7	55.7	56.0	23.5	26.9	34.1	35.1	39.1	30.8	34.8	42.8	49.5	49.8
MPSR [28]	41.7	43.1	51.4	55.2	61.8	24.4	–	39.2	39.9	47.8	35.6	–	42.3	48.0	49.7
SRR-FSD [35]	**47.8**	50.5	51.3	55.2	56.8	32.5	35.3	39.1	40.8	43.8	40.1	41.5	44.3	46.9	46.4
FSCE [23]	44.2	43.8	51.4	61.9	63.4	27.3	29.5	43.5	44.2	50.2	37.2	41.9	47.5	54.6	58.5
Meta-DETR* [32]	40.6	51.4	**58.0**	59.2	63.6	37.0	36.6	**43.7**	49.1	54.6	41.6	45.9	52.7	**58.9**	**60.6**
Ours	42.1	**52.2**	57.7	**59.5**	**64.3**	**37.3**	**37.1**	43.1	**49.5**	**55.1**	**42.4**	**46.0**	**54.3**	58.3	59.8

Table 2. FSOD performance (novel AP(%)) on COCO dataset. *indicates the result of the baseline,which is the first model to use Deformable DETR for few-shot

Method/Shot	Shot Number					
	10			30		
	AP	AP50(%)	AP75(%)	AP	AP50(%)	AP75(%)
FRCN-ft-full [30]	5.5	10.0	5.5	7.4	13.1	7.4
Meta-YOLO [11]	5.6	12.3	4.6	9.1	19.0	7.6
Meta R-CNN [30]	8.7	19.1	6.6	12.4	25.3	10.8
SSR-FSD [35]	11.3	23.0	9.8	14.7	29.2	13.5
CME [13]	15.4	24.6	16.4	16.9	28.0	17.8
DCNet [10]	12.8	23.4	11.2	18.6	32.6	17.5
FSCE [23]	11.1	–	9.8	15.3	–	14.2
Meta-DETR* [32]	19.0	30.5	19.7	22.2	35.0	22.8
Ours	**19.3**	**31.3**	**19.8**	**22.3**	**36.0**	**23.0**

The COCO dataset has many more categories than VOC dataset, but Table 2 proves that our method is still effective in improving the detection accuracy, and performs exceptionally well compared with other region-based methods under the metric AP0.50.

4.3 Ablation Analysis

In this section, we separately explore the influence of the self-supervised module and denoising module on the experimental results through experiments. All ablation experiments in this section, like most existing works, are based on Novel Set 1 of PASCAL VOC.

The Effect of the Self-supervised Module. We conduct ablation experiments to validate the effectiveness of the self-supervised module, without denoising module and team module.

Table 3. The effect of the self-supervised module.

Model/Shot	1	2	3	5	10
Baseline	40.6	51.4	**58.0**	59.2	63.6
self-supervised module	41.7	51.9	57.0	59.4	63.9
Ours	**42.1**	**52.2**	57.2	**59.5**	**64.3**

Table 3 shows the comparison results of the experiments. Taking 1-shot and 2-shot as an example, the accuracy is improved by 1.1% and 0.5% when using the self-supervised module compared to not using it. This shows that the self-supervised module is effective in improving the performance of the model.

The Weights of Self-supervised Branch. In our method, the weights of self-supervised branch is a very important hyperparameter. When the weights is small, the self-supervised branch is less effective. In the case of large weights, the self-supervised branch will have an excessive impact on the model, making the model more focused on the accuracy of auxiliary tasks and affecting the performance of small-sample target detection.

Table 4. The weights of self-supervised branch. Only the weights of self-supervised branch is different here, other settings are exactly the same.

λ_{ssl}/Shot	1	2	3	5	10
0.1	40.0	52.0	56.8	58.6	63.9
0.17	**42.1**	**52.2**	57.2	**59.5**	**64.3**
0.3	38.2	52.1	56.5	58.9	**64.3**
0.5	41.9	51.2	55.8	58.0	63.9
0.7	42.1	49.7	56.4	58.1	62.5

Table 4 shows the impact of different weights of self-supervised branch on model performance. We can see that when the weights of self-supervised branch is 0.17, the overall accuracy is the best.

Fig. 3. Visualize our 10-shot object detection on the boat class on the COCO dataset as an example.

4.4 Visualization

As shown in Fig. 3, we use the COCO dataset 10-shot as an example to visualize the boat in the novel class set. In Fig. 4, we visualize the results by taking the 1-shot of the VOC dataset 1-split as an example. "bus", "cow", "bird", and "motorbike" are categories in the novel class set, and the rest are in the base class set. We can see that our method can detect more novel classes of objects.

Fig. 4. Visualization of Our Model on the Novel Class Set: "Bus", "Cow", "Bird" and "Motorbike".

5 Conclusion

In this paper, we propose a multi-task-based self-supervised few-shot object detection model. We combine self-supervision with DETR in a multi-task manner for few-shot object detection. The self-supervised branch is used as an auxiliary task, and the few-shot object detection is used as the main task. We mainly focus on the detection performance of the main task. Our motivation is to use self-supervision to compensate for the number of samples in few-shot object detection. At the same time, We introduce a denoising module and team module to makes it more compatible with the few-shot object detection. Experimental results show that our model has better performance compared to other networks. We hope that our proposed method can be helpful for improving the accuracy of DETR as an end-to-end object detector in few-shot object detection.

References

1. Bar, A., et al.: Detreg: unsupervised pretraining with region priors for object detection. In: 2022 IEEE/CVF Conference on Computer Vision and Pattern Recognition (CVPR) (2022)
2. Carion, N., Massa, F., Synnaeve, G., Usunier, N., Kirillov, A., Zagoruyko, S.: End-to-End object detection with transformers, pp. 213–229 (2020)
3. Caron, M., Misra, I., Mairal, J., Goyal, P., Bojanowski, P., Joulin, A.: Unsupervised learning of visual features by contrasting cluster assignments. Le Centre pour la Communication Scientifique Directe - HAL - Université Paris Descartes (2020)
4. Chen, T., Kornblith, S., Norouzi, M., Hinton, G. A simple framework for contrastive learning of visual representations. Cornell University (2020)
5. Dai, Z., Cai, B., Lin, Y., Chen, J.. Up-detr: unsupervised pre-training for object detection with transformers. In: 2021 IEEE/CVF Conference on Computer Vision and Pattern Recognition (CVPR) (2021)
6. Everingham, M., Van Gool, L., Williams, C.K., Winn, J., Zisserman, A.: The pascal visual object classes (voc) challenge. Int. J. Comput. Vision **88**(2), 303–338 (2010)
7. Fan, Q., Zhuo, W., Tang, C.K., Tai, Y.W.: Few-shot object detection with attention-rpn and multi-relation detector. In: Proceedings of the IEEE/CVF Conference on Computer Vision and Pattern Recognition, pp. 4013–4022 (2020)
8. Girshick, R.: Fast r-cnn. In: Proceedings of the IEEE International Conference on Computer Vision, pp. 1440–1448 (2015)
9. Grill, J.-B., et al.: Bootstrap your own latent: a new approach to self-supervised learning. Le Centre pour la Communication Scientifique Directe - HAL - Diderot (2020)
10. Hu, H., Bai, S., Li, A., Cui, J., Wang, L.: Dense relation distillation with context-aware aggregation for few-shot object detection. In: 2021 IEEE/CVF Conference on Computer Vision and Pattern Recognition (CVPR) (2021)
11. Kang, B., Liu, Z., Wang, X., Yu, F., Feng, J., Darrell, T. Few-shot object detection via feature reweighting. In: Proceedings of the IEEE/CVF International Conference on Computer Vision, pp. 8420–8429 (2019)
12. Kolesnikov, A., Zhai, X., Beyer, L.: Revisiting self-supervised visual representation learning. In: 2019 IEEE/CVF Conference on Computer Vision and Pattern Recognition (CVPR) (2019)

13. Li, B., Yang, B., Liu, C., Liu, F., Ji, R., Ye, Q.: Beyond max-margin: class margin equilibrium for few-shot object detection. In: 2021 IEEE/CVF Conference on Computer Vision and Pattern Recognition (CVPR) (2021)
14. Lin, T.-Y., et al.: Microsoft COCO: common objects in context. In: Fleet, D., Pajdla, T., Schiele, B., Tuytelaars, T. (eds.) ECCV 2014. LNCS, vol. 8693, pp. 740–755. Springer, Cham (2014). https://doi.org/10.1007/978-3-319-10602-1_48
15. Liu, W., et al.: SSD: single shot multibox detector. In: Leibe, B., Matas, J., Sebe, N., Welling, M. (eds.) ECCV 2016. LNCS, vol. 9905, pp. 21–37. Springer, Cham (2016). https://doi.org/10.1007/978-3-319-46448-0_2
16. Meng, D., et al.: Conditional DETR for fast training convergence. In: 2021 IEEE/CVF International Conference on Computer Vision (ICCV) (2021)
17. Qiao, L., Zhao, Y., Li, Z., Qiu, X., Wu, J., Zhang, C.: Defrcn: decoupled faster r-cnn for few-shot object detection. In: Proceedings of the IEEE/CVF International Conference on Computer Vision, pp. 8681–8690 (2021)
18. Qiu, T., Zhou, L., Xu, W., Cheng, L., Feng, Z., Song, M.: Team-DETR: guide queries as a professional team in detection transformers (2023)
19. Redmon, J., Divvala, S., Girshick, R., Farhadi, A.: You only look once: unified, real-time object detection. In: Proceedings of the IEEE Conference on Computer Vision and Pattern Recognition, pp. 779–788 (2016)
20. Redmon, J., Farhadi, A.: Yolo9000: better, faster, stronger. In: Proceedings of the IEEE Conference on Computer Vision and Pattern Recognition, pp. 7263–7271 (2017)
21. Redmon, J., Farhadi, A.: Yolov3: an incremental improvement. arXiv preprint arXiv:1804.02767 (2018)
22. Ren, S., He, K., Girshick, R., Sun, J.: Faster r-cnn: towards real-time object detection with region proposal networks. Adv. Neural Inf. Process. Syst. **28** (2015)
23. Sun, B., Li, B., Cai, S., Yuan, Y., Zhang, C.: FSCE: few-shot object detection via contrastive proposal encoding. In: Proceedings of the IEEE/CVF Conference on Computer Vision and Pattern Recognition, pp. 7352–7362 (2021)
24. Wang, X., Huang, T.E., Darrell, T., Gonzalez, J.E., Yu, F.: Frustratingly simple few-shot object detection. arXiv preprint arXiv:2003.06957 (2020)
25. Wang, Y., Zhang, X., Yang, T., Sun, J.: Anchor detr: query design for transformer-based detector. In: Proceedings of the AAAI Conference on Artificial Intelligence, pp. 2567–2575 (2022)
26. Wang, Y. X., Ramanan, D., Hebert, M.: Meta-learning to detect rare objects. In: Proceedings of the IEEE/CVF International Conference on Computer Vision, pp. 9925–9934 (2019)
27. Wei, F., Gao, Y., Wu, Z., Qiu, J., Lin, S.: Aligning pretraining for detection via object-level contrastive learning. Cornell University (2021)
28. Wu, J., Liu, S., Huang, D., Wang, Y.: Multi-scale positive sample refinement for few-shot object detection. In: Vedaldi, A., Bischof, H., Brox, T., Frahm, J.-M. (eds.) ECCV 2020. LNCS, vol. 12361, pp. 456–472. Springer, Cham (2020). https://doi.org/10.1007/978-3-030-58517-4_27
29. Xiao, Y., Marlet, R.: IEEE Trans. Pattern Anal. Mach. Intell. (2022)
30. Yan, X., Chen, Z., Xu, A., Wang, X., Liang, X., Lin, L.: Meta r-cnn: towards general solver for instance-level low-shot learning. In: Proceedings of the IEEE/CVF International Conference on Computer Vision, pp. 9577–9586 (2019)
31. Yang, C., Wu, Z., Zhou, B., Lin, S.: Instance localization for self-supervised detection pretraining. In: 2021 IEEE/CVF Conference on Computer Vision and Pattern Recognition (CVPR) (2021)

32. Zhang, G., Luo, Z., Cui, K., Lu, S., Xing, E.P.: Meta-DETR: image-level few-shot detection with inter-class correlation exploitation. IEEE Trans. Pattern Anal. Mach. Intell. **45**, 12832–12843 (2022)
33. Zhang, H., et al.: Dino: Detr with improved denoising anchor boxes for end-to-end object detection (2022)
34. Zhang, W., Wang, Y.-X.: Hallucination improves few-shot object detection. In: Proceedings of the IEEE/CVF Conference on Computer Vision and Pattern Recognition, pp. 13008–13017 (2021)
35. Zhu, C., Chen, F., Ahmed, U., Shen, Z., Savvides, M.: Semantic relation reasoning for shot-stable few-shot object detection. In: Proceedings of the IEEE/CVF Conference on Computer Vision and Pattern Recognition, pp. 8782–8791 (2021)

CSTrack: A Comprehensive and Concise Vision Transformer Tracker

Yao Chen[1], Shuyan Ding[2(✉)], Jianhui Guo[1], Chen Yang[1], and Lunbo Li[1]

[1] School of Computer Science and Engineering, Nanjing University of Science and Technology, Nanjing 210094, China
{chenyao1019,guojianhui,kogenta,lunboli}@njust.edu.cn

[2] School of Electronic and Optical Engineering, Nanjing University of Science and Technology, Nanjing 210094, China
shuyanding@njust.edu.cn

Abstract. The attention mechanism has been widely applied in visual tracking tasks due to its remarkable ability to capture global dependencies. However, there are two issues in previous attention-based methods: redundancy of template information and inappropriate utilization of information streams. In this work, we propose a spatial positioning attention mechanism that addresses these issues by selective template feature enhancement and elimination of redundant information streams, respectively, significantly improving tracking accuracy and speed. Furthermore, previous trackers fail to focus on channels containing crucial target information within the template features and search region features. To tackle this, we introduce a channel focus attention mechanism to perform channel weight rescaling, which allows the tracker to concentrate on those target-related channels, improving its localization capability. Extensive experiments on four well-known datasets, GOT-10k, LaSOT, TrackingNet, and TNL2K, show that CSTrack outperforms all previous state-of-the-art trackers, running at over 70 FPS.

Keywords: Visual tracking · Vision Transformer · One-stream tracker

1 Introduction

Visual object tracking (VOT) is a fundamental task in computer vision, which aims to estimate the position of the target in the subsequent frames based only on its position in the first frame. Due to its balance of accuracy and speed, the Siamese framework has become the dominant tracking framework, represented by [1,16,25,29]. However, traditional Siamese-based trackers suffer from two notable drawbacks. First, the fusion of template features and search region features in traditional Siamese trackers relies on CNN-based methods, which inevitably confront the long-range dependency dilemma. This limitation hinders their ability to capture the necessary contextual information for accurate tracking. Second, two-stream trackers separate the processes of feature extraction and feature fusion, resulting in information loss and impacting the overall tracking

© The Author(s), under exclusive license to Springer Nature Singapore Pte Ltd. 2024
Q. Liu et al. (Eds.): PRCV 2023, LNCS 14436, pp. 120–132, 2024.
https://doi.org/10.1007/978-981-99-8555-5_10

performance. To address these issues, recent works [5,9,11,24,30,31], introduce attention mechanisms to VOT tasks, effectively getting rid of the long-range dependency dilemma and bringing about significant improvements in tracking performance. Furthermore, one-stream trackers like SimTrack [4] are proposed to mitigate information loss. These trackers perform feature extraction and feature fusion of template features and search region features simultaneously, thereby enhancing their ability to retain critical information for accurate tracking.

In VOT tasks, the template encompasses crucial information about the target, like its size and color. However, previous trackers blindly enhance template features, resulting in template information redundancy that hinders target identification and localization while also impacting tracking speed. Additionally, using the information stream from the search region to the template introduces useless background information from the search region into the template. This degrades the quality of template features and hampers the localization of the tracking target within the search region. As a result, we propose a spatial positioning attention mechanism to tackle the aforementioned issues. Concretely, we design the Gate module to control the template feature enhancement. By selectively applying template feature enhancement at specific feature fusion layers, we reduce template information redundancy and speed up tracking. Simultaneously, we remove the information stream from the search region to the template. This prevents interference from irrelevant background information in the search region, ultimately improving the localization capability of our tracker.

Moreover, as highlighted in SiamRPN++ [16], different feature channels represent distinct information. However, previous trackers treat all feature channels equally, which limits their ability to effectively identify and localize the target. In this work, we introduce a channel focus attention mechanism to perform feature channel weight rescaling, which enables the tracker to concentrate on target information-rich feature channels, improving tracking performance. Furthermore, we propose a one-stream tracker named CSTrack, which excels in handling complex tracking scenarios like background clutter.

Our contribution can be summarized as follows. (1) We propose a spatial positioning attention mechanism (SPA) to address the template feature information redundancy and the inappropriate use of information stream, improving the localization capability and tracking speed. (2) We introduce a channel focus attention mechanism (CFA) that allows our tracker to concentrate on channels that contain critical target information. (3) We design a one-stream tracker CSTrack, which incorporates both spatial and channel dimensions to locate the tracking target, significantly improving the tracking accuracy, while maintaining a high frame rate of 70 FPS. (4) Extensive experimental results on well-known datasets including GOT-10k, LaSOT, TNL2K, and TrackingNet, demonstrate that CSTrack suppresses all previous state-of-the-art trackers.

2 Related Work

Previous Trackers. The preliminary Siamese-based trackers are two-stream trackers, in which the feature extraction and feature fusion of the template and

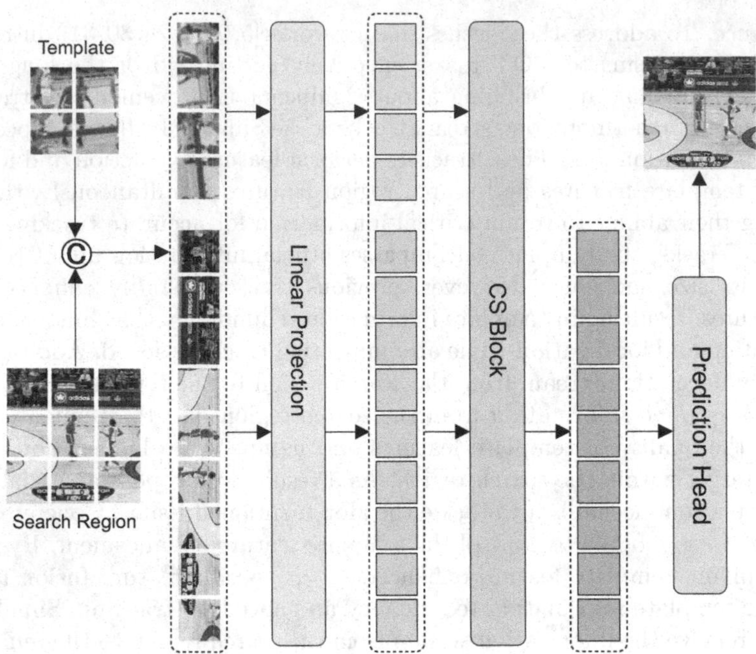

Fig. 1. The architecture of our tracker CSTrack.

search region are divided into two steps. SiamFC [1] is the first to apply the Siamese framework to VOT tasks, laying the foundation for subsequent research, like [2,6,10,16,25,29,33]. Due to the short-range property of CNNs, some works like [5,9,15,22,24,26,30,31], introduce attention mechanisms into VOT tasks, breaking the long-range dependency dilemma. However, in these trackers, the template and search region are fed into the feature fusion module after feature extraction, leading to information loss and limiting tracking performance. To address the limitations of two-stream trackers, Chen *et al.* propose a one-stream tracker SimTrack [4], which integrates feature extraction and feature fusion into a single module, significantly improving tracking performance. However, previous trackers blindly enhance template features and use inappropriate information stream, which restricts the localization capability.

Channel Attention Mechanism. Distinct feature channels represent different types of information, such as category and color. Consequently, it is crucial to assign varying levels of importance to these channels. In light of this, SENet [13] conducts feature channel weight rescaling through Squeeze (S) and Excitation (E) operations. SKNet [17] integrates channel information at different scales, taking into account the importance of each channel in a comprehensive manner. However, previous trackers treat all feature channels equally and cannot concentrate on the channels that are relevant to the target. To overcome this limitation, we employ a channel focus attention mechanism that realizes channel weight rescaling, which makes the tracker able to focus on the important feature channels, improving the anti-interference capability.

3 Method

3.1 Overview

As depicted in Fig. 1, our CSTrack adopts the prevailing one-stream architecture, where the feature extraction and fusion of the template image and the search region image are performed simultaneously. During tracking, the template image $t \in \mathbb{R}^{H_t \times W_t \times 3}$ and the search region image $s \in \mathbb{R}^{H_s \times W_s \times 3}$ are split into patches of size $P \times P$ and flattened, obtaining the patch sequence $t_p \in \mathbb{R}^{N_t \times (P^2 \cdot 3)}$ and $s_p \in \mathbb{R}^{N_s \times (P^2 \cdot 3)}$, where $N_t = H_t W_t / P^2$, $N_s = H_s W_s / P^2$. Next, the sequences t_p and s_p are sent into a linear projection layer to obtain the template feature embedding $E_t \in \mathbb{R}^{N_t \times d}$ and the search image feature embedding $E_s \in \mathbb{R}^{N_s \times d}$, where d means the feature embedding dimension. The feature tokens, E_t and E_s, are concatenated into a feature sequence with a length of $N_t + N_s$, which is subsequently fed into the CSBlock module for feature integration to produce the target localization feature $E_p \in \mathbb{R}^{N_s \times d}$. Finally, the feature E_p is sent into the prediction head to obtain the predicted position and size of the target. In brief, the tracking process can be formulated as follows.

$$
\begin{aligned}
[t_p; s_p] &= \text{SplitAndConcat}(t, s), \\
[E_t; E_s] &= \text{LinearProjection}([t_p; s_p]), \\
E_p &= \text{CSBlock}([E_t; E_s]), \\
[x, y, w, h] &= \text{PredictionHead}(E_p),
\end{aligned}
\tag{1}
$$

where [;] denotes the feature concatenation operation.

3.2 CSBlock

As illustrated in Fig. 2, the CSBlock module comprises three key components: spatial positioning attention (SPA), channel focus attention (CFA), and MLP module. The SPA module is applied to capture the spatial global feature dependencies and obtain the target information. The CFA module conducts feature channel weighting rescaling to highlight channels containing critical target-related information. Lastly, the MLP module is utilized to enhance the representation capability of both template features and search region features. The operation of the i-th layer of the CSBlock module can be expressed as follows.

$$
\begin{aligned}
[E_t^{i'}; E_s^{i'}] &= [E_t^i; E_s^i] + \text{CFA}(\text{LN}([E_t^i; E_s^i])), \\
[E_t^{i''}; E_s^{i''}] &= [E_t^{i'}; E_s^{i'}] + \text{SPA}(\text{LN}([E_t^{i'}; E_s^{i'}])), \\
[E_t^{i+1}; E_s^{i+1}] &= [E_t^{i''}; E_s^{i''}] + \text{MLP}(\text{LN}([E_t^{i''}; E_s^{i''}])),
\end{aligned}
\tag{2}
$$

where LN means the layer normalization operation.

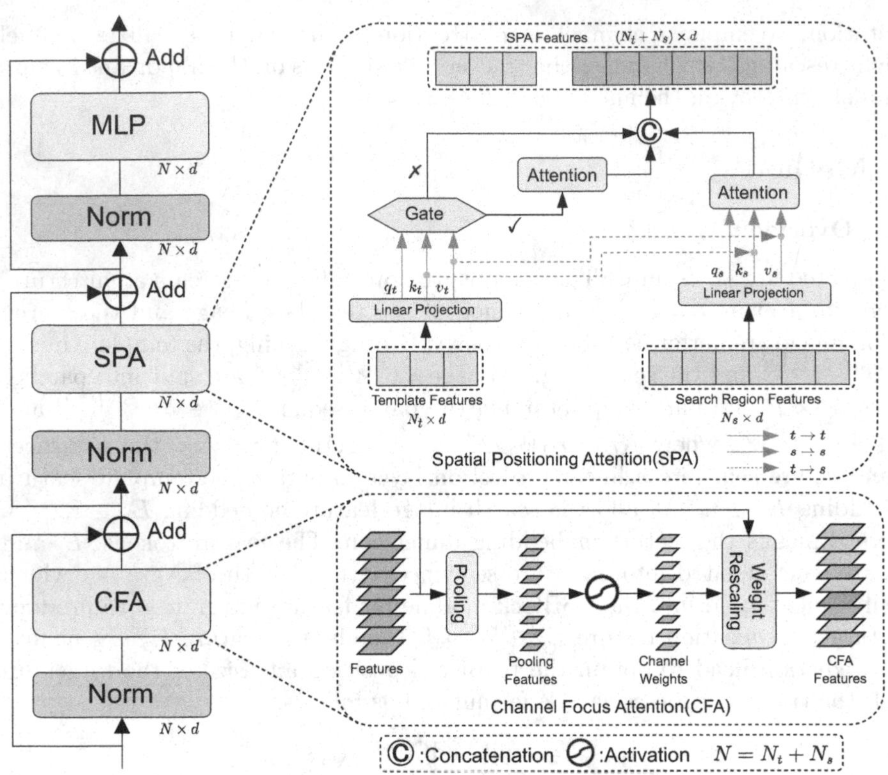

Fig. 2. The architecture of our CSBlock.

Spatial Positioning Attention. In VOT tasks, the attention mechanism plays a vital role in template feature enhancement, search region feature enhancement, and template-search region feature fusion. These tasks involve four distinct information streams: template feature enhancement ($t \rightarrow t$), search region feature enhancement ($s \rightarrow s$), and the fusion of template features and search region features ($t \rightarrow s$, $s \rightarrow t$). While previous trackers using four information streams achieve significant improvements in tracking performance, they also exhibit two notable drawbacks. Firstly, these trackers perform template feature enhancement with the information stream $t \rightarrow t$ at each layer. However, experimental results indicate that excessive template feature enhancement leads to template information redundancy, thereby degrading their quality. Since template information serves as the only ground truth in tracking, its quality directly influences tracking performance. Moreover, excessive template feature enhancement also slows down the tracking speed. Secondly, due to the presence of massive background information in the search region, the information stream $s \rightarrow t$ introduces unwanted background information into the template features, blurring the target information and hindering accurate tracking.

As displayed in Fig. 2, we propose the spatial positioning attention mechanism to address the aforementioned issues. Our approach differs from previous methods in that we selectively perform template feature enhancement ($t \rightarrow t$) only at specific feature layers. Through experiments and analysis, we find that template features in layer 0 directly represent the target information. Therefore, we only perform template feature enhancement in odd layers to preserve the integrity of layer 0 template features, prevent redundancy of template information, and improve both tracking accuracy and speed. The Gate module is employed to control the template feature enhancement process. Furthermore, we remove the information stream $s \rightarrow t$. This modification aims to reduce the interference caused by background information in the search region. In other words, the template features will only carry out feature enhancement operations.

The execution process of the SPA module is illustrated in Fig. 2. Initially, the template features E_t and the search region features E_s undergo a linear projection layer, obtaining the corresponding feature tokens q, k, and v. For the template features, the feature tokens k_t and v_t are transmitted to the search region branch for information fusion ($t \rightarrow s$). Subsequently, the Gate module determines whether template feature enhancement ($t \rightarrow t$) should be performed based on specific conditions. As for the search region features, the feature tokens k_s and v_s are concatenated with k_t and v_t, respectively, to conduct search region feature enhancement ($s \rightarrow s$) and information fusion ($t \rightarrow s$). Ultimately, the template features $E_t^{'}$ and the search region features $E_s^{'}$ are concatenated and passed to the subsequent module. In summary, the execution process of the SPA module of the i-th CSBlock is defined as follows (the i starts from 0).

$$\begin{aligned} q_t, k_t, v_t &= \text{LinearProjection}(E_t) \\ q_s, k_s, v_s &= \text{LinearProjection}(E_s) \\ \text{Gate}(E_t, i) &= \begin{cases} \text{Attention}(q_t, k_t, v_t), & i \text{ is odd} \\ E_t, & i \text{ is even} \end{cases} \\ E_t^{'} = \text{Gate}(E_t, i), &\ \ E_s^{'} = \text{Attention}(q_s, [k_t; k_s], [v_t; v_s]), \end{aligned} \tag{3}$$

Channel Focus Attention. As highlighted in SiamRPN++ [16], different feature channels represent distinct target information, such as category and shape. Effectively utilizing the channel information in template and search region features can improve the localization capability of the tracker. In contrast to previous trackers that ignore the feature channel information, inspired by SENet [13], we propose a channel focus attention to realize channel weight rescaling, focusing on the channels that represent the target information. Specifically, as depicted in Fig. 2, the features undergo a pooling operation along the spatial dimension. Subsequently, the weights of each channel are obtained through the sigmoid activation. The input features accomplish channel attention assignment in the weight rescaling module. Owing to the CFA module, critical channels can be focused on and tracking performance is significantly improved.

3.3 Prediction Head and Loss

We employ the prediction head module to estimate the position and size of the target. The feature token E_p is transformed into a 2D feature map and passed through three separate convolution branches, which are responsible for center classification, size regression, and offset regression, respectively. Specifically, the center classification branch is in charge of estimating the corresponding position of the center of the tracking target. The size regression branch is used to predict the width and height of the target. The offset regression branch is employed to compensate for discretization errors. Finally, the outputs of these branches are combined to derive the final predicted position and size of the target.

In the training stage, we employ the weighted focal loss to supervise the classification branch. The L1 loss and the generalized IoU loss [23] are used for the two regression branches, respectively. In short, the loss function is formulated as follows.

$$Loss = \lambda_{center} L_{focal} + \lambda_{iou} L_{iou} + \lambda_{L1} L_1, \quad (4)$$

where λ_{center}, λ_{iou} and λ_{L1} are trade-off weights to balance joint optimization.

4 Experiment

4.1 Implementation Details

Model. In our CSTrack, the size of the template image is 128×128 and the size of the search region image is 256×256. ViT-base [7] is adopted as the backbone network, which has 12 layers. We initialize the network parameters using the MAE [12] pre-training model, in order to converge quickly. The patch size $P \times P$ is 16×16, and the feature embedding dimension d is 768.

Training. The training sets of GOT-10k [14], LaSOT [8], TrackingNet [21], and COCO [18] datasets are used for model training. For the GOT-10k test set, we follow the default protocol to train the model using only the GOT-10k training set. The horizontal flip and brightness jittering are employed for data augmentation. We train our tracker using 2 RTX 3090 GPUs, with each GPU hosting 32 image pairs. The whole tracker is optimized with the AdamW [19] optimizer with the weight decay to 10^{-4}. The initial learning rate of the network is 4×10^{-5} and that of the rest parameters is 4×10^{-4}. The whole training process consists of 300 epochs with 60k image pairs per epoch. The learning rate decays by a factor of 10 after 240 epochs. For the GOT-10k dataset, we train only 100 epochs, with the learning rate decaying by a factor of 10 after 80 epochs.

4.2 Comparisons with the State-of-the-Art Trackers

GOT-10k. GOT-10k [14] is a large-scale dataset including more than 10,000 video sequences, with 180 test sequences. For the GOT-10k test set, we strictly follow the default protocol to train our model using only the GOT-10k training set. As illustrated in Table 1, owing to the selective template enhancement strategy, the tracking performance of our CSTrack significantly outperforms that of SimTrack [4] (AO: 1.0%, $SR_{0.5}$: 1.1%, and $SR_{0.75}$: 1.7%).

TrackingNet. TrackingNet [21] is a large-scale short-term dataset containing a large number of video sequences in the wild, of which 511 are test video sequences. As shown in Table 1, CSTrack achieves AUC (82.8%), P_{Norm} (87.4%), and P (81.2%), significantly exceeding the tracker CSWinTT [24] that utilizes information stream $s \rightarrow t$, proving the effectiveness of our method.

Table 1. State-of-the-art comparisons on TrackingNet [21], TNL2K [27], GOT-10k [14] and LaSOT [8]. The best two results are shown in red and **green** fonts.

Tracker	Source	GOT-10k [14]			TrackingNet [21]			TNL2K [27]		LaSOT [8]		
		AO	$SR_{0.5}$	$SR_{0.75}$	AUC	P_{Norm}	P	AUC	P	AUC	P_{Norm}	P
SiamFC [1]	ECCVW2016	34.8	35.3	9.8	57.1	66.3	53.3	29.5	28.6	33.6	42.0	33.9
SiamRPN++ [16]	CVPR2019	–	–	–	73.3	80.0	69.4	41.3	41.2	49.6	56.9	49.1
DiMP [2]	ICCV2019	61.1	71.7	49.2	74.0	80.1	68.7	44.7	43.4	56.9	65.0	56.7
KYS [3]	ECCV2020	63.6	75.1	51.5	74.0	80.0	68.8	44.9	43.5	55.4	63.3	–
PrDiMP [6]	CVPR2020	63.4	73.8	54.3	75.8	81.6	70.4	47.0	45.9	59.8	68.8	60.8
Ocean [33]	ECCV2020	61.1	72.1	47.3	–	–	–	38.4	37.7	56.0	65.1	56.6
SiamFC++ [29]	AAAI2020	56.9	69.5	47.9	75.4	80.0	70.5	38.6	36.9	54.4	62.3	54.7
STMTrack [10]	CVPR2021	64.2	73.7	57.5	80.3	85.1	76.7	–	–	60.6	69.3	63.3
AutoMatch [32]	ICCV2021	65.2	76.6	54.3	76.0	–	72.6	47.2	43.5	58.3	67.5	59.9
DTT [31]	ICCV2021	63.4	74.9	51.4	79.6	85.0	78.9	–	–	60.1	–	–
TrDiMP [26]	CVPR2021	67.1	77.7	58.3	78.4	83.3	73.1	–	–	63.9	73.0	61.4
STARK [30]	ICCV2021	68.0	77.7	62.3	81.3	86.1	–	52.5	–	66.0	75.5	70.8
TransT [5]	CVPR2021	67.1	76.8	60.9	81.4	86.7	**80.3**	50.7	51.7	64.9	73.9	69.0
SiamPW [25]	CVPR2022	64.4	76.7	50.9	–	–	–	–	–	55.8		57.0
CNNInMo [11]	IJCAI2022	–	–	–	72.1	–	–	42.2	41.9	53.9	61.6	53.9
SBT [28]	CVPR2022	66.4	77.3	59.2	–	–	–	–	–	65.9	–	70.0
UTT [20]	CVPR2022	67.2	76.3	60.5	79.7	–	77.0	–	–	64.6	-	67.2
TransInMo [11]	IJCAI2022	–	–	–	81.6	–	–	51.5	52.6	65.3	74.6	69.9
SLTrack [15]	ECCV2022	67.5	78.8	58.7	78.1	83.1	–	–	–	**66.4**	73.5	–
CIA [22]	ECCV2022	67.9	79.0	60.3	79.2	84.5	75.1	50.9	–	66.2	–	69.6
SparseTT [9]	IJCAI2022	69.3	**79.1**	63.8	81.7	86.6	79.5	–	–	66.0	74.8	70.1
CSWinTT [24]	CVPR2022	69.4	78.9	65.4	**81.9**	**86.7**	79.5	–	–	66.2	75.2	**70.9**
SimTrack [4]	ECCV2022	**69.8**	78.8	**66.0**	81.5	86.0	–	53.7	52.6	66.2	76.1	–
CSTrack	**Ours**	70.5	79.7	67.1	82.8	87.4	81.2	**53.3**	**52.5**	66.8	**75.9**	71.6

TNL2K. TNL2K [27] is a large-scale evaluation dataset comprising 700 test video sequences with various challenges like occlusion and deformation. As illustrated in Table 1, our CSTrack obtains AUC (53.3%) and P (52.5%). Since targets in TNL2K are often small, SimTrack [4] is able to track small targets better by using two template feature maps, the original-size template feature and the center-cropped template feature. However, our CSTrack utilizes the same size feature map in all layers. This is good for the tracking of large-size targets but not suitable for the identification and localization of small targets. This is the research direction for our future work.

LaSOT. LaSOT [8] is a large-scale long-term tracking dataset consisting of 280 long test sequences with an average of 2500 frames. As presented in Table 1, our CSTrack achieves AUC (66.8%), P_{Norm} (75.9%), and P (71.6%), demonstrating the effectiveness of the SPA module and the CFA module.

4.3 Ablation Study

Params, MACs and Speed. As shown in Table 2, the Params and MACs of our tracker are 93.5M and 29.1G, respectively. Notably, the tracking speed of our tracker is 70.6 FPS, which far exceeds the requirement for real-time tracking (over 35 FPS), demonstrating the help of the SPA module in tracking speed.

Table 2. Params, MACs and Speed of our designed tracker CSTrack.

Tracker	Params (M)	MACs (G)	Speed (FPS)
CSTrack	93.5	29.1	70.6

Table 3. The performance of different template feature enhancement strategies on the GOT-10k [14] dataset.

#	Layer	AO	$SR_{0.5}$	$SR_{0.75}$	FPS
1	All	69.3	79.1	65.2	68.4
2	None	67.7	77.0	62.2	74.5
3	0-th layer	66.7	76.0	61.6	73.4
4	Odd layers	**69.9**	**79.6**	66.0	74.3
5	Even layers	69.8	79.5	**66.4**	73.0
6	Every 3 layers	69.0	78.0	65.3	78.2
7	First half	69.5	79.5	66.4	71.2
8	Second half	68.5	78.2	64.7	76.9

Table 4. The performance of different usage of CFA on GOT-10k [14].

#	E		Layer		AO	$SR_{0.5}$	$SR_{0.75}$
	t	s	odd	even			
1	✓	✓	✓	✓	69.2	78.8	65.4
2	✓	✓	✓	–	**70.1**	**80.2**	**66.9**
3	✓	✓	–	✓	69.0	78.5	65.2
4	✓	–	✓	–	69.8	79.9	66.4

Table 5. The performance of SPA and CFA on GOT-10k [14] dataset.

#	Module		AO	$SR_{0.5}$	$SR_{0.75}$
	SPA	CFA			
1	–	–	69.3	79.1	65.2
2	✓	–	69.9	79.6	66.0
3	✓	✓	**70.1**	**80.2**	**66.9**

SPA Module. We explore the effect of performing template enhancement at different layers on tracking performance. As shown in results #1 and #2 in Table 3, omitting the template enhancement operation significantly degrades the tracking performance. This indicates that the interaction between template feature elements is crucial for robust tracking, highlighting the necessity of template enhancement. Results #2 and #3 indicate that the layer 0 template features contain essential original information about the target, and those information should be retained. Next, results #4, #5, and #6 demonstrate that performing spaced template enhancements can prevent template information redundancy and speed up tracking. Finally, results #7 and #8 reveal that template feature enhancement in the earlier layers yields greater benefits in improving performance compared to the later layers. This can be attributed to the fact that the features in the earlier layers directly represent template information. By enhancing these features, the information stream $t \rightarrow s$ can transmit discriminative template features to the search region branch for effective information fusion.

CFA Module. We investigate the effect of the CFA module on performance from two aspects. Firstly, we explore the usage of the CFA module at different positions. Experiments #1, #2, and #3 in Table 4 correspond to three usage positions: all layers, odd layers, and even layers, respectively. Since the template enhancement operation is performed only at the odd layer, using the CFA module only at the odd layer highlights the important channels and significantly enhances the subsequent feature enhancement. At the same time, utilizing the CFA module in all layers results in channel information redundancy, which hinders target localization. Furthermore, we explore the usage of the CFA module on different features. Results #2 and #4 demonstrate that both template features and search region features need to focus on the target-related feature channel, which facilitates target discrimination.

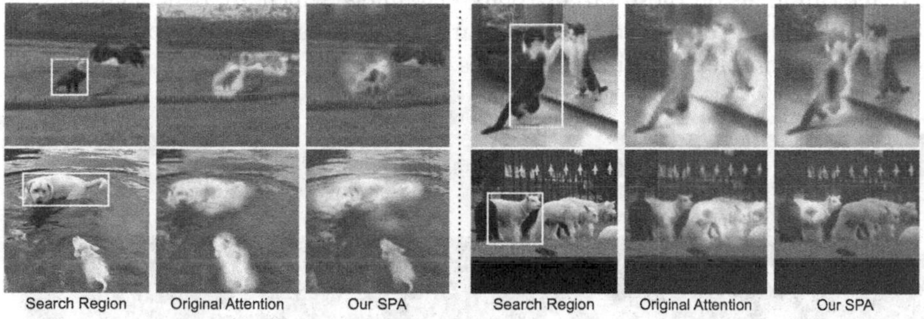

Fig. 3. Visualization of attention maps of the original attention mechanism and our proposed SPA mechanism. The red box denotes the tracking target. (Color figure online)

Module Ablation. We conduct experiments to investigate the effect of the SPA module and the CFA module on tracking performance. Results #1 and #2 in Table 5 demonstrate the effectiveness of the SPA module in addressing the issue of template information redundancy. By selectively enhancing template features and removing the information stream $s \rightarrow t$, the quality of template features is improved, leading to enhanced tracking performance. Furthermore, in experiment #3, we applied the CFA module to our tracker. The CFA module allows our tracker to focus on important feature channels, which significantly enhances the localization ability of the tracker. This improvement in localization ability further contributes to the overall enhancement of tracking performance.

4.4 Visualization of Attention Maps

To intuitively demonstrate the advantages of our proposed SPA mechanism, we visualize attention maps of its and the original attention mechanism. As depicted in Fig. 3, our SPA mechanism accurately distinguishes the tracking target from similar objects, whereas the original attention mechanism falls short in this regard. This improvement can be attributed to our selective template feature enhancement strategy, which effectively addresses the issue of template information redundancy. Meanwhile, the information stream $s \rightarrow t$ is removed to ensure the quality of the template features. In summary, our SPA mechanism exhibits enhanced robustness compared to the original attention mechanism.

4.5 Visualization of Tracking Performance

There are always many tracking challenges in real-world tracking scenarios, *e.g.*, motion blur and scale variations. Therefore, we visualize the tracking performance of our CSTrack and the state-of-the-art trackers CSWinTT [24] and SparseTT [9] in typical tracking challenges. In our approach, the useful information in template features is retained, improving the template quality. In addition, CSTrack can focus on those target-related channels, which are beneficial for target localization. As displayed in Fig. 4, our CSTrack is still able to robustly track the target in complex tracking challenges.

Fig. 4. Visualization of the tracking performance of our proposed CSTrack and state-of-the-art trackers CSWinTT [24] and SparseTT [9].

5 Conclusion

In this work, we propose a spatial positioning attention mechanism (SPA) to effectively address two key issues that existed in previous tracking methods: the redundancy of template information and the inappropriate use of information stream. The SPA module not only significantly improves tracking performance, but also accelerates tracking speed. In addition, we introduce a channel focus attention mechanism (CFA) that allows the tracker to focus on the channels containing critical target information in the template features and search region features, enhancing the target discrimination capability of the tracker. Further, we design a one-stream tracker CSTrack, which can cope well with complex tracking challenges. Extensive experiments demonstrate that our CSTrack noticeably outperforms previous state-of-the-art trackers, running at over 70 FPS.

Acknowledgements. This work was supported in part by National Natural Science Foundation of China under the Grants 61872187 and 62072246, in part by Natural Science Foundation of Jiangsu Province under the Grant BK20201306.

References

1. Bertinetto, L., Valmadre, J., Henriques, J.F., Vedaldi, A., Torr, P.H.S.: Fully-convolutional Siamese networks for object tracking. In: Hua, G., Jégou, H. (eds.) ECCV 2016. LNCS, vol. 9914, pp. 850–865. Springer, Cham (2016). https://doi.org/10.1007/978-3-319-48881-3_56
2. Bhat, G., Danelljan, M., Gool, L.V., Timofte, R.: Learning discriminative model prediction for tracking. In: ICCV, pp. 6182–6191 (2019)
3. Bhat, G., Danelljan, M., Van Gool, L., Timofte, R.: Know your surroundings: exploiting scene information for object tracking. In: Vedaldi, A., Bischof, H., Brox, T., Frahm, J.-M. (eds.) ECCV 2020. LNCS, vol. 12368, pp. 205–221. Springer, Cham (2020). https://doi.org/10.1007/978-3-030-58592-1_13
4. Chen, B., Li, P., Bai, L., Qiao, L., Shen, Q., Li, B.: Backbone is all your need: a simplified architecture for visual object tracking. In: Avidan, S., Brostow, G., Cissé, M., Farinella, G.M., Hassner, T. (eds.) Computer Vision – ECCV 2022. ECCV 2022. LNCS, vol. 13682, pp. 375–392. Springer, Cham (2022). https://doi.org/10.1007/978-3-031-20047-2_22
5. Chen, X., Yan, B., Zhu, J.: Transformer tracking. In: CVPR, pp. 8126–8135 (2021)
6. Danelljan, M., Gool, L.V., Timofte, R.: Probabilistic regression for visual tracking. In: CVPR, pp. 7183–7192 (2020)
7. Dosovitskiy, A., Beyer, L., Kolesnikov, A., Weissenborn, D., Zhai, X.: An image is worth 16 × 16 words: transformers for image recognition at scale. In: ICLR (2021)
8. Fan, H., et al.: LaSOT: a high-quality benchmark for large-scale single object tracking. In: CVPR, pp. 5374–5383 (2019)
9. Fu, Z., Fu, Z., Liu, Q., Cai, W., Wang, Y.: SparseTT: visual tracking with sparse transformers. In: IJCAI (2022)
10. Fu, Z., Liu, Q., Fu, Z., Wang, Y.: STMTrack: template-free visual tracking with space-time memory networks. In: CVPR, pp. 13774–13783 (2021)
11. Guo, M., Zhang, Z., Fan, H., Jing, L., Lyu, Y., Li, B.: Learning target-aware representation for visual tracking via informative interactions. In: IJCAI (2022)
12. He, K., Chen, X., Xie, S., Li, Y., Dollár, P., Girshick, R.: Masked autoencoders are scalable vision learners. In: CVPR, pp. 16000–16009 (2022)
13. Hu, J., Shen, L., Sun, G.: Squeeze-and-excitation networks. In: CVPR, pp. 7132–7141 (2018)

14. Huang, L., Zhao, X., Huang, K.: GOT-10k: a large high-diversity benchmark for generic object tracking in the wild. In: TPAMI, pp. 1562–1577 (2019)
15. Kim, M., Lee, S., Ok, J., Han, B., Cho, M.: Towards sequence-level training for visual tracking. In: ECCV, pp. 534–551 (2022)
16. Li, B., Wu, W., Wang, Q., Zhang, F., Xing, J., Yan, J.: SiamRPN++: evolution of Siamese visual tracking with very deep networks. In: CVPR, pp. 4282–4291 (2019)
17. Li, X., Wang, W., Hu, X., Yang, J.: Selective kernel networks. In: CVPR, pp. 510–519 (2019)
18. Lin, T.-Y., et al.: Microsoft COCO: common objects in context. In: Fleet, D., Pajdla, T., Schiele, B., Tuytelaars, T. (eds.) ECCV 2014. LNCS, vol. 8693, pp. 740–755. Springer, Cham (2014). https://doi.org/10.1007/978-3-319-10602-1_48
19. Loshchilov, I., Hutter, F.: Decoupled weight decay regularization. In: ICLR (2018)
20. Ma, F., et al.: Unified transformer tracker for object tracking. In: CVPR, pp. 8781–8790 (2022)
21. Müller, M., Bibi, A., Giancola, S., Alsubaihi, S., Ghanem, B.: TrackingNet: a large-scale dataset and benchmark for object tracking in the wild. In: Ferrari, V., Hebert, M., Sminchisescu, C., Weiss, Y. (eds.) ECCV 2018. LNCS, vol. 11205, pp. 310–327. Springer, Cham (2018). https://doi.org/10.1007/978-3-030-01246-5_19
22. Pi, Z., Wan, W., Sun, C., Gao, C., Sang, N., Li, C.: Hierarchical feature embedding for visual tracking. In: Avidan, S., Brostow, G., Cissé, M., Farinella, G.M., Hassner, T. (eds.) Computer Vision – ECCV 2022, ECCV 2022. LNCS, vol. 13682, pp. 428–445. Springer, Cham (2022). https://doi.org/10.1007/978-3-031-20047-2_25
23. Rezatofighi, H., Tsoi, N., Gwak, J.: Generalized intersection over union: a metric and a loss for bounding box regression. In: CVPR, pp. 658–666 (2019)
24. Song, Z., Yu, J., Chen, Y.P.P., Yang, W.: Transformer tracking with cyclic shifting window attention. In: CVPR, pp. 8791–8800 (2022)
25. Tang, F.: Ranking-based Siamese visual tracking. In: CVPR, pp. 8741–8750 (2022)
26. Wang, N., Zhou, W., Wang, J., Li, H.: Transformer meets tracker: exploiting temporal context for robust visual tracking. In: CVPR, pp. 1571–1580 (2021)
27. Wang, X., et al.: Towards more flexible and accurate object tracking with natural language: algorithms and benchmark. In: CVPR, pp. 13763–13773 (2021)
28. Xie, F.: Correlation-aware deep tracking. In: CVPR, pp. 8751–8760 (2022)
29. Xu, Y., Wang, Z., Li, Z., Yuan, Y.: SiamFC++: towards robust and accurate visual tracking with target estimation guidelines. In: AAAI, pp. 12549–12556 (2020)
30. Yan, B., Peng, H., Fu, J., Wang, D., Lu, H.: Learning spatio-temporal transformer for visual tracking. In: ICCV, pp. 10448–10457 (2021)
31. Yu, B., Tang, M., Zheng, L., Zhu, G., Wang, J., Feng, H.: High-performance discriminative tracking with transformers. In: ICCV, pp. 9856–9865 (2021)
32. Zhang, Z., Liu, Y., Wang, X., Li, B., Hu, W.: Learn to match: automatic matching network design for visual tracking. In: ICCV, pp. 13339–13348 (2021)
33. Zhang, Z., Peng, H., Fu, J., Li, B., Hu, W.: Ocean: object-aware anchor-free tracking. In: Vedaldi, A., Bischof, H., Brox, T., Frahm, J.-M. (eds.) ECCV 2020. LNCS, vol. 12366, pp. 771–787. Springer, Cham (2020). https://doi.org/10.1007/978-3-030-58589-1_46

Feature Implicit Enhancement via Super-Resolution for Small Object Detection

Zhehao Xu, Mengyin Liu, Chao Zhu(✉), Fang Zhou, and Xu-Cheng Yin

University of Science and Technology Beijing, Beijing 100083, China
xuzhehao-cn@outlook.com, blean@live.cn, {chaozhu,xuchengyin}@ustb.edu.cn, zhoufang@ies.ustb.edu.cn

Abstract. In recent years, object detection has made significant strides due to advancements in deep convolutional neural networks. However, the detection performance for small objects remains challenging. The visual information of small objects is easily confused with the background and even more likely to get lost in a series of downsampling operations due to the limited number of pixels, resulting in poor representations. In this paper, we propose a novel approach namely Feature Implicit Enhancement via Super-Resolution (FIESR) to learn more robust feature representations for small object detection. Our FIESR consists of two detection branches and requires two steps of training. Firstly, the detector learns the relationship between low-resolution and corresponding original high-resolution images to enhance the representations of small objects by minimizing a super-resolution loss between the two branches. Secondly, the detector is fine-tuned on original resolution images to fit extremely large objects. Additionally, our FIESR could be applied to various popular detectors such as Faster-RCNN, RetinaNet, FCOS, and DyHead. Our FIESR achieves competitive results on COCO dataset and is proved effective and flexible by extensive experiments.

Keywords: Small Object Detection · Super Resolution · COCO

1 Introduction

Small Object Detection (SOD), as a sub-field of generic object detection, which concentrates on detecting those objects with small size, is of great theoretical and practical significance in various scenarios. Although significant progress has been made in object detection in recent years, the performance of detecting small objects is often much lower than that of normal objects. One of the most important reasons is that as the scale of an object decreases, its appearance tends to become blurrier, making it easier to be confused with the background. Moreover, the visual information of small objects is more likely to get lost in a series of downsampling operations due to the limited number of pixels, resulting in poor representations. Take the famous COCO [1] dataset as an example, the

© The Author(s), under exclusive license to Springer Nature Singapore Pte Ltd. 2024
Q. Liu et al. (Eds.): PRCV 2023, LNCS 14436, pp. 133–145, 2024.
https://doi.org/10.1007/978-981-99-8555-5_11

Fig. 1. Comparisons of our FIESR and FCOS [2]. FCOS only detects large objects; our FIESR can detect small objects in step I, but might ignore some extremely large objetcs; after fine-tuning in step II, it can detect both small and large objects.

objects occupying an area less than or equal to 32^2 pixels come to the "Small" category. As shown in Fig. 1, small objects usually occupy only a small part of the image. Such poor-quality appearance provide insufficient information for a detector.

The most direct and effective approach to improve the performance of SOD is to use high-resolution images or feature maps. Thus, combined with super-resolution task, object detection task can utilize more visual information of small objects. Current super-resolution based approaches can be divided into image-level and feature-level styles. Some works [3–5] use a super-resolution network to preprocess the input images and some works [6–8] use a GAN [9] to reconstruct the RoI region of images or feature maps. However, these methods either introduce a large amount of computation from extra parameters, or cannot be applied to one-stage detectors without RoI operation.

There are two key observations: 1) Ref. [10,11] show that the performance of the detector can be further improved under the original structure and parameters with some training strategies. 2) The structure of the backbone and neck is similar in both two-stage and one-stage detectors, which is mostly agnostic to whether to use RoI operation. Inspired by these two observations, we propose a new small object detection method for feature implicit enhancement based on super-resolution (FIESR), which can be easily applied to both two-stage and one-stage detectors without introducing redundant computation and parameters during inference. Our method consists of two detection branches ($\mathcal{A}$ and $\mathcal{B}$ which is used only for training) and a super-resolution module (SRM) for the neck of branch $\mathcal{B}$. The purpose of our FIESR is to facilitate the detector in learning more robust feature representations by minimizing the distance between the super-resolution feature maps generated by SRM and the high-resolution feature maps generated by branch $\mathcal{A}$. The bottleneck of super-resolution module lies in generating features for small objects. To generate superior super-resolution features, the detector allocates more attention to small objects. Our method undergoes a two-step training process. In the step I, we utilize a pre-trained detector fed with original high-resolution images to supervise the training of final detector fed with low-resolution images. As shown in Fig. 1, although the detector is trained to successfully detect small objects, it might struggle to detect extremely large objects. To address this limitation, we employ the step II, where final detector focus on fitting these extremely large objects on original high-resolution images, while the capability of detecting small objects is mostly maintained.

Our contributions could be summarized as follows:

- We propose a novel detection method based on super-resolution, FIESR, to improve SOD performance. Our method utilizes high-resolution feature maps as supervision, enabling detectors to capture the differences in feature representations at high and low resolutions by minimizing a super-resolution loss. It significantly enhances the feature representations specifically for small objects.
- Our FIESR improves the detector performance without introducing extra parameters and computational overhead during inference. Moreover, it is generic and can be easily applied to various popular detectors.
- On the popular dataset COCO [1] for generic object detection, we verify the effectiveness of our FIESR on various detection architectures, including anchor-based one-stage, anchor-free one-stage and two-stage detectors.

2 Related Works

2.1 General Object Detection

Object detection can be mainly divided into two streams: two-stage and one-stage detection. On the one hand, two-stage object detection methods initially generate roughly localized regions of interest (RoIs) using a region proposal network. Subsequently, these RoIs are refined and classified more accurately by a detection head. Faster RCNN [12] is a classic two-stage detector.

On the other hand, one-stage object detection methods directly predict the class and location of objects from the feature maps. Furthermore, within the one-stage detection stream, there are anchor-based and anchor-free detection methods. RetinaNet [13], a one-stage and anchor-based detector, solves the class imbalance problem by introducing a focal loss. ATSS [14] improves the detection performance by optimizing the sampling strategy on the basis of retinanet. DyHead [15] introduces scale-aware, spatial-aware and task-aware attention to improve the performance of the detection head. While FCOS [2], a mainstream one-stage and anchor-free detector, predicts the width and height of the objects from the feature maps without predefined hyper-parameters of anchors. Although these detection methods are implemented in different ways, they all use the similar backbone and neck such as ResNet [16] and FPN [17].

2.2 Small Object Detection Based on Super-Resolution

Small object detection is a challenging computer vision task due to the lack of available information on small objects. There are four dominant research directions for small object detection: 1) data-augmentation; 2) scale-aware; 3) context-modeling; 4) super-resolution. In the following, we mainly introduce the detection method based on super-resolution.

Small object detection based on super-resolution are mainly divided into image-level and feature-level. The most straightforward image-level approach is using a super-resolution model to generate high-resolution images. Hu *et al.* [4] apply bilinear interpolation to generate high-resolution images. [5] uses the detector as the discriminator and the super-resolution module as the generator to jointly train the GAN [9] and the detector. There are two major problems with these methods. On the one hand, it takes a lot of time and computation for super-resolution model to generate high-resolution images. On the other hand, the super-resolution model may generate some non-object image parts that are not important for detection. Bai *et al.* [6] proposed SOD-MTGAN, an end-to-end model, which resolves these two problems to a certain extent by applying a super-resolution module on the RoIs instead of the whole image. However, it does not take the context information of RoIs into account. Some feature-level approaches were proposed, because the features contain contextual information after the information extraction via network layers. Perceptual GAN [7] is a notable approach focusing on the features of RoIs. Noh *et al.* [8] solved the problem by applying dilated convolutions to match the relative receptive fields of high-resolution feature maps and low-resolution feature maps. However, these methods are currently designed for two-stage detectors and cannot be readily applied to one-stage detectors.

Unlike these methods, our method assists the neck and backbone components to learn and enhance small object features by incorporating a super-resolution module during training. Additionally, the detector performs detection without super-resolution module during inference.

3 Methods

Most detectors have used FPN [17] or its variants to utilize the multi-scale semantic information. When an image $I \in \mathbb{R}^{3\times H\times W}$ is input to a detector, it generates multi-scale feature maps $P = \{P_l \in \mathbb{R}^{C\times H'\times W'}\}$. Here, l indicates the pyramid level and (H', W') is typically $\left(\lfloor \frac{H}{2^l} \rfloor, \lfloor \frac{W}{2^l} \rfloor\right)$ in a standard FPN implementation. Based on these multi-scale features, there are detection heads that perform classification prediction, bounding box height and width regression, and offset regression, respectively.

According to [10,11], FPN alone may not fully extract the information present in the image for the detection task, especially the information of small objects. To address this issue, we propose FIESR illustrated in Fig. 2, which utilizes super-resolution to assist FPN in extracting more information from small

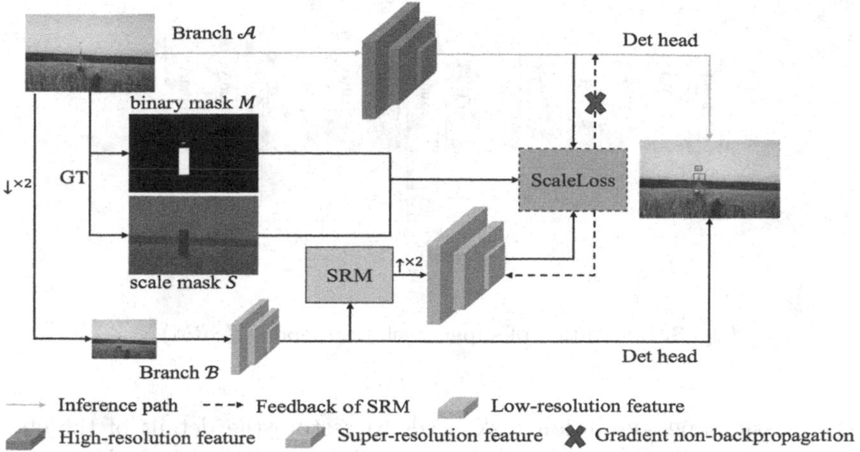

Fig. 2. The pipeline of the proposed small object detection (FIESR). Our FIESR consists of two detection branches and a SRM. ScaleLoss not only separates the foreground and background, but also enables the detector to allocate more attention to the information of small objects. During inference, FIESR only uses branch $\mathcal{A}$. The baseline can be mainstream type of detectors (*e.g.* Faster-RCNN [12], RetinaNet [13], or FCOS [2]).

objects, thereby generating improved feature representations. In the following sections, we will introduce our method in detail.

3.1 Overall Architecture

As shown in Fig. 2, our method consists of two detection branches ($\mathcal{A}$ and $\mathcal{B}$) and a super-resolution module (SRM). The original high-resolution (HR) image is input to branch $\mathcal{A}$ and low-resolution (LR) image is input to branch $\mathcal{B}$. By minimizing the distance between the super-resolution feature maps and high-resolution feature maps respectively generated from the two branches, our proposed framework enables the detector to learn more robust feature representations.

Given a HR image $I^{\text{hr}} \in \mathbb{R}^{3\times H\times W}$, we first downsample it to get corresponding LR image $I^{\text{lr}} \in \mathbb{R}^{3\times\frac{H}{2}\times\frac{W}{2}}$. Then, high-resolution feature maps $P^{\text{hr}} = \{P_l^{\text{hr}} \in \mathbb{R}^{C\times H'\times W'}\}$ is generated in branch $\mathcal{A}$, and low-resolution feature maps $P^{\text{lr}} = \{P_l^{\text{lr}} \in \mathbb{R}^{C\times\frac{H'}{2}\times\frac{W'}{2}}\}$ is generated in branch $\mathcal{B}$. Next, P^{lr} is fed to the SRM to recover the super-resolution feature maps $P^{\text{sr}} = \{P_l^{\text{sr}} \in \mathbb{R}^{C\times H'\times W'}\}$. SRM provides detectors with more detailed information during training by reconstructing super-resolution feature maps. According to [18], SRM is a residual bilinear module shown in Fig. 3.

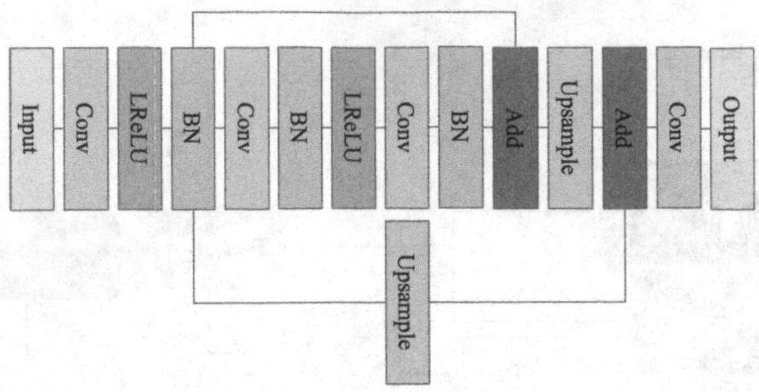

Fig. 3. Illustration of super-resolution module (SRM).

The classic super-resolution task needs to restore the details of the entire image, so the loss that treats all the pixels equally is often used for it. However, in object detection task, foreground objects are of greater importance. Additionally, large objects cover more pixels, resulting in their dominance over small objects in the loss calculation. This can potentially hinder the recovery of small object features. To effectively utilize the super-resolution task for assisting the detection task, we adopt a ScaleLoss, denoted as L_{fea}, that balances the loss of foreground and background as well as large and small objects. L_{fea} is defined as:

$$\begin{aligned} L_{fea}(P^{\mathrm{hr}}, P^{\mathrm{sr}}) = \alpha \sum_{c=1}^{C} \sum_{h=1}^{H} \sum_{w=1}^{W} M_{h,w} S_{w,h} (P^{\mathrm{sr}}_{c,h,w} - P^{\mathrm{hr}}_{c,h,w})^2 \\ + \beta \sum_{c=1}^{C} \sum_{h=1}^{H} \sum_{w=1}^{W} (1 - M_{h,w}) S_{w,h} (P^{\mathrm{sr}}_{c,h,w} - P^{\mathrm{hr}}_{c,h,w})^2 \end{aligned} \tag{1}$$

$$M_{h,w} = \begin{cases} 1, & (h,w) \in \text{ foreground} \\ 0, & (h,w) \in \text{ background} \end{cases} \tag{2}$$

where α, β are the weights of foreground and background, h, w are the horizontal and vertical coordinates of the feature maps. $M \in \{0,1\}^{H \times W}$ is defined as a binary mask to separate the foreground and background. S denotes the scale mask, which is set according to the area occupied by the object, the larger the area (h, w) belongs to, the smaller the value of its position. S is defined as:

$$S_{h,w} = \begin{cases} \frac{1}{H_r W_r}, & \text{if}(h,w) \in r \\ \frac{1}{\sum_{h=1}^{H} \sum_{w=1}^{W} (1 - M_{h,w})}, & \text{otherwise} \end{cases} \tag{3}$$

where H_r and W_r denote the height and width of the ground-truth box r.

3.2 Training

This section outlines the training process of our method. Two steps are involved, as shown in Algorithm 1.

Step I: The super-resolution task is employed to assist the detector in learning the ability to implicitly extract large-scale features of small objects. We utilize a pretrained detector D_p in branch $\mathcal{A}$ to supervise the training of the final detector D in branch $\mathcal{B}$ from scratches. Currently, branch $\mathcal{A}$ serves as a label generator exclusively for the SRM and its parameters are not updated during this process. The training loss in this step, denoted as L_1:

$$L_1(P^{\mathrm{lr}}, P^{\mathrm{hr}}, P^{\mathrm{sr}}) = L_{det}(P^{\mathrm{lr}}) + \lambda L_{fea}(P^{\mathrm{hr}}, P^{\mathrm{sr}}) \tag{4}$$

where λ is a hyper parameters to balance the loss. $L_{det}(P^{\mathrm{lr}})$ is the detection loss of P^{lr} as the input of the detection head. In this step, D learns more robust feature representations for small objects.

Algorithm 1. FIESR Algorithm Pipeline

---

inputs: HR image I, pre-trained detector D_p and initialized detector D
outputs: Trained detector D
(1). Step I:
Branch $\mathcal{A}$ use the parameters of D_p and is frozen
Branch $\mathcal{B}$ use the parameters of D
  for i in range(iterations) **do**
    (1). Downsample HR image to get LR image
    (2). Extract (P^{lr}, P^{hr}, P^{sr})
    (3). Train D by minimizing $L_1(P^{\mathrm{lr}}, P^{\mathrm{hr}}, P^{\mathrm{sr}})$ shown as Eq. 4
  end for
(2). Step II:
Branch $\mathcal{A}$ and $\mathcal{B}$ both use the shared parameters of D
  for i in range(iterations) **do**
    (1). Downsample HR image to get LR image
    (2). Extract (P^{hr}, P^{sr})
    (3). Train D by minimizing $L_2(P^{\mathrm{hr}}, P^{\mathrm{sr}})$ shown as Eq. 5
  end for

Step II: The purpose of this step is to adapt the final detector to large objects. At the same time, to preserve the ability learned in step I, branch $\mathcal{A}$ and $\mathcal{B}$ both use the shared parameters of D. Thus, different from step I, the detection head is now trained on original high-resolution (HR) images in branch $\mathcal{A}$. Note that we only allow the gradient of SRM propagate to branch $\mathcal{B}$ to ensure the forward learning of features. We denote the training loss as L_2:

$$L_2(P^{\mathrm{hr}}, P^{\mathrm{sr}}) = L_{det}(P^{\mathrm{hr}}) + \gamma L_{fea}(P^{\mathrm{hr}}, P^{\mathrm{sr}}) \tag{5}$$

where γ is a hyper parameters to balance the loss. $L_{det}(P^{\text{hr}})$ is the detection loss of P^{hr} as the input of the detection head.

Inference: Once training is done, the inference procedure is the same as a normal detector. Since the detector has the ability to generate better representations of small objects through two-step training, detection is performed directly on the original high-resolution image without SRM. This allows for more efficient and faster detection compared to other super-resolution based methods, as our method does not require additional computational cost for super-resolution processing.

Table 1. Ablation study on the generalization of our method when applying to popular object detection methods. '†' represents multi-scale training. '$\Diamond$' means that the baseline is first trained 12 epochs on the LR images, and then fine-tuned 12 epochs on the HR images without branch $\mathcal{B}$.

Method	Epochs	AP	AP_S	AP_M	AP_L
Faster-RCNN [12]	24	38.4	21.5	42.1	**50.3**
Faster-RCNN$^{\Diamond}$	12 + 12	38.7	23.0	42.4	49.5
+FIESR	12 + 12	**39.4**	**24.0**	**42.8**	49.7
RetinaNet [13]	24	37.4	20.0	40.7	49.7
RetinaNet$^{\Diamond}$	12 + 12	38.1	21.5	42.4	49.8
+FIESR	12 + 12	**38.7**	**22.7**	**42.7**	**50.1**
DyHead [15]	24	43.3	25.8	47.2	**57.0**
DyHead$^{\Diamond}$	12 + 12	44.0	28.0	47.9	56.7
+FIESR	12 + 12	**44.4**	**28.2**	**48.2**	**57.0**
FCOS† [2]	24	38.5	21.9	42.8	48.6
FCOS$^{\Diamond\dagger}$	12 + 12	39.0	23.4	42.8	**50.2**
+FIESR	12 + 12	**39.4**	**24.3**	**43.2**	49.6

4 Experiments and Details

4.1 Dataset and Details

We evaluate our method on the popular object detection COCO [1] dataset, which contains 80 object classes. We use the 120K train images for training and 5k val images for testing for all the experiments. Average Precision is adopted as evaluation metric, *i.e.*, mAP, AP_S, AP_M and AP_L. The last three measure performance with respect to objects with different scales.

We conduct experiments on different detection framework, including two-stage models [12,19], anchor-based one-stage models [13,15], and anchor-free one-stage models [2]. All experiments are conducted with mmdetection with Pytorch [20] framework. And the pre-trianed detector D_p is obtained by the mmdetection official site. All the detectors are trained in two steps, and each step trains 12 epochs with SGD optimizer, where the momentum is 0.9 and the weight decay is 0.0001. There are four hyper-paramters in the total training objectives, we set $\alpha = 0.0005$, $\beta = 0.00025$, $\lambda = 1.0$, $\gamma = 0.4$ for all the two-stage models and $\alpha = 0.001$, $\beta = 0.0005$, $\lambda = 1.0$, $\gamma = 0.4$ for all the one-stage models. Other settings are consistent with baseline.

4.2 Ablation Study

We conduct a series of ablation studies to demonstrate the effectiveness and efficiency of our method.

Table 2. Ablation study on the compatibility of our method when applying to different backbones on FCOS. DCN represents deformable convolution network [21]. ResNet-101 has more layers than ResNet-50.

Method	Backbone	Epochs	AP	AP_S	AP_M	AP_L
FCOS [2]	ResNet-50-DCN	24	42.5	25.1	46.3	55.9
FCOS$^\diamond$	ResNet-50-DCN	12 + 12	43.2	26.4	46.4	55.9
+FIESR	ResNet-50-DCN	12 + 12	**44.2**	**28.5**	**47.6**	**57.3**
FCOS [2]	ResNet-101	24	39.4	22.4	42.9	51.8
FCOS$^\diamond$	ResNet-101	12 + 12	40.6	24.2	44.7	**52.0**
+FIESR	ResNet-101	12 + 12	**41.4**	**26.5**	**45.1**	**52.0**

Table 3. Ablation study on different super-resolution loss and level outputs from FPN [17]. We use FCOS [2] with ResNet50 as baseline.

L1Loss	ScaleLoss	FPN1	FPN2	FPN3	AP	AP_S	AP_M	AP_L
					38.3	22.6	42.0	49.0
✓		✓	✓	✓	38.2	22.3	41.9	48.7
	✓	✓			38.3	23.9	41.5	48.7
	✓	✓	✓		**38.9**	23.7	**42.6**	49.5
	✓	✓	✓	✓	**38.9**	**24.0**	42.3	**49.7**

Generalization on Existing Object Detectors: We evaluate the generalization ability of our method by applying it to popular object detectors, such as Faster-RCNN [12], RetinaNet [13], FCOS [2] and DyHead [15]. These methods represent a wide variety of object detection frameworks (*e.g.*, two-stage vs. one-stage, anchor-based vs. anchor-free). Since our method is trained 12 epoches on LR images in step I, we set up a comparative experiment where the detector is trained 12 epoches on LR images and fine-tuned 12 epoches on HR images without our method. As shown in Table 1, our method outperforms baseline by 1.0$\sim$1.3 AP and 2.4$\sim$2.7 AP_S. Although the baseline we set is better than the original baseline, our method outperforms it by 0.4$\sim$0.7 AP and 0.2$\sim$1.2 AP_S. Even in multi-scale training, our method can achieve better results.

Cooperation with Different Backbones: We evaluate the compatibility of our method with different backbone. As shown in Table 2, we apply our method to FCOS [2] using ResNet-50-DCN [21] and ResNet-101 as backbone. Our method outperforms original baseline by 1.7 AP and 3.4 AP_S with ResNet-50-DCN backbone and by 2.0 AP and 4.1 AP_S with ResNet-101 backbone.

Comparison with Different Super-Resolution Loss and Level Outputs from FPN: We compare our method with different super-resolution loss (L1Loss or ScaleLoss for Eq. 1) and different level outputs from FPN. We use the largest three levels output features of FPN (1, 2, 3) as SRM input (1 is highest resolution feature map, 2 is second and so on). As shown in Table 3, while we using L1Loss, SRM cannot help detector to improve performance because of the imbalance between foreground and background, large objects and small objects. Moreover, we observe that the different levels of feature maps from FPN are added to SRM to improve the performance of the detector in different scales of objects, which matches the structure of FPN. When the three levels of feature maps are used as SMR input, the comprehensive performance of the detector is the best.

Comparison with Different Fine-Tuning strategies: We compare whether to use branch $\mathcal{B}$ in step II for fine-tuning. We experiment on FCOS [2] with

Table 4. Ablation study on different fine-tuning strategies.

Method	Backbone	Epochs	AP	AP_S	AP_M	AP_L
FCOS [2]$^\diamond$	ResNet-50	12 + 12	38.3	22.6	42.0	49.0
+FIESR/wo branch $\mathcal{B}$	ResNet-50	12 + 12	38.8	22.8	**42.6**	**49.7**
+FIESR	ResNet-50	12 + 12	**38.9**	**24.0**	42.3	**49.7**
FCOS [2]$^\diamond$	ResNet-50-DCN	12 + 12	43.2	26.4	46.4	55.9
+FIESR/wo branch $\mathcal{B}$	ResNet-50-DCN	12 + 12	43.9	27.9	47.4	**57.7**
+FIESR	ResNet-50-DCN	12 + 12	**44.2**	**28.5**	**47.6**	57.3

Table 5. Comparison with other object detectors using ResNet-50 on COCO minival set.

Method	Epochs	AP	AP_S	AP_M	AP_L
Cascade-RCNN [19]	20	41.0	22.7	44.4	54.3
FCOS [2]	12	38.7	22.9	42.5	50.1
RepPoints [22]	24	38.6	22.5	42.2	50.4
ATSS [14]	12	39.3	24.3	43.3	51.3
BorderDet [23]	12	41.4	23.6	45.1	54.6
Deformable DETR [24]	36	43.8	26.4	47.1	**58.0**
DyHead [15]	12	42.6	26.1	46.8	56.0
FIESR	12 + 12	**44.4**	**28.2**	**48.2**	57.0

ResNet-50 and ResNet-50-DCN. As shown in Table 4, step II with branch $\mathcal{B}$ achieves better performance than step II without branch $\mathcal{B}$, particularly when it comes to small objects, which validates the effectiveness of our FIESR.

4.3 Main Results

To verify the effectiveness performance of the proposed method, it is compared with other classic detection methods [2,14,15,19,22–24] in the literature. In this comparison, we choose DyHead [15] with ResNet-50 as baseline and apply our method to it, to make fair comparisons with other methods. As shown in Table 5, all the detectors use the same backbone (ResNet-50), and our method achieves 44.4 mAP, especially 28.2 AP_S for SOD which gains the best performance.

5 Conclusion

In this paper, we have proposed a novel detection framework based on super-resolution, FIESR, to improve the performance of small objects detection without introducing additional computational cost during inference. FIESR can be easily applied to both two-stage and one-stage detectors. By minimizing a super-resolution loss between the original high-resolution and super-resolution features, FIESR enables the detector to learn more robust feature representations, particularly for small objects. As future work, our method could be further improved in the following aspects: how to train the super-resolution module more efficiently, and how to design new modules on the detection head to assist the learning of the detector.

Acknowledgement.. This work was supported by National Key Research and Development Program of China(No. 2022ZD0119200), National Natural Science Foundation of China(No. 62072032 and 62076024), and National Science Fund for Distinguished Young Scholars(No. 62125601).

References

1. Lin, T.Y., et al.: Microsoft coco: Common objects in context. In: Fleet, D., Pajdla, T., Schiele, B., Tuytelaars, T. (eds.) Computer Vision - ECCV 2014, pp. 740–755. Springer International Publishing, Cham (2014). https://doi.org/10.1007/978-3-319-10602-1_48
2. Tian, Z., Shen, C., Chen, H., He, T.: FCOS: fully convolutional one-stage object detection. In: Proceedings of the IEEE/CVF International Conference on Computer Vision, pp. 9627–9636 (2019)
3. Haris, M., Shakhnarovich, G., Ukita, N.: Task-driven super resolution: object detection in low-resolution images. In: Mantoro, T., Lee, M., Ayu, M.A., Wong, K.W., Hidayanto, A.N. (eds.) ICONIP 2021. CCIS, vol. 1516, pp. 387–395. Springer, Cham (2021). https://doi.org/10.1007/978-3-030-92307-5_45
4. Hu, P., Ramanan, D.: Finding tiny faces. In: Proceedings of the IEEE Conference on Computer Vision and Pattern Recognition, pp. 951–959 (2017)
5. Rabbi, J., Ray, N., Schubert, M., Chowdhury, S., Chao, D.: Small-object detection in remote sensing images with end-to-end edge-enhanced GAN and object detector network. Remote Sens. **12**(9), 1432 (2020)
6. Bai, Y., Zhang, Y., Ding, M., Ghanem, B.: SOD-MTGAN: small object detection via multi-task generative adversarial network. In: Ferrari, V., Hebert, M., Sminchisescu, C., Weiss, Y. (eds.) ECCV 2018. LNCS, vol. 11217, pp. 210–226. Springer, Cham (2018). https://doi.org/10.1007/978-3-030-01261-8_13
7. Li, J., Liang, X., Wei, Y., Xu, T., Feng, J., Yan, S.: Perceptual generative adversarial networks for small object detection. In: Proceedings of the IEEE Conference on Computer Vision and Pattern Recognition, pp. 1222–1230 (2017)
8. Noh, J., Bae, W., Lee, W., Seo, J., Kim, G.: Better to follow, follow to be better: towards precise supervision of feature super-resolution for small object detection. In: Proceedings of the IEEE/CVF International Conference on Computer Vision, pp. 9725–9734 (2019)
9. Goodfellow, I., et al.: Generative adversarial nets. In: Advances in Neural Information Processing Systems, pp. 2672–2680. Curran Associates, Inc. (2014). https://www.papers.nips.cc/paper/5423-generative-adversarial-nets.pdf
10. Ni, Z., Yang, F., Wen, S., Zhang, G.: Dual relation knowledge distillation for object detection. arXiv preprint arXiv:2302.05637 (2023)
11. Guo, J., et al.: Distilling object detectors via decoupled features. In: Proceedings of the IEEE/CVF Conference on Computer Vision and Pattern Recognition, pp. 2154–2164 (2021)
12. Ren, S., He, K., Girshick, R., Sun, J.: Faster R-CNN: towards real-time object detection with region proposal networks. In: Cortes, C., Lawrence, N., Lee, D., Sugiyama, M., Garnett, R. (eds.) Advances in Neural Information Processing Systems, vol. 28. Curran Associates, Inc. (2015). https://www.proceedings.neurips.cc/paper/2015/file/14bfa6bb14875e45bba028a21ed38046-Paper.pdf
13. Lin, T.Y., Goyal, P., Girshick, R., He, K., Dollár, P.: Focal loss for dense object detection. In: 2017 IEEE International Conference on Computer Vision (ICCV), pp. 2999–3007 (2017). https://doi.org/10.1109/ICCV.2017.324
14. Zhang, S., Chi, C., Yao, Y., Lei, Z., Li, S.Z.: Bridging the gap between anchor-based and anchor-free detection via adaptive training sample selection. In: Proceedings of the IEEE/CVF Conference on Computer Vision and Pattern Recognition, pp. 9759–9768 (2020)

15. Dai, X., et al.: Dynamic head: unifying object detection heads with attentions. In: Proceedings of the IEEE/CVF Conference on Computer Vision and Pattern Recognition, pp. 7373–7382 (2021)
16. He, K., Zhang, X., Ren, S., Sun, J.: Deep residual learning for image recognition. In: 2016 IEEE Conference on Computer Vision and Pattern Recognition (CVPR), pp. 770–778 (2016). https://doi.org/10.1109/CVPR.2016.90
17. Lin, T.Y., Dollár, P., Girshick, R., He, K., Hariharan, B., Belongie, S.: Feature pyramid networks for object detection. In: 2017 IEEE Conference on Computer Vision and Pattern Recognition (CVPR), pp. 936–944 (2017). https://doi.org/10.1109/CVPR.2017.106
18. Cui, Z., et al.: Exploring resolution and degradation clues as self-supervised signal for low quality object detection. In: Avidan, S., Brostow, G., Cissé, M., Farinella, G.M., Hassner, T. (eds.)Computer Vision-ECCV 2022: 17th European Conference, Tel Aviv, Israel, October 23–27 2022, Proceedings, Part IX, vol. 13669, pp. 473–491. Springer,Cham (2022). https://doi.org/10.1007/978-3-031-20077-9_28
19. Cai, Z., Vasconcelos, N.: Cascade R-CNN: delving into high quality object detection. In: Proceedings of the IEEE Conference on Computer Vision and Pattern Recognition, pp. 6154–6162 (2018)
20. Paszke, A., et al.: Automatic differentiation in PyTorch (2017)
21. Dai, J., et al.: Deformable convolutional networks. In: Proceedings of the IEEE International Conference on Computer Vision, pp. 764–773 (2017)
22. Yang, Z., Liu, S., Hu, H., Wang, L., Lin, S.: RepPoints: point set representation for object detection. In: Proceedings of the IEEE/CVF International Conference on Computer Vision, pp. 9657–9666 (2019)
23. Qiu, H., Ma, Y., Li, Z., Liu, S., Sun, J.: BorderDet: border feature for dense object detection. In: Vedaldi, A., Bischof, H., Brox, T., Frahm, J.-M. (eds.) ECCV 2020. LNCS, vol. 12346, pp. 549–564. Springer, Cham (2020). https://doi.org/10.1007/978-3-030-58452-8_32
24. Zhu, X., Su, W., Lu, L., Li, B., Wang, X., Dai, J.: Deformable DETR: deformable transformers for end-to-end object detection. arXiv preprint arXiv:2010.04159 (2020)

Improved Detection Method for SODL-YOLOv7 Intensive Juvenile Abalone

Chengying Liu[1], Jun Yue[1(✉)], Guangjie Kou[1], Zhanming Zou[2], Zhenbo Li[3], and Changyi Dai[1]

[1] Ludong University, Yantai 264003, China
yuejuncn@sohu.com
[2] Weihai Ever Green Ocean Technology Co., Ltd., Weihai 264300, China
[3] China Agricultural University, Beijing 100091, China

Abstract. Achieving rapid and accurate detection of juvenile abalone is a prerequisite for estimating the number, density and size of juvenile abalone. Juvenile abalone are densely distributed in the breeding process, and the intra-class occlusion between each other is serious. Microorganisms in the water form inter-class occlusion for juvenile abalone, resulting in incomplete detection information. There is a lack of effective detection methods for juvenile abalone. To address the above problems, this paper proposed the SODL-YOLOv7 juvenile abalone detection method based on the establishment of the JAD (Juvenile abalone detection) dataset. First, the SODL backbone network for dense small target detection is proposed to improve the attention to small targets by incorporating null convolution kernels and pooling kernels with different sampling rates in the spatial null convolution and pooling layers; then, the ACBAM (Adaptive convolutional block attention module) is established to apply the adaptive pooling layer of channel space attention module, so that the network can pay more attention to the young abalone occlusion region and further improve the detection effect. Finally, the method ~~of~~ used in this paper was tested on the JAD dataset, with the results that the AP (average precision) reached 99.4%, an increase of 4.1% compared with the benchmark method YOLOv7, an increase of 9.2% compared with the instant-teaching method, and an increase of 2.2% compared with the TOOD method, therefore verifying the effectiveness of the method of this paper.

Keywords: abalone detection · occlusion detection · intra-class occlusion · inter-class occlusion · YOLOv7

1 Introduction

The nutritional value of abalone is extremely high and the market demand is large. However, due to the slow natural growth rate, it is difficult to meet the market demand. Artificial breeding is the main mode of abalone production. Juvenile abalone are small in size and large in number, and holding groups lead to serious obscuration, making it difficult to accurately monitor their numbers through manual observation. However, with the development of convolutional neural network technology, accurate and efficient

© The Author(s), under exclusive license to Springer Nature Singapore Pte Ltd. 2024
Q. Liu et al. (Eds.): PRCV 2023, LNCS 14436, pp. 146–157, 2024.
https://doi.org/10.1007/978-981-99-8555-5_12

juvenile abalone detection has become possible, effectively avoiding the limitations of manual detection.

Traditional visual object detection techniques rely on hand-designed feature extractors and shallow classifiers. Zhang et al. [1] extracted gray image features to detect vehicles accurately and rapidly. Chen et al. [2] used Histogram of Oriented Gradient (HOG) for flexible and accurate detection. Kumar et al. [3] demonstrated the feasibility of using machine learning and SIFT for disaster area identification. Arreola et al. [4] developed a UAV target recognition system based on Haar cascade classifier. Malepati et al. [5] improved facial image detection using Haar-like features. Kovač et al. [6] proposed a finger vein recognition algorithm based on adaptive Gabor filter and SIFT/SURF. However, these techniques only extract shallow features, leading to low accuracy and high false detection rates, limiting their application in abalone detection.

Deep learning techniques have provided a new solution for detecting juvenile abalone, significantly improving the accuracy of detecting small target objects. Han et al. [7] introduced a two-branch siamese network to calculate the similarity between image regions for detection, enabling the detection of small target objects with limited training samples. Other researchers have proposed methods such as QueryDet [8], Drone-YOLOv5 [9], Swin Transformer with YOLOv5 [10], slice-assisted hyperinference [11], and multiscale feature fusion [12] to improve small target detection. However, detecting juvenile abalone remains challenging due to their dense distribution, variable shape, and the issues of obscuration and interference, necessitating more effective detection methods.

This paper compares and analyzes abalone detection methods in intensive farming. Ye et al. [13] proposed an improved Faster R-CNN algorithm using VGG16 for feature extraction, achieving good detection results. However, VGG16's computational volume hampers realtime detection. Peng et al. [14] introduced the Piecewise Focal Loss (PFL) function for sample balancing, obtaining a 94% mean average precision (mAP). But further exploration needed for occlusion. Teh et al. [15] proposed an enhanced Mask R-CNN model with 97.48% accuracy, but its computational complexity is unsuitable for dense abalone detection.

YOLOv7 [16] target detection algorithm has relatively high detection accuracy and is suitable for small target detection. In this paper, we improve the YOLOv7 model and propose a SODL-YOLOv7 method for abalone detection during the nursery period based on the problems in the above literature and the characteristics of the task based on juvenile abalone detection. The components of this research include (1) Create JAD datasets; (2) Embedding the SODL small target detection network in the YOLOv7 backbone network to solve the problem of low detection rate and high false detection rate; (3) Introduce the improved ACBAM attention module to solve the intra-class inter-class occlusion problem.

2 Methods

The technical route of this study is shown in Fig. 1. In this paper, we proposed the SODL-YOLOv7 juvenile abalone detection method based on YOLOv7 [16], and the network structure diagram is shown in Fig. 2. Among them, the gray underlined part is the method proposed in this paper.

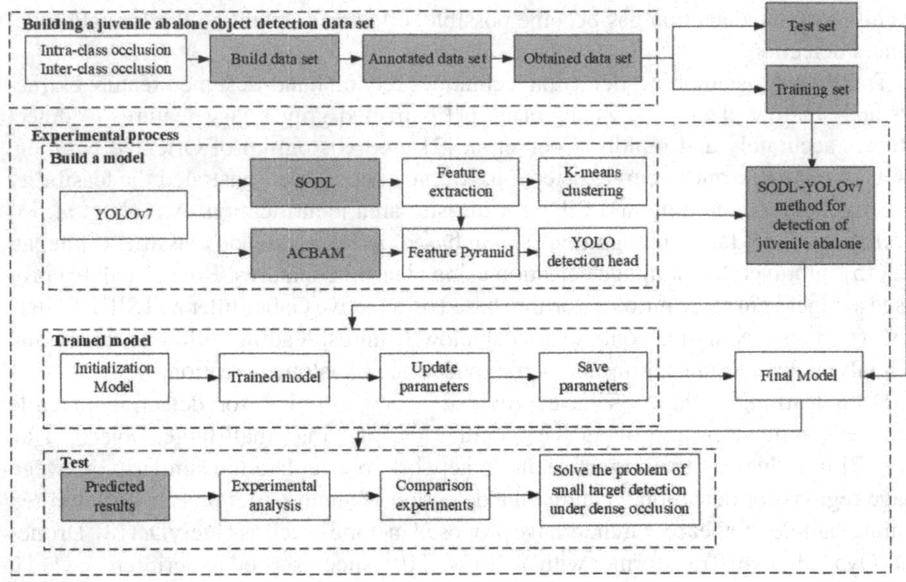

Fig. 1. Technology Road Map.

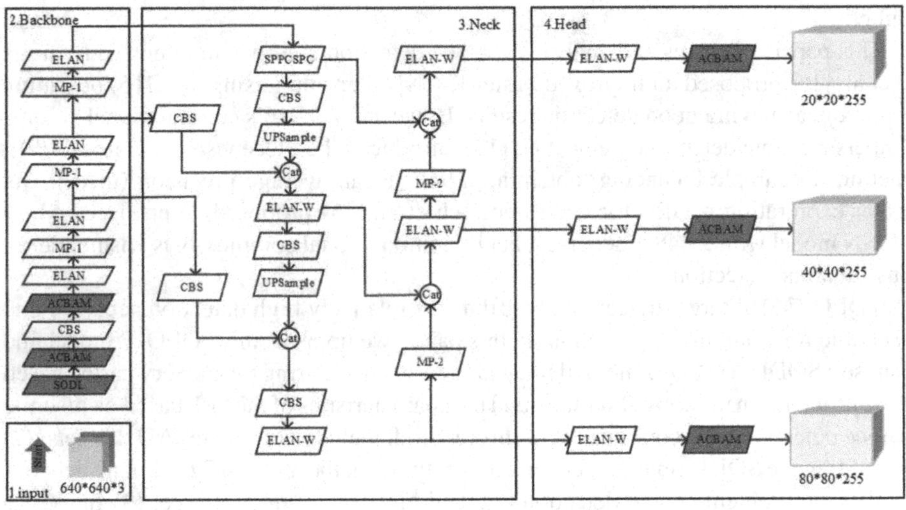

Fig. 2. SODL-YOLOv7 network structure.

2.1 SODL Small Target Detection Network

In YOLOv7, the backbone network consists of CSPDarknet53 [17], a deep convolutional neural network with 53 convolutional layers. To address the limitations of the network in detecting small targets like juvenile abalone, we propose modifying the YOLOv7 backbone by incorporating the SODL small target detection network. The SODL network replaces the traditional convolutional operation in YOLOv7 and offers several advantages. Firstly, it enhances the model's ability to capture details and features of small targets, improving detection accuracy and recall. Secondly, the SODL network utilizes atrous convolution to better understand the features of juvenile abalone at different scales and capture contextual information in the images. This helps handle the high variability in size and shape of juvenile abalone, reducing cases of missed and false detection. Additionally, the SODL network improves the model's understanding of the position and relationships of juvenile abalone in the image, leading to enhanced detection accuracy. The structure of the SODL small target detection network is depicted in Fig. 3.

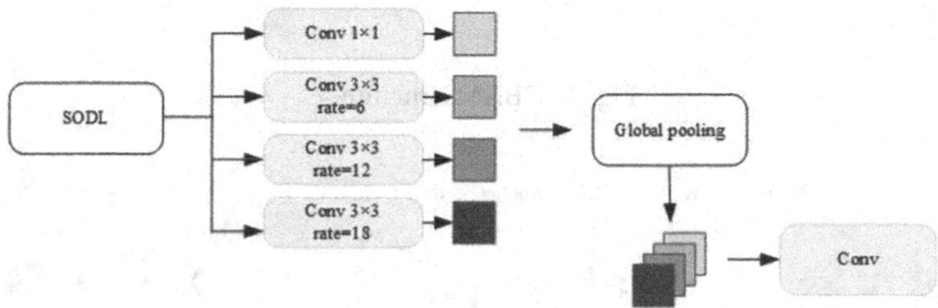

Fig. 3. SODL small target detection network structure.

The specific implementation process of SODL small target detection network is as follows: first, a 640 × 640 × 3 feature map is input to the network, and next, the input feature map is convolved using a void rate of 6, 12, and 18 to obtain multiple feature maps of different scales, as shown in Eq. (1).

$$fi = Conv_{3\times3}(x, ri) \tag{1}$$

where $Conv_{3\times3}$ denotes the 3 × 3 convolution operation, x is the input feature map, and r_i denotes the different void rates.

Then, the input feature map is pooled using the global pooling layer to obtain a global contextual feature: $f_4 = Pool_{global}(x)$.

Next, these feature maps at different scales are stitched together to obtain the multiscale feature maps: $f_{out} = Concat(f1, f2, f3, f4)$.

Finally, the multiscale feature map is convolved using a 1x1 convolution layer to obtain the output feature map: $y = Conv_{1\times1}(fout)$. Where $Conv_{1\times1}$ denotes the 1x1 convolution operation and y is the output feature map.

2.2 ACBAM Attention Module

This paper replaces the CBS attention mechanism in YOLOv7 with the improved CBAM attention mechanism, as shown in Fig. 4. The CBAM attention mechanism is improved by adding an adaptive pooling layer, and then the ACBAM attention mechanism is proposed, as shown in Fig. 5. The adaptive pooling layer allows different degrees of weighted pooling for each spatial location and each channel, thus enhancing the model's focus on the target. Applying ACBAM to juvenile abalone detection can improve the model's ability to understand juvenile abalone images, calculate attention weights in channel and spatial dimensions, thus making the network pay more attention to the important parts of the occluded objects and reduce the influence of the occluded parts on the results, thus achieving more accurate juvenile abalone detection.

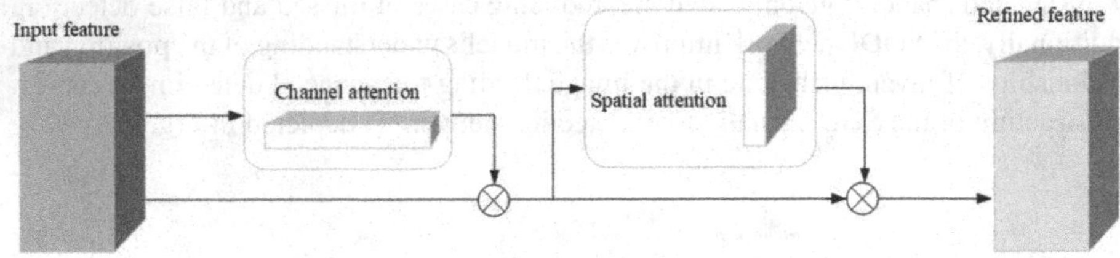

Fig. 4. CBAM structure.

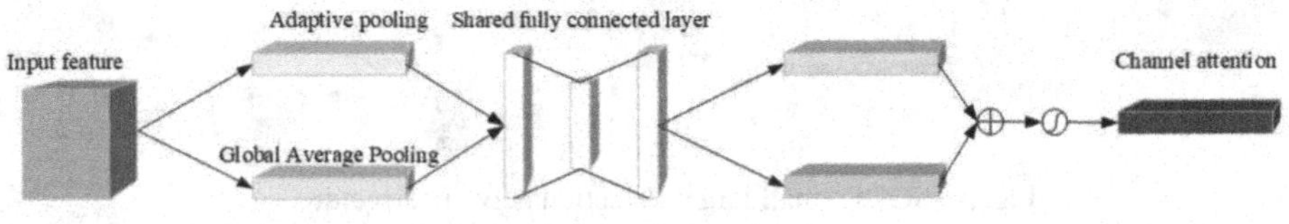

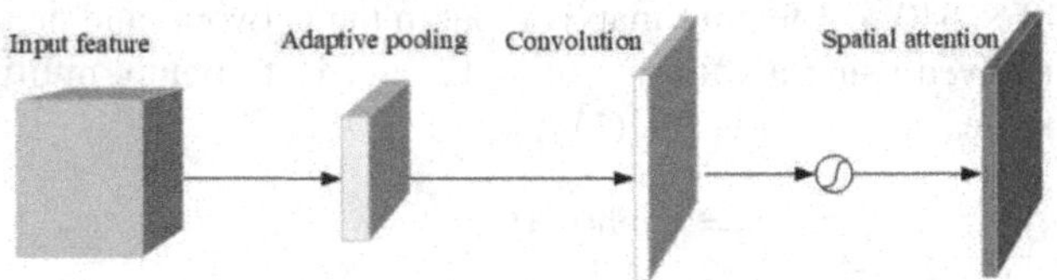

Fig. 5. Improved ACBAM structure.

First, the ACBAM module performs a dimensionality reduction operation on the input feature map M using an adaptive pooling layer to obtain the global average z of the channel dimensions, i.e.:

$$z = \frac{1}{H \times W} \sum_{i=1}^{H} \sum_{j=1}^{W} M_{i,j} \tag{2}$$

where, H,W denote the height and width respectively.

Then the pooling operation is performed on M using global average pooling to get the feature information of each channel. Next, a small fully connected neural network is used to compress M into a one-dimensional vector y, while entering a second fully connected layer to *Sigmoid* activate y and replicate it as the number of channels of M. Multiply the vectors activated by *Sigmoid* and the corresponding elements of M to obtain the weighted channel feature map M'. The specific formula is as follows:

$$M' = \sigma(Wv \cdot Sigmoid(Wg \cdot M)) \cdot M \tag{3}$$

where M is the channel feature map, Wg is the weight of the first fully connected layer, Wv is the weight of the second fully connected layer, and *Sigmoid* is the Sigmoid activation function.

Next, the ACBAM module takes the weighted channel feature map M' as the input feature map of this module and uses a 3 × 3 convolutional layer to extract the channel information of the feature map to obtain the weighted spatial feature map N'. An adaptive pooling operation is performed on the feature map N' to obtain a vector of size, and then, N' is passed through a convolution layer to obtain a weight vector s for calculating the importance in spatial dimensions, and then s is *Sigmoid* activated and replicated to the size of N. Multiplying this *Sigmoid* activated vector with the corresponding elements of N yields the weighted spatial feature map N''. The weighted channel feature map M' and the weighted spatial feature map N''' are stitched together as the output final feature map. The specific equation is shown as follows:

$$N'' = \sigma(Ws \cdot Sigmoid(Conv(Wp \cdot [N'; M']))) \cdot M' \quad (4)$$

where, Wp is the weight of the first convolution layer, Ws is the weight of the second fully connected layer, and *Sigmoid* is the Sigmoid activation function.

ACBAM improves the network's ability to capture juvenile abalone features, enhancing detection accuracy and recall. It adjusts channel and spatial feature importance for key parts like the head, eyes, and antennae. ACBAM enhances head feature representation, mitigating occlusion effects and promoting robust feature learning. This improves model generalization, enabling better detection in various breeding conditions and densities.

3 Experimental Results and Analysis

3.1 Experimental Data Preprocessing

In order to better solve the problem of dense distribution of juvenile abalone, we first went to Rongcheng Marine Farm for research to understand the growth environment of juvenile abalone and the attachment substrate style of juvenile abalone to ensure the accuracy of the experiment, and the culture environment is shown in Fig. 6.

Fig. 6. Environment of the farm.

In this paper, a JAD dataset was established based on the growth environment of juvenile abalone, and the image acquisition environment is shown in Fig. 7. This includes the necessary camera bracket for the high-definition camera, tile type attachment base, and juvenile abalone. The camera used was a 4K camera with the Reel Vision USB4KHDR01 for filming. In order to simulate the light conditions in the real environment, the photos in the dataset were taken at the same time to ensure constant light intensity, and the tile type attachment base underneath the juvenile abalone was used to imitate the natural growing environment of the juvenile abalone.

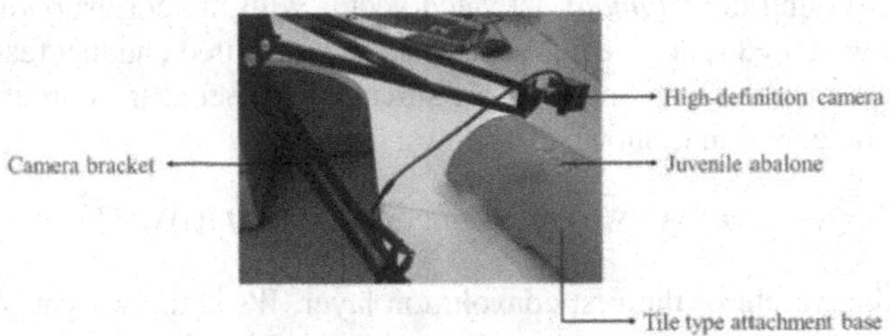

Fig. 7. Image acquisition environment.

The camera was placed at a distance of about 10.5 cm from the tiles, and a total of 2016 images were captured. The final number of images was 2000 after screening, which were then randomly divided into training and test sets according to the ratio of 8:1. The environment was consistent at the time of filming, as shown in Fig. 8, and there was a slight overlap between juvenile abalone. The dataset has a resolution of 3840*2140 pixels and is stored in "jpg" format. When the dataset is fed into the network, all images are scaled to 640 × 640 pixels, reducing training time and memory usage.

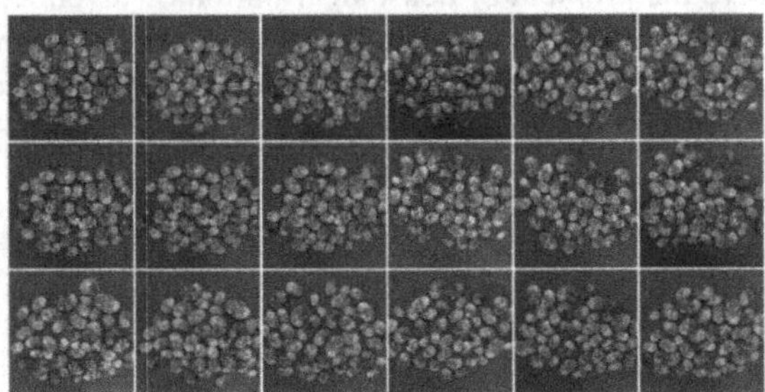

Fig. 8. Part of the juvenile abalone pictures.

The LabelImg labeling tool was used to label the 2000 images and generate.xml labeling files to output the juvenile abalone ID and coordinate position information. The final accuracy is then obtained by testing the test set against the weights trained from the training set. Some of the annotated images are shown in Fig. 9.

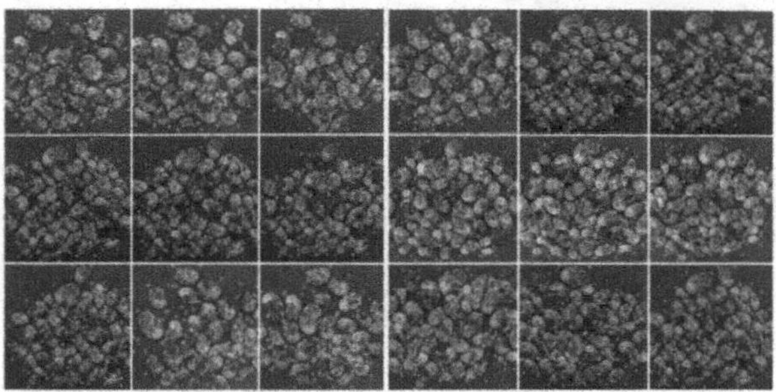

Fig. 9. Part of the labeled images.

3.2 Experimental Environment and Evaluation Index

Table 1. Software and hardware configuration.

Experimental environment	Configuration
Operating System	CentOS7
Processors	Intel(R) Xeon(R) Silver 4210
Graphics Processors	NVIDIA Corporation GP100GL
Graphics Processor Computing Platforms	CUDA 11.0, cuDNN 8.1
Frame	Pytorch
Compile Program	Pycharm, Anaconda
Operating System	Python 3.7

The hardware and software configurations in this paper are shown in Table 1. The model was trained under CentOS7 with NVIDIA Corporation GP100GL GPU and Intel(R) Xeon(R) Silver 4210 CPU. In this paper, the input image size is 640 × 640, epoch is 100, and batch_size size is set to 32. The model parameters of this paper are configured as shown in Table 2. The experiments in this paper are conducted on the JAD dataset, and the basic framework for the experiments is the YOLOv7 network, using an initial learning rate of 0.001.

All model training for the experiments in this paper was conducted on the training set of the JAD dataset, and all model testing was conducted on the test set of the JAD dataset.

Table 2. Model parameters.

Parameter Name	Configuration
Input Size	640 × 640
Learning Rate	0.001
Label Smoothing	0.01
Number of samples	2000
Category	1
Epoch	100
Batch_size	32

3.3 Experimental Results and Analysis

The loss of the training process is shown in Fig. 10. From the figure, we can see that the network fluctuates a lot in the first 5–10 epoch of training, and when the network is trained to nearly 50 epoch, it has tended to converge, and both the training and validation sets have good performance. Figure 11 shows the AP50 and AP75 values, and it can be seen that AP50 reaches 99.4% and AP75 reaches 93.6%, both of which have a high accuracy rate.

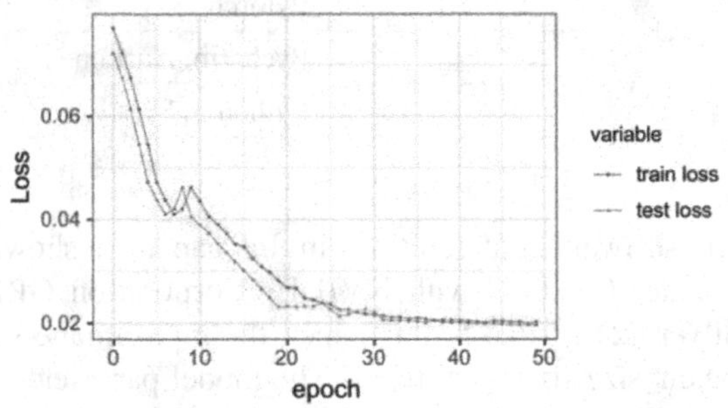

Fig. 10. Loss of training process.

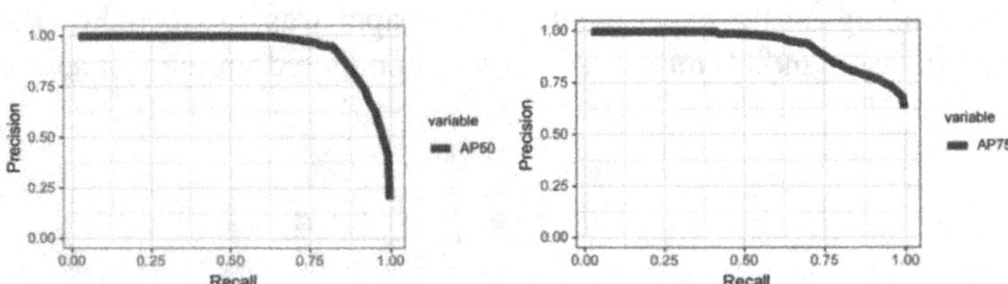

Fig. 11. Network model AP50 value and AP75 value.

In order to verify the effectiveness of the network proposed in this paper, the juvenile abalone detection results of the SODL-YOLOv7 model proposed in this paper were compared with those of the YOLOv7 model [16], the instant-teaching model [18], and the TOOD model [19] based on the JAD dataset, and the specific experimental comparison results are shown in Table 3.

Table 3. Comparison of the performance of different models.

Models	AP50(%)	Increase(%)	AP75(%)	Increase(%)
instant-teaching Model	90.2	2.2	80.3	11.5
TOOD Model	97.2	9.2	82.1	13.3
YOLOv7	95.3	4.1	90.4	3.2
SODL-YOLOv7 Model	**99.4**	--	**93.6**	--

As seen in Table 3, the SODL-YOLOv7 model in this paper improved AP50 by 4.1% and AP75 by 3.2% relative to the YOLOv7 model; Compared with the instant-teaching model, the SODL-YOLOv7 model in this paper has improved AP50 by 9.2% and AP75 by 13.3%; Compared with the TOOD model, the SODL-YOLOv7 model in this paper has improved AP50 by 2.2% and AP75 by 11.5%. The experimental results show that the detection results of the SODL-YOLOv7 model proposed in this paper are better than those of the YOLOv7 model, the instant-teaching model and the TOOD model when the identified juvenile abalone have similar color texture, dense individuals and certain occlusion. The visualization results of the specific performed target detection are shown in Fig. 12.

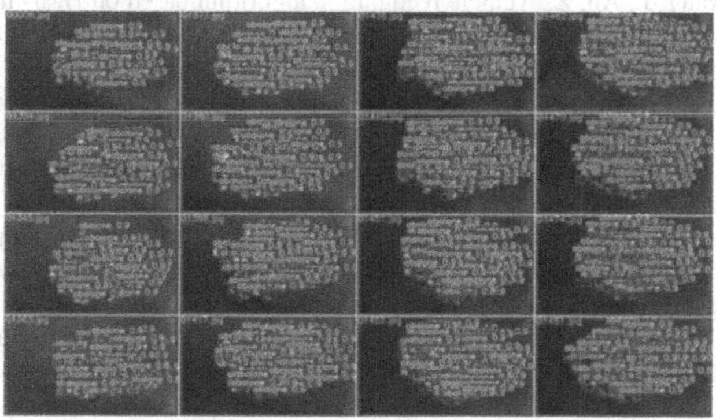

Fig. 12. Visualization results.

4 Conclusion

This paper designs a SODL-YOLOv7 network framework for helping to overcome the detection problem of juvenile abalone and to realize the detection of dense juvenile abalone based on the YOLOv7 algorithm. The JAD nursery abalone dataset was first constructed with a total of 2000 images. The SODL small target detection network and ACBAM attention module have been added to improve the detection accuracy and efficiency. The SODL small target detection network improves the detection accuracy of small targets by improving the YOLOv7 backbone network by adding multiscale adjustable convolution; By adding an adaptive pooling layer, the ACBAM attention module dynamically learns to adjust the pooling method based on the input feature maps, enabling the network to better adapt to feature maps of different sizes and focus more on the target region, which helps the network to better process and identify occluded objects, thus improving the accuracy and efficiency of detection. Finally the model achieves optimal results on the JAD dataset. The solution proposed in this paper can effectively improve the accuracy and efficiency of juvenile abalone detection and achieve good practical application. To lay the technical foundation for further operations that follow, such as abalone baiting estimation, abalone population estimation, and abalone density estimation, the model trained using this algorithm can be deployed into an embedded device with a camera to perform realtime detection of abalone.

Acknowledgements. This work is supported by Shandong Province Key R&D Program Project (2022TZXD005), Shandong Province Science and Technology SMEs Innovation Capacity Enhancement Project (2021TSGC1003), Yantai City Key R&D Program Project (2022XCZX079).

References

1. Zhang, L., Wang, J., An, Z.: Vehicle recognition algorithm based on Haar-like features and improved Adaboost classifier. J. Ambient Intell. Human. Comput. **14**(2), 807–815 (2023). https://doi.org/10.1007/s12652-021-03332-4
2. Chen, G.H., Ni, J., Chen, Z., et al.: Detection of highway pavement damage based on a CNN using grayscale and HOG features. Sensors **22**(7), 2455 (2022)
3. Kumar, V.S., Alemran, A., Gupta, S.K.: Extraction of sift features for identifying disaster hit areas using machine learning techniques. In: International Conference on Knowledge Engi-neering and Communication Systems (ICKES), Chickballapur, India, pp. 1–5 (2022)
4. Arreola, L., Gudiño, G., Flores, G.: Object recognition and tracking using Haar-like Features Cascade Classifiers: application to a quad-rotor UAV. In: 2022 8th International Conference on Control, Decision and Information Technologies (CoDIT). IEEE, vol. 1, pp. 45–50 (2022)
5. Malepati, H.R., Selvam, P.: Extraction of human facial behavioural expressions using Haar like features compared with Haar cascade classifier. In: AIP Conference Proceedings. AIP Publishing, vol. 2655, no. 1 (2023)
6. Kovač, I., Marak, P.: Finger vein recognition: utilization of adaptive gabor filters in the enhancement stage combined with sift/surf-based feature extraction. SIViP **17**(3), 635–641 (2023). https://doi.org/10.1007/s11760-022-02270-8
7. Han, G., Ma, J., Huang, S., et al.: Few-shot object detection with fully cross-transformer. In: Proceedings of the IEEE/CVF Conference on Computer Vision and Pattern Recognition, pp. 5321–5330 (2022)

8. Yang, C., Huang, Z., Wang, N.: QueryDet: cascaded sparse query for accelerating high-resolution small object detection. In: Proceedings of the IEEE/CVF Conference on Computer Vision and Pattern Recognition, pp. 13668–13677 (2022)
9. Li, K., Ou, O., Liu, G.B., Yu, Z.F., et al.: Target detection algorithm of remote sensing image based on improved YOLOv5. Comput. Eng. Appl. **59**(9), 207–214 (2023)
10. Gong, H., Mu, T., Li, Q., et al.: Swin-Transformer-enabled YOLOv5 with attention mechanism for small object detection on satellite images. Remote Sensing **14**(12), 2861 (2022)
11. Akyon, F.C., Altinuc, S.O., Temizel, A.: Slicing aided hyper inference and fine-tuning for small object detection. In: 2022 IEEE International Conference on Image Processing (ICIP). IEEE, pp. 966–970 (2022)
12. Zeng, N., Wu, P., Wang, Z., et al.: A small-sized object detection oriented multi-scale feature fusion approach with application to defect detection. IEEE Trans. Instrum. Measur. **71**, 1–14 (2022)
13. Ye, M., Li, J.: Abalone counting based on improved Faster R-CNN. In: 2022 2nd International Conference on Bioinformatics and Intelligent Computing, pp. 206–210 (2022)
14. Peng, F., Miao, Z., Li, F., et al.: S-FPN: a shortcut feature pyramid network for sea cucumber detection in underwater images. Expert Syst. Appl. **182**, 115306 (2021)
15. Hong, K.T., Abdullah, S.-S., Hasan, M.K., et al.: Underwater fish detection and counting using mask regional convolutional neural network. Water **14**(2), 222 (2022)
16. Wang, C.Y., Bochkovskiy, A., Liao, H.-M.: YOLOv7: Trainable bag-of-freebies sets new state-of-the-art for real-time object detectors. arXiv preprint arXiv:2207.02696 (2022)
17. Wang, C.Y., Liao, H.-M., Wu, Y.H., et al.: CSPNet: A new backbone that can enhance learning capability of CNN. In: Proceedings of the IEEE/CVF Conference on Computer Vision and Pattern Recognition Workshops, pp. 390–391 (2020)
18. Zhou, Q., Yu, C., Wang, Z., et al.: Instant-teaching: an end-to-end semi-supervised object detection framework. In: Proceedings of the IEEE/CVF Conference on Computer Vision and Pattern Recognition, pp. 4081–4090 (2021)
19. Feng, C., Zhong, Y., Gao, Y., et al.: TOOD: task-aligned one-stage object detection. In: 2021 IEEE/CVF International Conference on Computer Vision (ICCV). IEEE Computer Society, pp. 3490–3499 (2021)

MVP-SEG: Multi-view Prompt Learning for Open-Vocabulary Semantic Segmentation

Jie Guo[1], Qimeng Wang[2], Yan Gao[2], Xiaolong Jiang[2], Shaohui Lin[3,5], and Baochang Zhang[1,4](✉)

[1] Hangzhou Research Institute, School of General Engineering, Beihang University, Beijing, China
bczhang@buaa.edu.cn

[2] Xiaohongshu Inc., Beijing, China

[3] East China Normal University, Shanghai, China

[4] Nanchang Institute of Technology, Nanchang, China

[5] KLATASDS-MOE, Shanghai, China

Abstract. CLIP (Contrastive Language-Image Pretraining) is well developed for open-vocabulary zero-shot image-level recognition, while its applications in pixel-level tasks are less investigated, where most efforts directly adopt CLIP features without deliberative adaptations. In this work, we first demonstrate the necessity of image-pixel CLIP feature adaption, then provide Multi-View Prompt learning (MVP-SEG) as an effective solution to achieve image-pixel adaptation and to solve open-vocabulary semantic segmentation. Concretely, MVP-SEG deliberately learns multiple prompts trained by our Orthogonal Constraint Loss (OCLoss), by which each prompt is supervised to exploit CLIP feature on different object parts, and collaborative segmentation masks generated by all prompts promote better segmentation. Moreover, MVP-SEG introduces Global Prompt Refining (GPR) to further eliminate class-wise segmentation noise. Experiments show that the multi-view prompts learned from seen categories have strong generalization to unseen categories, and MVP-SEG+ which combines the knowledge transfer stage significantly outperforms previous methods on several benchmarks. Moreover, qualitative results justify that MVP-SEG does lead to better focus on different local parts.

Keywords: CLIP · Semantic segmentation · Open vocabulary

J. Guo and Q. Wang—Equal contribution.

Supplementary Information The online version contains supplementary material available at https://doi.org/10.1007/978-981-99-8555-5_13.

© The Author(s), under exclusive license to Springer Nature Singapore Pte Ltd. 2024
Q. Liu et al. (Eds.): PRCV 2023, LNCS 14436, pp. 158–171, 2024.
https://doi.org/10.1007/978-981-99-8555-5_13

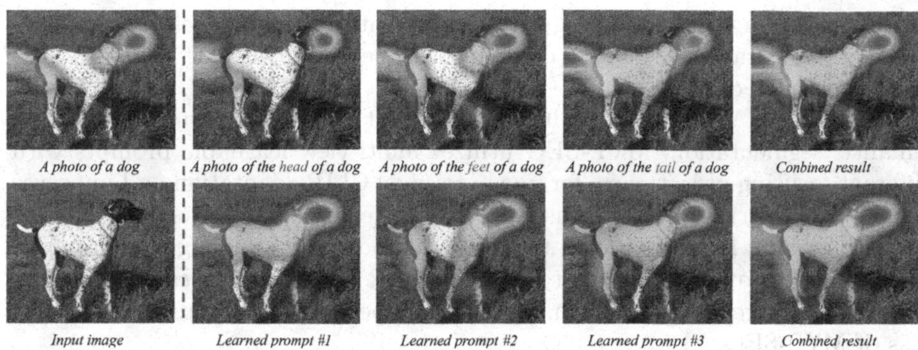

Fig. 1. In the first row, the carefully designed prompts effectively focus on the corresponding object parts and demonstrate superior localization compared to the left-most picture, which is the result of a single handcrafted prompt. The second row shows that using multi-view learnable prompts yields comparable results to the carefully designed prompts. Importantly, multi-view learnable prompts are flexible and can adapt to a wide range of objects without being limited to specific concepts like "head" or "tail".

1 Introduction

Open-vocabulary zero shot semantic segmentation [2,9] localizes objects from arbitrary classes, either seen or unseen during training time, with pixel-level masks. Compared to traditional segmentors working under closed-vocabulary setting [6,25], open-vocabulary methods find wider applications in image editing [14], view synthesis [18], and surveillance [23], while requiring to integrate zero-shot capability into the system.

Large-scale visual-language pre-training models such as CLIP [19] and ALIGN [10] are widely used to incorporate zero-shot capability into visual systems. CLIP, for example, trains on a large dataset of image-text pairs to embed semantic concepts into its parameters, serving as a knowledge base for downstream tasks.

MaskCLIP [25] pioneers the use of CLIP for open-vocabulary segmentation at the pixel level. It directly applies pretrained CLIP text features as a semantic classifier for pixel-level classification, without additional modifications. However, we believe that adapting CLIP from image to pixel can further improve segmentation performance. In Fig. 1, original CLIP prompt features (leftmost in the first row) primarily focus on the most distinctive parts of objects. This occurs because CLIP is pre-trained using image-level contrastive loss, which may lead to incomplete and partial segmentation.

We use prompt learning to adapt CLIP features from image to pixel while preserving zero-shot capability. Prompt learning [26] fine-tunes CLIP for downstream tasks by adjusting prompt parameters, avoiding the need for costly training. To improve image-to-pixel adaptation for segmentation, we propose using multiple prompts inspired by part-based representation [22]. As shown in the first row of Fig. 1, the carefully designed prompts capture different object parts, resulting in better coverage by combining the result of these prompts. On top of this, by switching carefully designed prompts with learnable ones, we get comparable segmentation results to carefully designed ones without relying on the prior knowledge of humans.

We propose MVP-SEG: Multi-View Prompt learning for open-vocabulary semantic segmentation. MVP-SEG utilizes multi-view learnable prompts with different focusing object parts for image-to-pixel CLIP adaptation and improves segmentation performance. To ensure prompt diversity, we introduce the Orthogonal Constraint Loss (OCLoss) to enforce orthogonality among the prompts. Additionally, the Global Prompt Refining (GPR) module reduces class-specific noise for improved segmentation. Extensive experiments validate the effectiveness of MVP-SEG.

In all, we propose three main contributions:

- We demonstrate that image-to-pixel adaptation is important for adopting exploiting CLIP's zero-shot capability in open-vocabulary semantic segmentation, and the proposed MVP-SEG successfully yields such adaptation with favorable performance gains.
- We design the OCLoss to build multi-view learnable prompts attending to different object parts so that collaboratively they yield accurate and complete segmentation. We also introduce GPR to eliminate class-wise segmentation noise.
- We conduct extensive experiments on three major benchmarks, MVP-SEG+ which combine MVP-SEG with the commonly used knowledge transfer stage reports state-of-the-art (SOTA) performance on all benchmarks and even surpass fully supervised counterparts on Pascal VOC and Pascal Context datasets.

2 Related Work

2.1 Vision-Language Models

Vision-language (VL) models pre-trained with web-scale image-text pairs infuse open-world knowledge into aligned textual and visual features, upon which zero-shot visual recognition [24] and generation [20] bloom in recent years. Recognition-purpose VL models can be categorized as two-stream [10,19] or one-stream [12] depending on whether multimodal inputs are processed separately or all together. Amongst, CLIP [19] is more popular in downstream applications and it is contrastively trained using 400 million image-text pairs. ALIGN [10] adopts an even larger dataset with 1.8 billion pairs but with considerable noise. For a zero-shot generation, BLIP-v2 [11] have shown outstanding generation quality in executing text-guided image generation. In this work, we build upon the zero-shot capability of CLIP to realize semantic segmentation in open vocabulary.

2.2 Zero-Shot Segmentation

Zero-shot segmentation performs pixel-level classification covering unseen classes during training. SPNet [21] and ZS3Net [2] are the representatives of discriminative [1] and generative [2,9] lines of work, the former projects visual embeddings towards semantic embeddings, while the later generates pixel-level features for the unseen using semantic embeddings. Following ZS3Net, fruitful generative methods are proposed [9] to mind the object-to-pixel feature gap considering structural consistency and uncertainty. Besides, self-training is widely adopted in zero-shot segmentation [2] to boost performance.

2.3 Prompt Learning

The idea of prompting originates from NLP (Nature Language Processing) and has been exploited for transferring VL-pretrained models to down-stream tasks in forms like "a photo of a [CLS]". Via prompting, one can eschew tuning huge VL models but use it as a fixed knowledge base, wherein only task-related information is elicited. Nonetheless, finding the optimal prompt is not trivial by hand, thus prompt learning [26] is proposed to automate this process with limited labeled data. CoOp [26] introduces continuous prompt learning such that a set of continuous vectors are end-to-end optimized via down-stream supervision [13], and CoOp applies learnable prompts on the text encoder of CLIP to replace sub-optimal hand-crafted templates.

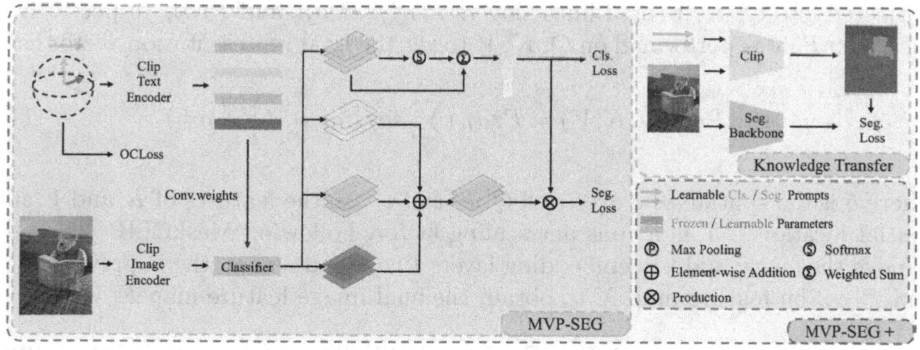

Fig. 2. The overview of MVP-SEG and MVP-SEG+. The MVP-SEG+ consists of MVP-SEG and a knowledge transfer stage. In MVP-SEG, we train multi-view prompts by Classification Loss, Segmentation Loss and Orthogonal Constraint Loss. The CLIP text and image encoders are fixed. In the knowledge transfer stage, the segmentation backbone is trained by the pseudo labels generated by MVP-SEG.

3 Method

In this section, we first give a brief introduction to the open-vocabulary semantic segmentation task, then we introduce our proposed MVP-SEG and MVP-SEG+ in detail.

3.1 Problem Definition

Semantic segmentation performs pixel-level classification to localize objects from different classes in the input image. Under traditional close-vocabulary settings, target classes form a finite set, and all classes are present in both training and test sets. Open-vocabulary semantic segmentation aims to generalize beyond finite base classes and segments objects from novel classes unseen during training. In order to quantitatively evaluate segmentation performance under this open-vocabulary setting, object classes in existing segmentation datasets are divided as seen and unseen subsets [25]. Specifically, models are trained on labeled seen classes and evaluated on both seen and unseen classes. Additionally, we also visualize segmentation results on arbitrary rare classes as in Fig. 1 in the Appendix to further validate our open-vocabulary capability.

3.2 MVP-SEG

The architecture of MVP-SEG is illustrated in Fig. 2. Note that MVP-SEG learns multiple prompts by training only on seen classes.

Multi-view Prompt Learning. In this stage, we adopt a two-stream network architecture separately employing an image and a text encoder. We modify a fixed CLIP image encoder as our vision encoder. In specifics, the original CLIP vision encoder adopts a transformer-style attention pooling layer *AttnPool* on the last feature map X to get the representation vector of the input image. X is projected into Q, K, V by 3 linear layers $Proj_q$, $Proj_k$ and $Proj_v$ respectively, then *AttnPool* is performed on Q, K, V to get the final representation vector as:

$$AttnPool(Q,K,V) = Proj_c(\sum_i softmax(\frac{(\bar{q}k_i^T)}{T})v_i)) \tag{1}$$

where $\bar{q}$ is the spatial-wise averaged Q. k_i and v_i are the features of K and V at spatial location i. T is a constant scaling factor. Following MaskCLIP [25], we remove the query and key embedding layers $Proj_q$ and $Proj_k$, then apply $Proj_v$ and $Proj_c$ on feature map X to obtain the final image feature map F, where

$$F = Proj_c(Proj_v(X)) \tag{2}$$

In practice, $Proj_c$ and $Proj_v$ are implemented as 1×1 convolution layers.

For the text encoder, we use a fixed CLIP text encoder to ensure feature alignment with the visual counterpart. For each class, we combine the class name with learnable prompts to form the inputs for the text encoder. We first initialize k + 1 prompts P, where $P = \{p_0, p_1, p_2, ...p_k\}$. p_0 is the global classification prompt and $p_1, p_2, ..p_k$ represent k segmentation prompts. Each prompt p_i contain 32 tokens, where $p_i \in \mathbb{R}^{32\times512}$. For class c, we concatenate each prompt with the class name as:

$$s_c^i = CONCAT(p_i, W(cls_c)) \tag{3}$$

here W denotes the word embedding function that maps cls_c to word embedding vectors. We feed each sentence s_c^i to the CLIP text encoder and get the text representation vector t_c^i, where t_c^0 and $t_c^i (i > 0)$ indicates global classification vector and segmentation vector of class c.

Each representation vector t_c^i is used as a classifier to perform pixel-level classification on the feature map F to get the mask map m_c^i, where m_c^0 is the global classification mask and $m_c^1, m_c^2 ... m_c^k$ are the segmentation masks.

The fused segmentation result m^f is computed by summing all segmentation masks as:

$$m^f = softmax(\tau_1 \sum_{i=1}^{k} m^i) \tag{4}$$

where τ_1 is a learnable scalar parameter.

In order to ensure different prompts result in masks attending to varied object parts and collectively they provide better segmentation, we design the Orthogonal Constraint Loss (OCLoss) to supervise prompt learning. For each segmentation prompt $p_i \in \mathbb{R}^{32 \times 512}$, we first average the token dimension to get its average vector p_i', where $p_i' \in \mathbb{R}^{1 \times 512}$, then applied the OCLoss which is formulated as:

$$L_{OC} = \sum_{i=1}^{k} \sum_{j=i+1}^{k} \frac{|p_i' \cdot p_j'|}{\|p_i'\| \|p_j'\|} \tag{5}$$

Global-Prompt Refinement. To incorporate CLIP's strong capability in zero-shot image classification into segmentation, we introduce Global-Prompt Refinement (GPR) module which uses a global classification prompt to obtain image classification scores and refine the segmentation mask by eliminating class-wise noises. The image-level classification score g_c is obtained by weighting and summing m_c^0 as

$$g_c = sigmoid(\frac{m_c^0 \cdot softmax(\gamma m_c^0)}{\tau_2}) \tag{6}$$

where τ_2 is a learnable scalar parameter and γ is a constant scale factor. The global classification loss is as:

$$L_{cls} = - \sum_{c} y_c log(g_c) \tag{7}$$

here y_c equals 1 if category c exists in this image, and else 0. The classification prompt tends to focus on the most class-discriminative object parts for better classification. More visualizations can be found in Fig. 3 in the Appendix.

The final segmentation mask m_c is obtained by multiplying the global classification score with fused segmentation masks:

$$m_c = m_c^f g_c \tag{8}$$

The segmentation loss is formulated as:

$$L_{seg} = - \sum^{H} \sum^{W} \sum_{c=1}^{C} m_c^* log(m_c) \tag{9}$$

where m_c^* is the ground-truth mask of category c. C is the number of segmentation classes. H and W denote the height and the width of the input image respectively. The overall loss function is obtained by summing the above losses as

$$L = \lambda_1 L_{seg} + \lambda_2 L_{cls} + \lambda_3 L_{OC} \tag{10}$$

Note that in the prompt learning stage, the image encoder and text encoder are fixed, only prompt vectors P and temperature τ_1, τ_2 are learnable, so that the image-text feature alignment within CLIP is preserved. In addition, our experiments also demonstrate that prompts trained on limited seen categories generalize well to unseen categories.

3.3 MVP-SEG+

We term the entire framework in Fig. 2 as MVP-SEG+, consisting of MVP-SEG and knowledge transfer stage.

Knowledge Transfer. The learned prompts in MVP-SEG are used to generate pseudo labels to transfer zero-shot knowledge from CLIP down to a segmentation network.

Following the application of MaskCLIP+ [25], the transfer learning stage contains two steps: pseudo-label training and self-training. At the pseudo-label training step, images are fed into MVP-SEG to generate the pseudo labels for the segmentation model. The classifier of segmentation model is replaced by the multi-view prompts of MVP-SEG to preserve the open-vocabulary capability. In self-training stage, the segmentation model in MVP-SEG+ starts to train itself with self-generated pseudo labels. For more details, please refer to MaskCLIP+ [25].

4 Experiments

To evaluate the efficacy of MVP-SEG and MVP-SEG+, we conduct comprehensive experiments on three widely-adopted benchmarks. In this section, we first analysis the performance and ablation study of MVP-SEG. Then, we compare MVP-SEG+ with the SOTA zero-shot segmentation method to show the effectiveness of our proposed method to adapt CLIP to pixel-level tasks. At last, we show the open-vocabulary ability of our method on unlabeled web images.

4.1 Datasets

Three semantic segmentation benchmarks are used in our experiment. PASCAL VOC 2012 [7] contains 1426 training images with 20 object classes and 1 background class. PASCAL Context [16] annotates PASCAL VOC 2010 data with segmentation masks of 10,103 images covering 520 classes, among which 59 common classes are used as foregrounds. COCO Stuff dataset [3] is an extension of the COCO dataset, 164,000 images covering 171 classes are annotated with segmentation masks. We follow the common zero-shot experimental setups as implemented in [25]. For PASCAL VOC, we ignore the background class, and *potted plant, sheep, sofa, train, tv monitor* are selected as 5 unseen classes while others are used as seen. For PASCAL Context, the background is not ignored and *cow, motorbike, sofa, cat, boat, fence, bird, tv monitor, keyboard, aeroplane* are unseen. For COCO Stuff, *frisbee, skateboard, cardboard, carrot, scissors, suitcase, giraffe, cow, road, wall concrete, tree, grass, river, clouds, playing field* are used as unseen classes.

4.2 Evaluation Metrics

We utilize the *mean intersection-over-union* (mIoU) on seen and unseen classes to evaluate semantic segmentation performance. Additionally, we also report the harmonic mean of seen and unseen mIoUs (hIoU).

4.3 Implementation Details

We adopt MMSegmentation[1] as our codebase. In the prompt learning stage, we use the encoders from CLIP-ResNet-50[2] to extract image and text features. We use 1 global classification and 3 multi-view segmentation prompts unless otherwise stated. All learnable prompts are implemented using unified context [26] with 32 context tokens. The weighting parameters λ_1, λ_2, λ_3 for L_{seg},L_{cls} and L_{OC} are empirically set to 1, 3, 100 respectively. The softmax scale factor γ

Fig. 3. Comparison of segmentation results between MVP-SEG+ and MaskCLIP+ [25]. The first row is the input image, the second row is the prediction result of MaskCLIP+ and the third row is our result.

[1] https://github.com/open-mmlab/mmsegmentation.
[2] https://github.com/openai/CLIP.

Table 1. Cross-dataset generalization test experiments. Columns represent different test sets, and rows represent the dataset used to learn prompts. Baseline indicates default prompts in MaskCLIP.

mIoU(U)	Pascal VOC (20)	Pascal Context (59)	COCO Stuff (171)
baseline	40.2	26.9	12.2
Pascal VOC (20)	45.9	29.8	13.9
Pascal Context (59)	47.8	35.7	19.5
COCO Stuff (171)	**50.0**	**38.7**	**19.9**

is set to 10. We use SGD optimizer to optimize learnable prompts with learning rate set to 2e−4 and 5e−4 weight decay. We also adopt linear warmup strategy with 1k warmup iters and 1e-3 warmup ratio. The prompt learning step takes 8k iterations with batch size 8. All the experiments are conducted using 4 T A100 GPUs.

In the knowledge transfer stage, we use the same settings as MaskCLIP [25] for fair comparison. We choose DeepLabv2 [4] as the segmentation backbone for PASCAL VOC and COCO Stuff and DeepLabv3+ for PASCAL Context. The training schedule is set to 20k/40k/80k for PASCAL VOC/PASCAL Context/COCO Stuff. The first 1/10 training iterations adopt MVP-SEG guided learning and the rest adopts self-training.

4.4 Ablation Studies on MVP-SEG

Comparison with Baseline. We adopt MaskCLIP [25] as our baseline. MaskCLIP feeds handcrafted prompts into the CLIP text encoder with 85 templates as described in [8].

We compare segmentation performance with MaskCLIP on the unseen classes to demonstrate the zero shot ability of MVP-SEG. As show in Table 2, MVP-SEG achieves significant improvements (+8.4% in COCO Stuff, +12.7% in PASCAL VOC) on the unseen classes over baseline method using only 4 learned multi-view prompts (1 global classification prompt and 3 multi-view segmentation prompts). This result shows that, multi-view learnable prompts effectively contribute to the improvement of CLIP performance on pixel-level tasks.

Table 2. Ablation study of results of MVP-SEG on COCO Stuff and PASCAL VOC dataset, we adopt MaskCLIP [25] as the baseline model for comparison. Note that the number of multiple prompts without GPR is 3.

Method	OCLoss	GPR	COCO Stuff				PASCAL VOC			
			mIoU(U)	mIoU(S)	mIoU	hIoU	mIoU(U)	mIoU(S)	mIoU	hIoU
Baseline	–	–	12.2	10.0	10.2	11.0	40.2	41.9	41.5	41.1
Single Prompt	✗	✗	15.8	15.1	15.2	15.5	41.8	46.9	45.6	44.2
Multiple Prompt	✗	✗	16.4	16.0	16.0	16.2	45.8	48.4	47.7	47.0
Multiple Prompt	✓	✗	19.9	16.1	16.4	17.9	45.9	51.1	49.8	48.3
Multiple Prompt	✓	✓	**20.6**	**17.7**	**18.0**	**19.2**	**52.9**	**53.4**	**53.2**	**53.1**

Study on Segmentation Prompts. As shown in Table 2, by replacing 85 manual prompts (as shown in the first row) with 1 learnable prompt (as in row 2), we get 3.6% performance improvement on unseen classes on the COCO Stuff dataset and 1.6% improvement on PASCAL VOC dataset. This result reveals the efficacy of learnable prompts for adapting CLIP features for segmentation. Increasing the number of learnable prompts is also beneficial for the performance. After deploying the OCLoss (as shown in row 4), the mIoU on unseen classes on COCO Stuff continues to increase from 16.4% to 19.9%, further indicating that our insight on multi-view prompt learning method is effective. We further test the optimal number of prompts with experiments on the COCO Stuff dataset. As depicted in Fig. 4 in the Appendix, the performance (on unseen classes using OCLoss) increases as the number of prompts increases and reaches the peak at 3 prompts, thus our default number for multi-view segmentation prompts is 3.

We also visualize text embeddings under different learnable prompts in MVP-SEG with t-SNE [15]. As illustrated in Fig. 5 in the Appendix, text embeddings of different prompts are scattered in the feature space, which proves that the learned multi-view segmentation prompts are diverse. Moreover, visualization in Fig. 1 shows that the learned prompts do contain high-level reasonable semantic information. More importantly, the segmentation results using different prompts tend to be complementary such that combining them together can yield more accurate and complete masks. For more visualizations please refer to Fig. 6 in the Appendix.

To further verify the advantages of multi-view learnable prompts, we also compare the segmentation performance of our prompts with manually designed ones. We select a set of animal-related classes including *cat, dog, horse, cow, bear, giraffe* from COCO Stuff for comparison. As stated in Sect. 1, We carefully hand-pick prompts related to body parts (*"the leg of {}", "the head of {}", and "the tail of {}"*) for animal-related segmentation. As shown in Table 1 in Appendix, manual prompts outperform the MaskCLIP baseline with 1.0% mIoU, indicating the feasibility of adopting multiple handcrafted prompts for segmentation. More importantly, we outperform both MaskCLIP and handcrafted prompts, showing the superiority of MVP-SEG.

The Influence of GPR. We use the GPR module to further refine segmentation masks with the image-level classification capability of CLIP. Row 5 of Table 2 shows that adding GPR can comprehensively improve performances on both seen and unseen classes. hIoU result on COCO Stuff dataset is improved from 17.9% to 19.2%, and on PASCAL VOC increases from 48.3% to 53.1%. Visualization results also show that GPR can improve segmentation by filtering out masks of false positive classes (please refer to Fig. 2 in the Appendix).

Table 3. Comparison with the SOTA methods. Depending on whether the unseen classes are visible during training, previous methods can be divided into inductive methods (top part) and transductive methods (middle part). Our method belongs to the latter. MVP-SEG+ and Fully Sup indicate the DeepLab models train by MVP-SEG pseudo label fully-supervised annotations respectively.

Method	ST	COCO Stuff				Pascal VOC				Pascal Context			
		mIoU(U)	mIoU(S)	mIoU	hIoU	mIoU(U)	mIoU(S)	mIoU	hIoU	mIoU(U)	mIoU(S)	mIoU	hIoU
SPNet [21]	✗	8.7	35.2	32.8	14.0	15.6	78.0	63.2	26.1	–	–	–	–
ZS3Net [2]	✗	9.5	34.7	33.3	15.0	17.7	77.3	61.6	28.7	12.7	20.8	19.4	15.8
CaGNet [9]	✗	12.2	35.5	33.5	18.2	26.6	78.4	65.5	39.7	18.5	24.8	23.2	21.2
SIGN [5]	✗	15.5	32.3	–	20.9	28.9	75.4	–	41.7	–	–	–	–
Joint [1]	✗	–	–	–	–	32.5	77.7	–	45.9	14.9	33.0	–	20.5
ZegFormer [6]	✗	33.2	36.6	–	34.8	63.6	86.4	–	73.3	–	–	–	–
SPNet [21]	✓	26.9	34.6	34.0	30.3	25.8	77.8	64.8	38.8	–	–	–	–
ZS3Net [2]	✓	10.6	34.9	33.7	16.2	21.2	78.0	63.0	33.3	20.7	27.0	26.0	23.4
CaGNet [9]	✓	13.4	35.5	33.7	19.5	30.3	78.6	65.8	43.7	–	–	–	–
SIGN [5]	✓	17.5	31.9	–	22.6	41.3	83.5	–	55.3	–	–	–	–
STRICT [17]	✗	30.3	35.3	34.9	32.6	35.6	82.7	70.9	49.8	–	–	–	–
MaskCLIP+ [25]	✗	54.7	–	39.6	45.0	86.1	–	88.1	87.4	66.7	–	48.1	53.3
MVP-SEG+	✓	**55.8**	**38.3**	**39.9**	**45.5**	**87.4**	**89.0**	**88.6**	**88.2**	**67.5**	**44.9**	**48.7**	**54.0**
Fully Sup [25]	✗	–	–	**39.9**	–	–	–	88.2	–	–	–	48.2	–

The Generalization of Multi-view Learnable Prompts. Experiment results in Table 2 show that learned prompts not only improve performances on seen classes but also significantly boost unseen classes as well. To further study the generalization ability of our learned prompts, we evaluate the performance by training prompts on one dataset but test on other datasets. As shown in Table 1, prompts trained on other datasets outperform the baseline by a large margin.

4.5 Comparison with State-of-the-Art

From Table 3, we can conclude: 1) Our method can effectively improve the performance of novel classes. MVP-SEG+ improves the previous SOTA method by 1.1%, 1.3% and 0.8% mIoU of unseen classes on COCO Stuff, Pascal VOC and Pascal Context respectively; 2) Our method consistently outperforms previous SOTA on all datasets (+0.3%, +0.5% and +0.6% hIoU); 3) Our proposed method can effectively adapt powerful CLIP to pixel level tasks so that the performance of MVP-SEG+ is competitive or even surpass fully supervised counterparts. To the best of our knowledge, this is the first time such a result has been achieved.

Visual comparison with MaskCLIP+, which is one of the SOTAs, is illustrated in Fig. 3, and our method obtains more complete masks compared, and false positive classes can be filtered. Furthermore, we also illustrate the performance of our method on web-crawled images which contain rare object classes that are not visible in any training set. As shown in Fig. 1 in the Appendix, MVP-SEG can correctly attend to rare objects such as Iron Man and Captain America, and MVP-SEG+ segments them with favorable accuracy.

5 Conclusion

In this work, we propose multi-view prompt learning (MVP-SEG) to settle open vocabulary semantic segmentation with pre-trained CLIP. At first, we demonstrate the efficacy of adapting pretrained CLIP model from image-to-pixel level for open-vocabulary segmentation, then introduce multi-view prompt learning, inspired by part-based representation, to convey this adaptation. Multi-view learnable prompts are optimized by our Orthogonal Constraint Loss (OCLoss) to ensure the part-wise attention of each prompt so that collaboratively they yield better segmentation results. In addition, we design Global Prompt Refining (GPR) adopting a global learnable prompt to remove class-wise noises and refine segmentation masks. MVP-SEG+ reports SOTA open-vocabulary segmentation performance on all three widely-used benchmarks and reveals superior results than fully-supervised counterparts on PASCAL VOC and PASCAL Context datasets.

Acknowledgement. This research was supported by Zhejiang Provincial Natural Science Foundation of China under Grant No. D24F020011, Beijing Natural Science Foundation L223024, the National Natural Science Foundation of China under Grant 62076016, the National Key Research and Development Program of China (Grant No. 2023YFC3300029) and "One Thousand Plan" projects in Jiangxi Province Jxsg2023102268. This work was also supported by National Natural Science Foundation of China (NO. 62102151), CCF-Tencent Open Research Fund, the Open Research Fund of Key Laboratory of Advanced Theory and Application in Statistics and Data Science, Ministry of Education (KLATASDS2305), the Fundamental Research Funds for the Central Universities.

References

1. Baek, D., Oh, Y., Ham, B.: Exploiting a joint embedding space for generalized zero-shot semantic segmentation. In: Proceedings of the IEEE/CVF International Conference on Computer Vision, pp. 9536–9545 (2021)
2. Bucher, M., Vu, T.H., Cord, M., Pérez, P.: Zero-shot semantic segmentation. In: Advances in Neural Information Processing Systems, vol. 32 (2019)
3. Caesar, H., Uijlings, J., Ferrari, V.: COCO-Stuff: thing and stuff classes in context. In: Proceedings of the IEEE Conference on Computer Vision and Pattern Recognition, pp. 1209–1218 (2018)
4. Chen, L.-C., Zhu, Y., Papandreou, G., Schroff, F., Adam, H.: Encoder-decoder with atrous separable convolution for semantic image segmentation. In: Ferrari, V., Hebert, M., Sminchisescu, C., Weiss, Y. (eds.) ECCV 2018. LNCS, vol. 11211, pp. 833–851. Springer, Cham (2018). https://doi.org/10.1007/978-3-030-01234-2_49
5. Cheng, J., Nandi, S., Natarajan, P., Abd-Almageed, W.: SIGN: spatial-information incorporated generative network for generalized zero-shot semantic segmentation. In: Proceedings of the IEEE/CVF International Conference on Computer Vision, pp. 9556–9566 (2021)
6. Ding, J., Xue, N., Xia, G.S., Dai, D.: Decoupling zero-shot semantic segmentation. In: Proceedings of the IEEE/CVF Conference on Computer Vision and Pattern Recognition, pp. 11583–11592 (2022)

7. Everingham, M., Van Gool, L., Williams, C.K.I., Winn, J., Zisserman, A.: The PASCAL Visual Object Classes Challenge 2012 (VOC 2012) Results (2012). http://www.pascal-network.org/challenges/VOC/voc2012/workshop/index.html
8. Gu, X., Lin, T.Y., Kuo, W., Cui, Y.: Open-vocabulary object detection via vision and language knowledge distillation. arXiv preprint arXiv:2104.13921 (2021)
9. Gu, Z., Zhou, S., Niu, L., Zhao, Z., Zhang, L.: Context-aware feature generation for zero-shot semantic segmentation. In: Proceedings of the 28th ACM International Conference on Multimedia, pp. 1921–1929 (2020)
10. Jia, C., et al.: Scaling up visual and vision-language representation learning with noisy text supervision. In: International Conference on Machine Learning, pp. 4904–4916. PMLR (2021)
11. Li, J., Li, D., Savarese, S., Hoi, S.: BLIP-2: bootstrapping language-image pre-training with frozen image encoders and large language models. arXiv preprint arXiv:2301.12597 (2023)
12. Li, L.H., Yatskar, M., Yin, D., Hsieh, C.J., Chang, K.W.: VisualBERT: a simple and performant baseline for vision and language. arXiv preprint arXiv:1908.03557 (2019)
13. Li, X.L., Liang, P.: Prefix-tuning: optimizing continuous prompts for generation. arXiv preprint arXiv:2101.00190 (2021)
14. Liu, X., et al.: Open-Edit: open-domain image manipulation with open-vocabulary instructions. In: Vedaldi, A., Bischof, H., Brox, T., Frahm, J.M. (eds.) Computer Vision-ECCV 2020: 16th European Conference, Glasgow, UK, 23–28 August 2020, Proceedings, Part XI 16, vol. 12356, pp. 89–106. Springer, Cham (2020). https://doi.org/10.1007/978-3-030-58621-8_6
15. Van der Maaten, L., Hinton, G.: Visualizing data using t-SNE. J. Mach. Learn. Res. **9**(11), 2579–2605 (2008)
16. Mottaghi, R., et al.: The role of context for object detection and semantic segmentation in the wild. In: Proceedings of the IEEE Conference on Computer Vision and Pattern Recognition, pp. 891–898 (2014)
17. Pastore, G., Cermelli, F., Xian, Y., Mancini, M., Akata, Z., Caputo, B.: A closer look at self-training for zero-label semantic segmentation. In: Proceedings of the IEEE/CVF Conference on Computer Vision and Pattern Recognition, pp. 2693–2702 (2021)
18. Qian, R., Li, Y., Xu, Z., Yang, M.H., Belongie, S., Cui, Y.: Multimodal open-vocabulary video classification via pre-trained vision and language models. arXiv preprint arXiv:2207.07646 (2022)
19. Radford, A., et al.: Learning transferable visual models from natural language supervision. In: International Conference on Machine Learning, pp. 8748–8763. PMLR (2021)
20. Ramesh, A., et al.: Zero-shot text-to-image generation. In: International Conference on Machine Learning, pp. 8821–8831. PMLR (2021)
21. Xian, Y., Choudhury, S., He, Y., Schiele, B., Akata, Z.: Semantic projection network for zero-and few-label semantic segmentation. In: Proceedings of the IEEE/CVF Conference on Computer Vision and Pattern Recognition, pp. 8256–8265 (2019)
22. Yu, F., Liu, K., Zhang, Y., Zhu, C., Xu, K.: PartNet: a recursive part decomposition network for fine-grained and hierarchical shape segmentation. In: Proceedings of the IEEE/CVF Conference on Computer Vision and Pattern Recognition, pp. 9491–9500 (2019)

23. Yu, L., Qian, Y., Liu, W., Hauptmann, A.G.: Argus++: robust real-time activity detection for unconstrained video streams with overlapping cube proposals. In: Proceedings of the IEEE/CVF Winter Conference on Applications of Computer Vision, pp. 112–121 (2022)
24. Yue, Z., Wang, T., Sun, Q., Hua, X.S., Zhang, H.: Counterfactual zero-shot and open-set visual recognition. In: Proceedings of the IEEE/CVF Conference on Computer Vision and Pattern Recognition, pp. 15404–15414 (2021)
25. Zhou, C., Loy, C.C., Dai, B.: Extract free dense labels from clip. In: Avidan, S., Brostow, G., Cissé, M., Farinella, G.M., Hassner, T. (eds.) Computer Vision-ECCV 2022: 17th European Conference, Tel Aviv, Israel, 23–27 October 2022, Proceedings, Part XXVIII, vol. 13688, pp. 696–712. Springer, Cham (2022). https://doi.org/10.1007/978-3-031-19815-1_40
26. Zhou, K., Yang, J., Loy, C.C., Liu, Z.: Learning to prompt for vision-language models. Int. J. Comput. Vision **130**(9), 2337–2348 (2022)

Context-FPN and Memory Contrastive Learning for Partially Supervised Instance Segmentation

Zheng Yuan, Weiling Cai[✉], and Chen Zhao

Nanjing Normal University, Nanjing 210097, China
caiwl@njnu.edu.cn

Abstract. Partially supervised instance segmentation aims to segment objects on both limited seen categories and novel unseen categories (without annotated masks), thereby eliminating expensive demands of mask annotation for new categories. Existing work mainly utilize the pipeline model of detection first and then segmentation, and explores how to provide more discriminative regions of interest for the class-agnostic mask head, but these methods do not perform well when faced with complex scenes. In this work, we propose a novel method, named CCMask, that combines Context Feature Pyramid Network (Context-FPN) and Memory Contrastive Learning Head (MCL Head) to achieve effective class-agnostic mask segmentation. Specifically, we introduce a Context-FPN to obtain context-rich feature map via context extraction module, which will benefit the subsequent task heads. In the MCL Head, we employ foreground/background query memory queue to store queries from recent training batches, this helps the MCL Head learns the general concepts of foreground and background. These strategies collectively contribute to improve the discrimination between foreground and background. Exhaustive experiments on COCO dataset demonstrate that our method achieves state-of-the-art results.

Keywords: Partially supervised instance segmentation · Contrastive learning · Feature Pyramid Network

1 Introduction

Instance segmentation is a fundamental task in computer vision, with wide-ranging applications in various domains such as human pose estimation, autonomous driving, and remote sensing image analysis. Current methods have achieved impressive results in this task by relying on abundant pixel-level annotated data. However, the high cost of mask annotation for new categories has limited instance segmentation in a narrow range of classes, hindering its further development. In contrast, bounding box annotation is easier to obtain and less expensive. As a result, recent research has proposed the task of partially supervised instance segmentation to address this issue. In this task, all object

© The Author(s), under exclusive license to Springer Nature Singapore Pte Ltd. 2024
Q. Liu et al. (Eds.): PRCV 2023, LNCS 14436, pp. 172–184, 2024.
https://doi.org/10.1007/978-981-99-8555-5_14

categories are divided into two sets: seen categories and unseen categories. Seen categories have instance mask annotations, while unseen categories only have bounding box annotations. The goal of this task is to take advantage of the partial supervision information to improve the mask prediction performance of the mask segmentation model in unseen categories.

Fig. 1. Visualization results of OPMask [1], ContrastMask [18] and our CCMask when an image contains multiple overlapping instances, the results show that our CCMask performs more precisely segmentation on both seen and unseen objects.

As the pioneering approach of this task, Mask^X R-CNN [10] designs a weight transfer function that transfers the parameters of each object category from the bounding box head to the partially supervised mask head. However, current mainstream methods utilizes class-agnostic mask head to separate foreground and background within each region of interest (RoI), instead of predicting a mask for per category. Therefore, the key to addressing this problem is to provide class-agnostic mask head with RoIs that has more distinctive foreground and background. ContrastMask [18] introduces an extra contrastive learning head to tackles this issue. In ContrastMask, it first generates queries and keys from foreground/background regions by utilizing the annotated masks of seen categories and pseudo masks of unseen categories as region priors. Specifically, in each mini-batch, it obtains a foreground/background query by averaging the features of all foreground/background regions, and meanwhile generates keys by random sampling from these regions. Then, it applies contrastive loss to pull together query and keys from foreground and push apart with background keys, and vice versa. Compared to other methods, ContrastMask achieves competitive segmentation performance by fully leveraging the information from both seen categories and unseen categories. However, as shown in Fig. 1, when an image contains multiple overlapping instances, ContrastMask often gets confused: 1) for large instances, it habitually produces defective masks, particularly for unseen categories; 2) it has difficulty in distinguishing foreground and background as well as differentiating between different instances. The reasons for this issue are that the RoIs lack contextual information from larger receptive fields and the queries in contrastive learning are difficult to represent the characteristic of the foreground/background because of the fact that they all come from the current batch.

To address these problems, in CCMask, we design a Context Feature Pyramid Network (Context-FPN) and Memory Contrastive Learning Head to collaboratively improve the learning capability of the class-agnostic mask segmentation model. In Context-FPN, we capture rich contextual information from different large receptive fields by using a module consist of multi-path dilated convolutional layers with different dilation rates, named as context extraction module (CEM). Then, we employ Content-Aware ReAssembly of FEatures (CARAFE) [16] instead of traditional upsampling operators to reduce information loss during the top-down pathway of the FPN [12]. By employing Context-FPN, we aggregate more discriminative features, facilitating learning of the following task heads. In the MCL Head, we maintain two memory query queues (QMQ) to store queries from the foreground and the background regions: the queries from the latest batch are enqueued, and the oldest are dequeued. The size of the queue is independent of the batch size, allowing them to be large. This enhances the diversity and representation of queries and enables our class-agnostic segmentation model to better learn the general concept of foreground and background. We perform comprehensive experiments to evaluate the effectiveness of our CCMask, and the results demonstrate that our approach achieves state-of-the-art performance on COCO dataset [13]. In summary, our CCMask main contributions are the following:

- We design a context feature pyramid network (Context-FPN), which captures abundant contextual information from various large receptive fields and compensate for the intrinsic defect in feature upsampling of FPN. It produces more reliable feature representations for the MCL Head and significantly improves the segmentation performance of our CCMask on large foregound objects.
- We maintain two query memory queues (QMQ) in our memory contrastive learning head (MCL Head), which are used to store the foreground/background queries from recent batches. It enhances the ability of the MCL Head to separate foreground from background.
- CCMask significantly surpasses all previous SOTA partially-supervised instances segmentation methods. Concretely, our method achieves 39.5 mAP for mask segmentation in the *nonvoc* $\rightarrow$ *voc* setting with the ResNet-50 [8] as backbone. Furthermore, we conduct extensive ablation experiments to analyze the impact of each of our contributions and the results prove that each of them is effective.

2 Related Work

Contrastive Learning. By introducing contrastive learning, various computer vision tasks such as image generation, image classification, semantic segmentation, and instance segmentation can be improved. Park et al. [15] proposes patch-wise contrastive learning to maximize the mutual information between the same location of input and generated images. By incorporating both pixel-pixel and pixel-region contrastive calculations, Wang et al. [17] fully exploits the semantic similarity between annotated pixels in the task of semantic segmentation. Yang et al. [20] proposes a contrastive learning-based few framework that integrates contrastive learning into both pretraining and meta-training stages to improve the few-shot learning image classification. C^2AM[19] proposes cross-image foreground-background contrastive learning for class-agnostic activation maps generation using unlabeled image data.

Feature Pyramid. Feature pyramid is a method used to deal with multi-scale problems and widely applied in object detection, instance segmentation and other fields. FPN [12] builds a feature pyramid through lateral connections and top-down pathway. PAFPN in PANet [14] improves its performance by adding an additional bottom-up pathway on FPN. Nas-FPN [5] use neural architecture search to discover a new and powerful feature pyramid structure. Based on FPN, FaPN [11] incorporates a feature alignment module and a feature selection module to generate multi-scale features for dense image prediction. A^2-FPN [9] improve multi-scale feature learning through attention-guided feature aggregation. Our Context-FPN aims to obtain contextual information from larger receptive fields.

Partially Supervised Instance Segmentation. Instance segmentation is one of the research hotspots in the computer vision, which aims to segment all object instances in an image. Mask R-CNN [7] extends Faster R-CNN [6] with a fully convolutional network branch for generating segmentation masks. Due to its excellent performance and extensibility, Mask R-CNN [7] has become an important benchmark model. In partially supervised instance segmentation, seen categories have both box and mask annotations while unseen categories only have box supervision. Due to the requirement of extensive mask annotations in instance segmentation tasks, partially supervised instance segmentation tasks have been proposed. $Mask^X$ R-CNN [10] addresses this problem by learning a weight transfer function that maps bounding box weights to mask weights. ShapeProp [22] learns salient regions from bounding box head and propagates them into an intermediate shape representation, which is a more accurate shape prior. CPMask [4] captures the underlying commonalities, including shape commonalities from seen categories and appearance commonalities from all categories, and generalizes them to unseen categories. OPMask [1] uses the object mask prior from bounding box head to help the class-agnostic mask head focus on foreground in each RoI. ContrastMask [18] achieves promising performance by integrating a contrastive learning module, which learns on both seen and unseen categories.

3 CCMask

In this section, we first introduce the overall framework of CCMask. Then, we elaborate the design details of the context feature pyramid network and the memory contrastive learning head. Finally, we explain the loss function of our segmentation model.

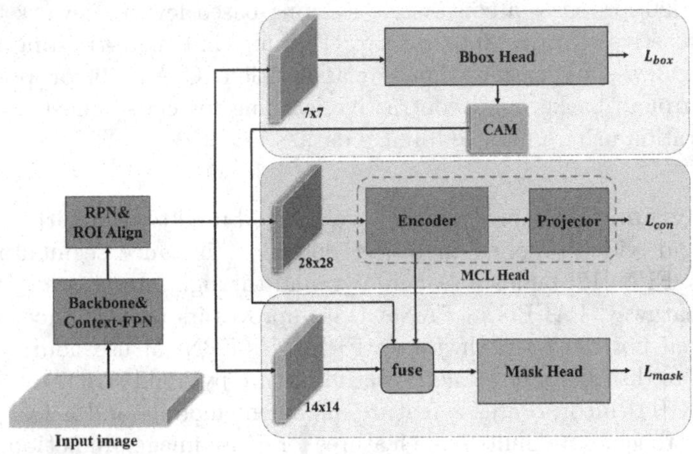

Fig. 2. The whole architecture of CCMask.

3.1 Overview

As illustrated in Fig. 2, our CCMask is built upon the architecture of Mask R-CNN [7], it replaces the original FPN [12] with the Context-FPN and adds an extra MCL Head. The Context-FPN can provide more discriminative features for subsequent heads. The MCL Head takes an RoI feature map as input and outputs an enhanced feature map. Finally, we mix the RoI feature map, Class Activation Map (CAM) [21] and the enhanced feature map from MCL Head as the input of the class-agnostic mask head, which can help the mask head predict a segmentation map. Next, we will depict each component of our model.

3.2 Context-FPN

FPN [12] is a classic framework designed to learn multi-scale feature representations. However, intrinsic defects in feature extraction and fusion inhibit FPN from further aggregating more discriminative features. To tackle these limitations in FPN, our Context-FPN introduces an additional context extraction module to extract more contextual information and employs CARAFE [16] as upsampling operator to mitigate the information loss caused by traditional upsampling operation during the feature fusion stage. These two approaches effectively address the shortcomings of FPN and enable it to generate more discriminative feature maps. The framework of our Context-FPN is illustrated in Fig. 3.

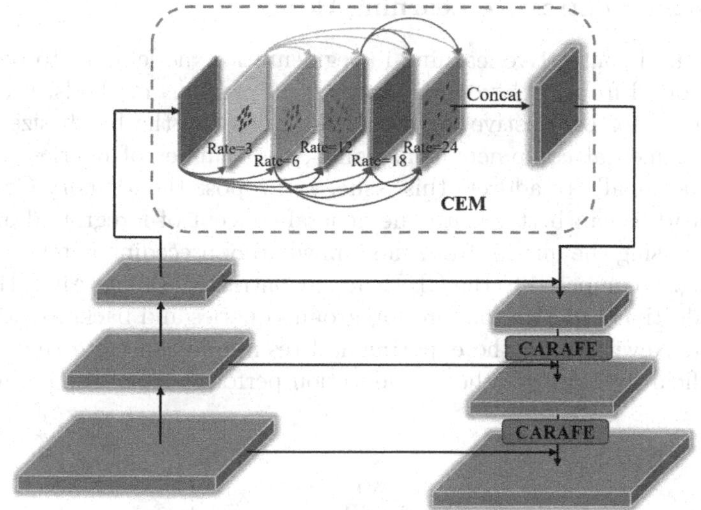

Fig. 3. The structure of our Context FPN.

Context Extract Module. CEM simply contains several convolutional layers and its input is the output of the last layer in the bottom-up pathway of FPN. Specifically, as shown in Fig. 3, CEM consists of multi-path dilated convolutional layers with different dilation rates: 3, 6, 12, 18, and 24 in our study. These convolutional layers can capture contextual information of various large receptive fields. Moreover, in order to enhance the perception of foreground objects boundary, these dilated convolution layers also incorporate deformable convolutions. In addition, to effectively fuse multi-scale information, we employ dense connections within the CEM, where the output of each dilated convolutional layer is concatenated with the input feature maps and fed into the next dilated convolutional layer.

Content-Aware RAassembly of FEatures. CARAFE is an effectively upsampling algorithm. On each location, CARAFE firstly predicts multiple reassembly kernels in a content-aware manner, and then reassembles the features inside a predefined nearby region via a weighted combination. Feature upsampling is then accomplished by rearranging the generated features as a spatial block. Compared to the traditional bilinear interpolation, CARAFE can aggregate more contextual information within a large receptive field. CARAFE also outperforms deconvolution as it can dynamically generate reassembly kernels based on local content, rather than applying the same kernel across the entire image.

3.3 Memory Contrastive Learning Head

The inclusion of contrastive learning in segmentation model aims to better separate foreground from background. However, in previous methods, the number of queries used for contrastive learning was limited by the batch size. Clearly, due to experimental equipment constraints, the number of queries per training batch was small. To address this issue, we propose the Memory Contrastive Learning head, it can better learn the general concept of foreground and background by reusing the queries from the immediately preceding batches. As illustrated in Fig. 4, compared to the CL Head of ContrastMask, our MCL Head only adds two additional queues to store foreground queries and background queries, respectively. Nevertheless, the experimental results demonstrate that our approach significantly improves the segmentation performance of the segmentation model.

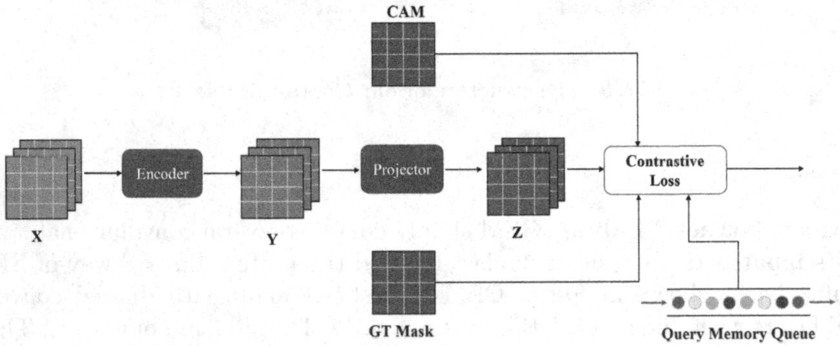

Fig. 4. The flowchart of our MCL Head, which includes an additional QMQ compared to the CL Head of ContrastMask.

Query Memory Queue. The introduction of QMQ decouples the number of queries from the batch size. Our queue size can be much larger than the batch size and can be flexibly set as a hyper-parameter. The queries in the queue are continuously updated, with query from the current mini-batch enqueued and query from the oldest mini-batch dequeued. Moreover, removing the query from the oldest mini-batch can be beneficial as they are the most outdated.

Memorial Queries Pixel-Region Contrastive Loss. The core design philosophy of our memorial queries pixel-region contrastive loss function is to involve the queries from both QMQ and current batch in the contrastive learning, thereby facilitating the learning of the MCL Head. In each mini-batch, we utilize the Ground-truth masks of seen categories and CAM [21] of unseen categories as region priors to indicate foreground and background separation. We get a foreground query by averaging the sum of the features of all foreground regions, and denoted as q^+. Similarly, we can obtain a q^- for the background regions. Then, we perform random sampling within these regions to obtain a set of foreground keys and a set of background keys, denoted as $\mathcal{K}^+$ and $\mathcal{K}^-$. We put the queries from current batch into the QMQ to replace the oldest queries. The length of the QMQ is N, and the number of q^+/q^- in a batch is n. The memorial queries pixel-region contrastive loss consists of two symmetrical formulations for foreground and background, and is defined as follows:

$$L_{con} = L^{q^+}_{\mathcal{K}^+,\mathcal{K}^-} + L^{q^-}_{\mathcal{K}^-,\mathcal{K}^+}, \tag{1}$$

$$\begin{aligned} L^{q^+}_{\mathcal{K}^+,\mathcal{K}^-} = -\frac{1}{N}\frac{1}{|\mathcal{K}^+|}\sum_{i=1}^{N}\sum_{k^+\in\mathcal{K}^+}\Big[sim(q_i^+,k^+)/\tau \\ -log\big(exp(sim(q_i^+,k^+)/\tau) + \sum_{k^-\in\mathcal{K}^-} exp(sim(q_i^+,k^-)/\tau)\big)\Big]. \end{aligned} \tag{2}$$

We describe in detail the loss function for foreground in Eq. (2), where $sim(\cdot,\cdot)$ denotes the cosine similarity and the τ is a temperature hyper-parameter. Similarly, we can obtain the loss function for background $L^{q}_{\mathcal{K}^+,\mathcal{K}^-}$.

3.4 Loss Function

The overall loss function of our CCMask as follows:

$$L = L_{box} + L_{mask} + \lambda L_{con}, \tag{3}$$

where the box detection loss L_{box} and the mask loss L_{mask} are inherited from Mask R-CNN. The contrastive loss L_{con} has been elaborately introduced in Eq. (1), and the λ is a weight parameter.

4 Experiments

4.1 Experimental Setup

We conduct our experiments on the COCO dataset [13]. In order to train on partially supervised setting, we split the 80 COCO clsasses into *voc* and *nonvoc* category subsets where "*voc*" categories are the 20 classes of the PASCAL VOC dataset [3] while "*nonvoc*" include the remaining 60 classes. Each time we select on subset as seen categories and the other subset as unseen categories. We train our model on COCO-train2017 and test on COCO-val2017. We adopt the typical metrics for instance segmentation to evaluate our model, including mAP, AP_{50}, AP_{75}, AP_S, AP_M and AP_L. These metrics are calculated on the unseen categories.

Implementation Details. We implement all experiments based on MMDetection [2]. All experiments are conducted with a batchsize of 8 over 4 RTX 3090 GPUs for 12 epochs. We use the ResNet [8] as backbone. The size of QMQ is 120. We linearly warmup the λ of L_{con} (Eq. (3)) from 0.25 to 1.0. Besides, the hyper-parameters in CARAFE [16] follow the default setting.

4.2 Experimental Results

In this subsection, we compare our CCMask to other recent methods for partially supervised instance segmentation.

Quantitative Results. The quantitative results on partially supervised setting are shown in Table 1. When using ResNet-50 as the backbone, our CCMask achieves new state-of-the-art performance of 39.5/34.7 mAP in the *nonvoc* $\rightarrow$ *voc* and *voc* $\rightarrow$ *nonvoc* settings, it even outperforms the previous models that uses ResNet-101 [8] as the backbone. Similarly, our method also achieves excellent performance adopting the ResNet-101 as the backbone, it outperforms OPMask [1] (the state-of-the-art method) by mAP in *nonvoc* $\rightarrow$ *voc* and *voc* $\rightarrow$ *nonvoc* settings, respectively.

Qualitative Results. To further demonstrate the contributions of these two improvement strategies, we visualize the segmentation results of our model in various scenarios: with and without Context-FPN, with and without MCL Head. As shown in Fig. 5, the results indicate that our two strategies can improve the segmentation performance of the model on both visible and invisible classes from different aspects.

4.3 Ablation Study

We conduct ablation experiments to verify the effects of main components in CCMask. The backbone network is ResNet-50. All experiments are conducted in the *nonvoc* $\rightarrow$ *voc* setting, and the results are evaluated on unseen categories.

Effectiveness of Context-FPN. The results are shown in Table 2. In our CCMask, by using CEM to capture contextual information from different large receptive fields, the performance is increased by 1.9 mAP compared with "CCMask *w/o* CEM". In addition, we performs another experiment, "CCMask *w/o* CARAFE", and the results also demonstrates the effectiveness of CARAFE.

Effectiveness of MCL Head. We next validate the design of our MCL Head. As shown in Table 2, the performance of CCMask significantly degrades by 5.2 mAP when removing the MCL Head. We then conduct a variant, "CCMask *w/o* QMQ": only use queries from current batch, it gets 37.0 mAP. This evidences the effectiveness of our MCL Head and the necessity of leveraging more queries during contrastive learning.

Table 1. Quantitative comparisons on COCO dataset. "*nonvoc* → *voc*" means that the categories in *nonvoc* is seen categories while the categories in *voc* is unseen categories.

Method	Backbone	*nonvoc* → *voc*						*voc* → *nonvoc*					
		mAP	AP_{50}	AP_{75}	AP_S	AP_M	AP_L	mAP	AP_{50}	AP_{75}	AP_S	AP_M	AP_L
Mask R-CNN [7]	ResNet-50 [8]	23.9	42.9	23.5	11.6	24.3	33.7	19.2	36.4	18.4	11.5	23.3	24.4
$Mask^X$ R-CNN [10]		28.9	52.2	28.6	12.1	29.0	40.6	23.7	43.1	23.5	12.4	27.6	32.9
CPMask [4]		–	–	–	–	–	–	28.8	46.1	30.6	12.4	33.1	43.4
ShapeProp [22]		34.4	59.6	35.2	13.5	32.9	48.6	30.4	51.2	31.8	14.3	34.2	44.7
OPMask [1]		36.5	62.5	37.4	17.3	34.8	49.8	31.9	52.2	33.7	16.3	35.2	46.5
ContrastMask [18]		35.1	60.8	35.7	17.2	34.7	47.7	30.9	50.3	32.9	15.2	34.6	44.3
CCMask		**39.5**	**64.0**	**41.7**	**20.3**	**39.3**	**51.7**	**34.7**	**55.7**	**36.8**	**18.3**	**38.7**	**50.3**
Mask R-CNN [7]	ResNet-101 [8]	24.7	43.5	24.9	11.4	25.7	35.1	18.5	34.8	18.1	11.3	23.4	21.7
$Mask^X$ R-CNN [10]		29.5	52.4	29.7	13.4	30.2	41.0	23.8	42.9	23.5	12.7	28.1	33.5
CPMask [4]		36.8	60.5	38.6	17.6	37.1	51.5	34.0	53.7	36.5	18.5	38.9	47.4
ShapeProp [22]		35.5	60.5	36.7	15.6	33.8	50.3	31.9	52.1	33.7	14.2	35.9	46.5
OPMask [1]		37.1	62.5	38.4	16.9	36.0	50.5	33.2	53.5	35.2	17.2	37.1	46.9
ContrastMask [18]		36.6	62.2	37.7	17.5	36.5	50.1	32.4	52.1	34.8	15.2	36.7	47.3
CCMask		**41.5**	**66.4**	**43.6**	**22.1**	**41.8**	**54.2**	**36.4**	**57.0**	**38.3**	**19.8**	**41.0**	**52.6**

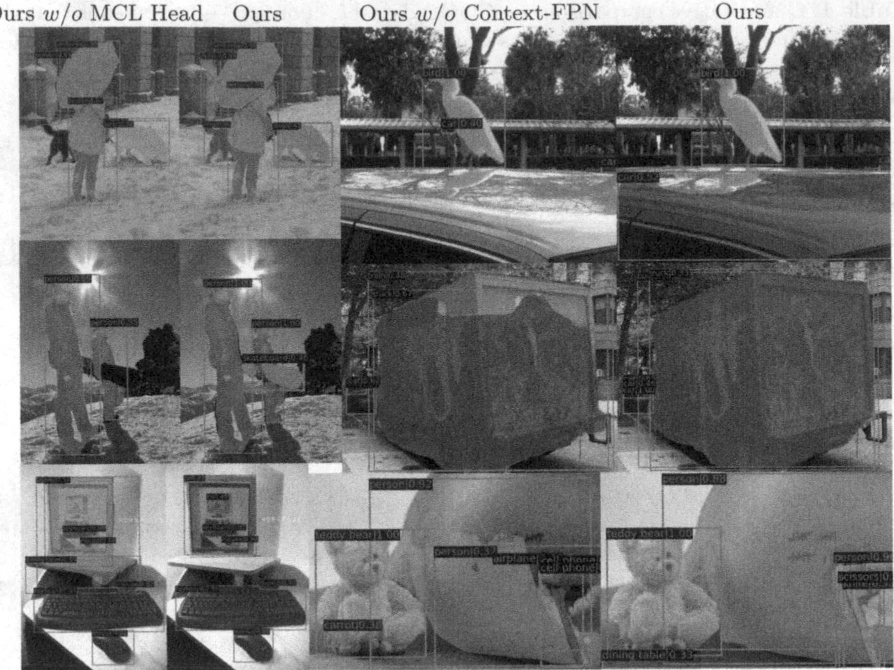

Fig. 5. Qualitative results on COCO dataset in *nonvoc* → *voc* setting. The results show that MCL Head can improve the model's ability to distinguish foreground and background, as well as different instances, and Context-FPN can improve the model's segmentation performance on large objects.

Table 2. Ablation on the impact of each component.

Method	*nonvoc* → *voc*					
	mAP	AP_{50}	AP_{75}	AP_S	AP_M	AP_L
CCMask	39.5	64.0	41.7	20.3	39.5	51.7
CCMask *w/o* CEM	37.6	61.6	39.5	19.4	37.8	49.4
CCMask *w/o* CARAFE	39.2	63.8	41.4	18.9	39.1	51.4
CCMask *w/o* MCL Head	34.3	60.7	33.9	17.3	33.7	47.1
CCMask *w/o* QMQ	37.0	60.9	38.6	19.1	36.4	48.7

5 Conclusion

We propose a novel approach CCMask for partially supervised instance segmentation. CCMask first utilizes Context-FPN to obtain feature maps with rich contextual information from different receptive fields. Then, it employs the MCL head to enhance the discrimination between foreground and background within

ROIs. Benefiting from these two effective strategies, our model achieves state-of-the-art results on the COCO dataset under the partially supervised setting.

References

1. Biertimpel, D., Shkodrani, S., Baslamisli, A.S., Baka, N.: Prior to segment: foreground cues for weakly annotated classes in partially supervised instance segmentation. In: Proceedings of the IEEE/CVF International Conference on Computer Vision, pp. 2824–2833 (2021)
2. Chen, K., Wang, J., Pang, J., et al.: MMDetection: Open MMLab detection toolbox and benchmark. arXiv preprint arXiv:1906.07155 (2019)
3. Everingham, M., Van Gool, L., Williams, C.K., Winn, J., Zisserman, A.: The Pascal visual object classes (VOC) challenge. Int. J. Comput. Vision **88**, 303–338 (2010)
4. Fan, Q., Ke, L., Pei, W., Tang, C.-K., Tai, Y.-W.: Commonality-parsing network across shape and appearance for partially supervised instance segmentation. In: Vedaldi, A., Bischof, H., Brox, T., Frahm, J.-M. (eds.) Computer Vision–ECCV 2020: 16th European Conference, Glasgow, UK, 23–28 August 2020, Proceedings, Part VIII 16. LNCS, vol. 12353, pp. 379–396. Springer, Cham (2020). https://doi.org/10.1007/978-3-030-58598-3_23
5. Ghiasi, G., Lin, T.Y., Le, Q.V.: NAS-FPN: learning scalable feature pyramid architecture for object detection. In: Proceedings of the IEEE/CVF Conference on Computer Vision and Pattern Recognition, pp. 7036–7045 (2019)
6. Girshick, R.: Fast R-CNN. In: Proceedings of the IEEE International Conference on Computer Vision, pp. 1440–1448 (2015)
7. He, K., Gkioxari, G., Dollár, P., Girshick, R.: Mask R-CNN. In: Proceedings of the IEEE International Conference on Computer Vision, pp. 2961–2969 (2017)
8. He, K., Zhang, X., Ren, S., Sun, J.: Deep residual learning for image recognition. In: Proceedings of the IEEE Conference on Computer Vision and Pattern Recognition, pp. 770–778 (2016)
9. Hu, M., Li, Y., Fang, L., Wang, S.: A2-FPN: attention aggregation based feature pyramid network for instance segmentation. In: Proceedings of the IEEE/CVF Conference on Computer Vision and Pattern Recognition, pp. 15343–15352 (2021)
10. Hu, R., Dollár, P., He, K., Darrell, T., Girshick, R.: Learning to segment every thing. In: Proceedings of the IEEE Conference on Computer Vision and Pattern Recognition, pp. 4233–4241 (2018)
11. Huang, S., Lu, Z., Cheng, R., He, C.: FAPN: feature-aligned pyramid network for dense image prediction. In: Proceedings of the IEEE/CVF International Conference on Computer Vision, pp. 864–873 (2021)
12. Lin, T.Y., Dollár, P., Girshick, R., He, K., Hariharan, B., Belongie, S.: Feature pyramid networks for object detection. In: Proceedings of the IEEE Conference on Computer Vision and Pattern Recognition, pp. 2117–2125 (2017)
13. Lin, T.Y., et al.: Microsoft COCO: common objects in context. In: Fleet, D., Pajdla, T., Schiele, B., Tuytelaars, T. (eds.) Computer Vision-ECCV 2014: 13th European Conference, Zurich, Switzerland, 6–12 September 2014, Proceedings, Part V 13, pp. 740–755. Springer, Cham (2014). https://doi.org/10.1007/978-3-319-10602-1_48
14. Liu, S., Qi, L., Qin, H., Shi, J., Jia, J.: Path aggregation network for instance segmentation. In: Proceedings of the IEEE Conference on Computer Vision and Pattern Recognition, pp. 8759–8768 (2018)

15. Park, T., Efros, A.A., Zhang, R., Zhu, J.Y.: Contrastive learning for unpaired image-to-image translation. In: Vedaldi, A., Bischof, H., Brox, T., Frahm, J.M. (eds.) Computer Vision-ECCV 2020: 16th European Conference, Glasgow, UK, 23–28 August 2020, Proceedings, Part IX 16. LNCS, vol. 12354, pp. 319–345. Springer, Cham (2020). https://doi.org/10.1007/978-3-030-58545-7_19
16. Wang, J., Chen, K., Xu, R., Liu, Z., Loy, C.C., Lin, D.: CARAFE: content-aware reassembly of features. In: Proceedings of the IEEE/CVF International Conference on Computer Vision, pp. 3007–3016 (2019)
17. Wang, W., Zhou, T., Yu, F., Dai, J., Konukoglu, E., Van Gool, L.: Exploring cross-image pixel contrast for semantic segmentation. In: Proceedings of the IEEE/CVF International Conference on Computer Vision, pp. 7303–7313 (2021)
18. Wang, X., Zhao, K., Zhang, R., Ding, S., Wang, Y., Shen, W.: ContrastMask: contrastive learning to segment every thing. In: Proceedings of the IEEE/CVF Conference on Computer Vision and Pattern Recognition, pp. 11604–11613 (2022)
19. Xie, J., Xiang, J., Chen, J., Hou, X., Zhao, X., Shen, L.: Contrastive learning of class-agnostic activation map for weakly supervised object localization and semantic segmentation. arXiv preprint arXiv:2203.13505 (2022)
20. Yang, Z., Wang, J., Zhu, Y.: Few-shot classification with contrastive learning. In: Computer Vision - ECCV 2022–17th European Conference, Tel Aviv, Israel, 23–27 October 2022, Proceedings, Part XX. LNCS, vol. 13680, pp. 293–309. Springer, Cham (2022). https://doi.org/10.1007/978-3-031-20044-1_17
21. Zhou, B., Khosla, A., Lapedriza, A., Oliva, A., Torralba, A.: Learning deep features for discriminative localization. In: Proceedings of the IEEE Conference on Computer Vision and Pattern Recognition, pp. 2921–2929 (2016)
22. Zhou, Y., Wang, X., Jiao, J., Darrell, T., Yu, F.: Learning saliency propagation for semi-supervised instance segmentation. In: Proceedings of the IEEE/CVF Conference on Computer Vision and Pattern Recognition, pp. 10307–10316 (2020)

A Dynamic Tracking Framework Based on Scene Perception

Jinpu Zhang, Ziwen Li, and Yuehuan Wang(✉)

School of Artificial Intelligence and Automation,
Huazhong University of Science and Technology, Wuhan, China
yuehwang@hust.edu.cn

Abstract. While recent large models have greatly improved tracking performance, not all scenes require a large and complex network. Dynamic networks can adapt the architecture to different inputs, leading to notable accuracy and computational efficiency. However, existing dynamic architectures and decision mechanisms designed for classification are not applicable to the tracking task. This paper proposes a dynamic tracking framework based on scene perception, named DynamicTrack. We classify tracking scenes into easy and hard categories, and propose a dynamic architecture with an easy-hard dual-branch to handle different scenes respectively. Unlike previous works in classification that selectively prune a subset of the backbone, complete execution of the entire backbone is necessary for tracking. Hence, we maintain two complete transformer backbones for the dual branches and vary the number of input tokens to achieve modeling at different granularities. Then, we propose a scene router that automatically selects the optimal branch for each input frame. The router directly assesses the scene complexity of features extracted by the easy branch for decision-making, without relying on the tracking head output. This enhances decision efficiency during dynamic inference. Moreover, we introduce two techniques that benefit DynamicTrack optimization, namely, the Gumbel-Softmax trick and cross-branch transmission (CBT). The former increases the stochasticity of decisions and prevents mode collapse into trivial solutions. The latter establishes information transmission between the two branches, facilitating discriminative power and learning efficiency. Extensive experiments on four benchmarks demonstrate that the proposed DynamicTrack achieves SOTA performance and accuracy-speed trade-offs.

Keywords: Object tracking · Dynamic network · Scene router

1 Introduction

Visual object tracking is a fundamental task in computer vision. It aims to estimate the position and shape of a given target in a video sequence. The continuous and arbitrary changes in targets and scenes pose challenges to learning an effective and efficient tracking model. Current mainstream trackers, such as

© The Author(s), under exclusive license to Springer Nature Singapore Pte Ltd. 2024
Q. Liu et al. (Eds.): PRCV 2023, LNCS 14436, pp. 185–197, 2024.
https://doi.org/10.1007/978-981-99-8555-5_15

Fig. 1. Examples of "easy" and "hard" images for the object tracking task.

Siamese-based [1,9,16,34] methods and transformer-based methods [4,6,29,30], have achieved remarkable success. However, these trackers aim to use a fixed feed-forward structure, i.e., static network, to generalize all scenes. The static network requires the same computational cost for all inputs during inference, potentially leading to redundancy in processing easy scenes and insufficiency in hard scenes. This static inference paradigm limits the model's representation power, efficiency, and interpretability.

Biological vision researches [14,22] suggest that humans can rapidly locate targets on the left "easy" images, while need more time to search targets on the right "hard" images, as shown in Fig. 1. Easy images usually refer to scenes with singular composition, simple texture and strong target-to-background contrast. Hard images have opposite characteristics and may be accompanied with distractors. The human brain processes images of different complexities at varied speeds. This property motivates the dynamic tracking network, which adaptively selects the inference architecture based on the scene complexity.

Dynamic networks have been extensively studied in the classification task. A common paradigm is to automatically select a subset of a multi-stage backbone using early exiting [23] or layer skipping [24,27] mechanisms. However, the tracking task depends on multi-stage features to model targets of arbitrary categories and scales. Discarding a portion of features would lead to a significant performance degradation [5,16]. Thus, it is necessary to completely execute the entire backbone, and existing dynamic architectures are not applicable to the tracking task. Furthermore, dynamic classification networks usually add a head network, typically a fully connected layer classifier, at the intermediate layer for adaptive decision-making [12,23]. While the tracking head includes matching, classification and regression subnetworks. Incorporating a tracking head for decisions brings large computational redundancy.

To address the above issues, we propose a dynamic tracking framework based on scene perception, named DynamicTrack. First, we design a dynamic architecture with an easy-hard dual-branch to handle different scenes respectively. The dual branch contains two complete transformer trackers with the same structure but different number of input tokens. Representing the image as less tokens is sufficient for many easy scenes and enjoys high computational efficiency. While increasing input tokens enables a more fine-grained representation, which can adapt to hard scenes but incurs a higher computational burden. To balance

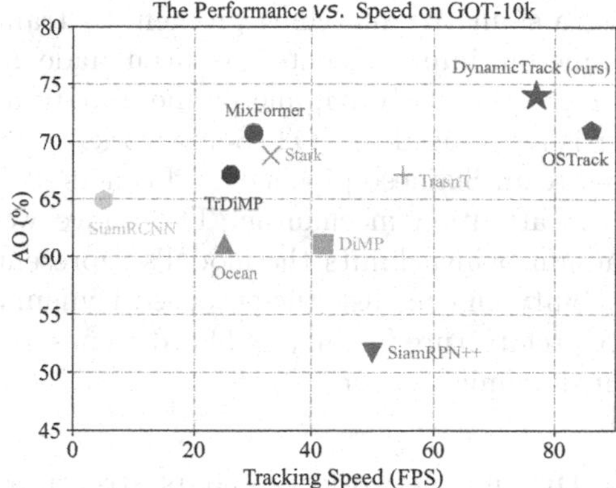

Fig. 2. A comparison of AO and speed of SOTA trackers on GOT-10k test set. Our DynamicTrack achieves SOTA performance and accuracy-speed trade-offs (74% AO, 77FPS).

accuracy and efficiency, we design a scene router that adaptively selects the optimal inference route for each input frame. The router directly assesses the scene complexity of features extracted by the easy branch to determine whether to continue with the easy branch or switch to the hard branch. This process avoids using the tracking head and significantly improves decision efficiency during dynamic inference. Finally, we introduce two techniques advantageous to DynamicTrack optimization. 1) We employ the Gumbel-Softmax trick [15] to increase the stochasticity of decisions. It prevents mode collapse into trivial solutions, i.e., always selecting the more accurate hard branch. 2) We design a cross-branch transmission (CBT) module to transmit the large receptive field context from the easy branch to the hard branch. It facilitates the discriminative power and learning efficiency of the hard branch.

The main contributions of this work are as follows:

- We propose a dynamic tracking framework based on scene perception, named DynamicTrack. It customizes an easy-hard dual-branch network to handle different scenes respectively.
- We propose a scene router to perceive the complexity of tracking scenes and achieve adaptive decision-making. Moreover, we introduce the Gumbel-Softmax trick and a CBT module to facilitate optimization.
- Extensive experiments on four benchmarks prove SOTA performance and accuracy-speed trade-offs of our method.

2 Related Work

Visual Object Tracking. In the early development of VOT, DCF-based methods [2,7,11] are dominant trackers due to their favorable ability in modeling target appearance variation. With the development of deep learning, SiamFC [1] formulates tracking as a similarity matching problem by training on large-scale image pairs. Later, numerous improvements have been made, including backbone design [16,33], scale regression [9,16,34] and online update mechanism [10,31]. Recently, transformer is introduced to VOT. Some works [4,26,29] embed transformers into the two-stream Siamese pipeline, and others [3,6,30] adopt a one-stream pipeline by the attention mechanism. The above works all adopt the static inference paradigm, which limits the model's representation power, efficiency, and interpretability. In contrast, the proposed DynamicTrack customizes the dynamic inference architecture for easy and hard scenes respectively, yielding a unified and efficient dynamic tracker.

Dynamic Network Dynamic networks can adapt structures or parameters to the input during inference. Branchynet [23] introduces the early exiting strategy, allowing easy samples to be output at shallow exits without executing deeper layers. MSDNet [12] adopts a multi-scale architecture and dense connections to improve the joint optimization of multiple classifiers. SkipNet [27] and Conv-AIG [24] propose a layer skipping strategy. The network depth can be adapted on the fly by skipping the calculation of intermediate layers. DVT [28] cascades multiple transformers with increasing numbers of tokens and activates them sequentially to achieve dynamic inference. However, these methods are all designed specifically for the classification task and are not applicable to object tracking.

DynamicDet [18] proposes a dynamic architecture for object detection, including two identical detectors and a router. Our method shares a similar spirit. Differently, we customize two trackers with different number of input tokens, motivated by the redundancy of existing models in numerous easy scenes. Moreover, our decision mechanism and optimization strategies are more suitable for the tracking task and can flexibly generalize multiple tracking benchmarks.

3 Method

This section presents the proposed DynamicTrack. As shown in Fig. 3(a), DynamicTrack consists of an easy-hard dual-branch network and a scene router. The easy-hard dual-branch network cascades two VIT-based trackers OSTrack [30] with increasing number of tokens of the template and search region. The template and search image pairs are first split into a small number of tokens and fed into the easy branch encoder. Then the router determines the inference route based on the output search region tokens. If the scene is classified as "Easy", the router selects the tracker head of the easy branch for prediction. Otherwise the router switches to the hard branch, where the original image pairs are split into more tokens and processed by the encoder with CBT and the corresponding tracker head. CBT transmits the priors of the easy branch to the hard branch to facilitate learning. In the following section, we depict the proposed easy-hard dual-branch network and the scene router in detail.

3.1 Easy-Hard Dual-Branch Network

Token Representation. The input is a pair of template $z \in \mathbb{R}^{3\times H_z \times W_z}$ and search region $x \in \mathbb{R}^{3\times H_x \times W_x}$. They are first split and flattened into patch sequences $z_p \in \mathbb{R}^{N_z \times (3 \cdot P^2)}$ and $x_p \in \mathbb{R}^{N_x \times (3 \cdot P^2)}$, where P is the patch size and $N_z = H_z W_z / P^2$, $N_x = H_x W_x / P^2$ are the number of patches. A linear projection maps z_p and x_p to token embeddings. Learnable position embeddings are then added to the token embeddings. Finally, the template tokens $\boldsymbol{Z}$ and search tokens $\boldsymbol{X}$ are concatenated as $\boldsymbol{H}^0 \in \mathbb{R}^{N \times D}$, where $N = N_z + N_x$ is the number of tokens and D is the token dimension. The concatenated $\boldsymbol{H}^0$ are fed into the encoder with L layers.

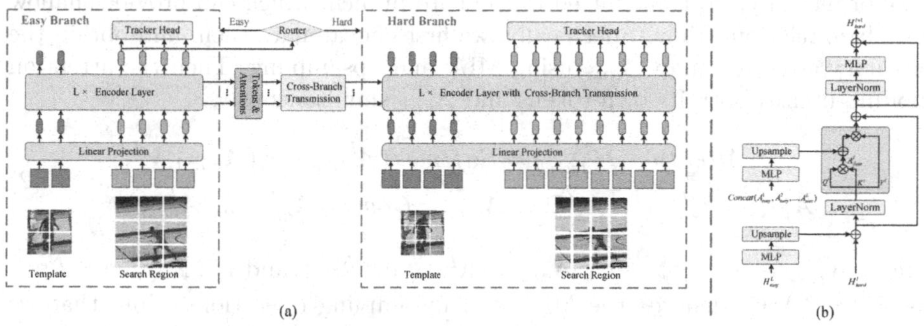

Fig. 3. (a) An Overview of our DynamicTrack. The template and search image pairs are first split into a small number of tokens and fed into the easy branch encoder. Then the router determines the inference route based on the output search region tokens. Easy images are directly output through the tracker head, while hard images are split into more tokens and enter the hard branch. (b) The structure of the encoder with CBT in the hard branch.

For the easy branch, we set the patch size $P_{easy} = 32$ and denote the input tokens as $\boldsymbol{H}^0_{easy} \in \mathbb{R}^{N_{easy} \times D}$. For the hard branch, we set the patch size $P_{hard} = 16$ and denote the input tokens as $\boldsymbol{H}^0_{hard} \in \mathbb{R}^{N_{hard} \times D}$, $N_{hard} = 4 \cdot N_{easy}$. Since the computational cost grows quadratically with respect to the token number, the easy branch is highly efficient in processing easy images. While the hard branch represents the input as more tokens, achieving a more fine-grained representation of hard images.

Encoder with Cross-Branch Transmission. Since the two branches have the same structure and training objective, the hard branch can use the tokens and relations in the easy branch as priors to improve learning efficiency. Moreover, the easy branch has a larger receptive field, which can provide global context to facilitate discriminative power. Thus we propose the cross-branch transmission (CBT) module to transmit these priors from easy branch to hard branch.

The structure of encoder with CBT is shown in Fig. 3(b). For token transmission, we leverage the tokens $\boldsymbol{H}_{easy}^{L}$ output by the final encoder layer in the easy branch as it has the most discriminative power. We use a MLP and upsampling to align $\boldsymbol{H}_{easy}^{L}$ with the input tokens $\boldsymbol{H}_{hard}^{l}$ in each layer of the hard branch, then add them by element,

$$\boldsymbol{H}_{fuse}^{l} = f_l(\boldsymbol{H}_{easy}^{l}) + \boldsymbol{H}_{hard}^{l}, \qquad l \in \{1, ..., L-1\} \tag{1}$$

where $f_l : \mathbb{R}^{N_{easy} \times D} \rightarrow \mathbb{R}^{N_{hard} \times D}$ denotes the MLP and upsampling operations. Here we upsample the tokens of the template and the search region separately. Given the fused tokens $\boldsymbol{H}_{fuse}^{l}$, query $\boldsymbol{Q}^{l}$, key $\boldsymbol{K}^{l}$ and value $\boldsymbol{V}^{l}$ are generated to calculate the attention. The multi-level attention maps $\boldsymbol{A}_{easy}^{1}, ..., \boldsymbol{A}_{easy}^{L}$ in the easy branch are also transmitted to the hard branch, which can provide shallow and deep relation priors. Specifically, we first concatenate them and obtain the auxiliary attention map $\boldsymbol{A}_{aux}$ using MLP and upsampling. Then the attention module utilizes both its own tokens and $\boldsymbol{A}_{aux}$ simultaneously,

$$\begin{aligned} Attention(\boldsymbol{H}_{fuse}^{l}) &= Softmax(\boldsymbol{A}_{fuse}^{l} + r_l(\boldsymbol{A}_{aux}))\boldsymbol{V}^{l}, \\ \boldsymbol{A}_{fuse}^{l} = \boldsymbol{Q}^{l}(\boldsymbol{K}^{l})^{\top}/\sqrt{D}, &\quad \boldsymbol{A}_{aux} = Concat(\boldsymbol{A}_{easy}^{1}, ..., \boldsymbol{A}_{easy}^{L}) \end{aligned} \tag{2}$$

where $\boldsymbol{A}_{fuse}^{l} \in \mathbb{R}^{N_{hard} \times N_{hard}}$, $\boldsymbol{A}_{aux} \in \mathbb{R}^{L \times N_{easy} \times N_{easy}}$ and $r_l : \mathbb{R}^{L \times N_{easy} \times N_{easy}} \rightarrow \mathbb{R}^{N_{hard} \times N_{hard}}$ denotes the MLP and upsampling operations. Note that we split $\boldsymbol{A}_{aux}$ into $[\boldsymbol{A}_{zz}, \boldsymbol{A}_{zx}, \boldsymbol{A}_{xz}, \boldsymbol{A}_{xx}]$ for upsampling separately, where $\boldsymbol{A}_{zx}$ is the relation between the template and the search region and the rest are similar. This way prevents different kinds of relations affecting each other during upsampling. Finally, a feed-forward network (FFN) are followed to obtain the output tokens $\boldsymbol{H}_{hard}^{l+1}$ for the next layer.

3.2 Scene Router

Not all tracking scenes require a fine-grained hard branch, and the inference architecture should depend on the input image. Therefore, we design an adaptive router based on scene perception, that is, a simple yet effective decision-maker for the dynamic tracker. As shown in Fig. 3(a), the scene router receives the search region tokens $\boldsymbol{X}_{easy}^{L}$ of the final encoder layer in the easy branch. The search region contains current tracking scene information, including composition, texture and target-to-background contrast. We first pool $\boldsymbol{X}_{easy}^{L}$ along the spatial dimension to obtain a scene vector. Then a MLP maps this scene vector into a two-dimensional decision vector $\boldsymbol{s}$ corresponding to the probabilities of selecting easy and hard branches, respectively. The computational burden of our router can be neglected as it involves only two nonlinear transformations on the vector. Moreover, the scene router directly evaluates the tokens. If the hard branch is selected, the tracker head of the easy branch is skipped for efficient inference.

Gumbel-Softmax Optimization. We employ the IOU metric, a general evaluation criterion in tracking, as the supervision for optimizing the router. An intuitive attempt is to directly calculate the cross-entropy loss between the IOU and the decision vector $\boldsymbol{s}$. However, this hard optimization tends to mode collapse, i.e., the router will always select the most accurate hard branch as it yields a higher IOU. To mitigate this problem, we use the Gumbel-Softmax [15] trick to introduce stochasticity by adding noise to $\boldsymbol{s}$. Concretely, given a distribution with (two) class probabilities $\boldsymbol{s} = \{s_1, s_2\}$, gumbel sampling can be written as,

$$\arg \max_{k \in \{1,2\}} (\log s_k + g_k) \tag{3}$$

where g_k is noise sample drawn from Gumbel distribution. The Gumbel-Softmax defines a continuous, differentiable approximation by replacing the argmax with a softmax,

$$\boldsymbol{y} = Softmax((\log s_k + g_k)/\tau) \tag{4}$$

where τ is the temperature of the softmax. The training loss of the router is the cross-entropy between the Gumbel-Softmax vector $\boldsymbol{y} = \{y_1, y_2\}$ and IOU,

$$L_{router} = -IOU \cdot y_1 - (1 - IOU) \cdot y_2 \tag{5}$$

By incorporating the Gumbel-Softmax trick, we can perform probabilistic sampling based on the decision scores. Consequently, when the decision scores of the easy and hard branches are close, the router has a probability to select the easy branch rather than exhibiting a consistent bias towards the hard branch.

4 Experiments

4.1 Implementation Details

Network Architecture. The implementation of DynamicTrack is developed by two OSTrack [30] trackers. We remove the candidate elimination module to maintain a consistent number of tokens in each encoder layer, enabling the utilization of CBT. The tracker head consists of a classification branch, a local offset branch and a box size branch. The template and search region resolution are 128×128 and 256×256 respectively. The temperature τ in Eq. 4 is set to 1. The speed is measured on a single 3080Ti GPU.

Training Procedure. The training procedure comprises two steps. In step 1, we train the easy-hard dual-branch network. The training loss is the summation of the losses of the two branches, where the loss calculation of each branch follows OSTrack [30]. In step 2, we freeze the parameters of the dual branch and only utilize Eq. 5 to train the scene router. Both stages are trained with AdamW on LaSOT [8], GOT-10k [13], TrackingNet [21] and COCO [17]. The training set in step 1 follows OSTrack. The step2 training takes 5 epochs, with each containing 60000 pairs. The batch size is 4 and the learning rate is 4×10^{-6}.

Table 1. SOTA comparisons on four tracking benchmarks. The top three results are highlight with red, green and blue fonts, respectively.

Method	LaSOT			TrackingNet			UAV123		GOT-10k		
	AUC	P_{norm}	P	AUC	P_{norm}	P	AUC	P	AO	SR_{50}	SR_{75}
SiamFC [1]	33.6	42	33.9	57.1	66.3	53.3	49.4	72.5	34.8	35.3	9.8
ECO [7]	32.4	33.8	30.1	55.4	61.8	49.2	52.5	74.1	31.6	30.9	11.1
SiamRPN++ [16]	49.6	56.9	49.1	73.3	80	69.4	64.2	84	51.7	61.6	32.5
DiMP [2]	56.9	65	56.7	74	80.1	68.7	64.2	84.9	61.1	71.7	49.2
SiamRCNN [25]	64.8	72.2	–	81.2	85.4	80	–	–	64.9	72.8	59.7
Ocean [34]	56	65.1	56.6	69.2	79.4	68.7	62.1	82.3	61.1	72.1	47.3
AutoMatch [32]	58.3	–	59.9	76	–	72.6	64.4	83.8	65.2	76.6	54.3
TrDiMP [26]	63.9	–	61.4	78.4	83.3	73.1	67	87.6	67.1	77.7	58.3
TransT [4]	64.9	73.8	69	81.4	86.7	80.3	68.1	87.6	67.1	76.8	60.9
STARK [29]	67.1	77	–	82	86.9	–	69.2	88.2	68.8	78.1	64.1
MixFormer-22k [6]	69.2	78.7	74.7	83.1	88.1	81.6	70.4	91.8	70.7	80	67.8
UTT [19]	64.6	-	67.2	79.7	–	77	–	–	67.2	76.3	60.5
SimTrack-B [3]	69.3	78.5	–	82.3	86.5	–	69.8	89.6	68.6	78.9	62.4
OSTrack-256 [30]	69.1	78.7	75.2	83.1	87.8	82	68.3		71	80.4	68.2
DynamicTrack	70	78.9	76.2	83.8	88	82.6	69.9	89.1	74	83.6	71.3

4.2 Comparison with State-of-the-arts

We compare our DynamicTrack with recent SOTA trackers on four tracking benchmarks.

GOT-10k. GOT-10k [13] test set contains 180 videos covering a wide range of common challenges in tracking. Following the official requirements, we only use the GOT-10k training set to train our models. We report the average overlap (AO) and success rate (SR_{50}, SR_{75}) in Table 1. DynamicTrack outperforms other SOTA one-stream trackers OSTrack-256 [30] and MixFormer-22k [6] by 3% and 3.3% in AO, respectively. The SR_{75} of DynamicTrack reaches 71.3%, outperforming OSTrack-25 by 3.1%. This proves the excellent discriminative power and localization accuracy of our method in various scenes. Moreover, Fig. 2 illustrates the performance and speed comparison. Our method achieves a good balance between accuracy and inference speed.

LaSOT. LaSOT [8] is a challenging large-scale long-term tracking benchmark, containing 280 videos for testing. Methods are ranked by AUC, normalized precision (P_{norm}) and precision (P). As reported in Table 1, DynamicTrack achieves the best performance in terms of AUC, normalization precision and precision. DynamicTrack performs slightly better than MixFormer-22k [6], getting 0.8% AUC improvement. Besides, the inference speed of DynamicTrack (77FPS) is 2× faster than MixFormer-22k (30FPS).

TrackingNet. TrackingNet [21] contains 511 testing sequences, which covers diverse target classes. Table 1 shows DynamicTrack gets the best AUC of 83.8%, surpassing OSTrack-256 [30] and MixFormer-22k [6] by 0.7%.

UAV123. UAV123 [20] is a large-scale aerial tracking benchmark involving 123 challenging sequences with more than 112K frames. Table 1 shows our method achieves competitive results (rank at second) compared with the previous SOTA trackers. The AUC of DynamicTrack is slightly lower than that of MixFormer-22k [6], i.e., 69.9 vs 70.4. We believe that the variations of targets and scenes are drastic in aerial tracking conditions. Thus MixFormer with an online update mechanism is more adaptable.

Table 2. The analysis of the dynamic architecture and decision mechanism. The reported GFLOPs are the average GFLOPs on the corresponding dataset.

	GOT-10k			LaSOT		
	AO	GFLOPs	FPS	AUC	GFLOPs	FPS
① Base	71.1	29	74	68.7	29	74
② Easy branch	65.4	**7.4**	**102**	65.4	**7.4**	**102**
③ Hard branch w/o CBT	71.6	36	60	69.3	36	60
④ Hard branch	73.4	37.1	51	**70.7**	37.1	51
⑤ Router w/o gumbel	73.5	36.5	53	70.3	35.6	56
⑥ Router	**74**	27	77	70	25.2	80

4.3 Ablation Study and Analysis

Table 2 analyzes the effect of each component in the proposed DynamicTrack.

Dynamic Architecture. Our baseline (①) is the original OSTrack without the candidate elimination module. The easy branch (②) has excellent efficiency, with only 1/4 GFLOPs of ①. Moreover, using easy branch for all inputs also yields good performance, i.e., 65.4 on GOT-10k and 65.4 on LaSOT, outperforming most Siamese-based trackers in Table 1. This proves that the easy branch is sufficient in processing many simple scenes. The hard branch (④) achieves better performance at the cost of computation and speed, thus adapting to complex scenes. It improves 8% AO on GOT-10k and 5.3% AUC on LaSOT with 5× GFLOPs compared with easy branch (④ vs ②). When the scene router (⑥) is introduced for adaptive decision, DynamicTrack can achieve an accuracy-speed trade-off. Compared with baseline (①), the performance gains are 2.9% and 1.3%, and the GFLOPs are reduced by 6.9% and 13.1% on GOT-10k and LaSOT, respectively. Moreover, we find that DynamicTrack performs even higher AO than only using hard branch on GOT-10k (⑥ vs ④). This can potentially be attributed to that over-processing in certain simple scenes is detrimental.

Gumbel-Softmax Trick. The router w/o Gumbel-Softmax always selects the hard branch, whose accuracy, GFLOPs and FPS are similar to only using hard branch (⑤ vs ④). While the router w/ Gumbel-Softmax can prevent this trivial solution and effectively improve inference efficiency, reducing 26% and 29.2% GFLOPs on GOT-10k and LaSOT respectively (⑥ vs ⑤). This proves the necessity of Gumbel-Softmax for adaptive decision-making.

CBT. Hard branch w/o CBT is optimized from scratch without any priors, resulting in slight improvements compared to the baseline. (③ vs ①). The introduction of CBT can significantly improve performance (④ vs ③). This proves that the information of easy branch is also valid for hard branch.

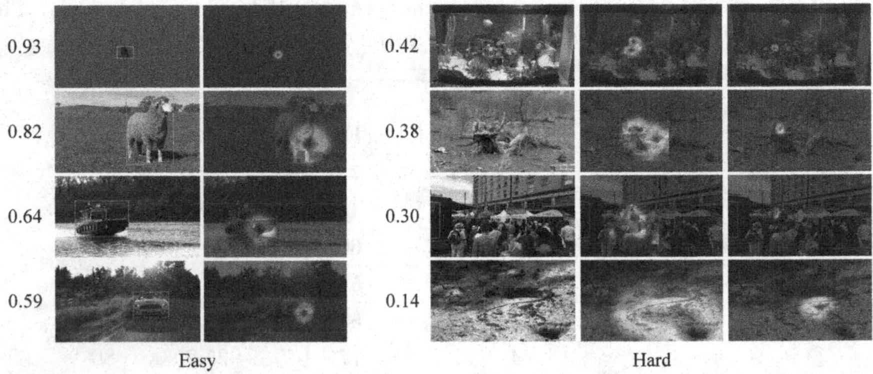

Fig. 4. Visualization of heatmaps for easy and hard images. **1st col:** easy images, **2nd col:** easy branch's heatmaps on easy images, **3rd col:** hard images, **4th col:** easy branch's heatmaps on hard images, **5th col:** hard branch's heatmaps on hard images. The numbers indicate the 0-dim of the decision vectors, and a smaller one indicates a harder scene.

Visualization of Scene Perception. Figure 4 visualizes the heatmaps of our method on easy and hard images, respectively. The easy images (1st col) usually contain singular composition, clean background and strong target-to-background contrast. Our scene router selects the easy branch to process these scenes and the heatmaps (2nd col) can effectively focus on targets. While hard images (3rd col) have the opposite characteristics and may be accompanied with distractors. The easy branch confuses these targets (4th col) and the router will select the hard branch to represent them accurately (5th col).

5 Conclusion

This paper proposes a dynamic tracking framework based on scene perception, DynamicTrack. We first design an easy-hard dual-branch network to support dynamic inference. It customizes the inference architecture for different scenes by varying the number of tokens in the transformer backbone. Then we propose a scene router to perceive the complexity of scenes and determine the inference route. Furthermore, we introduce two tricks to facilitate DynamicTrack optimization. Extensive experiments demonstrate the SOTA performance and the accuracy-speed trade-off of our method.

References

1. Bertinetto, L., Valmadre, J., Henriques, J.F., Vedaldi, A., Torr, P.H.S.: Fully-convolutional Siamese networks for object tracking. In: Hua, G., Jégou, H. (eds.) ECCV 2016. LNCS, vol. 9914, pp. 850–865. Springer, Cham (2016). https://doi.org/10.1007/978-3-319-48881-3_56
2. Bhat, G., Danelljan, M., Gool, L.V., Timofte, R.: Learning discriminative model prediction for tracking. In: Proceedings of the IEEE International Conference on Computer Vision, pp. 6182–6191 (2019)
3. Chen, B., et al.: Backbone is all your need: a simplified architecture for visual object tracking. In: Avidan, S., Brostow, G., Cissé, M., Farinella, G.M., Hassner, T. (eds.) Proceedings of the European Conference on Computer Vision. LNCS, vol. 13682, pp. 375–392. Springer, Cham (2022). https://doi.org/10.1007/978-3-031-20047-2_22
4. Chen, X., Yan, B., Zhu, J., Wang, D., Yang, X., Lu, H.: Transformer tracking. In: Proceedings of the IEEE Conference on Computer Vision and Pattern Recognition, pp. 8126–8135 (2021)
5. Cheng, S., et al.: Learning to filter: Siamese relation network for robust tracking. In: Proceedings of the IEEE/CVF Conference on Computer Vision and Pattern Recognition, pp. 4421–4431 (2021)
6. Cui, Y., Jiang, C., Wang, L., Wu, G.: MixFormer: end-to-end tracking with iterative mixed attention. In: Proceedings of the IEEE Conference on Computer Vision and Pattern Recognition, pp. 13608–13618 (2022)
7. Danelljan, M., Bhat, G., Shahbaz Khan, F., Felsberg, M.: ECO: efficient convolution operators for tracking. In: Proceedings of the IEEE Conference on Computer Vision and Pattern Recognition, pp. 6638–6646 (2017)
8. Fan, H., et al.: LaSOT: a high-quality benchmark for large-scale single object tracking. In: Proceedings of the IEEE Conference on Computer Vision and Pattern Recognition, pp. 5374–5383 (2019)
9. Guo, D., Wang, J., Cui, Y., Wang, Z., Chen, S.: SiamCAR: Siamese fully convolutional classification and regression for visual tracking. In: Proceedings of the IEEE Conference on Computer Vision and Pattern Recognition, pp. 6269–6277 (2020)
10. Guo, Q., Feng, W., Zhou, C., Huang, R., Wan, L., Wang, S.: Learning dynamic Siamese network for visual object tracking. In: Proceedings of the IEEE International Conference on Computer Vision, pp. 1763–1771 (2017)
11. Henriques, J.F., Caseiro, R., Martins, P., Batista, J.: High-speed tracking with kernelized correlation filters. IEEE Trans. Pattern Anal. Mach. Intell. **37**(3), 583–596 (2014)
12. Huang, G., Liu, S., van der Maaten, L., Weinberger, K.Q.: Multi-scale dense networks for resource efficient image classification. In: International Conference on Learning Representations (2018)

13. Huang, L., Zhao, X., Huang, K.: GOT-10k: a large high-diversity benchmark for generic object tracking in the wild. IEEE Trans. Pattern Anal. Mach. Intell. (2019)
14. Hubel, D.H., Wiesel, T.N.: Receptive fields, binocular interaction and functional architecture in the cat's visual cortex. J. Physiol. **160**(1), 106 (1962)
15. Jang, E., Gu, S., Poole, B.: Categorical reparameterization with Gumbel-Softmax. arXiv preprint arXiv:1611.01144 (2016)
16. Li, B., Wu, W., Wang, Q., Zhang, F., Xing, J., Yan, J.: SiamRPN++: evolution of Siamese visual tracking with very deep networks. In: Proceedings of the IEEE Conference on Computer Vision and Pattern Recognition, pp. 4282–4291 (2019)
17. Lin, T.-Y., et al.: Microsoft COCO: common objects in context. In: Fleet, D., Pajdla, T., Schiele, B., Tuytelaars, T. (eds.) ECCV 2014. LNCS, vol. 8693, pp. 740–755. Springer, Cham (2014). https://doi.org/10.1007/978-3-319-10602-1_48
18. Lin, Z., Wang, Y., Zhang, J., Chu, X.: DynamicDet: a unified dynamic architecture for object detection. In: Proceedings of the IEEE Conference on Computer Vision and Pattern Recognition, pp. 6282–6291 (2023)
19. Ma, F., et al.: Unified transformer tracker for object tracking. In: Proceedings of the IEEE/CVF Conference on Computer Vision and Pattern Recognition, pp. 8781–8790 (2022)
20. Mueller, M., Smith, N., Ghanem, B.: A benchmark and simulator for UAV tracking. In: Leibe, B., Matas, J., Sebe, N., Welling, M. (eds.) ECCV 2016. LNCS, vol. 9905, pp. 445–461. Springer, Cham (2016). https://doi.org/10.1007/978-3-319-46448-0_27
21. Müller, M., Bibi, A., Giancola, S., Alsubaihi, S., Ghanem, B.: TrackingNet: a large-scale dataset and benchmark for object tracking in the wild. In: Ferrari, V., Hebert, M., Sminchisescu, C., Weiss, Y. (eds.) ECCV 2018. LNCS, vol. 11205, pp. 310–327. Springer, Cham (2018). https://doi.org/10.1007/978-3-030-01246-5_19
22. Murata, A., Gallese, V., Luppino, G., Kaseda, M., Sakata, H.: Selectivity for the shape, size, and orientation of objects for grasping in neurons of monkey parietal area AIP. J. Neurophysiol. **83**(5), 2580–2601 (2000)
23. Teerapittayanon, S., McDanel, B., Kung, H.T.: BranchyNet: fast inference via early exiting from deep neural networks. In: International Conference on Pattern Recognition, pp. 2464–2469. IEEE (2016)
24. Veit, A., Belongie, S.: Convolutional networks with adaptive inference graphs. In: Ferrari, V., Hebert, M., Sminchisescu, C., Weiss, Y. (eds.) ECCV 2018. LNCS, vol. 11205, pp. 3–18. Springer, Cham (2018). https://doi.org/10.1007/978-3-030-01246-5_1
25. Voigtlaender, P., Luiten, J., Torr, P.H., Leibe, B.: Siam R-CNN: visual tracking by re-detection. In: Proceedings of the IEEE Conference on Computer Vision and Pattern Recognition, pp. 6578–6588 (2020)
26. Wang, N., Zhou, W., Wang, J., Li, H.: Transformer meets tracker: exploiting temporal context for robust visual tracking. In: Proceedings of the IEEE Conference on Computer Vision and Pattern Recognition, pp. 1571–1580 (2021)
27. Wang, X., Yu, F., Dou, Z.-Y., Darrell, T., Gonzalez, J.E.: SkipNet: learning dynamic routing in convolutional networks. In: Ferrari, V., Hebert, M., Sminchisescu, C., Weiss, Y. (eds.) ECCV 2018. LNCS, vol. 11217, pp. 420–436. Springer, Cham (2018). https://doi.org/10.1007/978-3-030-01261-8_25
28. Wang, Y., Huang, R., Song, S., Huang, Z., Huang, G.: Not all images are worth 16×16 words: dynamic transformers for efficient image recognition. Adv. Neural. Inf. Process. Syst. **34**, 11960–11973 (2021)

29. Yan, B., Peng, H., Fu, J., Wang, D., Lu, H.: Learning spatio-temporal transformer for visual tracking. In: Proceedings of the IEEE International Conference on Computer Vision, pp. 10448–10457 (2021)
30. Ye, B., Chang, H., Ma, B., Shan, S., Chen, X.: Joint feature learning and relation modeling for tracking: a one-stream framework. In: Avidan, S., Brostow, G., Cissé, M., Farinella, G.M., Hassner, T. (eds.) Proceedings of the European Conference on Computer Vision. LNCS, vol. 13682, pp. 341–357. Springer, Cham (2022). https://doi.org/10.1007/978-3-031-20047-2_20
31. Zhang, L., Gonzalez-Garcia, A., Weijer, J.V.D., Danelljan, M., Khan, F.S.: Learning the model update for Siamese trackers. In: Proceedings of the IEEE International Conference on Computer Vision, pp. 4010–4019 (2019)
32. Zhang, Z., Liu, Y., Wang, X., Li, B., Hu, W.: Learn to match: automatic matching network design for visual tracking. In: Proceedings of the IEEE International Conference on Computer Vision, pp. 13339–13348 (2021)
33. Zhang, Z., Peng, H.: Deeper and wider Siamese networks for real-time visual tracking. In: Proceedings of the IEEE Conference on Computer Vision and Pattern Recognition, pp. 4591–4600 (2019)
34. Zhang, Z., Peng, H., Fu, J., Li, B., Hu, W.: Ocean: object-aware anchor-free tracking. In: Vedaldi, A., Bischof, H., Brox, T., Frahm, J.-M. (eds.) ECCV 2020. LNCS, vol. 12366, pp. 771–787. Springer, Cham (2020). https://doi.org/10.1007/978-3-030-58589-1_46

HPAN: A Hybrid Pose Attention Network for Person Re-Identification

Ruohong Huan(✉), Tianya Chen, Ziwei Zhan, Peng Chen, and Ronghua Liang

College of Computer Science and Technology, Zhejiang University of Technology, Hangzhou, Zhejiang, China

huanrh@zjut.edu.cn

Abstract. To address the difficulty in expressing the correlation between different local features extracted by the current person re-identification feature extraction methods, and the challenge of effectively integrating local features with global features, a Hybrid Pose Attention Network (HPAN) for person re-identification is proposed. In HPAN, the high-resolution network HRNet-W32 serves as the backbone for person re-identification and pose estimation, extracting global features and local key point heatmaps of the human images, and then generating local key point features. Self-attention is used to extract the correlation between each local key point feature, generating local pose features. Furthermore, a Hybrid Pose and Global Feature Fusion (HPGFF) module is adopted to fuse the global features and local pose features, creating integrated features. To evaluate, we conduct experiments on five publicly available datasets, and HPAN has all achieved competitive or state-of-the-art results.

Keywords: Re-identification · Pose Estimation · Local Feature · Feature Correlation · Fusion

1 Introduction

Person Re-Identification (Re-ID) is a significant field and research hotspot in computer vision. The goal of person Re-ID is to automatically locate all pedestrian images of a given query object across multiple non-overlapping cameras. In practical applications, combining person Re-ID with techniques like pedestrian detection and pedestrian tracking to form a person Re-ID system can be applied in fields such as video surveillance, criminal investigation, unmanned supermarkets, and others.

In recent years, due to the development of deep learning and the enhancement of computer hardware, person Re-ID has made significant progress [1–4]. Particularly in feature extraction, deep neural networks often extract discriminative features [5–8]. Regarding the features extracted by neural networks, we can divide them into global features and local features according to the size of the region. Global features can represent the holistic attributes of pedestrians with good invariance, but are easily affected by the background. For similar appearances, global features are also difficult to distinguish subtle differences. Therefore, it is challenging to use them alone in person Re-ID tasks.

© The Author(s), under exclusive license to Springer Nature Singapore Pte Ltd. 2024
Q. Liu et al. (Eds.): PRCV 2023, LNCS 14436, pp. 198–211, 2024.
https://doi.org/10.1007/978-981-99-8555-5_16

Local features are extracted from specific regions in the image and contain rich fine-grained information. Many person Re-ID methods are based on local features to identify pedestrian identities [9–13]. For example, in PCB [14], Sun et al. horizontally sliced the whole pedestrian picture and then extracted features from each slice separately. In DSR [15], He et al. divided the entire pedestrian picture into lots of equal-sized blocks, using the set distance between the blocks to identify pedestrian identities. In VPM [16], Sun et al. performed self-supervised learning on the whole image and part of the image after being divided into blocks, learning local features in the image through self-comparison.

In person Re-ID methods, it is a common method to use pose estimation networks to extract local information for expressing pedestrians [8, 17–21]. The HORe-ID [18] model proposed by Wang et al. used a graph structure and adaptive graph convolution layer to learn local and edge features of pedestrians. The PGFA [17] model based on the pose model had two branches, one generated key point information and heatmaps and the other branch extracted global features. The PFD [19] method used ViT [22] and matching mechanism to extract and separate pose and patch features.

Despite these methods having extracted local features of pedestrians, the correlation among these local features is minor, and they are insensitive to large-scale action changes and scaling [23]. In recent years, person Re-ID methods based on transformers have placed more emphasis on the correlation among local features. In ViT [22], Dosovitskiy et al. equally divided the entire image into lots of patches of the same size, and used multi-head attention to get the correlation between the patches.

Although local features can represent fine-grained information, a large number of unordered, unrelated local features lack unified global guidance. Therefore, many current studies often combine local features with global features [24–27]. For example, in transformer-based methods, PFD [19] and TransReID [28] computed loss for the collections of global and local features separately, achieving better results than ViT [22] that only used a 'cls' token. In PGFA [17], Miao et al. combined global features and pose-guided features by concatenation for identity prediction.

Although the above methods have realized the fusion of local features and global features, some of them calculate losses for global features and local features separately [19, 28], some provide additional clues for extracting local features through global features, and some just do simple calculations on global features and local features [13, 27]. There still lacks an effective mechanism to integrate the coarse-grained global features and the fine-grained local features, so that the fused features have both sufficient detail information of local features and global guidance of global features.

To address the difficulty in expressing the correlation between each local feature in the local feature extraction methods of current person Re-ID, and the challenge of effectively fusing local features and global features, this paper proposes a Hybrid Pose Attention Network (HPAN) for person Re-ID.

The main contributions include: (1) the high-resolution network HRNet-W32 serves as the backbone in HPAN for person Re-ID and pose estimation to extract global features of the human image and local key point heatmaps, and then generate local key point features. (2) Self-attention mechanism is used to extract the correlation between each local key point feature and generate local pose features. (3) Hybrid Pose and Global

Feature Fusion (HPGFF) module is proposed to fuse global features and local pose features to generate the fused features. (4) For evaluation, we test on five publicly available datasets, Market-1501, DukeMTMC, MSMT17, CUHK03, and Occ-duke. The experimental results demonstrate our HPAN can achieve competitive or State-Of-The-Art (SOTA) performance.

2 The Proposed Method

The network structure of HPAN is shown in Fig. 1, which includes two sub-networks, a pose estimation sub-network and a person Re-ID sub-network. Both of these sub-networks adopt HRNet-W32 as the backbone network. First, we input the preprocessed pedestrian images into the two sub-networks. The pose estimation sub-network generates S pose key point heatmaps $\{H_1, H_2, \ldots, H_s\}$, and the person Re-ID sub-network generates the global feature map F_g of the person image. The global feature map F_g and the S key point heatmaps $\{H_1, H_2, \ldots, H_s\}$ are multiplied element-wise, yielding S local key point feature maps $\left\{F_{p1}, F_{p2}, \ldots, F_{pS}\right\}$. After passing through the self-attention module, we obtain a local pose feature map F_{hp}. The global feature map F_g and the local pose feature map F_{hp} are fused through the HPGFF module, resulting in the final feature map F_h. Then, Global Average Pooling (GAP) is implemented, and a feature vector is obtained as the input for similarity measurement.

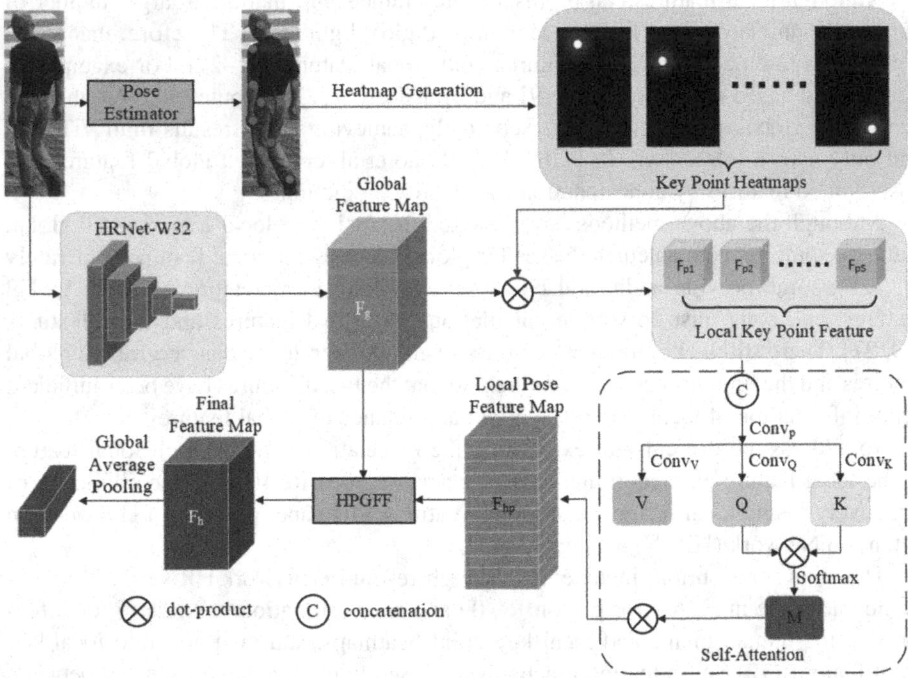

Fig. 1. The general workflow of HPAN

2.1 Local Key Point Features

For the pose estimation sub-network, we output the high-resolution feature map, and a 1×1 convolution is performed to obtain a key point heatmap with 13 channels. For the person Re-ID sub-network, we generate the corresponding feature maps $\{X_{128}, X_{256}, X_{512}, X_{1024}\}$ through the bottleneck corresponding to the input channel number. These feature maps are then upsampled so that the corresponding W and H of each feature map are equal, and these feature maps are concatenated to obtain the final global feature map F_g with 1920 channels. We perform a 1×1 convolution on the global feature map and reduce its dimensionality to get a feature map with 256 channels, which is then point-multiplied with the heatmap to obtain the 13 local key point features $\{F_{p1}, F_{p2}, \ldots, F_{p13}\}$.

2.2 Self-Attention

After obtaining the local feature maps of different key points, each feature map reflects the feature information of the corresponding key point. We use self-attention to associate and aggregate information from each key point. For the feature maps that contain each key point, we concatenate them to obtain $concat(F_{p1}, F_{p2}, \ldots, F_{p13})$, and use a 1×1 convolution $Conv_p$ to generate the feature map F_p containing each key point, as shown in Eq. (1),

$$F_p = Conv_p(concat(F_{p1}, F_{p2}, \ldots, F_{p13})) \tag{1}$$

Here, $Conv_p$ is a 1×1 convolution, $concat$ is the concatenation of feature maps in the channel dimension, and $\{F_{p1}, F_{p2}, \ldots, F_{p13}\}$ is the collection of each key point feature map.

F_p does not contain the association between key points. In order to capture the correlation between each key point, we input F_p into two 1×1 convolution layers $Conv_Q$ and $Conv_K$ to obtain two new feature maps, and then transform them into matrices Q and K of size $N \times C$ and $C \times N$. Here, N represents all pixels in each channel, as shown in Eq. (2),

$$N = H \times W \tag{2}$$

Here, H and W represent the height and width of the feature map, C represents the number of channels of the feature map. The relation mapping M is generated by Q multiplied by K and normalized by *SoftMax*, as follows,

$$M = SoftMax(QK) \tag{3}$$

A point (i, j) in M represents the correlation between the i-th pixel and the j-th pixel. We input F_p into another 1×1 convolution $Conv_V$ to generate a feature map and re-transform it into a matrix V of size $N \times C$, then multiply it with M to integrate the relation mapping into the original V. Then we get the feature map, and perform residual connection with F_p. Through a 1×1 convolution $Conv_{hp}$, we obtain the final local pose feature map F_{hp}, as shown in Eq. (4),

$$F_{hp} = Conv_{hp}(VM + F_p) \tag{4}$$

2.3 Hybrid Pose and Global Feature Fusion (HPGFF)

Based on the concept of the Asymmetric Fusion Non-local Block (AFNB) [29], the HPGFF module, as shown in Fig. 2, includes two inputs, a global feature map F_g and a local pose feature map F_{hp}. The sizes of these two feature maps are $C \times H \times W$, where C is the number of channels in the feature map, and H and W are the length and width of the feature map, respectively. First, we transform F_{hp} and F_g into three nonlinear mapping features, as shown in Eq. (5),

$$F_q = W_q \cdot F_{hp}, F_{k'} = PPM_k(W_k \cdot F_g), F_{v'} = PPM_v(W_v \cdot F_g) \quad (5)$$

where W_q, W_k, W_v are three 1×1 convolutions. PPM is pyramid pooling [30] used to reduce the computational overhead of the module itself. The obtained $F_q \in \mathbb{R}^{C'' \times (H \times W)}$, $F_{k'} \in \mathbb{R}^{C' \times S}$, $F_{v'} \in \mathbb{R}^{S \times C'}$, where C'' is the number of channels after the W_q convolution, S is the pixel of pyramid pooling, C' is the number of channels after W_k, W_v convolutions, numerically equal to $H \times W$. Then we calculate the relationship of pixels in F_q and $F_{k'}$ by matrix operation, normalize the relationship matrix M by *SoftMax*, and convert the pixels in the relationship matrix M between 0 and 1, as shown in Eq. (6),

$$M = SoftMax(F_k'^T \cdot F_q) \quad (6)$$

The obtained $M \in S \times (H \times W)$, then we multiply it by $F_{v'}$ that has been through pyramid pooling to get F_c, as shown in Eq. (7),

$$F_c = M \cdot F_{v'} \quad (7)$$

Each pixel in F_c reflects the weight of the corresponding F_{hp} in F_g. These weights are selected from all pixels in F_g. The final output result or fusion feature F_h obtained by convolution $Conv_h$ is shown in Eq. (8),

$$F_h = Conv_h(concat(F_c, F_{hp})) \quad (8)$$

F_h is taken as the final output feature map, and after global average pooling, a feature vector is obtained as the input for similarity measurement.

2.4 Loss Function

The task of person Re-ID is decomposed into two tasks, person classification and person similarity measurement. We use the cross-entropy function and triplet loss function based on hard sample mining respectively as the loss functions for these two tasks. The loss function of the overall task is the sum of these two loss functions, as shown in Eq. (9),

$$L = L_{crossEntropy} + L_{tri} \quad (9)$$

Here, $L_{crossEntropy}$ represents the cross-entropy loss function, and L_{tri} represents the triplet loss function.

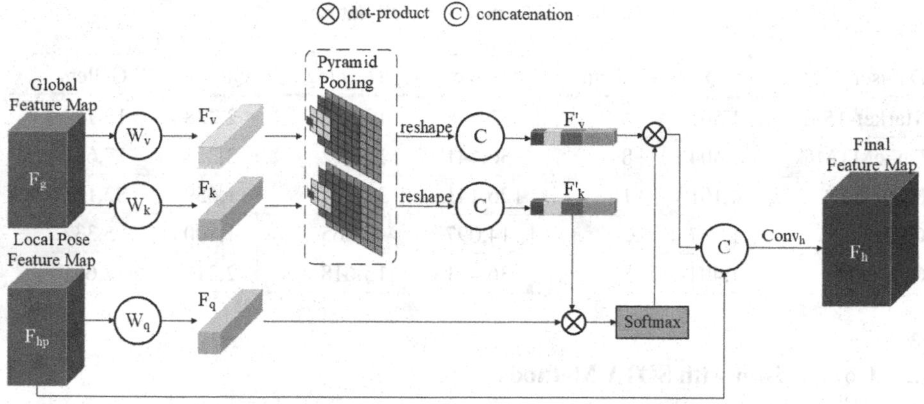

Fig. 2. Structure diagram of the HPGFF module

2.5 Training Strategy

We resize all images to a resolution of 256 × 128 and apply random horizontal flipping and random erasing for data augmentation in the training set. The backbone network, HRNet-W32, is pretrained on ImageNet before further training. Each training batch size is 64, where we randomly select 16 pedestrian identities, each with 4 images. For the optimizer, we use Adam to optimize our network, with the weight decay parameter set to 0.0005. During training, we set the epoch to 600, with an initial learning rate of 0.0008. The learning rate lr for each epoch is shown in Eq. (10) [21],

$$lr = \begin{cases} 8 \times 10^{-6} + (8 \times 10^{-3} - 8 \times 10^{-6}) \times epoch \div 20 & epoch \leq 20 \\ 8 \times 10^{-3} & 20 < epoch \leq 80 \\ 8 \times 10^{-3} \div 2^{(epoch-80)\%40+1} & 80 < epoch \leq 360 \\ 3.125 \times 10^{-6} & 360 < epoch \leq 600 \end{cases} \quad (10)$$

3 Experiments

3.1 Datasets and Evaluation Metrics

Datasets. We conduct experiments on five publicly available datasets, Market-1501 [31], DukeMTMC [32], MSMT17 [33], CUHK03 [34], and Occluded-DukeMTMC (Occ-Duke) [17]. The information about these datasets in shown in Table 1.

Evaluation Metrics. We utilize Rank-1 and mean Average Precision (mAP) for fair comparison. All experiments are performed in the single query setting.

Table 1. Information about datasets

Dataset	#ID	#cam	#image	Training	Query	Gallery
Market-1501	1,501	6	32,668	12,936	2,228	19,732
DukeMTMC	1,404	8	36,441	16,522	2,228	17,661
MSMT17	4,101	15	126,441	32,621	11,659	82,161
CUHK03	1,467	2	14,097	7,365	1,400	5,332
Occ-Duke	1,404	8	36,441	15,618	2,210	17,661

3.2 Comparison with SOTA Methods

Comparisons on Unoccluded Datasets. We evaluate our HPAN model across four unoccluded datasets, Market-1501, DukeMTMC, MSMT17, and CUHK03, and compare with various SOTA methods in Table 2. From Table 2, we can observe that the method we proposed achieves competitive results. Specifically, for the Market-1501 dataset, our Rank-1 is only about 0.1% lower than the SAN method, but our mAP is 2.5% higher, which is currently the highest. Compared with the PFD method which is also based on pose estimation, our method improves Rank-1 by 0.5% and mAP by 0.9%. Our method achieves the SOTA performance in terms of Rank-1 and mAP on the DukeMTMC dataset, with 91.8% and 84.5% respectively. Meanwhile, our Rank-1 surpasses the PFD method by 1.2%, and mAP is improved by 2.3%. Our method also achieves good performance on MSMT17 and CUHK03 datasets. Our method achieves 84.7% Rank-1 and 64.1% mAP on MSMT17. Although the mAP is 1% and 3.3% lower than PFD and TransReID methods respectively, our Rank-1 achieves a competitive result, improving by 2% compared to PFD. Our method achieves 85.6% Rank-1 and 84.1% mAP on CUHK03. For the Rank-1, our model is only 1.2% lower than the SCSN method, but our mAP is 0.1% higher than it, which is currently the highest.

Comparisons on Occluded Dataset. We also evaluate our HPAN model on occluded dataset Occ-Duke. As shown in Table 3, we categorize the baselines into two categories, CNN-based methods and Transformer-based methods. We can observe that our method achieves a significant improvement on the Occ-Duke dataset, achieving 72.9% Rank-1 and 62.0% mAP. Compared with the PFD method, our method significantly improves Rank-1 by 5.2% and mAP by 1.9%. It can be seen that although our method is designed for holistic person Re-ID tasks, it still achieves competitive results on the occluded pedestrian dataset, which proves the robustness of our proposed method.

Table 2. Performance comparison on the holistic Re-ID datasets Market-1501, DukeMTMC, MSMT17 and CUHK03. *represents the second best current result, and the best result is shown in bold.

Method	Market-1501		DukeMTMC		MSMT17		CUHK03	
	Rank-1	mAP	Rank-1	mAP	Rank-1	mAP	Rank-1	mAP
PCB [14]	92.3	77.4	81.8	66.1	–	–	–	–
DSR [15]	83.6	64.3	–	–	–	–	–	–
BOT [11]	94.1	85.7	86.4	76.4	–	–	–	–
VPM [16]	93.0	80.8	83.6	72.6	–	–	–	–
SAN [12]	**96.1**	88.0	87.9	75.7	79.2	55.7	–	–
OSNet [9]	–	–	–	–	78.7	52.9	–	–
DG-Net [35]	–	–	–	–	77.2	52.3	–	–
CBN [10]	–	–	–	–	72.8	42.9	–	–
Circle [1]	94.2	84.9	–	–	76.3	–	–	–
RGA-SC [21]	–	–	–	–	80.3	57.5	81.1	77.4
BDB [36]	–	–	–	–	–	–	79.4	76.7
DSA-reID [37]	–	–	–	–	–	–	78.9	75.2
SCSN [38]	–	–	–	–	–	–	**86.8**	84.0*
ISP [39]	–	–	–	–	–	–	76.5	74.1
ADC + 2O-IB [26]	–	–	–	–	–	–	80.6	79.3
MVPM [4]	91.4	80.5	83.4	70.0	71.3	46.3	–	–
SFT [24]	93.4	82.7	86.9	73.2	73.6	47.6	–	–
CAMA [25]	94.7	84.5	85.8	72.9	–	–	–	–
IANet [13]	94.4	83.1	87.1	73.4	75.5	46.8	–	–
FED [20]	95.0	86.3	89.4	78.0	–	–	–	–
PGFA [17]	91.2	76.8	82.6	65.5	–	–	–	–
HORe-ID [18]	94.2	84.9	86.9	75.6	–	–	–	–
PFD [19]	95.5	89.6*	90.6	82.2*	82.7	65.1*	–	–
PAT [3]	95.4	88.0	88.8	78.2	–	–	–	–
DeiT [2]	94.4	86.6	89.3	78.9	81.9	61.4	–	–
ViT [22]	94.7	86.8	88.8	79.3	81.8	61.0	–	–

(continued)

Table 2. (*continued*)

Method	Market-1501		DukeMTMC		MSMT17		CUHK03	
	Rank-1	mAP	Rank-1	mAP	Rank-1	mAP	Rank-1	mAP
TransReID [28]	95.2	88.9	90.7*	82.0	**85.3**	**67.4**	–	–
HPAN (ours)	96.0*	**90.5**	**91.8**	**84.5**	84.7*	64.1	85.6*	**84.1**

Table 3. Performance comparison on the occluded Re-ID dataset Occluded-DukeMTMC. *represents the second best current result, and the best result is show in bold.

Method (CNN-based)	Rank-1	mAP	Method (Transformer-based)	Rank-1	mAP
PCB [14]	42.6	33.7	PAT [3]	64.5	53.6
RE [27]	40.5	30.0	DeiT [2]	60.6	53.1
FD-GAN [5]	40.8	–	ViT [22]	60.5	53.1
DSR [15]	40.8	30.4	TransRe-ID [28]	66.4	59.2
PGFA [17]	51.4	37.3			
PVPM [6]	47.0	37.7			
ISP [39]	62.8	52.3			
HORe-ID [18]	55.1	43.8			
MoS [7]	61.0	49.2			
OAMN [8]	62.6	46.1			
PFD [19]	67.7	60.1*			
FED [20]	68.1*	56.4			
HPAN (ours)	**72.9**	**62.0**			

3.3 Ablation Studies

Ablation experiments are conducted to fully verify the effectiveness of the modules in HPAN, and the results are shown in Table 4. The methods for the ablation experiments include, M1(only the global feature map F_g is used), M2 (neither self-attention nor HPGFF are used. The obtained local key point feature map undergoes 1×1 convolution to obtain the local pose feature map, and the local pose feature map is fused with the global feature map using a residual connection), M3 (only self-attention is used. The local pose feature map is obtained using self-attention, and fused with the global feature map F_g using a residual connection), and M4 (HPAN).

Table 4. Ablation Studies on Market-1501, DukeMTMC and CUHK03

Method	Market-1501		DukeMTMC		CUHK03	
	Rank-1	mAP	Rank-1	mAP	Rank-1	mAP
M1	95.4	88.5	89.4	81.2	84.2	81.8
M2	95.5	89.1	90.0	82.3	84.3	82.2
M3	95.7	89.9	91.0	82.9	84.5	82.7
M4	96.0	90.5	91.8	84.5	85.6	84.1
Δ(M3-M1)	0.3	1.4	1.6	1.7	0.3	0.9
Δ(M3-M2)	0.2	0.8	1.0	0.6	0.2	0.5
Δ(M4-M3)	0.3	0.6	0.8	1.6	1.1	1.4
Δ(M4-M1)	0.6	2.0	2.4	3.3	1.4	2.3

As can be observed from Table 4, we find compared with M1 and M2, the method with self-attention (M3) improves both mAP and Rank-1. This demonstrates that using self-attention can extract the correlation between key points and learn the correlation of local features of each key point, which results in better performance. Compared with the method of only using the self-attention (M3), the method using both the self-attention and HPGFF (M4) shows an extra increase in mAP and Rank-1. It demonstrates the effectiveness of the HPGFF module. By the HPGFF, fusing the global features with local pose features, the global features can guide the local features, resulting in a final feature that is more discriminative. In short, the ablation experiments on the three datasets quantitatively demonstrate the effectiveness of the self-attention and the HPGFF module.

3.4 Visualization of Attention Maps

To further analyze the difference between the global feature map and the fused feature map, we visualize the attention maps of the features of our model, using the Gram-CAM tool to identify areas that the network considers important. Figure 3 shows the comparison of attention maps of global features extracted by the backbone network HRNet-W32 and the fused features obtained by HPAN. In the subfigures of Fig. 3, the left image is the original image, the middle is the global feature attention map, and the right is the fused feature attention map obtained by HPAN. From the figure, we can see that in addition to global features, HPAN can also focus on discriminative local human key point features.

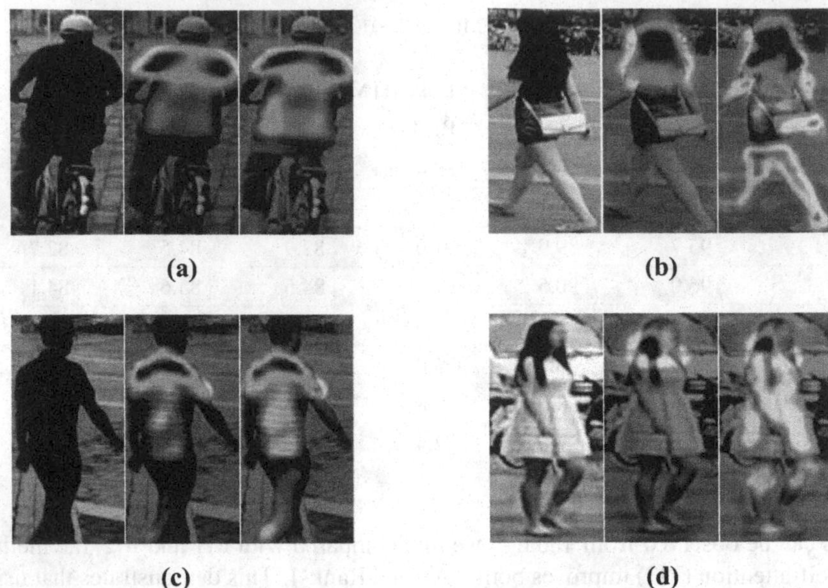

Fig. 3. Visualization of attention map

4 Conclusion

This paper presents a person Re-ID method HPAN, based on hybrid pose attention. The self-attention within it can mine and generate local pose features from the associations among extracted local key point features. The fusion module HPGFF combines the extracted global features with the local pose features to generate more distinguishable and robust features. Our HPAN model shows superior performance on five popular datasets, surpassing SOTA methods.

Acknowledgement. This work is supported by the National Natural Science Foundation of China (grant number 62276237, 62036009, U1909203), Zhejiang Province Basic Public Welfare Research Program Project (grant number LTGY23F020006), and Zhejiang Provincial Natural Science Foundation of China (grant number LDT23F0202, LDT23F02021F02).

References

1. Sun, Y., et al.: Circle loss: a unified perspective of pair similarity optimization. Presented at the Proceedings of the IEEE/CVF Conference on Computer Vision and Pattern Recognition (2020)
2. Training data-efficient image transformers & distillation through attention. https://proceedings.mlr.press/v139/touvron21a. Accessed 16 June 2023
3. Li, Y., He, J., Zhang, T., Liu, X., Zhang, Y., Wu, F.: Diverse part discovery: occluded person re-identification with part-aware transformer. Presented at the Proceedings of the IEEE/CVF Conference on Computer Vision and Pattern Recognition (2021)

4. Sun, H., Chen, Z., Yan, S., Xu, L.: MVP Matching: a maximum-value perfect matching for mining hard samples, with application to person re-identification. Presented at the Proceedings of the IEEE/CVF International Conference on Computer Vision (2019)
5. FD-GAN: Pose-guided Feature Distilling GAN for Robust Person Re-identification. https://proceedings.neurips.cc/paper/2018/hash/c5ab0bc60ac7929182aadd08703f1ec6-Abstract.html. Accessed 16 June 2023
6. Gao, S., Wang, J., Lu, H., Liu, Z.: Pose-guided visible part matching for occluded person ReID. Presented at the Proceedings of the IEEE/CVF Conference on Computer Vision and Pattern Recognition (2020)
7. Matching on Sets: Conquer Occluded Person Re-identification Without Alignment | Proceedings of the AAAI Conference on Artificial Intelligence. https://ojs.aaai.org/index.php/AAAI/article/view/16260. Accessed 16 June 2023
8. Chen, P., et al.: Occlude Them all: occlusion-aware attention network for occluded person re-ID. Presented at the Proceedings of the IEEE/CVF International Conference on Computer Vision (2021)
9. Zhou, K., Yang, Y., Cavallaro, A., Xiang, T.: Omni-scale feature learning for person re-identification. Presented at the Proceedings of the IEEE/CVF International Conference on Computer Vision (2019)
10. Rethinking the Distribution Gap of Person Re-identification with Camera-Based Batch Normalization | SpringerLink. https://link.springer.com/chapter/https://doi.org/10.1007/978-3-030-58610-2_9. Accessed 16 June 2023
11. Luo, H., Gu, Y., Liao, X., Lai, S., Jiang, W.: Bag of tricks and a strong baseline for deep person re-identification. Presented at the Proceedings of the IEEE/CVF Conference on Computer Vision and Pattern Recognition Workshops (2019)
12. Jin, X., Lan, C., Zeng, W., Wei, G., Chen, Z.: Semantics-aligned representation learning for person re-identification. In: Proceedings of the AAAI Conference on Artificial Intelligence, vol. 34, pp. 11173–11180 (2020). https://doi.org/10.1609/aaai.v34i07.6775
13. Hou, R., Ma, B., Chang, H., Gu, X., Shan, S., Chen, X.: Interaction-and-aggregation network for person re-identification. Presented at the Proceedings of the IEEE/CVF Conference on Computer Vision and Pattern Recognition (2019)
14. Sun, Y., Zheng, L., Yang, Y., Tian, Q., Wang, S.: Beyond Part Models: Person Retrieval with Refined Part Pooling (and A Strong Convolutional Baseline). Presented at the Proceedings of the European Conference on Computer Vision (ECCV) (2018)
15. He, L., Liang, J., Li, H., Sun, Z.: Deep spatial feature reconstruction for partial person re-identification: alignment-free approach. Presented at the Proceedings of the IEEE Conference on Computer Vision and Pattern Recognition (2018)
16. Sun, Y., Xu, Q., Li, Y., Zhang, C., Li, Y., Wang, S., Sun, J.: Perceive where to focus: learning visibility-aware part-level features for partial person re-identification. Presented at the Proceedings of the IEEE/CVF Conference on Computer Vision and Pattern Recognition (2019)
17. Miao, J., Wu, Y., Liu, P., Ding, Y., Yang, Y.: Pose-guided feature alignment for occluded person re-identification. Presented at the Proceedings of the IEEE/CVF International Conference on Computer Vision (2019)
18. Wang, G., et al.: High-order information matters: learning relation and topology for occluded person re-identification. Presented at the Proceedings of the IEEE/CVF Conference on Computer Vision and Pattern Recognition (2020)
19. Wang, T., Liu, H., Song, P., Guo, T., Shi, W.: Pose-guided feature disentangling for occluded person re-identification based on transformer. Proceedings of the AAAI Conference on Artificial Intelligence, vol. 36, pp. 2540–2549 (2022). https://doi.org/10.1609/aaai.v36i3.20155

20. Wang, Z., Zhu, F., Tang, S., Zhao, R., He, L., Song, J.: Feature erasing and diffusion network for occluded person re-identification. Presented at the Proceedings of the IEEE/CVF Conference on Computer Vision and Pattern Recognition (2022)
21. Zhang, Z., Lan, C., Zeng, W., Jin, X., Chen, Z.: Relation-aware global attention for person re-identification. Presented at the Proceedings of the IEEE/CVF Conference on Computer Vision and Pattern Recognition (2020)
22. Dosovitskiy, A., et al.: An Image is Worth 16x16 Words: Transformers for Image Recognition at Scale. http://arxiv.org/abs/2010.11929 (2021). https://doi.org/10.48550/arXiv.2010.11929
23. Zhang, X., et al.: AlignedReID: Surpassing Human-Level Performance in Person Re-Identification. http://arxiv.org/abs/1711.08184 (2018). https://doi.org/10.48550/arXiv.1711.08184
24. Luo, C., Chen, Y., Wang, N., Zhang, Z.: Spectral feature transformation for person re-identification. Presented at the Proceedings of the IEEE/CVF International Conference on Computer Vision (2019)
25. Yang, W., Huang, H., Zhang, Z., Chen, X., Huang, K., Zhang, S.: Towards rich feature discovery with class activation maps augmentation for person re-identification. Presented at the Proceedings of the IEEE/CVF Conference on Computer Vision and Pattern Recognition (2019)
26. Zhang, A., Gao, Y., Niu, Y., Liu, W., Zhou, Y.: Coarse-to-fine person re-identification with auxiliary-domain classification and second-order information bottleneck. Presented at the Proceedings of the IEEE/CVF Conference on Computer Vision and Pattern Recognition (2021)
27. Zhong, Z., Zheng, L., Kang, G., Li, S., Yang, Y.: Random erasing data augmentation. In: Proceedings of the AAAI Conference on Artificial Intelligence, vol. 34, pp. 13001–13008 (2020). https://doi.org/10.1609/aaai.v34i07.7000
28. He, S., Luo, H., Wang, P., Wang, F., Li, H., Jiang, W.: TransReID: transformer-based object re-identification. Presented at the Proceedings of the IEEE/CVF International Conference on Computer Vision (2021)
29. Zhu, Z., Xu, M., Bai, S., Huang, T., Bai, X.: Asymmetric non-local neural networks for semantic segmentation. Presented at the Proceedings of the IEEE/CVF International Conference on Computer Vision (2019)
30. Zhao, H., Shi, J., Qi, X., Wang, X., Jia, J.: Pyramid scene parsing network. Presented at the Proceedings of the IEEE Conference on Computer Vision and Pattern Recognition (2017)
31. Zheng, L., Shen, L., Tian, L., Wang, S., Wang, J., Tian, Q.: Scalable person re-identification: a benchmark. Presented at the Proceedings of the IEEE International Conference on Computer Vision (2015)
32. Zheng, Z., Zheng, L., Yang, Y.: Unlabeled samples generated by GAN improve the person re-identification baseline in vitro. Presented at the Proceedings of the IEEE International Conference on Computer Vision (2017)
33. Wei, L., Zhang, S., Gao, W., Tian, Q.: Person transfer GAN to bridge domain gap for person re-identification. Presented at the Proceedings of the IEEE Conference on Computer Vision and Pattern Recognition (2018)
34. Li, W., Zhao, R., Xiao, T., Wang, X.: DeepReID: deep filter pairing neural network for person re-identification. Presented at the Proceedings of the IEEE Conference on Computer Vision and Pattern Recognition (2014)
35. Zheng, Z., Yang, X., Yu, Z., Zheng, L., Yang, Y., Kautz, J.: Joint discriminative and generative learning for person re-identification. Presented at the Proceedings of the IEEE/CVF Conference on Computer Vision and Pattern Recognition (2019)
36. Dai, Z., Chen, M., Gu, X., Zhu, S., Tan, P.: Batch DropBlock network for person re-identification and beyond. Presented at the Proceedings of the IEEE/CVF International Conference on Computer Vision (2019)

37. Zhang, Z., Lan, C., Zeng, W., Chen, Z.: Densely semantically aligned person re-identification. Presented at the Proceedings of the IEEE/CVF Conference on Computer Vision and Pattern Recognition (2019)
38. Chen, X., et al.: Salience-guided cascaded suppression network for person re-identification. Presented at the Proceedings of the IEEE/CVF Conference on Computer Vision and Pattern Recognition (2020)
39. Zhu, K., Guo, H., Liu, Z., Tang, M., Wang, J.: Identity-guided human semantic parsing for person re-identification. In: Vedaldi, A., Bischof, H., Brox, T., and Frahm, J.-M. (eds.) Computer Vision – ECCV 2020. pp. 346–363. Springer International Publishing, Cham (2020). https://doi.org/10.1007/978-3-030-58580-8_21

SpectralTracker: Jointly High and Low-Frequency Modeling for Tracking

Yimin Rong[1,2], Qihua Liang[1,2(✉)], Ning Li[1,2], Zhiyi Mo[1,2,3], and Bineng Zhong[1,2]

[1] Key Laboratory of Education Blockchain and Intelligent Technology, Ministry of Education, Guangxi Normal University, Guilin 541004, China
[2] Guangxi Key Laboratory of Multi-source Information Mining and Security, Guangxi Normal University, Guilin 541004, China
qhliang@gxnu.edu.cn
[3] Guangxi Key Laboratory of Machine Vision and Intelligent Control, Wuzhou University, Wuzhou 543002, China

Abstract. Recently, a considerable number of top-performing Transformer based trackers have been proposed. However, most of them mainly focus on utilizing low-frequency information from a spatial-spectral analysis perspective, limiting their performance in complicated scenes. To address this problem, we propose a spectral tracker that explores how to capture high and low-frequency information for robust tracking jointly. Specifically, we design a novel dual-spectral information extraction and aggregation module (DSM) consisting of a high and low-frequency branch to capture and combine complementary frequency information of a Transformer effectively. Firstly, we divide the local window in the high-frequency branch to focus on more fine-grained high-frequency information. Then, in the low-frequency branch, we apply AvgPooling with a low-pass effect on a Transformer to amplify its low-frequency information. Furthermore, we design a shared MLP strategy to polarize the dual-frequency branching to high and low-frequency information attention. Finally, we utilize an MLP to complementarily fuse high and low-frequency information for frequency domain modeling. Comprehensive experiments on five tracking benchmarks (i.e., GOT-10k, TrackingNet, LaSOT, UAV123 and TNL2K) show that our spectral tracker achieves better performance than the state-of-the-art trackers.

Keywords: Object Tracking · Vision Transformer · Spectral Domain · Single-stage Backbone

1 Introduction

Visual tracking aims to estimate the future state of target object in sequential video frames, given its initial state. This is a highly challenging task in computer

© The Author(s), under exclusive license to Springer Nature Singapore Pte Ltd. 2024
Q. Liu et al. (Eds.): PRCV 2023, LNCS 14436, pp. 212–224, 2024.
https://doi.org/10.1007/978-981-99-8555-5_17

Fig. 1. Comparison of the attention weight maps estimated by our proposed SpectralTrack and existing state-of-the-art Transformer-based visual tracker OSTrack [7].

vision due to factors such as heavy occlusion, abrupt changes, interference from similar objects and large deformation, among others. Although many trackers [1–6] have been proposed in the past few years and achieved high performance on existing benchmark datasets, these trackers almost invariably ignore the role of spectral domain information in the tracking process.

The traditional image processing approaches mainly include two types: spatial and spectral. Coincidentally, most studies focus more on exploiting spatial domain information to extract effective feature representations at the spatial level for object tracking. In the earlier years, the classic Siamese-based trackers [1–4,8] only extracted the spatial level features to model the entire architecture. With the introduction of Transformer from NLP to the visual domain, Transformer-based trackers [5,6,9,10] have led the modeling of trackers in a new direction. However, most trackers still focus only on spatial domain-level modeling to interpret Transformer. Recently, a growing body of research has indicated that attention mechanisms are more adept at capturing low-frequency signals but are slightly less effective at capturing high-frequency signals. It is mentioned in [11,12] that the global shapes and structures of the scene or object in the image represents the low-frequency signal, and conversely, local edges and textures are within the realm of high-frequency signals. Although the interpretation in the two recent Transformer-based works [7,13] is still based on the spatial domain, spectral domain-related signals exist in their feature information structure. However, these signals merely represent ambiguous frequency domain information that is difficult to control. This ambiguity can be attributed to the co-existence of high and low-frequency modeling within Transformer, without considering the strength of the spectral modeling capability.

Through the above analysis, we propose a Transformer-based spectral tracker, namely SpectralTracker, which utilizes explicit high and low-frequency signals of spectral domain to model the whole tracking process. Specifically, Dual-Spectral Module (DSM) is the core component of our method. It comprises two seemingly independent but interconnected spectral modeling branches that can capture and model the high and low-frequency feature information between the target template and the search region. In addition, we build our Spectral-

Tracker backbone with Patch Embedding and stacked multiple blocks of DSM, and finally supplement it with a simple convolution-based prediction head to form our entire tracking framework. By adopting this design, we can extract and aggregate the frequency domain features from both the target template and the search region, which effectively resolves the following problems in the tracking process: Firstly, the Transformer model tends to exhibit weaker high-frequency modeling capability, which results in difficulties in capturing local details among neighboring regions, and prevent the full utilization of its global high-frequency modeling ability. Secondly, filling all layers of the model with low-frequency information degrades high-frequency information, such as local textures, and weakens the overall modeling capability of the Transformer. Comparison of the attention weight maps estimated is shown in Fig. 1.

Our main contributions are three-fold: (1) We propose a novel Dual-Spectral information extract and aggregate Module (DSM). This allows for efficiently extracting the high and low-frequency signals of the spectral domain simultaneously in the DSM's high and low-frequency branches. (2) We propose a simple tracking backbone network with Dual-Spectral Module (DSM) as the core component to model the spectral features. Furthermore, we have designed a shared MLP strategy to enhance attention to high and low-frequency information at the same position. (3) Our tracker achieves comparable results to state-of-the-art trackers on multiple tracking benchmarks.

2 Related Work

2.1 Visual Tracking

Numerous trackers [1–4,8,14,15], which belong to Siamese trackers, use AlexNet or ResNet as the backbone of shared weights to extract the features of template and search region. A correlation operation is applied in these trackers as an important aggregation module. In recent years, with Transformer has been widely used in the field of computer vision, resulting in the emergence of several excellent trackers based on Transformer. TransT [5], STARK [6], TREG [9], TrDiMP [10], TrTr [16], MixFormer [13] and OSTrack [7] take innovative advantage of Transformer and achieve impressive performance. However, the theoretical support of both Siamese trackers and Transformer trackers is based on spatial domain modeling, which invariably focuses on the spatial information in the image and even only adds the temporal domain information between sequences to improve the tracking effect, while the frequency domain information always be ignored. In particular, the one-stream compact trackers MixFormer and OSTrack perform fuzzy frequency modeling, resulting in the loss of global high and local low-frequency information.

2.2 Frequency Modeling in Visual Transformer

As pointed out in [17], ViT [18] and its variants are more proficient at capturing global low-frequency spectral domain information, such as shapes and lines, than local high-frequency information, such as object edges and textures, within a scene. IFormer [12] expresses the same view that ViT's self-attention is a global operation for information exchange between patch tokens, which is better at capturing global information. However, this does not mean that ViT cannot model local high-frequency information, as demonstrated by LITv2 [11], which can also capture fine-grained local high-frequency information by combining the corresponding means with ViT. CvT [19] is a variant of ViT that combines convolutional operations, which are good at capturing local high-frequency spectral information. Therefore, we propose a Dual-Spectral Module that utilizes a combination of operations, such as attention window division and AvgPooling, to simultaneously model high and low-frequency features extraction and aggregation in a compact architecture. Our proposed SpectralTracker is a simple, compact, and efficient spectral-based tracker capable of capturing both local and global spectral information.

3 Method

In this section, we will describe our compact, dual-spectral tracking framework, named SpectralTracker. In the following, we will elaborate on our Module with Dual-Spectral information extraction and aggregation in Sect. 3.1. Then, we introduce the whole SpectralTracker in detail in Sect. 3.2 and Sect. 3.3.

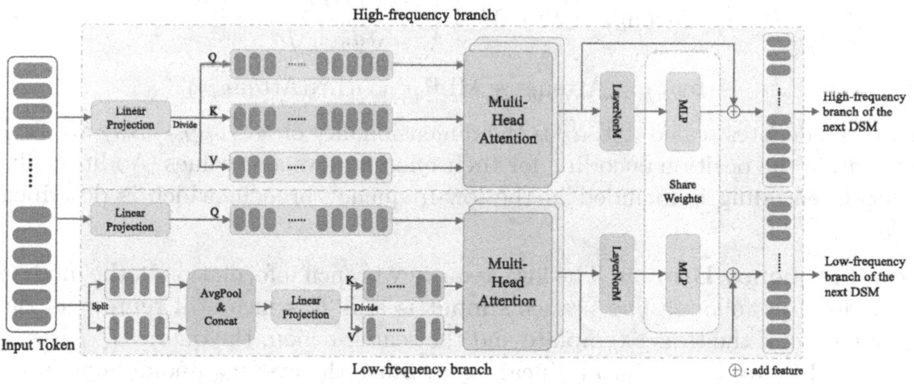

Fig. 2. Dual-Spectral Module (DSM) is the core component of our SpectralTracker, which consists of a high-frequency branch, a low-frequency branch and a shared MLP.

3.1 Dual-Spectral Module

Being the core design of our SpectralTracker, the Dual-Spectral Module (DSM) is designed to extract parallelly high and low-frequency features of token sequences and simultaneously model spectral information between them. In contrast to the traditional Attention mechanism [20], DSM performs mutually independent attention operations to model high and low-frequency features on the concatenated token sequences. This dual-spectral mechanism can be effectively implemented by having high and low-frequency branches in the module. As shown in Fig. 2, the attention-based Dual spectral mechanism could be implemented efficiently via the following process:

High-Frequency Branch. Given a concatenated token sequences T, which is concatenated by template token sequences $T_z \in \mathbb{R}^{N_z \times N}$ and search token sequences $T_x \in \mathbb{R}^{N_x \times N}$. N_z and N_x are the numbers of patches of template and search region, respectively, and N represents the dimensional size of the space of a high-dimensional feature projection.

To implement modeling of local high-frequency information, we reshape and divide the token sequences T into a local attention window of size 1×1 firstly, then map the token sequences on multiple feature spaces to generate feature map queries, keys and values. Formally, we generate them by a linear projection, reshape them to $\mathbb{R}^{3 \times h \times (N_z+N_x) \times N'}$, then divide them into $Q_{hf}, K_{hf}, V_{hf} \in \mathbb{R}^{h \times (N_z+N_x) \times N'}$, where Q_{hf}, K_{hf}, V_{hf} are the queries, keys, values of the high-frequency branch, h is the number of heads and $N' = N/3h$. Local window division is reflected in $\mathbb{R}^{3 \times h \times (N_z/n+N_x/n) \times N' \times n}$, where n is the number of local attention windows. The formulation of our high-frequency branch is as follows:

$$Q_{hf}, K_{hf}, V_{hf} = Div(\mathrm{FC}(T)). \tag{1}$$

$$\mathrm{Atten}_{hf} = SoftMax\left(\frac{Q_{hf} \cdot K_{hf}^T}{\sqrt{d_h}}\right) V_{hf}. \tag{2}$$

$$\mathrm{Atten}_{hf} = \mathrm{Atten}_{hf} + \mathrm{MLP}_{shared}(\mathrm{LN}(\mathrm{Atten}_{hf})) \tag{3}$$

where $(\cdot)$ denotes scaled dot, d_h is the dimensionality of K_{hf}. Q_{hf}, K_{hf} and V_{hf} all contain the position encoding for their queries, keys and values. Additionally, position encoding is included in the low-frequency branch, which is described below.

Low-Frequency Branch. The low-frequency branch information in the module is modeled as follows: The branch's input is still T, which still represents the information of the target template and the search region, the queries Q_{lf} is still computed by feeding T into a linear layer. Since the average pooling operation is a low-pass filter and to maximize the attention mechanism's ability to capture low-frequency signals better, we process the input token sequences T through an AvgPooling operation to generate low-frequency keys K_{lf} and values V_{lf} that all are distinct from the other branch. In this way, the interference of high-frequency signals is weakened. As a result, the branch can not only focus on modeling low-frequency information but also reduce the computation complexity of attention.

For the computation of keys K_{lf} and values V_{lf}, we first split the input token sequences T into sequences belonging to the target template (T_z) and the search region (T_x). We then embed the two token sequences using AvgPooling and concatenate the resulting outputs. Finally, we feed the concatenated output into a linear layer:

$$Q_{lf} = \mathrm{FC}(T). \tag{4}$$

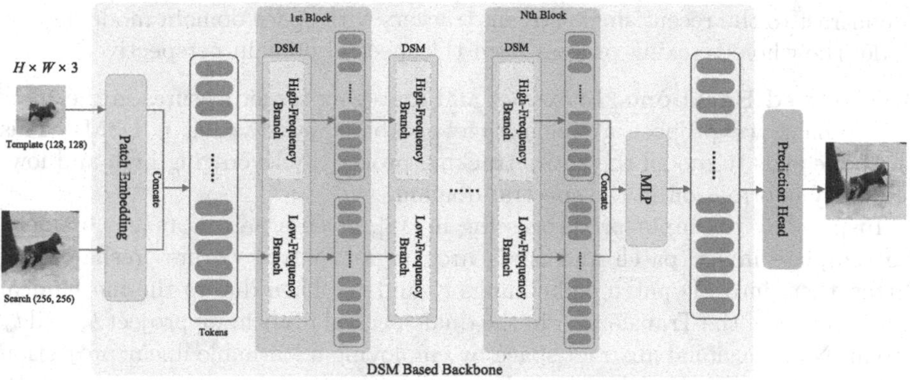

Fig. 3. The overall architecture of our SpectralTracker consists of a backbone based on DSM and a prediction head, which is an end-to-end tracking architecture.

$$T_{lo} = \text{Concat}\left[\text{AvgPool}\left(T_z\right), \text{AvgPool}\left(T_x\right)\right]. \tag{5}$$

$$K_{lf}, V_{lf} = Div(\text{FC}(T_{lo})). \tag{6}$$

where $K_{lf}, V_{lf} \in \mathbb{R}^{h\times(Pz+Px)\times N}$, P is the resolution of the target template and search region token sequences after the AvgPooling operation. Similarly, queries, keys and values also contain their position encoding. Besides, $Atten_{lf}$ is calculated in the same way as $Atten_{hf}$.

It is worth noting that the inputs of the first DSM and subsequent DSMs in the backbone are different, as illustrated in Fig. 2.

Shared MLP in Both Branches. Throughout the module, since the parallel high and low-frequency branches are unconnected in both input and output, they are entirely independent. Suppose the DSM with these two independent branches is applied to the whole tracking architecture. In that case, it is difficult to achieve simultaneous attention to the high and low-frequency signals, as demonstrated in the ablation experiments below.

In the point cloud [21], shared MLP transforms and extracts features for each point simultaneously. Inspired by this, we designed a shared MLP in both branches. Therefore, the high and low-frequency information at the same location of the two branches should be transformed and extracted synchronously instead of being considered as two completely independent branches. It provides a more expressive frequency domain feature for the complementary fusion of the MLP at the backbone end.

3.2 Dual-Spectral for Tracking

Overall Architecture. The overall architecture is depicted in Fig. 3. The core idea of SpectralTracker is to explore how to jointly capture high and low-frequency information for robust tracking. SpectralTracker consists of a DSM-based backbone and a prediction head based on a fully convolutional network.

In contrast to the recent single-stream trackers with space domain modeling, we model the whole tracking process from the spectral domain perspective.

DSM-Based Backbone. To extract and aggregate target features in the spectral domain, we propose a single-stage backbone by stacking n DSMs. This backbone aims to model the entire tracking process by leveraging high and low-frequency information in the spectral domain.

Inspired by the input pre-processing in [18], given a pair of images, namely, the template image patch and the search region patch, we first reshape and flatten them into 2D patch embeddings t_z and t_x. Then due to the mechanical requirements of the Transformer in the dual-spectral branch, we project t_z and t_x into an N-dimensional abstract space by employing a learnable linear projection with parameter E. Finally, we add trainable position embeddings P_z and P_x after the projection of token sequences are completed. The relevant equations are expressed as follows:

$$T_z = [t_z E] + P_z = [(\tau_z^1 E + \rho_z^1), \cdots, (\tau_z^{N_z} E + \rho_z^{N_z})], \tag{7}$$

$$T_x = [t_x E] + P_x = [(\tau_x^1 E + \rho_x^1), \cdots, (\tau_x^{N_x} E + \rho_x^{N_x})]. \tag{8}$$

where the size of outputs T_z and T_x are $N_z \times N$ and $N_x \times N$, respectively, the token sequences T_z and T_x are concatenated as T, as mentioned in Sect. 3.1.

The token sequences T will be fed into the first DSM's high and low-frequency branches to generate the corresponding spectral domain features, e.g., $T_{hf}^1, T_{lf}^1 \in \mathbb{R}^{(N_z+N_x)\times N}$. These features will be fed into the high and low-frequency branches of the subsequent N-1 DSMs in parallel and independently. At the final DSM, the obtained dual-frequency information, T_{hf}^n and T_{lf}^n will be concatenated together, and an MLP layer will be applied to achieve the transformation and merge modeling of the high and low-frequency feature information. Finally, the final spectral domain feature of the entire backbone is output and fed into the prediction head.

3.3 Prediction Head and Total Loss

Similar to [7], we estimate the bounding box of the tracked object directly by a full convolution-based prediction head and compute the classification map, local bias, and bounding box size using several stacked Conv-BN-ReLU layers. For the loss calculation, we employ the Gaussian weighted focal loss (referred as $\mathcal{L}_{cls}$) for the classification task. For the localization task, we adopt $\mathcal{L}_1$ loss. We use the generalized IoU loss (denoted as $\mathcal{L}_G$) for bounding box regression. Finally, our model is trained in an end-to-end fashion, and the total loss function follows:

$$\mathcal{L}_{total} = \lambda_1 \mathcal{L}_{cls} + \lambda_2 \mathcal{L}_1 + \lambda_3 \mathcal{L}_G \tag{9}$$

where $\lambda_1 = 1$, $\lambda_2 = 5$, $\lambda_3 = 2$ are the weight to balance the contributions of each loss function.

4 Experiments

4.1 Implementation Details

We adopt the training splits of the COCO [22], LaSOT [23], GOT-10k [24] and TrackingNet [25] as the training dataset. Our tracker is implemented using Python 3.7 and PyTorch 1.9.0. SpectralTracker training is conducted on 2 T V100 GPUs and the mini-batch size is set to 32, resulting in a total batch size of 64. We optimize the network using AdamW [28] with an initial learning rate of 4×10^{-5} for the DSM-based backbone, 4×10^{-4} for the rest, and weight decay set to 1×10^{-4}. For the full data training, we set the total training epochs to 300, sampled 60,000 image pairs per epoch, and adjusted the learning rate downward by a factor of 10 after 240 epochs. For the GOT-10k training, we set 100 epochs and adjusted the learning rate after 80 epochs.

4.2 State-of-the-Art Comparison

To verify the validity of the proposed models, we compare them with state-of-the-art trackers on five different benchmarks.

Table 1. Performance comparisons with state-of-the-art trackers on the test set of LaSOT [23], TrackingNet [25] and GOT-10k [24]. The best two results are shown in red and blue fonts.

Method	Source	GOT-10k [24]			TrackingNet [25]			LaSOT [23]		
		AO	$SR_{0.5}$	$SR_{0.75}$	AUC	P_{norm}	P	AUC	P_{norm}	P
SiamPRN++ [3]	CVPR19	51.7	61.6	32.5	73.3	80.0	69.4	49.6	56.9	49.1
DiMP [26]	ICCV19	61.1	71.7	49.2	74.0	80.1	68.7	56.9	65.0	56.7
Ocean [27]	ECCV20	61.1	72.1	47.3	–	–	–	56.0	65.1	56.6
TrDiMP [10]	CVPR21	67.1	77.7	58.3	78.4	83.3	73.1	63.9	–	61.4
TransT [5]	CVPR21	67.1	76.8	60.9	81.4	86.7	80.3	64.9	73.8	69.0
AutoMatch [28]	ICCV21	65.2	76.6	54.3	76.0	–	72.6	58.3	–	59.9
STARK [6]	ICCV21	68.8	78.1	64.1	82.0	86.9	–	67.1	77.0	–
KeepTrack [29]	ICCV21	67.1	77.2	70.2	–	–	–	–	–	–
UTT [30]	CVPR22	67.2	76.3	60.5	79.7	–	77.0	64.6	–	67.2
CSWinTT [31]	CVPR22	69.4	78.9	65.4	81.9	86.7	79.5	66.2	75.2	70.9
AiATrack [32]	ECCV22	69.6	63.2	80.0	82.7	87.8	80.4	69.0	79.4	73.8
MixFormer-22k [13]	CVPR22	70.7	80.0	67.8	83.1	88.1	81.6	69.2	78.7	74.7
SimTrack-B/16 [33]	ECCV22	69.4	78.0	64.3	82.5	87.0	80.4	69.6	78.6	74.1
OSTrack-256 [7]	ECCV22	71.0	80.4	68.2	83.1	87.8	82.0	69.1	78.7	75.2
SpectralTracker	Ours	73.5	82.9	69.8	83.4	88.0	82.1	68.9	78.2	74.5

Table 2. Comparison with state-of-the-art on UAV123 [34] and TNL2K [35] in terms of success score (AUC). The best two results are shown in red and blue fonts, respectively.

Method	SiamBAN [4]	STMTracker [36]	ToMP-50 [37]	TransT [5]	STARK [6]	MixFormer-1k [13]	TransInMo* [38]	OSTrack-256 [7]	SimTrack-B/16 [33]	SpectralTracker (Ours)
UAV123 [34]	63.1	64.7	66.9	68.1	68.2	68.7	69.0	68.3	69.8	70.2
TNL2K [35]	41.0	–	–	50.7	–	–	52.0	54.3	54.8	55.8

GOT-10k. The GOT-10k [24] is a large-scale dataset containing over 10,000 videos of real-world moving objects. Notably, GOT-10k proposes a protocol requiring the tracker to train only with its training set. In our study, we also followed this protocol to train our model and submit them to the official platform for evaluation. As shown in Table 1, we compared our tracker's average overlap (AO) and success rates (SR0.5 and SR0.75) with those of other state-of-the-art trackers and achieved a new state-of-the-art performance.

TrackingNet. TrackingNet [25] is a recently released large-scale short-term tracking dataset that provides a containing 511 video sequences in the test set. Our evaluation results on the testing dataset of TrackingNet are shown in Table 1. Our proposed method gets a success score (AUC) of 83.4%, a normalized precision score (P_{Norm}) of 88.0%, a precision score (P) of 82.1%.

UAV123. UAV123 [34] is a large-scale aerial tracking benchmark involving 123 challenging sequences with more than 112K frames, which are captured from low-altitude UAVs. As shown in Table 2, our SpectralTracker has achieved the best performance.

LaSOT. The LaSOT [23] is a large-scale benchmark with high-quality annotations, consisting of 1400 challenging sequences with an average video length is over 2500 frames. Compared with the top-performing trackers, as shown in Table 1, our model has a good performance in challenging videos.

TNL2K. We also evaluate our tracker on TNL2K [35] benchmark, which includes 700 video sequences. We evaluate our tracker on the dataset as shown in Table 2. It can be observed that SpectralTracker performs the best among all compared trackers.

4.3 Ablation Studies

In order to explore the role of the components of the proposed tracker, we have carried out comprehensive ablation studies on the GOT-10k dataset. First, we discuss the use of high and low-frequency branches with or without discussion, then explore the effectiveness of shared MLPs. Finally, we discuss the number of DSM that make up the backbone. The comparison results are shown in Table 3.

High and Low-Frequency Branches. Comparing experimental results between high-frequency modeling (experiment ①) and low-frequency modeling (experiment ②), it is apparent that the former outperforms the latter significantly. The reason for this difference lies in the ability of the high-frequency branch to capture local details of the target object, while the low-frequency branch captures the overall shape or structure of the scene or target. However, when comparing the results of experiments ①, ② and ⑥, it is found that the joint modeling of high-frequency and low-frequency components yields better results than either of the two alone. The joint modeling approach significantly improves the limitations of single-frequency modeling and spatial-domain modeling.

Table 3. Ablation studies on the variants of our tracker in GOT-10k benchmark [24].

	Methods			Number of DSM			Got-10k		
	High-Frequency Branch	Low-Frequency Branch	Shared MLP	8	10	12	AO	$SR_{0.5}$	$SR_{0.75}$
①	✓					✓	0.708	0.801	0.662
②		✓				✓	0.587	0.686	0.464
③	✓	✓				✓	0.701	0.801	0.644
④	✓	✓	✓	✓			0.703	0.802	0.651
⑤	✓	✓	✓		✓		0.715	0.808	0.669
⑥	✓	✓	✓			✓	0.735	0.829	0.698

Shared MLP. Based on the excellent effect of experiment ①, experiments ③ introducing the low-frequency branch led to a decrease in performance, which proves that independent high-frequency and low-frequency branches cannot further improve tracking performance. By observing experiments ①, ③ and ⑥, introducing the shared MLP strategy can significantly improve the performance of dual spectral tracking. With this strategy, ensuring that the independent dual-spectral branches will model at the same position is easier, further improving the effect.

Number of DSMs. Compared to experiments ④, ⑤ and ⑥, the overall performance of the tracker gradually improves with an increase in the number of DSMs. However, as the number of DSMs used in the backbone increases, training and testing of the structure become slower, and the model becomes larger and more cumbersome, and lead to a relatively less pronounced improvement. Therefore, we do not desire to sacrifice other performance factors to pursue better results and limit the number of DSMs to 12.

5 Conclusion

In this work, we proposed a novel, simple, Transformer-based tacking framework, which fully explores the joint capture of high and low-frequency signals of the spectral domain for robust tracking. We designed a Dual-Spectrum Module (DSM) that allows the extraction and aggregation of information in the frequency domain and only consists of high-frequency and low-frequency branches. Furthermore, we introduced a shared MLP to guide the dual-spectral branch to intensify high and low-frequency attention in a more polarized manner at the same location. Extensive experiments on five visual tracking benchmarks demonstrate that our SpectralTracker obtains state-of-the-art performance.

Acknowledgment. This work was supported by the Project of Guangxi Science and Technology (No. 2022GXNSFDA035079), the National Natural Science Foundation of China (No. 61972167 and U21A20474), the Guangxi "Bagui Scholar" Teams for Innovation and Research Project, the Guangxi Collaborative Innovation Center of Multi-source Information Integration and Intelligent Processing, the Guangxi Talent Highland Project of Big Data Intelligence and Application, and the Research Project of Guangxi Normal University (No. 2022TD002).

References

1. Bertinetto, L., Valmadre, J., Henriques, J.F., Vedaldi, A., Torr, P.H.S.: Fully-convolutional Siamese networks for object tracking. In: Hua, G., Jégou, H. (eds.) ECCV 2016. LNCS, vol. 9914, pp. 850–865. Springer, Cham (2016). https://doi.org/10.1007/978-3-319-48881-3_56
2. Li, B., Yan, J., Wu, W., Zhu, Z., Hu, X.: High performance visual tracking with Siamese region proposal network. In: CVPR (2018)
3. Li, B., Wu, W., Wang, Q., Zhang, F., Xing, J., Yan, J.: SiamRPN++: evolution of Siamese visual tracking with very deep networks. In: CVPR (2019)
4. Chen, Z., Zhong, B., Li, G., Zhang, S., Ji, R.: Siamese box adaptive network for visual tracking. In: CVPR (2020)
5. Chen, X., Yan, B., Zhu, J., Wang, D., Yang, X., Lu, H.: Transformer tracking. In: CVPR (2021)
6. Yan, B., Peng, H., Fu, J., Wang, D., Lu, H.: Learning spatio-temporal transformer for visual tracking. In: ICCV, pp. 10428–10437 (2021)
7. Ye, B., Chang, H., Ma, B., Shan, S.: Joint feature learning and relation modeling for tracking: a one-stream framework. In: Avidan, S., Brostow, G., Cissé, M., Farinella, G.M., Hassner, T. (eds.) ECCV. LNCS, vol. 13682. Springer, Cham (2022). https://doi.org/10.1007/978-3-031-20047-2_20
8. Guo, D., Wang, J., Cui, Y., Wang, Z., Chen, S.: SiamCAR: Siamese fully convolutional classification and regression for visual tracking. In: CVPR (2020)
9. Cui, Y., Jiang, C., Wang, L., Wu, G.: Target transformed regression for accurate tracking. arXiv Computer Vision and Pattern Recognition (2021)
10. Wang, N., Zhou, W., Wang, J., Li, H.: Transformer meets tracker: exploiting temporal context for robust visual tracking. In: Computer Vision and Pattern Recognition (2021)
11. Pan, Z., Cai, J., Zhuang, B.: Fast vision transformers with HiLo attention. CoRR (2022)
12. Si, C., Yu, W., Zhou, P., Zhou, Y., Wang, X., Yan, S.: Inception transformer. CoRR (2022)
13. Cui, Y., Jiang, C., Wang, L., Wu, G.: MixFormer: end-to-end tracking with iterative mixed attention. In: CVPR, pp. 13598–13608. IEEE (2022)
14. Zhu, Z., Wang, Q., Li, B., Wu, W., Yan, J., Hu, W.: Distractor-aware Siamese networks for visual object tracking. In: Ferrari, V., Hebert, M., Sminchisescu, C., Weiss, Y. (eds.) ECCV 2018. LNCS, vol. 11213, pp. 103–119. Springer, Cham (2018). https://doi.org/10.1007/978-3-030-01240-3_7
15. Zheng, Y., Zhong, B., Liang, Q., Tang, Z., Ji, R., Li, X.: Leveraging local and global cues for visual tracking via parallel interaction network. IEEE Trans. Circuits Syst. Video Technol. **33**(4), 1671–1683 (2022)
16. Zhao, M., Okada, K., Inaba, M.: TrTr: visual tracking with transformer. arXiv preprint arXiv:2105.03817 (2021)

17. Park, N., Kim, S.: How do vision transformers work? In: ICLR (2022)
18. Dosovitskiy, A., et al.: An image is worth 16 × 16 words: transformers for image recognition at scale. In: ICLR (2021)
19. Wu, H., et al.: CVT: introducing convolutions to vision transformers. In: ICCV, pp. 22–31. IEEE (2021)
20. Vaswani, A., et al.: Attention is all you need. In: Neural Information Processing Systems (2017)
21. Qi, C.R., Su, H., Mo, K., Guibas, L.J.: PointNet: deep learning on point sets for 3D classification and segmentation. In: CVPR (2017)
22. Lin, T.-Y., et al.: Microsoft COCO: common objects in context. In: Fleet, D., Pajdla, T., Schiele, B., Tuytelaars, T. (eds.) ECCV 2014. LNCS, vol. 8693, pp. 740–755. Springer, Cham (2014). https://doi.org/10.1007/978-3-319-10602-1_48
23. Fan, H., et al.: LaSOT: a high-quality benchmark for large-scale single object tracking. In: CVPR (2019)
24. Huang, L., Zhao, X., Huang, K.: GOT-10k: A large high-diversity benchmark for generic object tracking in the wild. IEEE Trans. Pattern Anal. Mach. Intell. **43**(5), 1562–1577 (2021)
25. Müller, M., Bibi, A., Giancola, S., Alsubaihi, S., Ghanem, B.: TrackingNet: a large-scale dataset and benchmark for object tracking in the wild. In: Ferrari, V., Hebert, M., Sminchisescu, C., Weiss, Y. (eds.) ECCV 2018. LNCS, vol. 11205, pp. 310–327. Springer, Cham (2018). https://doi.org/10.1007/978-3-030-01246-5_19
26. Bhat, G., Danelljan, M., Gool, L.V., Timofte, R.: Learning discriminative model prediction for tracking. In: ICCV (2019)
27. Zhang, Z., Peng, H., Fu, J., Li, B., Hu, W.: Ocean: object-aware anchor-free tracking. In: Vedaldi, A., Bischof, H., Brox, T., Frahm, J.-M. (eds.) ECCV 2020. LNCS, vol. 12366, pp. 771–787. Springer, Cham (2020). https://doi.org/10.1007/978-3-030-58589-1_46
28. Zhang, Z., Liu, Y., Wang, X., Li, B., Hu, W.: Learn to match: automatic matching network design for visual tracking. In: ICCV (2021)
29. Mayer, C., Danelljan, M., Paudel, D.P., Gool, L.V.: Learning target candidate association to keep track of what not to track. In: ICCV, pp. 13424–13434. IEEE (2021)
30. Ma, F., et al.: Unified transformer tracker for object tracking. In: CVPR (2022)
31. Song, Z., Yu, J., Chen, Y.P., Yang, W.: Transformer tracking with cyclic shifting window attention. In: CVPR (2022)
32. Gao, S., Zhou, C., Ma, C., Wang, X., Yuan, J.: AiATrack: attention in attention for transformer visual tracking. In: Avidan, S., Brostow, G., Cissé, M., Farinella, G.M., Hassner, T. (eds) Computer Vision – ECCV 2022. LNCS, vol. 13682, pp. 146–164. Springer, Cham (2022). https://doi.org/10.1007/978-3-031-20047-2_9
33. Chen, B., et al.: Backbone is all your need: a simplified architecture for visual object tracking. In: Avidan, S., Brostow, G., Cissé, M., Farinella, G.M., Hassner, T. (eds.) ECCV. LNCS, vol. 12356. Springer, Cham (2022). https://doi.org/10.1007/978-3-030-58621-8_6
34. Mueller, M., Smith, N., Ghanem, B.: A benchmark and simulator for UAV tracking. In: Leibe, B., Matas, J., Sebe, N., Welling, M. (eds.) ECCV 2016. LNCS, vol. 9905, pp. 445–461. Springer, Cham (2016). https://doi.org/10.1007/978-3-319-46448-0_27
35. Wang, X., et al.: Towards more flexible and accurate object tracking with natural language: algorithms and benchmark. In: CVPR (2021)
36. Fu, Z., Liu, Q., Fu, Z., Wang, Y.: STMTrack: template-free visual tracking with space-time memory networks. In: CVPR (2021)

37. Mayer, C., et al.: Transforming model prediction for tracking. In: CVPR (2022)
38. Guo, M., et al.: Learning target-aware representation for visual tracking via informative interactions. In: Raedt, L.D. (ed.) IJCAI (2022)

DiffusionTracker: Targets Denoising Based on Diffusion Model for Visual Tracking

Runqing Zhang, Dunbo Cai, Ling Qian, Yujian Du(✉), Huijun Lu, and Yijun Zhang

China Mobile (Suzhou) Software Technology Co., Ltd., Suzhou 215153, China
zhangrunqing_yewu@cmss.chinamobile.com, zrq1993@bupt.cn

Abstract. The problem of background clutter (BC) is caused by distractors in the background that resemble the target's appearance, thereby reducing the precision of visual trackers. We consider these similar distractors as noise and formulate a denoising task to solve the visual tracking problem. We propose a target denoising method based on a diffusion model for visual tracking, referred to as DiffusionTracker, which introduces the diffusion model to distinguish between targets and noise (distractors). Specifically, we introduce a reverse diffusion model to eliminate noisy distractors from the proposal candidates generated by the Siamese tracking backbone. To handle the difficulty that distractors do not strictly conform to a Gaussian distribution, we incorporate Spatial-Temporal Weighting (STW) to integrate spatial correlation and noise decay time information, mitigating the impact of noise distribution on denoising effectiveness. Experimental results demonstrate the effectiveness of the proposed method, with DiffusionTracker achieving a precision of 64.0% on BC sequences and a success rate of 63.8% on BC sequences from the LaSOT test datasets, representing improvements of 11.7% and 10.2% respectively over state-of-the-art trackers. Furthermore, our proposed method can be seamlessly integrated as a plug-and-play module with cutting-edge tracking algorithms, significantly improving the success rate for tracking task in background clutter scenarios.

Keywords: Visual Tracking · Diffusion Model · Siamese Network · Deep Learning · Denoising

1 Introduction

Visual tracking is one of the fundamental tasks in computer vision, widely used in the civil and military fields [8, 30]. In real-world applications, there are distractors similar to the target in appearance around the target to be tracked, which is known as the background clutter (BC) problem as shown in Fig. 1. This problem causes the tracker to fail to distinguish between the target and the distractors (Fig. 2).

© The Author(s), under exclusive license to Springer Nature Singapore Pte Ltd. 2024
Q. Liu et al. (Eds.): PRCV 2023, LNCS 14436, pp. 225–237, 2024.
https://doi.org/10.1007/978-981-99-8555-5_18

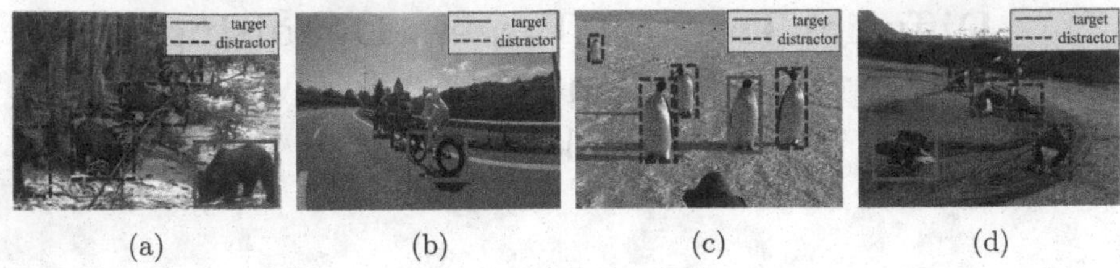

Fig. 1. The distractors (yellow) in (a) bear-17, (b) bicycle-7, (c) bird-2, and (d) bird-15 from the LaSOT dataset. (Color figure online)

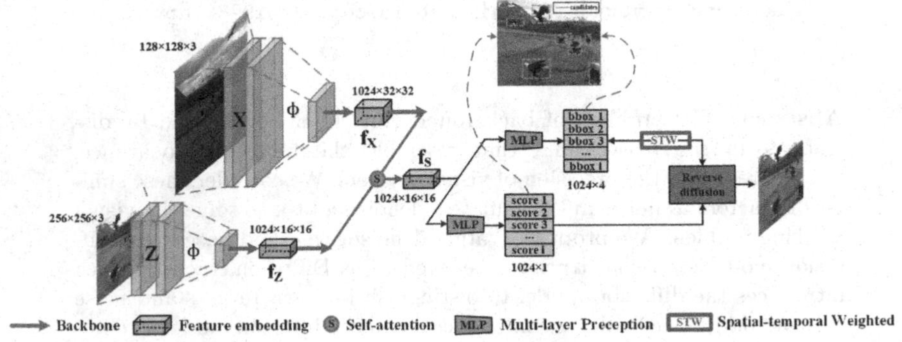

Fig. 2. The framework of our DiffusionTracker.

Current state-of-the-art trackers have improved the extraction of feature embeddings by employing deeper network structures, attention mechanisms, and search area constraints [4,15,28], which rely on target appearance information to enhance tracking performance. However, these methods overlook the potential spatial probability distribution relationship between distractors and the target, leading to difficulties to distinguish objects in the background that are similar to the target, resulting in tracking loss.

This paper proposes a method that addresses the tracking task by converting it into a denoising problem. Our approach utilizes a diffusion model to remove distractors, which are treated as noise, in the background that share a similar appearance with the target. Specifically, we first construct a tracking pipeline to generate candidates that contain targets and distractors, each assigned a confidence score. Then, we train the diffusion model to explore the potential probability distribution of the distractors and remove them during the backward diffusion process. By leveraging the spatial probability distribution relationship between the distractors and the target, our method achieves more accurate target tracking in the presence of background clutter, while utilizing a Spatial-temporal Weighted (STW) strategy. Our contributions are summarized as:

1. The proposed DiffusionTracker introduces reverse diffusion to gradually eliminate noisy candidates for distractor discrimination, which improves the tracking success rate in scenarios containing background clutter.

2. Based on spatial-temporal weighting, the proposed tracker utilizes spatial distribution information and noise decay time to enhance the confidence of distractors, enabling noise samples that do not meet the Gaussian distribution to also benefit from reverse diffusion.
3. The proposed method achieved superior performance on four public datasets: LaSOT, GOT-10k, TrackingNet, and OTB-2015. The proposed method can be combined with cutting-edge tracking algorithms as a module to improve the tracking performance in scenes with background clutter.

2 Related Works

In this section discusses the algorithms that are closely related to this paper, including the visual tracking based on Siamese Network and diffusion model.

2.1 Visual Tracking Based on Siamese Network

Siamese Network and Self-correlation Networks are two popular approaches in visual tracking that have gained significant attention due to their effectiveness and unique characteristics.

Siamese Networks revolutionize tracking task by employing end-to-end networks that extract adaptive cross-correlation features to discriminate between the target and the background. SiamRPN [16], a notable method in this category, improves tracking efficiency by introducing bounding box regression based on anchors instead of multi-scale estimation. Building upon SiamRPN, SiamRPN++ [15] enhances feature extraction using ResNet as the backbone network, resulting in more robust feature maps. Another approach, Target-aware deep tracking (TADT) [18], incorporates attention mechanisms based on Siamese networks to select effective feature channels. Based on the Siamese Network, methods based on the Transformer utilized the self-attention layer instead of the cross-correlation layers, such as Transformer tracking (TransT) [4], effectively incorporating semantic information such as target posture and spatial self-structure. This technique significantly enhances feature extraction by capturing rich semantic information.

Despite the enhancement of feature embeddings that describe the appearance of objects, these tracking algorithms have not fully utilized the distribution information of similar distractors and the target. As a result, the success rate of tracking in background clutter scenarios still requires further improvement.

2.2 Diffusion Model

Diffusion models progressively add noise on image data and then learn to recover the original data. These models add noise in a smooth manner and estimate the score function to guide the denoising process.

Diffusion models have been extensively utilized in various image processing tasks, such as Image Super Resolution, Semantic Segmentation, and Anomaly

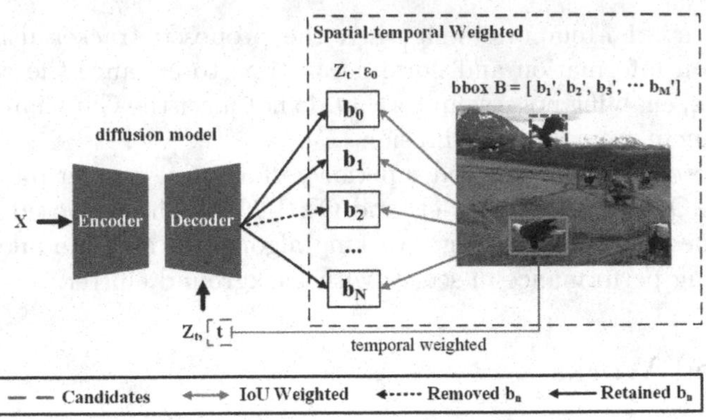

Fig. 3. The Reverse Diffusion process for time t during inference.

Detection [1,3,17,20]. For image resolution tasks, methods like Super-Resolution via Repeated Refinement (SR3) [23] and Cascaded Diffusion Model (CDM) [12] employ diffusion models to achieve high-quality results in super-resolution and inpainting. Some approaches [21,24] utilize pre-trained autoencoders to shift the diffusion process to the latent space for efficient training. In semantic segmentation, generative pre-training with diffusion models has proven effective in leveraging learned representations for accurate label utilization. Methods like Decoder Denoising Pretraining (DDeP) [2] and ODISE [27] have demonstrated promising results in label-efficient semantic segmentation. Diffusion models also excel in anomaly detection by modeling normal data and reconstructing healthy approximations. AnoDDPM [26] and DDPM-CD [10] utilize diffusion models to detect anomalies in input images, outperforming adversarial training-based alternatives.

While diffusion models have found widespread applications, their utilization in detection or tracking tasks remains challenging. In this context, our paper represents the first attempt to apply a diffusion model in a tracking task, serving as an academic exploration of generative algorithms.

3 Method

In this section, we introduce our DiffusionTrack to distinguish targets from the distractors, including the architecture, training process and inference process in Sect. 3.1, Sect. 3.1 and Sect. 3.1, respectively.

3.1 Architecture

The network architecture encompasses a primary tracking framework network and a reverse diffusion model with Spatial-temporal Weighted.

Tracking Framework. The tracking framework network, which is based on a Siamese network, is utilized for extracting image features and generating multiple proposal bounding boxes, including input heads, feature embeddings, and prediction heads.

Fig. 4. Some examples of the datasets for experiments.

The Siamese network contains two input heads, including the template branch for the template image patch **Z** and anther head for the search region patch **X**. Both the template patch and the search region patch are then reshaped into dimensions of $3 \times 128 \times 128$ and $3 \times 256 \times 256$, respectively.

For feature embeddings extraction, we modify the backbone network, utilizing a modified version of ResNet50 [11] with the convolution stride of the down-sampling layer in the fourth stage to 1 to enhance feature resolution for the tracking task. The output of the fourth stage in ResNet50 is set as feature output $\mathbf{f_Z}$ and $\mathbf{f_X}$ for both two branches. In order to capture the correlation between the template feature map $\mathbf{f_Z}$ and the search region feature map $\mathbf{f_X}$, a self-attention layer is employed to generate a feature map $\mathbf{f_S}$ that encapsulates semantic information.

The prediction head comprises two branches to predicts 1024 coordinates: a classification branch for confidence scores and a bounding box regression branch. Each branch consists of a multi-layer perceptron (MLP) with three linear layers and a ReLU activation function.

Reverse Diffusion Model. The reverse diffusion model f_θ with the learnable parameter θ is employed to discern candidate boxes resembling the target object from proposal bounding boxes containing noise, as shown in Fig. 3. The distractors are removed, while the vaild boxes are retained.

Feature Encoder. To avoid extracting high-level features at each step of the reverse diffusion process, the feature encoding network shares parameters with the tracking framework network. Furthermore, during each frame of tracking, image features $\mathbf{f_S}$ are extracted in a single pass. In the reverse diffusion process, noisy boxes z_T are generated by utilizing Gaussian random sampling based on the previous frame's target position, where T is the total number of the reverse diffusion steps. Corresponding features are then extracted from the image features for each bounding box.

Tracking Decoder. In the Reverse Diffusion model, the tracking decoder is used to find out the noise ε_0 from the noisy boxes z_t, where $t \in [1, 2, ..., T]$ is the reverse step. The tracking decoder separates the boxes from the encoder into valid boxes and noise. The noise boxes are removed, and the valid boxes are retained as shown in Fig. 3. Similar with Sparse R-CNN, there are regression and classification heads in the tracking decoder. The output features obtained from the feature encoder are fed into the detection head of the decoder, resulting in regression and classification results for the boxes.

Spatial-Temporal Weighted. Considering that real distractors may not strictly follow a Gaussian distribution, we introduce the spatial-temporal weighted (STW) strategy. We utilize the proposal boxes from the tracking framework as the candidates. The boxes from the Reverse Diffusion Model serve as weighted reference samples for noise removal. Since the Reverse Diffusion Model and proposal boxes do not strictly correspond on a one-to-one basis, we have designed two weighting methods based on spatial and temporal aspects. Firstly, in terms of spatial weighting, we calculate the Intersection over Union (IoU) between the predicted noisy box z_t from the Reverse Diffusion Model and the proposal box B to measure their correlation. A higher IoU indicates a higher likelihood of the proposal box being noise. Secondly, in terms of temporal weighting, during the reverse diffusion process, boxes that disappear earlier (corresponding to smaller t) are more likely to be the distractor.

3.2 Training Process

Giving a group of noisy boxes $z_t = [b_0, b_1, ..., b_n, ..., b_N], n \in [1, 2, ..., N]$, we utilize the reverse process to distinguish the distractors (noise) ε_0 and the target $(z_t - \varepsilon_0)$. b_n is the coordinate of the bounding box, including the center position, weight, and height. The purpose of the training process is learning a network f_θ to estimate ε_0 for each step t.

We model the denoising process as the reverse diffusion process. Given a group of noisy boxes z_t and the input image $\mathbf{X}$, the objective function can be represented as:

$$L_{train} = \frac{1}{2}||f_\theta(z_t, t, \mathbf{X}) - \varepsilon_0||^2 \quad (1)$$

where $\hat{\varepsilon}_0 = f_\theta(z_t, t, x)$ is the forecast value of ε_0. z_t, ε_0 is unknown for each step t. We can obtain these samples during the forward process of the diffusion model. Given a input image $\mathbf{X}$ with target's bounding box z_0 obeyed distribution $q(z_0)$, the noisy boxes groups $z_1, z_2, ..., z_t, ..., z_T$ are obtained by cumulatively adding Gaussian noise $\varepsilon_0 \sim N(0, I)$:

$$z_t = \sqrt{\alpha_t} z_{t-1} + \sqrt{1-\alpha_t}\varepsilon_0 \tag{2}$$

where the variance of Gaussian distribution is $\beta_t \in (0,1)$, and $\alpha = 1-\beta_t$. According to the linear properties of Gaussian distribution, z_T can be represented by z_0, ε_0, and $\{\alpha_t\}$:

$$z_T = \bar{\alpha}_T^{\frac{1}{2}} z_0 + (1-\bar{\alpha}_T)^{\frac{1}{2}}\varepsilon_0 \tag{3}$$

where $\bar{\alpha}_T = \prod_{t=0}^{T} \alpha_t$. During the forward process, $\{z_t\}$ from each step and ε_0 are recorded as the input and the label in Eq.(1).

3.3 Inference Process

During reverse diffusion process, the noisy boxes z_T is restored to z_0 step by step. According to Bayes Rule, the distribution of z_{t-1} can be represented as:

$$q(z_{t-1}|z_t, z_0) = q(z_t|z_{t-1}, z_0)\frac{q(z_{t-1}|z_0)}{q(z_t|z_0)} \tag{4}$$

where $q(z_t|z_{t-1}, z_0)$, $q(z_{t-1}|z_0)$, and $q(z_t|z_0)$ can be calculated as:

$$q(z_{t-1}|z_0) = \sqrt{\bar{\alpha}_{t-1}} z_0 + \sqrt{1-\bar{\alpha}_{t-1}}\varepsilon_0 \sim N(\sqrt{\bar{\alpha}_{t-1}} z_0, 1-\bar{\alpha}_{t-1}) \tag{5}$$

$$q(z_t|z_0) = \sqrt{\bar{\alpha}_t} z_0 + \sqrt{1-\bar{\alpha}_t}\varepsilon_0 \sim N(\sqrt{\bar{\alpha}_t} z_0, 1-\bar{\alpha}_t) \tag{6}$$

$$q(z_t|z_{t-1}, z_0) = \sqrt{\alpha_t} z_{t-1} + \sqrt{1-\alpha_t}\varepsilon_0 \sim N(\sqrt{\bar{\alpha}_t} z_{t-1}, 1-\bar{\alpha}_t) \tag{7}$$

Substituting Eqs. (4), (5), and (6) into Eq. (3), we can obtain:

$$q(z_{t-1}|z_t, x_0) \propto \exp(-\frac{1}{2}((\frac{\alpha_t}{\beta_t} + \frac{1}{1-\bar{\alpha}_{t-1}} z_{t-1}^2 - (\frac{2\sqrt{\alpha_t}}{\beta_t} z_t + \frac{2\sqrt{\bar{\alpha}_t}}{1-\bar{\alpha}_{t-1}} z_0) + C(z_t, z_0))) \tag{8}$$

where $C(z_t, z_0)$ is the set of terms in the Eq. (8) that are not related to z_{t-1}. According to the form of the Gaussian probability density function, the Eq. (9) can be balanced by completing the square, and the mean μ_{z_t} and variance $\sigma_{z_t}^2$ can be calculated as:

$$\mu_{z_t} = \frac{\sqrt{\alpha_t}(1-\bar{\alpha}_{t-1})}{1-\bar{\alpha}_t} z_t + \frac{\sqrt{\bar{\alpha}_{t-1}}\beta_t}{1-\bar{\alpha}_t} z_0 \tag{9}$$

$$\sigma_{z_t}^2 = \frac{1}{\frac{\alpha_t}{\beta_t} + \frac{1}{1-\bar{\alpha}_{t-1}}} \tag{10}$$

By substituting Eq. (2) into Eq. (9), we obtain:

$$\mu_{z_t} = \frac{1}{\sqrt{\alpha_t}}(z_t - \frac{\beta_t}{\sqrt{(1-\bar{\alpha}_t)}}\hat{\varepsilon}_0) \tag{11}$$

During the inference process, $\hat{\varepsilon}_0$ can be predicted by the diffusion network f_θ.

Table 1. Ablation Study on LaSOT dataset.

Method	diffusion model	STW	Success - All (%)	Success - only BC (%)
Baseline			64.9	57.9
DiffusionTracker-	✓		63.7	59.3
DiffusionTracker	✓	✓	**67.3**	**63.8**

4 Experiments

To validate the effectiveness of the proposed method, we carry out extensive experiments comparing the state-of-the-art methods on four challenging datasets, including LaSOT [9], GOT-10k [14], TrackingNet [19], and OTB-2015 [25], as shown in Fig. 4.

In this section, firstly, we provide the implementation details of the proposed method in Sect. 4.1. Then, we conduct the ablation study, general datasets evaluation on four datasets, tracking attributes evaluation, and compatibility experiment are conducted in Sect. 4.2, Sect. 4.3, Sect. 4.4 and Sect. 4.5, respectively.

4.1 Implementation Details

The training sets of TrackingNet, LaSOT, and GOT-10k are used to train the proposed tracking method. The training samples consist of image pairs from the same sequence with common data augmentation techniques, such as translation and brightness correction. The backbone branch is initialized with ResNet50, pre-trained on the ImageNet dataset [22]. The proposed model is trained using AdamW, with the learning rate of the backbone set to 1e−5 and the learning rate of the other branches set to 1e−4. The training process and subsequent experiments were conducted on a computer equipped with V100 GPUs.

4.2 Ablation Study

The proposed tracking methods take advantage of both the reverse diffusion and the STW. We perform an ablation study from these two modules on the LaSOT test dataset. The experimental results are listed in Table 1. The baseline method TransT [4] is based on the self-attention module.

The advantages of the proposed method are obvious both in terms of position precision and success rate. Compared with the baseline method TransT [4], the proposed method achieves higher precision and success rate utilizing the reverse diffusion and STW. The utilization of the reverse diffusion model enables effective differentiation of distractors. However, the distractors do not strictly adhere to a Gaussian distribution, which limits the effectiveness of the diffusion model. By incorporating the STW module, our approach employs spatial-temporal confidence weighting, allowing non-Gaussian background objects to benefit from the guidance of the diffusion model. Through the integration of the diffusion model and STW module, our method achieves more accurate target detection, resulting in an improved overlap rate between the detection box and ground truth.

4.3 General Datasets Evaluation

We compare the proposed method against the state-of-the-art tracking methods, including TransT [4], SiamFC++ [28], SiamMask [13], JCAT [29], MixFormer [5], TADT [18], ASRCF [6], ECO [7], and SiamRPN++ [15]. The overall performance (%) on LaSOT [9] and other datasets is shown in Fig. 5 and Table 2.

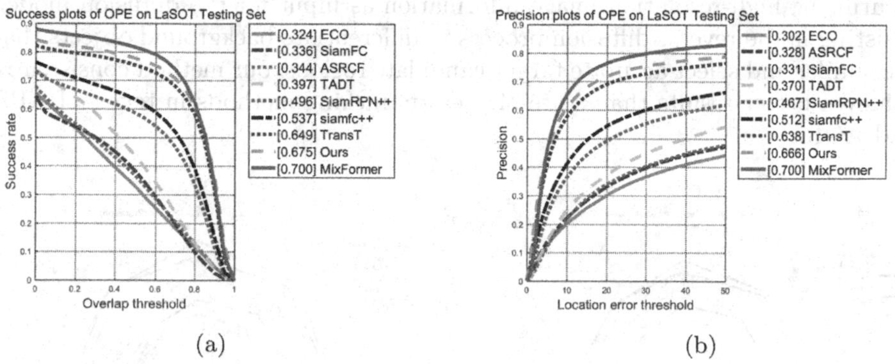

(a) (b)

Fig. 5. Quantitative ablation study on LaSOT test dataset.

In the LaSOT test dataset, the occurrence of background clutter presents a similar phenomenon to the presence of distractors in the background. As shown in Fig. 5, the proposed method outperforms the baseline with a precision of 66.5% and a success rate of 67.3%. This improvement in performance can be attributed to the effective enhancement of the tracker's discriminative capability achieved through the utilization of the reverse diffusion module, leading to an enhanced success rate in tracking.

Table 2. General Datasets Evaluation on OTB-2015, GOT-10k, and TrackingNet.

	OTB-2015		GOT-10k		TrackingNet		
Method	APE	AOR	AO	$SR_{0.5}$	Precision	NormPrecision	Success
SiamFC++ [28]	91.5	68.3	59.5	69.5	75.4	70.4	80.0
SiamRPN++ [15]	90.3	69.6	51.7	61.5	73.3	69.4	80.0
TADT [18]	78.4	65.0	36.7	38.9	54.0	67.1	59.3
JCAT [29]	87.0	63.5	66.6	76.3	74.6	68.8	78.3
ECO [7]	–	69.1	31.6	30.9	56.1	48.9	62.1
ASRCF [6]	92.2	69.2	31.3	31.7	30.2	15.3	27.3
TransT [4]	–	69.4	72.3	82.4	81.4	86.7	80.3
SiamMask [13]	–	–	51.4	58.7	66.4	77.8	72.5
Ours	**92.4**	**70.4**	**72.5**	**82.7**	**81.6**	**87.2**	**80.9**

Other datasets such as GOT-10k [14], TrackingNet, and OTB-2015 are also employed for comparison in Table 2. Specitially, for GOT-10k, AO means average overlap rate and $SR_{0.5}$ means success rate. The parameters 0.5 is the threshold of the success rate. Leveraging spatial-temporal and distractors distribution information, the proposed method effectively extracts more precise features for challenging targets, yielding a superior performance with 81.6% precision for TrackingNet and 92.4% APE for OTB-2015. The proposed method is capable of utilizing high-resolution image information as input for the diffusion model, assisting in the reverse diffusion process to differentiate background objects that are similar and select accurate target candidate regions. our method consistently achieves better results than state-of-the-art tracking methods in terms of APE and AOR.

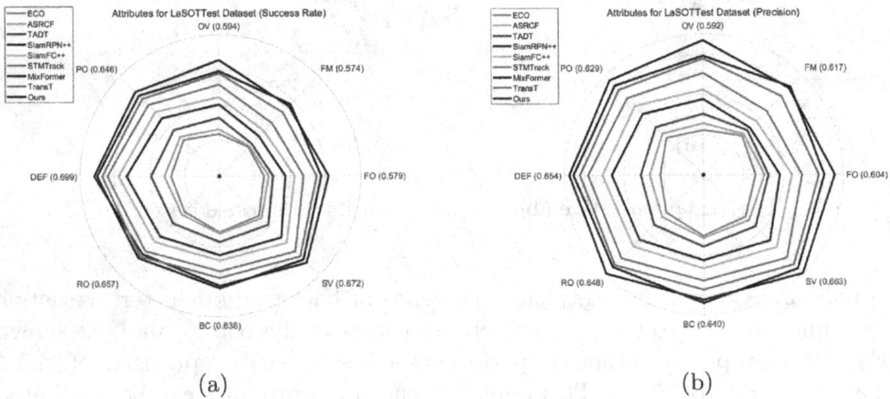

(a) (b)

Fig. 6. Quantitative experiment on the LaSOT test dataset.

4.4 Attributes Evaluation

We compare our method with the state-of-the-art tracking methods on LaSOT test datasets with different attributes, including Partial Occlusion (PO), Background Clutter (BC), Out-of-view (OV), Fast Motion (FM), Full Occlusion (FO), Scale Variation (SV), Rotation (RO), and Deformation (DEF).

Figure 6 illustrates the superior performance of our method compared to the baseline approach across various challenging attributes. What's more, Fig. 7 allows a qualitative comparison between the performance of the proposed method and the baseline algorithm TransT [4] on the sequences containing background clutter from LaSOT test dataset. Benefiting from the reverse diffusion and STW modules, our proposed method achieves a tracking success rate of 63.8% and a tracking accuracy of 64.0%.

4.5 Compatibility Experiment

We combine the proposed method as a pluggable module with state-of-the-art tracking algorithms. By utilizing this approach, we can differentiate targets from distractors in the proposal candidates provided by cutting-edge tracking algorithms. Our experiments are conducted on the LaSOT test dataset, which includes the BC attribute. As shown in Table 3 and Fig. 7, the proposed method can be effectively integrated with tracking algorithms that utilize the proposal candidate strategy, leading to enhanced tracking performance over 4% in background clutter scenarios.

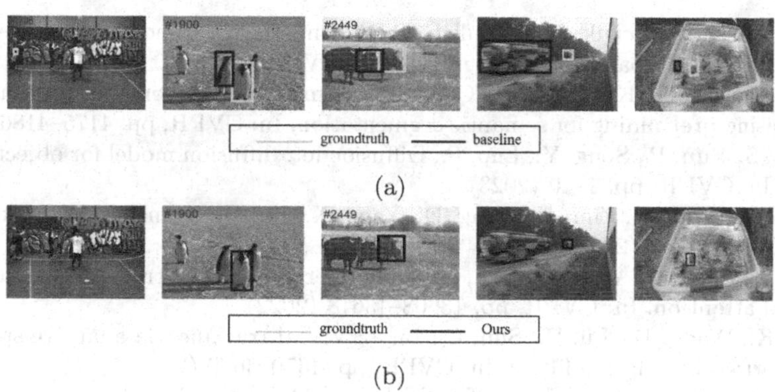

Fig. 7. Quantitative experiment on the LaSOT test dataset.

Table 3. Compatibility Experiment on the sequences containing BC problem in LaSOT test dataset.

Method	Precision (%)	Success (%)	Precision (+ours) (%)	Success (+ours) (%)
SiamFC++ [28]	46.7	48.0	48.1 ↑	52.9 ↑
SiamRPN++ [15]	42.6	44.9	45.7 ↑	49.3 ↑
ASRCF [6]	31.5	32.6	35.2 ↑	37.8 ↑

5 Conclusion

This paper proposed a visual tracking method for the background clutter problem, utilizing the Siamese network and diffusion model. Firstly, reverse diffusion is employed, treating the process of distinguishing the target from background interferences as a denoising process. Secondly, the spatial-temporal weighted module was introduced to support the evaluation of background interferences that do not conform to a Gaussian distribution, determining whether they are noise. According to the experimental results, the proposed method achieves a high success rate while preserving high positional precision for visual tracking.

In the future, to further mitigate the impact of the probability distribution of distractors on the proposed method, we will combine feedback reinforcement learning techniques to model the actual distribution of distractors and incorporate it into the inference process of reverse diffusion. This enhancement could further improve the tracking success rate.

Acknowledgment. This work was fully supported by Applied Basic Research Foundation of China Mobile (NO. R23100TM and R23103H0).

References

1. Batzolis, G., Stanczuk, J., Schönlieb, C.B., Etmann, C.: Conditional image generation with score-based diffusion models. In: CVPR, pp. 1–10 (2021)
2. Brempong, E.A., Kornblith, S., Chen, T., Parmar, N., Minderer, M., Norouzi, M.: Denoising pretraining for semantic segmentation. In: CVPR, pp. 4175–4186 (2022)
3. Chen, S., Sun, P., Song, Y., Luo, P.: Diffusiondet: diffusion model for object detection. In: CVPR, pp. 1–10 (2023)
4. Chen, X., Yan, B., Zhu, J., Wang, D., Yang, X., Lu, H.: Transformer tracking. In: CVPR, pp. 1–11 (2021)
5. Cui, Y., Jiang, C., Wang, L., Wu, G.: Mixformer: end-to-end tracking with iterative mixed attention. In: CVPR, pp. 13608–13618 (2022)
6. Dai, K., Wang, D., Lu, H., Sun, C., Li, J.: Visual tracking via adaptive spatially-regularized correlation filters. In: CVPR, pp. 4670–4679 (2019)
7. Danelljan, M., Bhat, G., Khan, F.S., Felsberg, M., et al.: Eco: efficient convolution operators for tracking. In: CVPR, pp. 3–14 (2017)
8. Dou, Y., Li, T., Li, L., Zhang, Y., Li, Z.: Tracking the research on ten emerging digital technologies in the AECO industry. J Constr. Eng. Manag. **149**(3), 03123003 (2023)
9. Fan, H., et al.: Lasot: a high-quality benchmark for large-scale single object tracking. In: CVPR, pp. 5374–5383 (2019)
10. Gedara Chaminda Bandara, W., Gopalakrishnan Nair, N., Patel, V.M.: Remote sensing change detection (segmentation) using denoising diffusion probabilistic models. In: CVPR, pp. arXiv-2206 (2022)
11. He, K., Zhang, X., Ren, S., Sun, J.: Deep residual learning for image recognition. In: CVPR, pp. 770–778 (2016)
12. Ho, J., Saharia, C., Chan, W., Fleet, D.J., Norouzi, M., Salimans, T.: Cascaded diffusion models for high fidelity image generation. J. Mach. Learn. Res. **23**(47), 1–33 (2022)
13. Hu, W., Wang, Q., Zhang, L., Bertinetto, L., Torr, P.H.: Siammask: a framework for fast online object tracking and segmentation. PAMI **45**(3), 3072–3089 (2023)
14. Huang, L., Zhao, X., Huang, K.: Got-10k: a large high-diversity benchmark for generic object tracking in the wild. In: PAMI, p. 1 (2019)
15. Li, B., Wu, W., Wang, Q., Zhang, F., Xing, J., Yan, J.: SiamRPN++: evolution of Siamese visual tracking with very deep networks. In: CVPR, pp. 4282–4291 (2019)
16. Li, B., Yan, J., Wu, W., Zhu, Z., Hu, X.: High performance visual tracking with Siamese region proposal network. In: CVPR, pp. 8971–8980 (2018)
17. Li, H., et al.: SRDIFF: single image super-resolution with diffusion probabilistic models. Neurocomputing **479**, 47–59 (2022)

18. Li, X., Ma, C., Wu, B., He, Z., Yang, M.H.: Target-aware deep tracking. In: CVPR, pp. 1369–1378 (2019)
19. Müller, M., Bibi, A., Giancola, S., Alsubaihi, S., Ghanem, B.: TrackingNet: a large-scale dataset and benchmark for object tracking in the wild. In: Ferrari, V., Hebert, M., Sminchisescu, C., Weiss, Y. (eds.) ECCV 2018. LNCS, vol. 11205, pp. 310–327. Springer, Cham (2018). https://doi.org/10.1007/978-3-030-01246-5_19
20. Özbey, M., et al.: Unsupervised medical image translation with adversarial diffusion models. In: CVPR, pp. 1–10 (2022)
21. Rombach, R., Blattmann, A., Lorenz, D., Esser, P., Ommer, B.: High-resolution image synthesis with latent diffusion models. In: CVPR, pp. 10684–10695 (2022)
22. Russakovsky, O., et al.: Imagenet large scale visual recognition challenge. IJCV **115**(3), 211–252 (2015)
23. Saharia, C., Ho, J., Chan, W., Salimans, T., Fleet, D.J., Norouzi, M.: Image super-resolution via iterative refinement. In: PAMI (2022)
24. Vahdat, A., Kreis, K., Kautz, J.: Score-based generative modeling in latent space. In: NIPS, vol. 34, pp. 11287–11302 (2021)
25. Wu, Y., Lim, J., Yang, M.H.: Object tracking benchmark. PAMI **37**(9), 1–1 (2015)
26. Wyatt, J., Leach, A., Schmon, S.M., Willcocks, C.G.: AnodDPM: anomaly detection with denoising diffusion probabilistic models using simplex noise. In: CVPR, pp. 650–656 (2022)
27. Xu, J., Liu, S., Vahdat, A., Byeon, W., Wang, X., De Mello, S.: Open-vocabulary panoptic segmentation with text-to-image diffusion models. In: CVPR, pp. 2955–2966 (2023)
28. Xu, Y., Wang, Z., Li, Z., Yuan, Y., Yu, G.: SiamFC++: towards robust and accurate visual tracking with target estimation guidelines. In: AAAI, pp. 12549–12556 (2020)
29. Yang, Y., Gu, X.: Joint correlation and attention based feature fusion network for accurate visual tracking. IEEE Trans. Image Process. **32**, 1705–1715 (2023)
30. Zhang, J., Yang, X., Wang, W., Guan, J., Ding, L., Lee, V.C.: Automated guided vehicles and autonomous mobile robots for recognition and tracking in civil engineering. Autom. Constr. **146**, 104699 (2023)

Instance-Proxy Loss for Semi-supervised Learning with Coarse Labels

Hongyan Wu[1], Qinghai Miao[1], Haiyun Guo[1,2](✉), Min Huang[1], and Jinqiao Wang[1,2,3,4]

[1] School of Artificial Intelligence, University of Chinese Academy of Sciences, Beijing 100049, China
wuhongyan19@mails.ucas.ac.cn, {miaoqh,huangm}@ucas.ac.cn
[2] Foundation model research center, Institute of automation, Chinese Academy of Sciences, Beijing 100864, China
{haiyun.guo,jqwang}@nlpr.ia.ac.cn
[3] Peng Cheng Laboratory, Shenzhen 518066, China
[4] Wuhan AI Research, Wuhan 430073, China

Abstract. Objects are often organized in a hierarchy where coarse-grained categories are comprised of subordinate fine-grained classes. Comparing with the fine-grained labels, the coarse-grained labels are much affordable to obtain. The coarse-grained labels can boost the semi-supervised learning (SSL) by offering extra regularization on the feature space of finer-grained recognition. However, coarse-grained labels are ignored by most of works in SSL. An intuitive way to utilize the coarse labels for SSL is to impose an extra coarse-grained categorization constraint, which will cause the class confusion between fine-grained categories belonging to the same coarse-grained category thus is sub-optimal for SSL. In this paper, we present an instance-proxy loss (IPL) to boost the separability of the fine-grained classes within the same coarse-grained class, as well as keep the intra-class feature space of coarse-grained classes compact. Specifically, IPL includes instance-level loss and proxy-level loss to impose constraints on both instance-to-instance and instance-to-proxy relations. Our approach outperforms the state-of-the-art methods on three benchmark datasets, showing significant improvement with small proportion of fine-grained labels, e.g., it brings 10.14% accuracy improvement on CUB-200-2011 with 15% of labeled data.

Keywords: Semi-supervised Learning · Fine-grained Visual Categorization · Instance-level loss · Proxy-level loss

1 Introduction

Recently, the impressive progress of deep learning highly depends on the large-scale labeled datasets. However, labeling can be a time-consuming work requiring

---

Supplementary Information The online version contains supplementary material available at https://doi.org/10.1007/978-981-99-8555-5_19.

© The Author(s), under exclusive license to Springer Nature Singapore Pte Ltd. 2024
Q. Liu et al. (Eds.): PRCV 2023, LNCS 14436, pp. 238–250, 2024.
https://doi.org/10.1007/978-981-99-8555-5_19

expert knowledge, especially in fine-grained visual classification (FGVC) task, which aims at distinguishing subordinate categories of some general classes like birds [23], cars [11,26] and aircrafts [16]. FGVC has attracted much research attention due to plenty of applications, while it is challenging to collect datasets with sufficient annotations in many practical FGVC problems.

Semi-supervised learning (SSL) is a compelling technique to reduce the dependence on labeled data by utilizing limited number of labeled data and large amounts of unlabeled data. Recent years have witnessed some progress in dealing with the label scarcity issue of FGVC in SSL framework. Objects are inherently organized into the hierarchical structure in which fine categories are grouped into coarse categories. The coarse-grained labels are an effective alternative to reducing annotation cost, as they can be easily obtained from non-expert. But only relatively few works [8,21] exploit coarse-grained labels in SSL. This paper focuses on the recently proposed task setting [21] where the coarse-grained labels are available for training while only small proportion of the fine-grained labels can be used.

An intuitive way to leverage coarse-grained labels is to train two separate classifiers for fine level and coarse level. Although the supervision of coarse level does work by helping distinguish similar fine-grained classes from different coarse-grained classes (Fig. 1 (a)-top), it reduces the intra-class variance and squeezes the feature space within the same coarse-grained class, eliminating the discrimination between fine-grained classes (Fig. 1 (a)-bottom). This contradiction issue also happens in unsupervised learning with coarse-grained labels. ANCOR [3] alleviates this issue by presenting an angular normalization to maximize the synergy between contrastive InfoNCE loss and coarse-grained classification loss. While it pushes the query away from its corresponding coarse-grained proxy when imposing constraint on positive pair, which may diminish the effectiveness of synergy. (In contrastive loss, a query and a key form a positive pair if they are data-augmented versions of the same image and the key is positive key. It forms a negative pair otherwise, and the key is negative key.) In this paper, we aim to mitigate this contradiction issue in semi-supervised learning with coarse-grained labels by introducing instance-proxy loss (IPL), which takes good points of both instance-level and proxy-level losses and corrects the defect of angular normalization. It not only keeps the compactness within coarse-grained classes (Fig. 1 (b)-top), but also facilitates the separability of fine-grained classes belonging to the same coarse-grained class (Fig. 1 (b)-bottom).

Our proposed IPL incorporates instance-level loss and proxy-level loss to impose constraints on both instance-to-instance and instance-to-proxy relations. The instance-level loss introduces coarse-grained constraint into supervised contrastive loss [10]. (Supervised contrastive loss treats a representation of an image as query. Positive keys are the representations which belong to the same class of query, and negative keys are from different classes of query. Positive pair means query and key from the same class, negative pair are opposite.) Instance-level loss restricts positive keys and negative keys in the same coarse-grained category, and increases the distance between negative pairs, providing distinction between

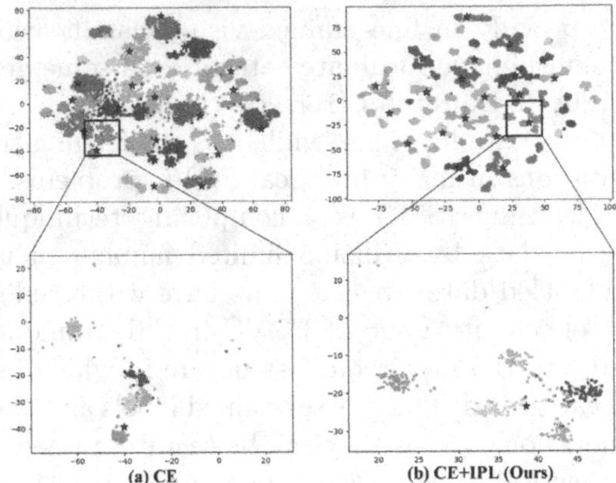

Fig. 1. The tSNE visualization of learned representation. We train two classifiers supervised by 15% fine-grained labels and 100% coarse-grained labels. Stars are proxies for coarse-grained classes (weights of coarse-level classifier). For the top images, points of different colors means features from different coarse-grained classes. For the bottom images, points of different colors means features of different fine-grained classes. (a) CE: it is training with cross-entropy (CE) losses for fine level and coarse level. (a)-top depicts feature distribution of all categories. (a)-bottom shows feature distribution of coarse-grained class 11. (b) CE+IPL (Ours): it is training with two CE loss and IPL. (b)-top depicts feature distribution of all categories. (b)-bottom shows the feature distribution of coarse-grained class 11. Our method can not only keeps the compactness within coarse-grained classes (top), but also boosts the separability of fine-grained classes belonging to the same coarse-grained class (bottom). (Color figure online)

fine-grained classes within the same coarse-grained class. Meanwhile, to retain the compact feature space of coarse-grained class, it still pulls query close to its coarse-grained proxy (coarse-level classifier weights), which improves ANCOR by removing the angular normalization of positive pair, and further applies it to semi-supervised scenario to obtain instance-level loss. For the proxy-level loss, it is proposed to further explore the underultilized instance-to-proxy relations of unlabeled data. It encourages query to be close to the proxy of its fine-grained class (weights of fine-level classifier) and far away from those of different fine-grained classes, which further ensures separation of fine-grained classes.

To summarize, our contributions are as follows: (1) we present IPL, to improve performance of SSL with coarse labels, by facilitating separability of fine-grained classes belonging to the same coarse-grained class, as well as maintaining the compactness within coarse-grained class. Specifically, IPL is comprised of instance-level loss and proxy-level loss, leveraging both instance-to-instance and instance-to-proxy relations to enable a more separable feature space of fine-grained classes with coarse supervision loss. (2) our method achieves state of the art (SOTA) performance on three benchmark datasets, CIFAR100 [12], CUB-200-2011 [23], and Semi-iNat [20], showing significant improvement with small proportion of labeled data.

2 Related Work

Fine-Grained Feature Learning with Coarse-Grained Labels. Recently, researchers delve into weakly supervised learning with coarse-grained labels. [28] gives a theoretical guarantee for learning fine-grained representations with instance loss when coarse-grained labels are available. However, [28] neglects the contradiction between instance loss and coarse-grained classification objective. ANCOR alleviates the problem by introducing an angular normalization module into contrastive InfoNCE loss. Our instance-level loss is inspired by ANCOR, but we improve ANCOR and apply it into the task of SSL with coarse-grained labels and fine-grained labels of certain proportion. Besides, we also proposed a proxy-level loss to exploit the relations between instance and proxy to further facilitate the task, which is considered by relatively few works. Specifically, HIERMATCH [8] proposes to combine SSL methods with label hierarchy exploration method [4] which trains a set of independent classifiers for different class hierarchies with disentangled features to fed, and each classifier uses its feature as well as the features from finer classifiers to predict while gradient is stopped from flowing to the finer level. Hierarchical supervised loss (HL) [21] utilizes the unlabeled data with its coarse-level labels with CE loss, and obtains the prediction of each coarse-grained class by summing probabilities of its subordinate fine-grained classes. Both HL and HIERMATCH tackle the task by conquering two separate problems, i.e. SSL and the utilization of coarse-grained labels. Differently, we focus on how to combine SSL and coarse-grained labels, and propose to embed coarse-grained constraint into SSL framework.

Semi-supervised Learning. Recent advances in self-supervised learning and semi-supervised learning drive newly proposed methods to make a combination of them [15,17,24,25,31]. Among them, Self-Tuning [25] is the most related work. It presents a supervised constrastive loss named PGC to mitigate the confirmation bias induced by model's overfitting to false pseudo-labels. Our work is different from Self-Tuning as following: (1) We target at a different task, SSL with coarse-grained labels and fine-grained labels of certain proportion. (2) PGC loss uses fixed number of negative keys from each different fine-grained category, and it can only be applied to datasets with small class number as the size of its negative keys increases with the class number, resulting in an out of memory of GPU. However, our instance-level loss restricts the negative keys within the same coarse-grained class, and always use constant quantity of negative keys, which is applicable to large-scale datasets.

3 Method

The framework of SSL with the proposed IPL is illustrated in Fig. 2. It consists of: (1) an encoder $\mathcal{B} : I \to q \in \mathbb{R}^d$, which is attached by a global average pooling. (2) a momentum encoder $\mathcal{E} : I \to k \in \mathbb{R}^d$, which is similar to $\mathcal{B}$, but updates with exponential moving average of weights from $\mathcal{B}$. (3) a coarse-grained class linear classifier $\mathcal{C} : q \to p_c \in \mathbb{R}^C$ with W_c denoting its weights. (4) a fine-grained class

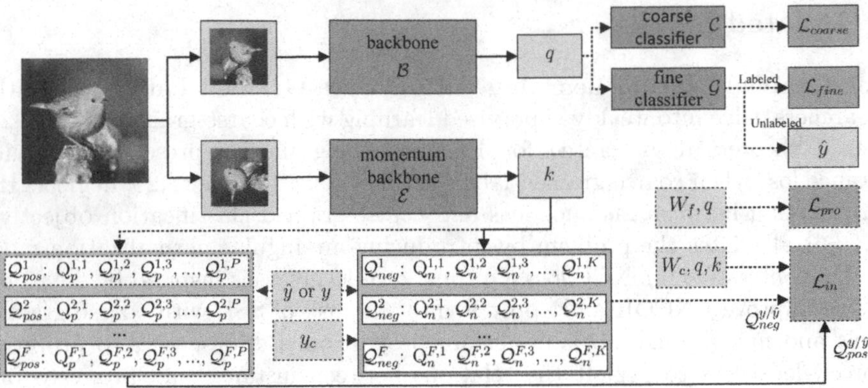

Fig. 2. An overview of our method. The image is augmented twice and fed into model $\mathcal{B}$ and its momentum-updated model $\mathcal{E}$ to generate q and k, respectively. q is input into coarse classifier $\mathcal{C}$ to compute the coarse-level CE loss, $\mathcal{L}_{coarse}$, and fine classifier $\mathcal{G}$ to get $\mathcal{L}_{fine}$ for labeled data or generate fine pseudo label $\hat{y}$ for unlabeled data. k is stored in the queue for positive keys of fine-grained class y or $\hat{y}$, and the queue for negative keys of one of other fine-grained classes with the same coarse-grained class of k. IPL consists of $\mathcal{L}_{in}$ and $\mathcal{L}_{pro}$: $\mathcal{L}_{in}$ uses $q, k, \mathcal{Q}_{neg}^{y/\hat{y}}, \mathcal{Q}_{pos}^{y/\hat{y}}$ as well as weights of $\mathcal{C}$, W_c, and $\mathcal{L}_{pro}$ utilizes q and weights of $\mathcal{G}$, W_f. The combination of them helps learn a feature space where different subordinate fine-grained classes are separable, and the interiors of coarse-grained classes are compact.

linear classifier $\mathcal{G} : q \rightarrow p_f \in \mathbb{R}^F$ with weights W_f. (5) a set of queues for placing the positive keys: $\{\mathcal{Q}_{pos}^i\}_{i=1}^F$, where $\mathcal{Q}_{pos}^i \in \mathbb{R}^{d\times P}$ with P positive keys for each fine-grained class i. (6) a queue set: $\{\mathcal{Q}_{neg}^i\}_{i=1}^F$ for negative keys. $\mathcal{Q}_{neg}^i \in \mathbb{R}^{d\times K}$ with K negative keys, which stores the keys from other fine-grained class within the same coarse-grained class of i. (7) the supervised CE losses: $\mathcal{L}_{coarse}$ and $\mathcal{L}_{fine}$. (8) our IPL, which is described below. Here, we will give a description of the training process.

For clarification, the process focuses on an unlabeled image I^u with coarse-grained label y_c^u. I^u is augmented twice to generate I_q^u and I_k^u, then they are fed into the corresponding encoders to extract $q^u = \mathcal{B}(I_q^u)$ and $k^u = \mathcal{E}(I_k^u)$. By inputing q^u into fine-level classifier $\mathcal{G}$, fine-grained pseudo-label $\hat{y} = \text{argmax}_f \mathcal{G}(q^u)$ is obtained. Our positive keys are comprised of k^u and keys in $\mathcal{Q}_{pos}^{\hat{y}}$, and negative keys are from $\mathcal{Q}_{neg}^{\hat{y}}$. For the losses, the supervised CE loss for coarse level: $\mathcal{L}_{coarse}$ are computed based on $p^u = \mathcal{C}(q^u)$ and y_c^u, and IPL is applied to positive keys $\{k^+\}$, negative keys $\{k^-\}$, q^u, positive proxy $W_f^{\hat{y}}$ (the $\hat{y}$-th row of W_f), and negative proxies $\{W_f^-\}$ (the other rows of W_f except $\hat{y}$-th). As for a labeled image I with fine-grained label y and coarse-grained label y_c, we can simply replace $\hat{y}$ with y to get losses $\mathcal{L}_{coarse}$, IPL. And we compute $\mathcal{L}_{fine}$ with $p_f = \mathcal{G}(q)$ and y, which is only for labeled data. Note that these queues are iteratively and progressively updated by replacing the oldest keys with the newly-generated ones.

3.1 Instance-Level Loss

We first review the ANCOR method, then improve it, and introduce it into SSL to attain our instance-level loss.

Revisit of ANCOR. ANCOR is presented to induce synergy between coarse-level supervised CE loss and contrastive InfoNCE loss. The coarse-grained supervision loss forces feature embedding q (query) to shift towards the proxy of coarse-grained class y_c, presented by $W_c^{y_c}$. And all features of class y_c will collapse to $W_c^{y_c}/||W_c^{y_c}||$, which conflicts with the contrastive loss attempting to push features in y_c away from each other. To mitigate this contradiction, ANCOR defines an angular normalization, which is applied in InfoNCE loss. It can be formulated as:

$$\mathcal{L}_{acon} = -\log\frac{\exp(\hat{\mathcal{A}}(q,k^+,y_c)/\tau)}{\exp(\hat{\mathcal{A}}(q,k^+,y_c)/\tau)+\sum_j\exp(\hat{\mathcal{A}}(q,k_j^-,y_c)/\tau)} \tag{1}$$

where $\hat{\mathcal{A}}(x,z,y) = \mathcal{A}(x,y)\cdot\mathcal{A}(z,y)$, $\mathcal{A}(x,y) = \frac{x-W_c^y}{||x-W_c^y||}$. And x and W_c^y are normalized, $x = x/||x||$ and $W_c^y = W_c^y/||W_c^y||$. (All features occur below are normalized) $\mathcal{A}$ is the angular normalization module. q is a query representation, k^+ is a positive key, and k_j^- is a negative key. τ is a temperature hyper-parameter.

Improvement of ANCOR. By expanding the formula of $\hat{\mathcal{A}}$, we give some insights into how ANCOR works and improve it. $\hat{\mathcal{A}}$ can be rewritten as: $\hat{\mathcal{A}}(x,z,y) = \frac{x\cdot z - x\cdot W_c^y - W_c^y\cdot z+1}{||x-W_c^y||\cdot||z-W_c^y||}$. Then, we find that $\mathcal{L}_{acon}$ minimizes $\hat{\mathcal{A}}(q,k_j^-,y_c)$, meaning that $q\cdot k_j^-$ is minimized, and $q\cdot W_c^{y_c}$ is maximized at the same time. The core of ANCOR is to pull query to its coarse-grained proxy when pushing negative pairs away from each other. However, ANCOR still can be improved. Since it also maximizes $\hat{\mathcal{A}}(q,k^+,y_c)$, the $q\cdot k^+$ is maximized and $q\cdot W_c^{y_c}$ minimized meanwhile. It suggests that the constraint imposed on positive pair can push query away from its coarse-grained proxy, which hinders the maintaining of a compact feature space within coarse-grained class. Therefore, we discard the angular normalization of positive pair. And we also remove the $||q-W_c^y||\cdot||k_j^- - W_c^y||$ in angular normalization of negative pair to balance the scale of negative pairs and positive pair. Then, the loss become:

$$\mathcal{L}_{incon} = -\log\frac{\exp(q\cdot k^+/\tau)}{\exp(q\cdot k^+/\tau)+\sum_j\exp(\mathcal{IN}(q,k_j^-,y_c)/\tau)} \tag{2}$$

where

$$\mathcal{IN}(x,z,y) = x\cdot z - x\cdot W_c^y - W_c^y\cdot z + 1 \tag{3}$$

Instance-Level Loss. We apply the $\mathcal{IN}$ operation in Eq.(3) to SSL framework to obtain our instance-level loss:

$$\mathcal{L}_{in} = -\frac{1}{P+1}\sum_{p=0}^{P}\log\frac{\exp(q\cdot k_p^+/\tau)}{Pos+Neg} \tag{4}$$

where $Pos = \sum_{i=0}^{P} \exp(q \cdot k_i^+/\tau)$, $Neg = \sum_{j=1}^{K} \exp(\mathcal{IN}(q, k_j^-, y_c)/\tau)$. And k_0^+ is the differently-augmented view of q. k_1^+, k_2^+, $\cdots$, k_P^+ are positive keys in fine-grained class y. k_j^- is negative keys belonging to other fine-grained classes in coarse-grained class y_c.

Our instance-level loss pull the query from the same fine-grained class to be more close and push the query away from negative keys which are from the same coarse-grained class. And it can boost the discrimination between different fine-grained classes in the same coarse-grained class. In the meanwhile, $\mathcal{IN}$ can keep the compactness within the same coarse-grained class by pulling query to its coarse-grained proxy.

3.2 Proxy-Level Loss

We further present a proxy-level loss to leverage the relationship between instance and proxies. CE loss is effective to encourage the similarity of data and positive proxy (the fine-grained proxy the data belongs to), while in SSL it is easily misled by false pseudo-labels, making the network overfit to these incorrect predictions. For our work, if we simply use CE loss to delve into the instance-to-proxy relations, the fine-grained proxies will suffer from these false pseudo-labels badly, due to that they are only supervised by CE loss. And the rest of the model can alleviate the reliance on CE loss by supervision of $\mathcal{L}_{in}$. Therefore, our proxy-level loss stops the gradient of CE loss flowing to these fine-grained proxies in implementation. And it can be written as: $\mathcal{L}_{pro} = -\log \frac{\exp(q \cdot W_f^y/\tau)}{\exp(q \cdot W_f^y/\tau) + \sum_{j=1}^{F-1} \exp(q \cdot W_f^j/\tau)}$, where $W_f^F = W_f^y$. W_f^y is the positive proxy, and W_f^j is the negative proxies (the proxies of fine-grained classes other than y). Different from $\mathcal{L}_{in}$, it pulls q towards its corresponding fine-grained proxy, as well as pushes q away from other proxies.

3.3 Instance-Proxy Loss

Combining $\mathcal{L}_{in}$ and $\mathcal{L}_{pro}$ primarily ($\mathcal{L}_{in} + \mathcal{L}_{pro}$) will result in slow convergence or sub-optimization when conflict between supervised contrastive and CE loss occurs, according to [7]. Therefore, we incorporate our losses referring to Paco [7], which adjusts the intensity of them and forbid conflict. And our IPL is:

$$\mathcal{L}_{ipl} = \sum_{p=0}^{P+1} -w_p \log \frac{\exp(q \cdot z_p/\tau)}{Pos^2 + Neg^2} \tag{5}$$

$$Pos^2 = \sum_{i=0}^{P+1} \exp(q \cdot z_i/\tau), \quad Neg^2 = \sum_{j=1}^{K+F-1} \exp(T(q, z_j, y_c)/\tau)$$

$$w_p = \begin{cases} \alpha, & p \in [0, P] \\ 1.0, & p = P+1 \end{cases}, \quad z_p = \begin{cases} k_p^+, & p \in [0, P] \\ W_f^p, & p = P+1 \end{cases},$$

$$T(q, z_j, y_c) = \begin{cases} \mathcal{IN}(q, k_j^-, y_c), & j \in [1, K] \\ q \cdot W_f^j & j \in [K+1, K+F-1] \end{cases}$$

Concretely, α is a hyper-parameter in $(0, 1)$ to trade off the performance of $\mathcal{L}_{in}$ and $\mathcal{L}_{pro}$. k_p^+ and k_j^- are positive keys and negative key. W_f^p (or W_f^y) and W_f^j are positive fine-grained proxy and negative fine-grained proxy. And $\mathcal{L}_{ipl}$ is scaled by $\frac{1}{\sum_{p=0}^{P+1} w_p}$ in implementation. Our **final loss** is: $\mathcal{L} = \mathcal{L}_{ipl} + \mathcal{L}_{coarse} + \mathcal{L}_{fine}$ ($\mathcal{L}_{fine}$ is only available with labeled data).

Table 1. Statistics of datasets. The number of train images includes that of labeled and unlabeled train images.

Dataset	CUB-200-2011	CIFAR100	Semi-iNat
# Coarse classes	37	20	8
# Fine classes	200	100	810
# Labeled train images	900/1.8K/3K	400/1K/2.5K	9.7K
# Train images	6K	50K	101K
# Test images	6K	10K	16.2K

Table 2. Error rates (%) ↓ on *CIFAR100* with 400, 2500 and 10000 labeled data.

Type	Method	#Labels		
		400	2.5k	10k
Baselines	Fine−	82.56	57.94	35.41
	Fine−Coarse	44.49	33.48	25.16
	Fine+	19.30	19.30	19.30
SSL	Π-model [13]	–	57.25	37.88
	Pseudo-Labeling [14]	–	57.38	36.21
	Mean Teacher [22]	–	53.91	35.83
	MixMatch [2]	67.61	39.94	28.31
	UDA [27]	59.28	33.13	24.50
	ReMixMatch [1]	44.28	27.43	23.03
	FixMatch [18]	49.95	28.64	23.18
	SimPLE [9]	–	–	21.89
	BAM-UDA [15]	40.30	–	21.70
	FlexMatch [30]	39.94	26.49	21.90
	Ours	**35.51**	**26.26**	21.61

Table 3. Classification accuracy (%) ↑ on CUB-200-2011 (ResNet-50 pre-trained).

Type	Method	Label Proportion		
		15%	30%	50%
Baselines	Fine−	45.25	59.68	70.12
	Fine-Coarse	63.00	71.28	77.16
	Fine+	82.02	82.02	82.02
SSL	Π-model [13]	45.20	56.20	64.07
	Pseudo-Labeling [14]	45.33	62.02	72.30
	Mean Teacher [22]	53.26	66.66	74.37
	UDA [27]	46.90	61.16	71.86
	FixMatch [18]	44.06	63.54	75.96
	SimCLRv2 [6]	45.74	62.70	71.01
	Self-Tuning [25]	64.17	75.13	80.22
SSL+Coarse	Resnet50+HL [21]	64.45	71.51	77.05
	Pseudo-Labeling+HL [21]	63.76	72.23	77.55
	Ours	**74.31**	**77.09**	**81.12**

4 Experiments

Datasets. We perform our experiments on three datasets: (1) CUB-200-2011 [23]. Similar to self-tuning [25], we make the labeled proportion of training data range from 15% to 50%. The coarse-grained labels are obtained from [5]. (2) CIFAR100 [12]. Since it is a benchmark SSL datasets, we conduct experiments on it to compare with SSL methods. (3) Semi-iNat [20]. It is collected by the semi-supervised challenge at the FGVC8 workshop [20]. There are 7 levels in Semi-iNat, we primarily conduct our experiments on Phylum level to compare with SOTA methods. And extensive exploration of other levels is presented in **supplementary material**. Our Table 1 illustrates some details of CUB-200-2011, CIFAR100, and Semi-iNat. Besides, experiments on combination with self-supervised method are also conducted in **supplementary material**.

Implementation Details. We use ResNet50 and Wide ResNet28-8 (WRN-28-8) [29] (only for CIFAR100 to compare with SOTA methods) as our backbones, and the dimension of the output of encoder $\mathcal{B}$ are $d = 2048$ and $d = 512$, respectively. We use cosine-annealing with base learning rate of 0.001 and weight decay of 0.0001 to train ResNet50. For WRN-28-8, the learning rate and weight decay are 0.03 and 0.001. Empirically, the temperature τ is set as 0.2 for CIFAR100 and CUB-200-2011, and 0.6 for Semi-iNat. The positive queue size P is 32, and the negative queue size K is 1000 for CIFAR100, and 250 for CUB-200-2011 and Semi-iNat. The α is set 0.5 on CIFAR100, and 0.01 on the other.

Baselines. Here, we give some baselines used in all experiments. (1) Fine+. The backbone network $\mathcal{B}$ along with a fine-level classifier($\mathcal{G}$) is trained with all labeled data from fine level. It can be a upper-bound of our method. (2) Fine−. It is similar to Fine+, except that it is only trained with a subset of labeled data in fine level, which is a lower-bound of our method. (3) Fine-Coarse. It consists of the backbone network $\mathcal{B}$, a fine classifier($\mathcal{G}$) and a coarse classifier($\mathcal{C}$). The labels of data is comprised of a subset of fine-grained labels, and all labels of coarse-grained classes.

Table 4. Classification accuracy (%) ↑ on Semi-iNat (ResNet-50 pre-trained on ImageNet).

Type	Method	Acc
Baselines	Fine−	42.65
	Fine-Coarse	44.52
	Fine+	86.00
SSL	Pseudo-Labeling [14]	40.40
	Self-Training [19]	42.40
	MoCo [19]	41.70
	MoCo + Self-Training [19]	42.60
	FixMatch [18]	44.10
SSL+Coarse	HL [21]	46.60
	Pseudo-Labeling+HL [21]	44.90
	FixMatch+HL [21]	47.90
	Self-Training+HL [21]	44.80
	MoCo + Self-Training+HL [21]	45.80
	Ours	**49.11**

Table 5. Ablation studies on CIFAR100 with 400 labels in fine-level. (ResNet-50 pre-trained)

Type	Method	Acc
Baselines	Fine−	36.50
	Fine-Coarse	61.82
	Fine+	84.55
Instance-level	$\mathcal{L}_{in}$ w/o $\mathcal{IN}$+	65.35
	$\mathcal{L}_{in}$ w/o $\mathcal{IN}$	64.92
	$\mathcal{L}_{in}$ w/o $\mathcal{IN}$ w $\mathcal{A}$	66.40
	$\mathcal{L}_{in}$	68.70
Proxy-level	$\mathcal{L}_{pro}$	66.14
Combination	$\mathcal{L}_{in} + \mathcal{L}_{pro}$	63.84
	$\mathcal{L}_{ipl}$	**69.13**

4.1 Comparison to SOTA Methods

CIFAR100. CIFAR100 is a popular dataset for SSL methods, and we conduct experiments on it to verify the superiority of our approach. The results of

CIFAR100 is illustrated in Table 2. As can be seen from Table 2, our method outperforms all the SSL methods across 15% and 30% label proportions, especially in the results of 400 labels, where we surpass the best SSL method by 4.43%. We also observe that our method will gain much more performance improvement with fewer proportion of labeled data.

CUB-200-2011. CUB-200-2011 is a popular dataset in FGVC, and the results of experiments is in Table 3. It also shows consistent advantages of our method over Self-Tuning (best SSL methods) across various label proportions, with obvious boosting of 10.14% in label proportion of 15%. It is noteworthy that the size of negative keys utilized by Self-Tuning is twenty-five times of ours, suggesting that even much larger and more diverse negative keys are not sufficient to the performance gap brought by our method by leveraging coarse-grained labels. Comparing with the methods of SSL+Coarse, our approach performs much better than them.

Semi-iNat. Semi-iNat is a newly proposed dataset for FGVC tasks with SSL method. And we are capable of comparing to SOTA methods which also use coarse labels to promote SSL ('SSL+Coarse' in Table 4) on Semi-iNat. The results are suggested in Table 4. Our method achieves about 4.59% excess accuracy over 'Fine-Coarse', and provides better performance than any other SSL and SSL+Coarse methods. It demonstrates that our method can be also applied to large-scale datasets.

4.2 Ablation Study

We conduct ablation studies on the effect of our losses (in Table 5).

The Effectiveness of Instance-Level Loss. For the instance-level loss, we compare some variants: (1) $\mathcal{L}_{in}$ w/o $\mathcal{IN}+$. We remove $\mathcal{IN}$ operation of $\mathcal{L}_{in}$ in Eq. (4), and use negative keys from all other fine-grained categories. The Neg in Eq. (4), become $Neg = \sum_{j=1}^{(F-1)K} \exp(q \cdot k_j^- / \tau)$. For a fair comparison, we introduce 10 keys from each fine-grained category to obtain 990 negative keys. (There are 1000 negative keys in $\mathcal{L}_{in}$.) (2) $\mathcal{L}_{in}$ w/o $\mathcal{IN}$. It is our $\mathcal{L}_{in}$ in Eq. (4) without $\mathcal{IN}$ operation, and the Neg in Eq. (4) is $Neg = \sum_{j=1}^{K} \exp(q \cdot k_j^- / \tau)$. (3) $\mathcal{L}_{in}$ w/o $\mathcal{IN}$ w $\mathcal{A}$. We include it here to demonstrate the improvement of our $\mathcal{IN}$ based loss over angular normalization $\mathcal{A}$ based loss. The loss is below: $\mathcal{L}_{ain} = -\frac{1}{P+1} \sum_{p=0}^{P} \log \frac{\exp(\hat{\mathcal{A}}(q, k_p^+, y_c)/\tau)}{Pos^3 + Neg^3}$, $Pos^3 = \sum_{i=0}^{P} \exp(\hat{\mathcal{A}}(q, k_i^+, y_c)/\tau)$, $Neg^3 = \sum_{j=1}^{K} \exp(\hat{\mathcal{A}}(q, k_j^-, y_c)/\tau)$.

We draw following conclusion on CIFAR100 for the instance-level: (1) without $\mathcal{IN}$, the performance of $\mathcal{L}_{in}$ w/o $\mathcal{IN}+$ is better than $\mathcal{L}_{in}$ w/o $\mathcal{IN}$. It may result from that comparing with using negative keys within the same coarse class ($\mathcal{L}_{in}$ w/o $\mathcal{IN}$), using negative keys from all other fine-grained classes ($\mathcal{L}_{in}$ w/o $\mathcal{IN}+$) can diminish the constraint the loss imposes to disperse same coarse-grained class elements in feature space. (2) $\mathcal{L}_{in}$ w/o $\mathcal{IN}$ w $\mathcal{A}$ and $\mathcal{L}_{in}$ can bring performance improvement over $\mathcal{L}_{in}$ w/o $\mathcal{IN}$, proving that $\mathcal{IN}$ and $\mathcal{A}$ can mitigate the strong intensity of negative keys belonging to the same coarse-grained class, and pull query towards its coarse proxy to hold the compact feature space of coarse-grained classes. (3)$\mathcal{L}_{in}$ surpasses $\mathcal{L}_{in}$ w/o $\mathcal{IN}$ w $\mathcal{A}$ by 2.3%, showing our $\mathcal{IN}$ is more effective in pulling query close to its coarse-grained proxy.

The Effectiveness of IPL. There are some findings on the rest: (1) $\mathcal{L}_{pro}$ and $\mathcal{L}_{in}$ can boost the accuracy over the 'Fine-Coarse' method, by exploiting the instance-to-proxy and instance-to-instance relations, respectively. (2) IPL can further facilitate the results by providing an synergy between $\mathcal{L}_{pro}$ and $\mathcal{L}_{in}$, which is demonstrated by that IPL achieves a much higher accuracy than $\mathcal{L}_{in} + \mathcal{L}_{pro}$. (3) $\mathcal{L}_{ipl}$ boosts the performance of 'Fine-Coarse' method by a large margin, which verifies our method can leverage the coarse-grained labels effectively by making the fine-grained classes in the same coarse-grained class separable and maintaining the intra-class feature space in coarse level compact.

Hyper-Parameter Analysis of α. We search α in [0.005, 0.9]. And α is 0.5 on CIFAR100, 0.01 on CUB-200-2011 and Semi-iNat. We use pretrained Resnet50 as backbone. When α is (0.005, 0.01, 0.05, 0.1, 0.5, 0.9), the accuracy(%) is: CIFAR100 with 400 labels (68.15, 67.64, 68.78, 69.02, 69.13, 68.29), CUB-200-2011 with 15% label rate (73.73, 74.31, 73.43, 72.80, 70.95, 70.81), Semi-iNat (48.77, 49.11, 48.14, 48.48, 47.48, 47.69). The results suggest the accuracy is fluctuating within a relatively small range as α changes, and it is always above the baseline 'Fine-Coarse' in Table 2, Table 3 and Table 4.

5 Conclusion

Coarse-grained labels can mitigate the reliance on labeled fine-grained labels which is expensive to obtain. However, introducing the coarse-level supervision will result in the suppression of intra-class diversity and inseparability of subordinate fine-grained classes. To diminish this challenge, this paper proposes IPL to discriminate the different fine-grained classes from the same coarse-grained class, and retain the compactness within coarse-grained classes. IPL consists of instance-level loss and proxy-level loss to exploit the rich relationship between distance and instance, as well as instance and proxy. Experiments on CIFAR100, CUB-200-2011 and Semi-iNat demonstrate our superiority over both SSL and 'SSL+Coarse' methods.

Acknowledgement. This work was supported by National Key R & D Program of China under Grant No.2021ZD0110400, National Natural Science Foundation of China (No.62276260, 62002356, 62271485, 62076235) and Zhejiang Lab (No.2021KH0AB07).

References

1. Berthelot, D., et al.: Remixmatch: semi-supervised learning with distribution alignment and augmentation anchoring. arXiv preprint arXiv:1911.09785 (2019)
2. Berthelot, D., Carlini, N., Goodfellow, I., Papernot, N., Oliver, A., Raffel, C.A.: Mixmatch: a holistic approach to semi-supervised learning. In: Advances in Neural Information Processing Systems, vol. 32 (2019)
3. Bukchin, G., et al.: Fine-grained angular contrastive learning with coarse labels. In: Proceedings of the IEEE/CVF Conference on Computer Vision and Pattern Recognition, pp. 8730–8740 (2021)
4. Chang, D., Pang, K., Zheng, Y., Ma, Z., Song, Y.Z., Guo, J.: Your "flamingo" is my "bird": fine-grained, or not. In: Proceedings of the IEEE/CVF Conference on Computer Vision and Pattern Recognition, pp. 11476–11485 (2021)
5. Chen, T., Wu, W., Gao, Y., Dong, L., Luo, X., Lin, L.: Fine-grained representation learning and recognition by exploiting hierarchical semantic embedding. In: Proceedings of the 26th ACM International Conference on Multimedia, pp. 2023–2031 (2018)
6. Chen, T., Kornblith, S., Swersky, K., Norouzi, M., Hinton, G.: Big self-supervised models are strong semi-supervised learners. In: NeurIPS (2020)
7. Cui, J., Zhong, Z., Liu, S., Yu, B., Jia, J.: Parametric contrastive learning. In: Proceedings of the IEEE/CVF International Conference on Computer Vision, pp. 715–724 (2021)
8. Garg, A., Bagga, S., Singh, Y., Anand, S.: Hiermatch: leveraging label hierarchies for improving semi-supervised learning. In: Proceedings of the IEEE/CVF Winter Conference on Applications of Computer Vision, pp. 1015–1024 (2022)
9. Hu, Z., Yang, Z., Hu, X., Nevatia, R.: Simple: similar pseudo label exploitation for semi-supervised classification. In: Proceedings of the IEEE/CVF Conference on Computer Vision and Pattern Recognition, pp. 15099–15108 (2021)
10. Khosla, P., et al.: Supervised contrastive learning. In: Larochelle, H., Ranzato, M., Hadsell, R., Balcan, M.F., Lin, H. (eds.) Advances in Neural Information Processing Systems, vol. 33, pp. 18661–18673. Curran Associates, Inc. (2020). https://proceedings.neurips.cc/paper/2020/file/d89a66c7c80a29b1bdbab0f2a1a94af8-Paper.pdf
11. Krause, J., Stark, M., Deng, J., Fei-Fei, L.: 3D object representations for fine-grained categorization. In: 2013 IEEE International Conference on Computer Vision Workshops, pp. 554–561 (2013). https://doi.org/10.1109/ICCVW.2013.77
12. Krizhevsky, A., Hinton, G., et al.: Learning multiple layers of features from tiny images (2009)
13. Laine, S., Aila, T.: Temporal ensembling for semi-supervised learning. In: ICLR (2017)
14. Lee, D.H., et al.: Pseudo-label: the simple and efficient semi-supervised learning method for deep neural networks. In: Workshop on Challenges in Representation Learning, ICML, vol. 3, p. 896 (2013)
15. Loh, C., et al.: On the importance of calibration in semi-supervised learning. arXiv abs/2210.04783 (2022)
16. Maji, S., Kannala, J., Rahtu, E., Blaschko, M., Vedaldi, A.: Fine-grained visual classification of aircraft. Technical report (2013)
17. Mugnai, D., Pernici, F., Turchini, F., Del Bimbo, A.: Fine-grained adversarial semi-supervised learning. ACM Trans. Multimedia Comput. Commun. Appl. (TOMM) **18**(1s), 1–19 (2022)

18. Sohn, K., et al.: Fixmatch: simplifying semi-supervised learning with consistency and confidence. In: Advances in Neural Information Processing Systems, vol. 33, pp. 596–608 (2020)
19. Su, J.C., Cheng, Z., Maji, S.: A realistic evaluation of semi-supervised learning for fine-grained classification. In: Proceedings of the IEEE/CVF Conference on Computer Vision and Pattern Recognition, pp. 12966–12975 (2021)
20. Su, J.C., Maji, S.: The semi-supervised inaturalist challenge at the fgvc8 workshop (2021)
21. Su, J.C., Maji, S.: Semi-supervised learning with taxonomic labels. In: British Machine Vision Conference (2021)
22. Tarvainen, A., Valpola, H.: Mean teachers are better role models: weight-averaged consistency targets improve semi-supervised deep learning results. In: NeurIPS (2017)
23. Wah, C., Branson, S., Welinder, P., Perona, P., Belongie, S.: The Caltech-UCSD birds-200-2011 dataset (2011)
24. Wang, W., Lin, L., Fan, Z., Liu, J.: Semi-supervised learning for mars imagery classification and segmentation. ACM Trans. Multimed. Comput. Commun. Appl. **19**(4), 1–23 (2023)
25. Wang, X., Gao, J., Long, M., Wang, J.: Self-tuning for data-efficient deep learning. In: International Conference on Machine Learning, pp. 10738–10748. PMLR (2021)
26. Wu, H., Guo, H., Miao, Q., Huang, M., Wang, J.: Graph neural networks based multi-granularity feature representation learning for fine-grained visual categorization. In: Þór Jónsson, B., et al. (eds.) MMM 2022. LNCS, vol. 13142, pp. 230–242. Springer, Cham (2022). https://doi.org/10.1007/978-3-030-98355-0_20
27. Xie, Q., Dai, Z., Hovy, E., Luong, T., Le, Q.: Unsupervised data augmentation for consistency training. In: Advances in Neural Information Processing Systems, vol. 33, pp. 6256–6268 (2020)
28. Xu, Y., Qian, Q., Li, H., Jin, R., Hu, J.: Weakly supervised representation learning with coarse labels. In: Proceedings of the IEEE/CVF International Conference on Computer Vision, pp. 10593–10601 (2021)
29. Zagoruyko, S., Komodakis, N.: Wide residual networks. CoRR abs/1605.07146 (2016)
30. Zhang, B., et al.: Flexmatch: boosting semi-supervised learning with curriculum pseudo labeling. In: Advances in Neural Information Processing Systems, vol. 34, pp. 18408–18419 (2021)
31. Zhao, J., Liu, X., Zhao, W.: Balanced and accurate pseudo-labels for semi-supervised image classification. ACM Trans. Multimed. Comput. Commun. Appl. **18**(3s), 1–18 (2022)

FAFVTC: A Real-Time Network for Vehicle Tracking and Counting

Zhiwen Wang[1], Kai Wang[2], and Fei Gao[1(✉)]

[1] Zhejiang University of Technology, HangZhou 310023, China
feig@zjut.edu.cn

[2] Zhejiang Institute of Metrology, HangZhou 310018, China

Abstract. In a complex traffic environment, the detection and association of moving objects can easily lead to tracking errors. This work proposes a novel attention mechanism called MCSA, which integrates multi-spectral attention and spatial attention. Additionally, a fast and anchor-free real-time vehicle tracking and counting model named FAFVTC is constructed. MCSA is used for extracting the features of moving objects, while FAFVTC is able to better detect and associate these objects. The effectiveness of the FAFVTC method is verified on the UA-DETRAC dataset. FAFVTC outperforms existing techniques with a 1.3 improvement in the PR-MOTA metric and a 2.16 improvement in the MOTA metric. The average tracking speed achieved is 27.9 FPS. The experimental results demonstrate that the proposed approach enables fast and accurate vehicle tracking and counting.

Keywords: Vehicle tracking · Attention mechanism · Vehicle counting

1 Introduction

In recent years, the development of advanced computer vision technology has made it possible to track and count vehicles efficiently and accurately. MOT is a fundamental task in computer vision. As one of the key technologies within ITS, vehicle tracking provides essential information for various applications such as traffic flow estimation, vehicle monitoring, and road condition analysis. However, in practical applications, vehicle tracking and analysis based on traffic surveillance videos is not a simple task due to factors such as occlusions, lighting variations, weather changes, and low video resolutions.

Traditional MOT algorithms include KCF [1], MDP, JPDA, and Particle Filter. However, these methods suffer from significant prediction errors and poor robustness against occlusions and similar motion interferences. Recent research in this field has primarily focused on Joint Detection and Tracking (JDT). Prominent examples of such approaches are FairMOT [2] and FAFMOTS [3], which aim to reduce processing time and achieve real-time tracking capabilities.

This work proposes a single-stage multi-object detection, tracking, and counting method called FAFVTC. It combines MCSA attention for object

© The Author(s), under exclusive license to Springer Nature Singapore Pte Ltd. 2024
Q. Liu et al. (Eds.): PRCV 2023, LNCS 14436, pp. 251–264, 2024.
https://doi.org/10.1007/978-981-99-8555-5_20

identification, localization, and feature extraction, while also replacing the data association method. FAFVTC demonstrates excellent performance on datasets including congested scenes, achieving high accuracy and real-time capabilities. As a result, its deployment in ITS holds great potential. In summary, the contributions of this work are as follows:

(1) The vehicle tracking and counting framework FAFVTC was proposed, which combines multi-object detection, data association, and vehicle counting modules within a unified framework.
(2) The MCSA attention mechanism was proposed, which integrates multi-spectral channel attention and spatial attention. It addresses the deficiency of feature information in existing attention methods and fully utilizes low-frequency component information.
(3) A vehicle counting method was proposed which was validated on the UA-DETRAC dataset and real-world surveillance videos. The experiments demonstrate that the proposed FAFVTC enables the rapid and accurate completion of vehicle tracking and counting tasks.

2 Related Work

Vehicle Detection And Tracking: Existing MOT methods can be categorized into different research directions, such as Tracking-by-Detection (TBD) and Joint Detection and Tracking (JDT).

The TBD method separates object detection and tracking, which prevents the attainment of globally optimal results. SORT [4], DeepSORT [5], ByteTrack [6], and TADAM [7] are examples. Li et al. [8] represent data association as a graph optimization problem. STURE [9] utilizes a motion prediction network that incorporates temporal features for dynamic position prediction.

The JDT method integrates motion object detection and tracking into a unified framework. FairMOT [2], CenterTrack [10], and TraDeS [11]are all based on the CenterNet architecture and utilize anchor-free detection methods for object detection. FAFMOTS [3] based on FairMOT, integrates the idea of instance segmentation. FAMNet [12] enhances feature representation, affinity models, and multi-dimensional assignment, optimizing these three components jointly. LGM-Tracker [13] solves the vehicle tracking problem solely from a motion perspective.

Attention Mechanism: The attention mechanism enhances the model's ability to focus on relevant information, and selectively attend to specific parts of the input data. It can be categorized into channel-wise, spatial-wise, and hybrid attention mechanisms. SENet [14] enhances the model's representation capability by dynamically adjusting the weights of each channel. CBAM [15], based on SENet, combines channel attention and spatial attention along two independent dimensions. LCT [16] discovers a negative correlation between global context and attention values and models it using linear transformations. GCT [17] captures global background information from input images to improve the accuracy of model predictions. FCANet [18] dynamically focuses on different frequency channels to enhance the representation of important frequency information. CFCANet [19] improves the distribution of DCT frequency components in FCANet. EANet [20] implicitly considers the correlation between different samples.

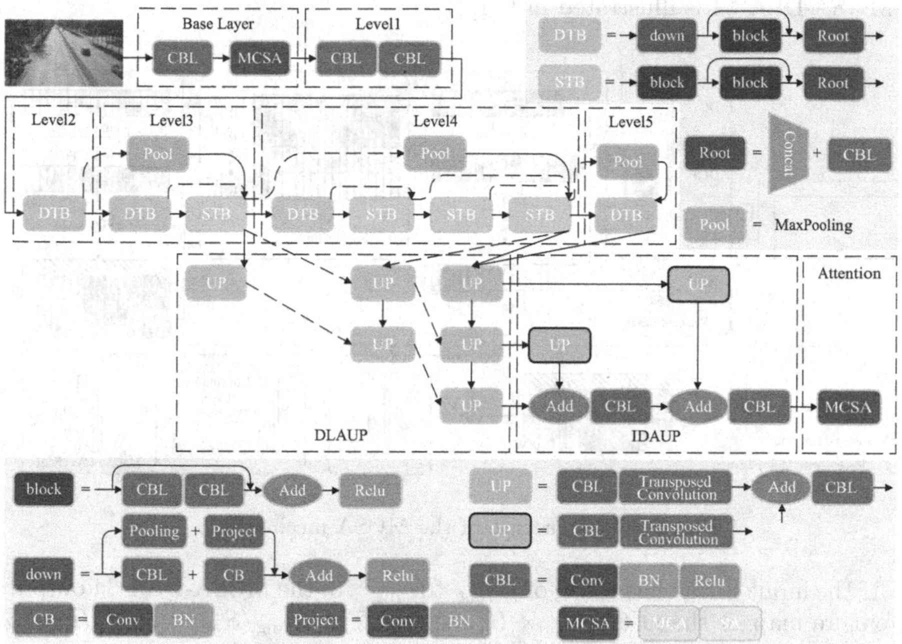

Fig. 1. The architecture of MCSA-DLA-34.

Vehicle Counting: Vehicle counting based on surveillance videos typically involves three components: vehicle detection, tracking, and trajectory processing. The challenges in it lie in the presence of motion blur, vehicle occlusion, and target scale variations. Amato et al. [21] proposed a real-time method for assessing vehicle numbers on highways or in parking lots. Zhang et al. [22] designed a traffic surveillance system based on Mask R-CNN, including modules for vehicle counting, vehicle type detection, and vehicle speed estimation. Gomaa et al. [23] introduced a method based on YOLOv2 and feature point motion analysis. Ciampi et al. [24] presented an approach for handling overlapping regions in multi-camera setups to estimate vehicle numbers in parking lots. Xu et al. [25] proposed a CityCam-to-Edge collaborative learning framework.

3 Method

3.1 Backbone Network

The DLA-34 is obtained by incorporating DLA [26] into ResNet-34 and replacing the regular convolutions in the DLA upsampling modules with Deformable

Convolution (DCN). The DCN can better adapt to the shape and position changes of moving objects, thereby improving the accuracy of motion object recognition and localization. Incorporating the MCSA attention into DLA-34, we propose MCSA-DLA-34. MCSA-DLA-34 is employed as the backbone network to enhance the feature extraction capability of the network. The structure of MCSA-DLA-34 is illustrated in Fig. 1.

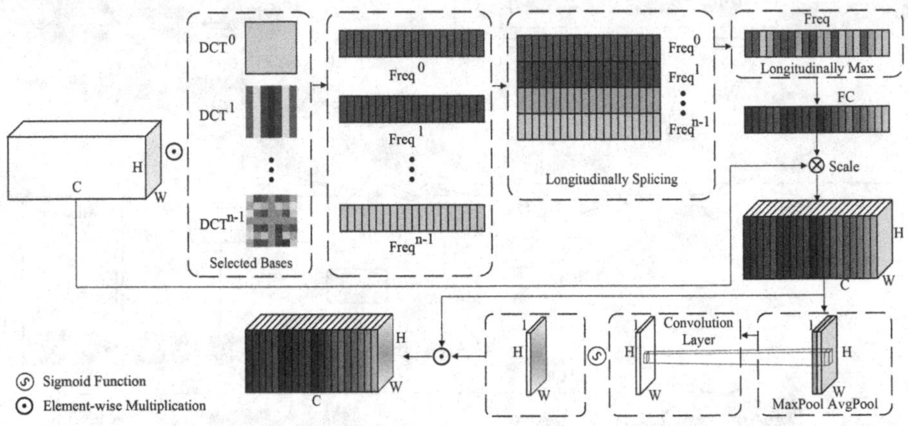

Fig. 2. The structure of the MCSA mechanism.

If the input image has a size of $W_{\text{img}} \times H_{\text{img}} \times 3$, the MCSA-DLA-34 outputs a feature map of shape $C \times W \times H$, where $W = W_{\text{img}}/S$ and $H = H_{\text{img}}/S$. Here, C and S are hyperparameters, where C represents the number of feature channels, and S represents the stride.

3.2 Multi-spectral Channel and Spatial Attention (MCSA)

The FAFVTC incorporates the MCSA-DLA-34 as its backbone network, which integrates the MCSA attention. The predict heads of FAFVTC also utilize the MCSA attention. MCSA integrates the multi-spectral channel attention and spatial attention. It dynamically focuses on different frequency channels and spatial information based on their importance, effectively enhancing the representation of important information. The structure of MCSA is illustrated in Fig. 2.

In FCANet, the MCA divides the input image X along the channel dimension into n channel blocks and assigns a 2D DCT frequency component index to each channel block. Although the approach of allocating frequency components by groups, as in MCA, introduces information from different frequency component indices, each channel block only utilizes one frequency component index, neglecting the other frequency component indices within the block and utilizing only a partial portion of the low-frequency information.

Considering that low-frequency components contain vital information, such as basic object structure, it is necessary to fully exploit the complete information of low-frequency components to enhance features. Inspired by CFCANet, we can assign the corresponding 2D DCT frequency component indices to the entire feature map. To fully leverage the complete information of low-frequency components, in MCSA, the input image $X \in \mathbb{R}^{C\times H\times W}$ is not divided into channel blocks. Instead, it is directly multiplied by n 2D DCT frequency component indices, where C represents the number of channels, and H and W represent the height and width of the feature map, respectively. $Freq^i$ denotes the result of element-wise multiplication between the i-th 2D DCT frequency component index and the input X. $Freq^i$ can be calculated using Eq. (1).

$$Freq^i = \sum_{h=0}^{H-1}\sum_{w=0}^{W-1} X_{:,h,w} \cos\left(\frac{\pi h}{H}\left(u_i + \frac{1}{2}\right)\right)\cos\left(\frac{\pi w}{W}\left(v_i + \frac{1}{2}\right)\right) \tag{1}$$

where $[u_i, v_i]$ corresponds to the i-th 2D DCT frequency component index of X. $i \in \{0, 1, \ldots, n-1\}, h \in \{0, 1, \ldots, H-1\}, w \in \{0, 1, \ldots, W-1\}$.

By using Eq. (2), we vertically concatenate all $Freq^i$ values to obtain a $C \times n$ matrix. Taking the maximum value of each column in the matrix yields the most prominent feature value for each channel, resulting in a multi-spectral vector $Freq$.

$$Freq = \mathrm{MF}_{\max}\left(\mathrm{cat}\left(\left[Freq^0, Freq^1, \ldots, Freq^{n-1}\right]\right)\right) \tag{2}$$

where $\mathrm{MF}_{\max}$ denotes the maximum value taken for each column of the matrix.

After obtaining the multi-spectral vector $Freq$, a fully connected layer learning process is performed to obtain the channel attention mechanism. The entire channel attention mechanism can be represented by Eq. (3).

$$MC\_att = sigmoid(FC(Freq)) \tag{3}$$

To capture the spatial relationships of features, spatial attention was incorporated after channel attention. This spatial attention complements channel attention. The entire MCSA mechanism can be represented by Eq. (4).

$$MCSA\_att = sigmoid\left(f^{7\times 7}([AvgPool(MCattF); MaxPool(MCattF)])\right) \tag{4}$$

where $f^{7\times 7}$ represents the convolution operation with a filter size of 7×7. $MCattF$ represents the output features of the multi-spectral channel attention.

3.3 Data Association

FAFVTC achieves satisfactory detection results and the Kalman filter exhibits high prediction accuracy, which can replace Re-ID for long-term association between moving objects, thereby achieving improved tracking speed. Inspired by ByteTrack, FAFVTC adopts a layered online data association approach.

Figure 3 illustrates the data association process. By detecting video frames, detection boxes D and their confidence scores $Score_D$ are obtained. Based on the given high-score box threshold τ_{high} and low-score box threshold τ_{low}, the boxes are classified into high-score boxes and low-score boxes. If $Score_D$ is higher than

τ_{high}, it is classified as a high-score box. If $Score_D$ is higher than τ_{low} but lower than τ_{high}, it is classified as a low-score box.

All high-score bounding boxes detected in the first frame are initialized as tracks. The Kalman filter is used to predict the new positions of the tracks in the next frame. Starting from the second frame, a matching process is conducted between high-score bounding boxes and existing tracks based on the IoU distance threshold $Score_{\text{IoU}}$. If the IoU distance is less than $Score_{\text{IoU}}$, the high-score bounding box is considered unmatched and is initialized as a new track. After matching the high-score bounding boxes, a similar matching process is performed between low-score bounding boxes and unmatched tracks based on the $Score_{\text{IoU}}$ value. Unmatched tracks are retained for 30 frames.

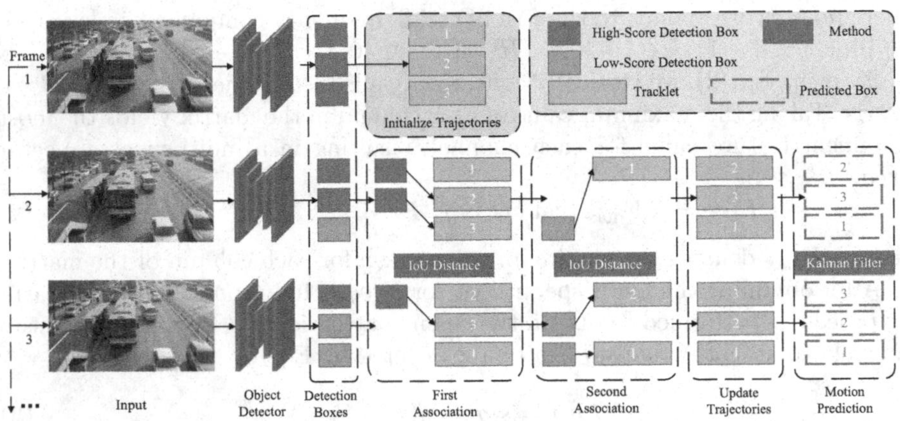

Fig. 3. The schematic diagram of data association.

3.4 Vehicle Counting

To meet the requirements of high counting accuracy and real-time performance, this work adopts a method that combines virtual detection lines with object tracking. As shown in Fig. 4, a single virtual detection line L is placed on the lane. The tracking algorithm extracts the motion trajectories T of the vehicle centers. When a vehicle's motion trajectory T intersects with the detection line L, the total count of vehicles is incremented by 1. The motion trajectory T starts when the motion object enters the detection area and ends when the motion object leaves the detection area. If the detection line is set at a distant location, the counting accuracy may be compromised due to poor detection of small targets. Therefore, when setting the detection line, it is crucial to choose a position that maximizes the detection of vehicle features to enhance the reliability of vehicle counting. To evaluate the counting accuracy, the Matching Accuracy (MA) is used as a metric, which is defined in Eq. (5).

$$MA = right/(right + error) \tag{5}$$

where *right* represents the number of correctly detected vehicles, while *error* represents the number of vehicles that were either incorrectly detected or missed.

To capture high-quality images of well-tracked moving objects within a trajectory, this work introduces the concept of Image Quality Quantification Score IQS. The vehicle image with the highest IQS score within a trajectory is preserved. The IQS score for the vehicle image c_0 is calculated using Eqs. (6)–(8).

$$Q_1 = \frac{\sqrt{\frac{1}{w_0 \times h_0}\sum_{a=1}^{w_0}\sum_{b=2}^{h_0}(G_{a,b} - G_{a,b-1})^2 + \frac{1}{w_0 \times h_0}\sum_{a=2}^{w_0}\sum_{b=1}^{h_0}(G_{a,b} - G_{a-1,b})^2}}{\sqrt{\frac{1}{w_0 \times h_0}\sum_{a=1}^{w_0}\sum_{b=2}^{h_0}(255)^2 + \frac{1}{w_0 \times h_0}\sum_{a=2}^{w_0}\sum_{b=1}^{h_0}(255)^2}} \tag{6}$$

$$Q_2 = (w_0 \times h_0)/(w_1 \times h_1 + w_0 \times h_0) \tag{7}$$

$$IQS = (Q_1 + Q_2 + Q_3 * 2)/4 \tag{8}$$

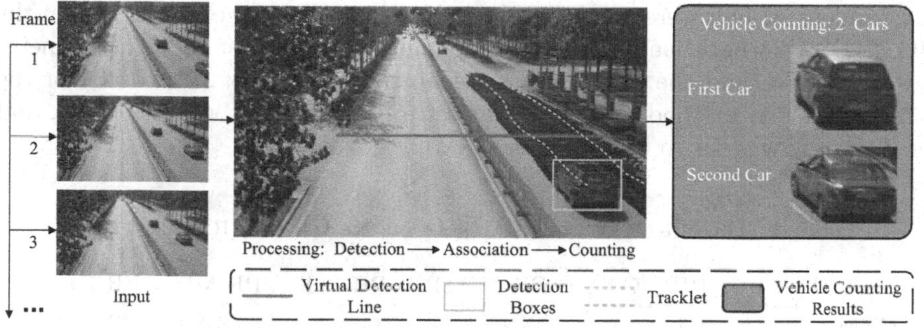

Fig. 4. The schematic diagram of vehicle counting.

where $G_{a,b}$ denotes the grayscale value of the pixel at the b-th row and a-th column of c_0. Q_1 represents the normalized spatial frequency of c_0, $Q_1 \in [0, 1]$. Q_2 represents the proportional relationship in terms of pixel size between c_0 and the first vehicle image c_1 in the trajectory T, $Q_2 \in (0, 1)$. w_i and h_i respectively represent the width and height of the bounding box of the vehicle image c_i. Q_3 represents the confidence score of the vehicle detection box, $Q_3 \in [0, 1]$.

4 Experiments

4.1 Datasets and Metrics

The UA-DETRAC dataset [27] consists of video sequences captured from various urban scenes and different traffic surveillance cameras, comprising a total of 100 video sequences. During the experimental process, 60 sequences were used for training, while 40 sequences were used for testing. In order to adapt to the original data input format of FairMOT, the data from the UA-DETRAC dataset was converted to the annotation format of the MOTChallenge dataset.

The performance evaluation on the dataset utilized metrics integrated with the precision-recall (PR) curve. The evaluation was conducted using the official tool, DETRAC-toolkit-test-trk. Standard tracking metrics including PR-MOTA and PR-MOTP were reported. PR-MOTA combines precision, recall, and MOTA to provide a comprehensive assessment of tracking performance.

4.2 Implementation Details

The training phase was conducted on two NVIDIA GeForce RTX 3090 GPUs. The input image size was adjusted to 1088×608 pixels. The model was initialized using pre-trained parameters from the COCO dataset. The FAFVTC model was trained for 35 epochs using the Lamb optimizer. With an initial learning rate of 1e−4, a batch size of 32 was set, and the learning rate was reduced to 1e-5 after 20 epochs. Data augmentation techniques such as random rotation, color jittering, and random scaling were employed during training to reduce overfitting. In the MCSA-DLA-34 backbone network, the parameter C was set to 256, parameter S was set to 4, and the output feature map size was $256 \times 272 \times 152$. For the data association method, the τ_{high} was set to 0.6, the τ_{low} was set to 0.1, and the $Score_{\mathrm{IoU}}$ was set to 0.2.

Table 1. The PR-MOTA metric results on the UA-DETRAC test set.

Method	PR-MOTA ↑	PR-MOTP ↑	PR-MT ↑	PR-ML ↓	PR-IDs ↓
CEM	5.1	35.2	3.0%	35.3%	267
H^2T	12.4	35.7	14.8%	19.4%	852
CMOT	12.6	36.1	16.1%	18.6%	285
GOG	14.2	**37.0**	3.0%	35.3%	3335
IOUT	16.1	**37.0**	13.9%	19.9%	2308
V-IOU	17.7	36.4	17.4%	18.8%	364
FAMNet [12]	19.8	36.7	17.1%	18.2%	617
FairMOT [2]	20.5	30.3	14.1%	**17.2%**	226
FAFVTC(ours)	**21.8**	24.9	**19.5%**	23.3%	**118**

4.3 Comparison Experiments

In the comparison experiments, separate experiments were conducted to evaluate the performance of vehicle detection and tracking as well as vehicle counting for FAFVTC. The performance of different models on the UA-DETRAC dataset was compared. For the experiments, FairMOT [2] and FAFVTC utilized their respective detection outputs. From Table 1, it can be seen that the FAFVTC method achieved higher PR-MOTA metric compared to FAMNet [12] and FairMOT, indicating superior tracking accuracy of FAFVTC over FAMNet and FairMOT. The reason behind this lies in the fact that FAMNet and FairMOT exhibit lower feature extraction and data association capabilities compared to FAFVTC.

FAFVTC outperforms FairMOT with a 1.3 improvement in the PR-MOTA metric, demonstrating superior multi-object detection and tracking performance. The runtime speed of FAFVTC is 27.9 FPS, making it suitable for real-time applications. Figure 5 visualizes the detection and tracking results of FAFVTC. The experiments demonstrate the accurate tracking of vehicles achieved by FAFVTC.

We achieved the best results in the PR-MOTA metric but slightly lagged in the PR-MOTP metric. This is due to performance trade-offs. Our tracking method prioritizes tracking consistency and maintaining accurate track identities to effectively handle occluded and congested scenes. This might come at the cost of precise object spatial localization. Experiments show that our tracking method minimizes identity switches and ensures consistent trajectories, avoiding unstable and fragmented tracking paths, which is beneficial for tracking and counting tasks.

Fig. 5. Visualization of partial results on the UA-DETRAC test set. Frame1, Frame2, and Frame3 refer to video frames with a 10-frame interval. Ignored Region refers to the ignored areas annotated by the official guidelines.

Table 2. The experimental results of vehicle counting on the UA-DETRAC test set. Right indicates correct counting, while Error represents incorrect counting.

Video	Right	Error	MA	Video	Right	Error	MA	Video	Right	Error	MA
39031	41	0	100.00%	40742	25	2	92.59%	40852	24	0	100.00%
39051	29	1	96.67%	40743	50	1	98.04%	40853	11	1	91.67%
39211	17	0	100.00%	40761	23	0	100.00%	40854	40	0	100.00%
39271	45	0	100.00%	40762	35	1	94.59%	40855	23	0	100.00%
39311	33	4	89.19%	40763	9	0	100.00%	40863	18	0	100.00%
39361	53	1	98.15%	40771	34	0	100.00%	40864	47	1	97.92%
39371	30	1	96.77%	40772	23	2	92.00%	40891	25	0	100.00%
39401	79	5	94.05%	40773	16	0	100.00%	40892	22	0	100.00%
39501	19	0	100.00%	40774	6	0	100.00%	40901	14	0	100.00%
39511	13	0	100.00%	40775	23	1	95.83%	40902	73	2	97.33%
40701	38	0	100.00%	40792	8	0	100.00%	40903	36	0	100.00%
40711	20	0	100.00%	40793	45	3	93.75%	40904	11	0	100.00%
40712	53	0	100.00%	40851	16	0	100.00%	40905	23	1	95.83%
40714	27	0	100.00%								
								Average	1177	28	97.68%

Table 3. The results of the vehicle counting experiments in real-world scenarios.

Video	Right	Error	MA	Video	Right	Error	MA	Video	Right	Error	MA
1001	39	0	100.00%	2002	51	3	94.44%	3002	36	1	97.30%
1002	29	1	96.67%	2003	45	1	97.83%	4001	37	2	94.87%
2001	27	0	100.00%	3001	54	1	98.18%	4002	26	1	96.30%
								Average	344	10	97.18%

Table 4. The results of the ablation experiments, where AM indicates the type of attention mechanism used.

Method	AM	Re-ID	BYTE	MOTA ↑	IDF1 ↑	MT ↓	ML ↓	FP ↓	FN ↓	IDs ↓
Baseline	×	✓	×	78.66	83.41	1680	108	18764	124313	1163
Method1	MCSA	✓	×	80.17	83.80	1742	90	18960	113500	1528
Method2	MCA	×	✓	80.35	84.96	1740	95	20555	111861	359
Method3	MCSA	×	✓	80.82	85.66	1797	88	23508	105689	396
Method4	MCSA	✓	✓	81.01	84.89	1795	86	22060	105412	885

Furthermore, this work conducted experiments to evaluate the vehicle counting performance of FAFVTC. From Table 2, it can be observed that FAFVTC achieved an average counting accuracy of 97.68% on the UA-DETRAC test set across various scenarios. Following the difficulty categorization by Wen et al. [27], in some simple scenarios such as MVI_39361 and MVI_40712, where vehicles are minimally occluded, the counting accuracy is relatively high. However, in challenging scenarios like MVI_39311 and MVI_40762, where vehicles are nearby,

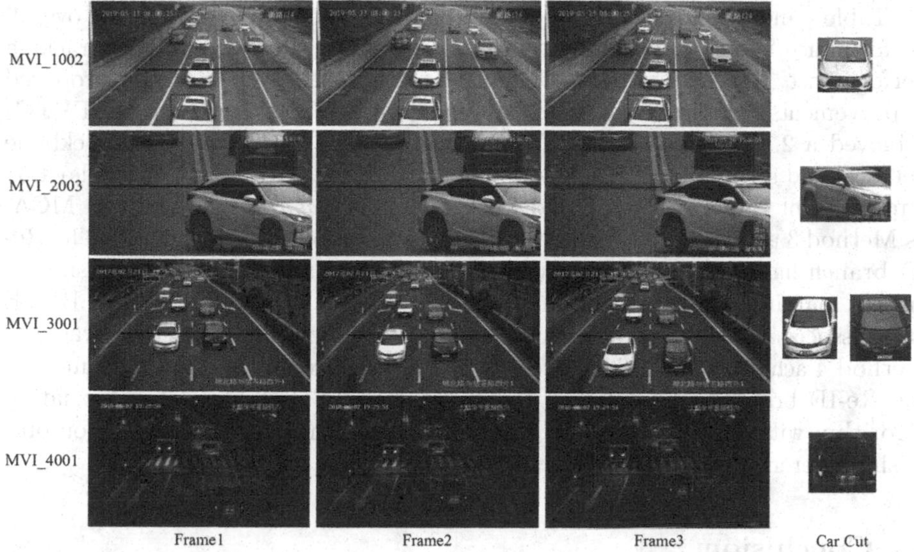

Fig. 6. The visualizations of experimental results in real-world scenarios. Frame1, Frame2, and Frame3 refer to video frames with a temporal interval of 10 frames.

occlusion between vehicles occurs, and low brightness in nighttime videos, resulting in a decrease in counting accuracy.

In this work, experiments were conducted to evaluate the vehicle counting performance of FAFVTC in real-world scenarios. The videos were captured in cities such as Lishui, Wenzhou, Wuxi, and Shenzhen, encompassing both daytime and nighttime scenes. From Table 3, it can be observed that the FAFVTC method achieves an average counting accuracy of 97.18% in real-world scenarios. Figure 6 visualizes the counting results of FAFVTC in real-world scenarios. The experiments demonstrate the accurate vehicle counting capability of the proposed FAFVTC in real-world scenarios.

Table 5. Comparison of inference speeds.

Method	Re-ID	FPS↑
Baseline + MCSA + BYTE	✓	26.55
Baseline + MCSA + BYTE	×	27.86

4.4 Ablation Study

This work conducted an ablation study on the vehicle detection and tracking performance of FAFVTC under different constraints. In the ablation experiments, FairMOT was used as the baseline model, DLA-34 served as the backbone network, and the Adam optimizer was used for training.

Table 4 and Table 5 present the impact of different components on the overall performance, with the MOT metrics used to evaluate the detection and tracking performance. The experiments demonstrate the effectiveness of the proposed improvements. Compared to the baseline (FairMOT), Method 3 (FAFVTC) achieved a 2.16 increase in the MOTA metric. The MCSA-DLA-34 backbone network had a significant impact on model accuracy, as Method 1 showed a 1.51 improvement in MOTA compared to the baseline. MCSA outperformed MCA, as Method 3 achieved a 0.47 increase in MOTA compared to Method 2. The Re-ID branch had a minimal impact on model performance, as Method 4 showed only a marginal improvement in MOTA compared to Method 3. The BYTE data association method played a crucial role in improving model accuracy, as Method 4 achieved a 0.84 increase in MOTA compared to Method 1. Removing the Re-ID branch and replacing the original DeepSORT-like data association algorithm with BYTE improved the model's inference speed without compromising inference accuracy.

5 Conclusion

n complex traffic environments, the misdetection and occlusion of moving objects can easily lead to tracking errors. In this work, we investigate the problem of efficient vehicle object tracking and counting and propose the MCSA attention mechanism along with the MCSA-DLA-34 backbone network, which effectively utilizes frequency and spatial domain information. Building upon MCSA-DLA-34, we introduce the FAFVTC framework for vehicle tracking and counting. By replacing the data association method and removing the Re-ID branch, we improve the model's inference speed without compromising accuracy. Experimental results on the UA-DETRAC dataset and surveillance videos captured in real-world scenarios demonstrate the effectiveness of our approach. The proposed FAFVTC method offers a solution for fast and accurate online vehicle tracking and counting, with potential applications. In the future, we will design a lighter and faster model to further enhance the accuracy and efficiency of online vehicle tracking and counting.

Acknowledgment. This work is being supported by Zhejiang Provincial Science and Technology Planning Key Project of China under Grant No. 2021C03129 and 2021C01194, and the Eagle Plan of Zhejiang Provincial Administration for Market Regulation under Grant No. CY2022339.

References

1. Henriques, J.F., et al.: High-speed tracking with kernelized correlation filters. TPAMI **37**(3), 583–596 (2014)
2. Zhang, Y., et al.: Fairmot: on the fairness of detection and re-identification in multiple object tracking. IJCV **129**(11), 3069–3087 (2021)

3. Li, S., et al.: FAFMOTS: a fast and anchor free method for online joint multi-object tracking and segmentation. In: ISMARW, pp. 465–470 (2022)
4. Bewley, A., et al.: Simple online and realtime tracking. In: ICIP, pp. 3464–3468 (2016)
5. Wojke, N., et al.: Simple online and realtime tracking with a deep association metric. In: ICIP, pp. 3645–3649 (2017)
6. Zhang, Y., et al.: Bytetrack: multi-object tracking by associating every detection box. In: Avidan, S., Brostow, G., Cissé, M., Farinella, G.M., Hassner, T. (eds.) ECCV 2022. LNCS, vol. 13682, pp. 1–21. Springer, Cham (2022). https://doi.org/10.1007/978-3-031-20047-2_1
7. Guo, S., et al.: Online multiple object tracking with cross-task synergy. In: CVPR, pp. 8136–8145 (2021)
8. Li, W., et al.: Simultaneous multi-person tracking and activity recognition based on cohesive cluster search. CVIU **214**, 103301, 1–13 (2022)
9. Wang, H., et al.: STURE: spatial-temporal mutual representation learning for robust data association in online multi-object tracking. CVIU **220**, 1–10 (2022)
10. Zhou, X., Koltun, V., Krähenbühl, P.: Tracking objects as points. In: Vedaldi, A., Bischof, H., Brox, T., Frahm, J.-M. (eds.) ECCV 2020. LNCS, vol. 12349, pp. 474–490. Springer, Cham (2020). https://doi.org/10.1007/978-3-030-58548-8_28
11. Wu, J., et al.: Track to detect and segment: an online multi-object tracker. In: CVPR, pp. 12352–12361 (2021)
12. Chu, P., et al.: Famnet: joint learning of feature, affinity and multi-dimensional assignment for online multiple object tracking. In: ICCV, pp. 6172–6181 (2019)
13. Wang, G., et al.: Track without appearance: Learn box and tracklet embedding with local and global motion patterns for vehicle tracking. In: ICCV, pp. 9876–9886 (2021)
14. Hu, J., et al.: Squeeze-and-excitation networks. In: CVPR, pp. 7132–7141 (2018)
15. Woo, S., Park, J., Lee, J.-Y., Kweon, I.S.: CBAM: convolutional block attention module. In: Ferrari, V., Hebert, M., Sminchisescu, C., Weiss, Y. (eds.) ECCV 2018. LNCS, vol. 11211, pp. 3–19. Springer, Cham (2018). https://doi.org/10.1007/978-3-030-01234-2_1
16. Ruan, D., et al.: Linear context transform block. In: AAAI, vol. 34, no. 4, pp. 5553–5560 (2020)
17. Ruan, D., et al.: Gaussian context transformer. In: CVPR, pp. 15129–15138 (2021)
18. Qin, Z., et al.: Fcanet: frequency channel attention networks. In: ICCV, pp. 783–792 (2021)
19. Su, B., et al.: CFCAnet: a complete frequency channel attention network for SAR image scene classification. In: IEEE J-STARS, vol. 14, pp. 11750–11763 (2021)
20. Guo, M.H., et al.: Beyond self-attention: external attention using two linear layers for visual tasks. TPAMI **45**(5), 5436–5447 (2022)
21. Amato, G., et al.: Counting vehicles with deep learning in onboard UAV imagery. In: ISCC, pp. 1–6 (2019)
22. Zhang, B., et al.: A traffic surveillance system for obtaining comprehensive information of the passing vehicles based on instance segmentation. TITS **22**(11), 7040–7055 (2021)
23. Gomaa, A., et al.: Faster CNN-based vehicle detection and counting strategy for fixed camera scenes. MTA **81**(18), 25443–25471 (2022)

24. Ciampi, L., et al.: Multi-camera vehicle counting using edge-AI. ESWA **207**, 117929, 1–9 (2022)
25. Xu, H., et al.: Efficient CityCam-to-edge cooperative learning for vehicle counting in ITS. TITS **23**(9), 16600–16611 (2022)
26. Yu, F., et al.: Deep layer aggregation. In: CVPR, pp. 2403–2412 (2018)
27. Wen, L., et al.: UA-DETRAC: a new benchmark and protocol for multi-object detection and tracking. CVIU **193**, 102907, 1–9 (2020)

Ped-Mix: Mix Pedestrians for Occluded Person Re-identification

Shang Gao[1,2], Chenyang Yu[1], Pingping Zhang[1], and Huchuan Lu[1,2(✉)]

[1] Dalian University of Technology, Dalian, China
yuchenyang@mail.dlut.edu.cn, {zhpp,lhchuan}@dlut.edu.cn
[2] NingBo Institute of Dalian University of Technology, Ningbo, China

Abstract. Occluded person re-identification is a very challenging task due to the interference of occluding objects. Most existing approaches concentrate on modifying the network architecture to facilitate the extraction of more distinctive local features or render the network less sensitive to occlusions. However, it is easy to fail when encountering previously unseen occlusions or when other humans act as occluders, due to the limited occlusion variance in the training set. In this paper, we propose a data augmentation method that blends the target pedestrian with other pedestrians to simulate non-target pedestrian occlusion. Furthermore, we propose a non-target suppression (NTS) loss to reduce the information flow from the occluded region to the final embedding, where the occluded region can be easily obtained from the augmentation. Experimental results demonstrate that this simple augmentation technique yields significant performance improvements in the task of occluded person re-identification.

Keywords: Occluded person re-identification · Data Augmentation · Non-Target Suppression Loss

1 Introduction

Person re-identification (ReID) is the task of finding and matching the same person across multiple non-overlapping cameras. This task has received a lot of attention in recent years because of its wide applications in surveillance systems. The performance of ReID has been significantly improved thanks to the abundant training data provided by large-scale benchmarks [18,25,30] and the powerful representation capabilities of deep learning techniques [6,10,22]. However, the occlusion problem caused by various obstacles such as vehicles, road signs, or other pedestrians is still a challenging issue and widely occurs in real application scenarios.

Many network structures that incorporate attention mechanisms are proposed [16,17,33]. While, some methods use human parsing and keypoint

---

Supplementary Information The online version contains supplementary material available at https://doi.org/10.1007/978-981-99-8555-5_21.

© The Author(s), under exclusive license to Springer Nature Singapore Pte Ltd. 2024
Q. Liu et al. (Eds.): PRCV 2023, LNCS 14436, pp. 265–277, 2024.
https://doi.org/10.1007/978-981-99-8555-5_21

estimation to align different human parts [7,19,21,23]. However, such methods are limited by the fact that the current datasets for ReID usually exhibit limited variance in occlusion, which reduces the robustness when encountering new occlusion types. Occlusion can be divided into two types [24]: non-pedestrian occlusion and non-target pedestrian occlusion. As shown in Fig. 1, it can be seen that the situation of occlusion is very complex and varied.

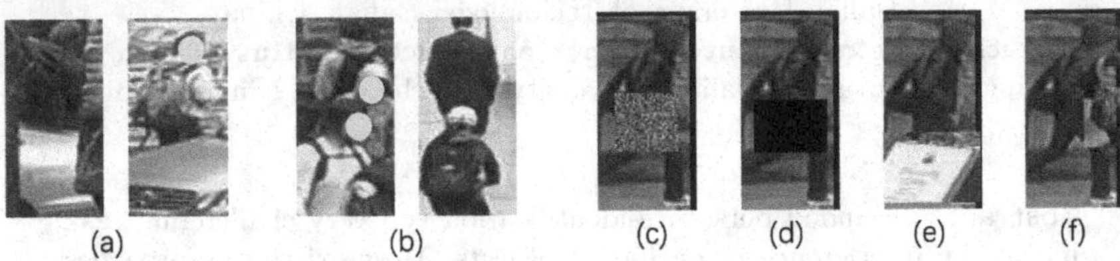

Fig. 1. Illustration of various occlusion scenarios and different data augmentation methods: (a) Non-pedestrian occlusion, (b) Non-target pedestrian occlusion, (c) Random erasing [32], (d) Cutout [5], (e) NPO [24], (f) The proposed Ped-Mix.

Data augmentation is an effective method to enlarge the training data and enhance the model's robustness, which can complement the above methods. Most of the data augmentation methods used in occluded person ReID can be regarded as CutMix [28] family methods, which randomly select a region from the input image and then replace the selected region with another image patch. As shown in Fig. 1, random erasing [32] substitutes the original region with random values; Cutout [5] sets the replacement area to zero; while NPO [24] occludes the selected region with a random pre-cropped background patch. Among these augmentation methods, most of them only account for the case of non-pedestrian occlusion, and only the method proposed in [24] considers the case of non-target pedestrian occlusion. It simulates multi-pedestrian images by diffusing characteristics of non-target pedestrians to the original features. This method cannot guarantee the rationality of the diffusion, resulting in a very limited performance improvement, and it requires extra computational overhead.

In this paper, to address the occlusion problem caused by non-target pedestrians, we propose a novel data augmentation approach, Ped-Mix. It can be easily implemented to various network structures. In occluded person ReID, it is generally assumed that the pedestrian closest to the center of the bounding box is the target pedestrian, while any other pedestrians in the same image are considered as occlusion or interference. Based on this assumption, we design our data augmentation method. In the training phase, an image undergoes the following three operations: 1) randomly select another sample within the batch; 2) randomly shift the selected image and crop out the overlapping part with its original area; 3) randomly mask the cropped patch and replace the overlapping regions of the target image with the cropped patch. This method enables the simulation of realistic occlusion between pedestrians during the training process, while the mask operation preserves the model from only extracting the most obvious visible parts. The experiments show that using this simple data augmentation method can significantly boost performance.

Furthermore, the features of the target person should not obtain any information from the occluded area. Therefore, we attempt to use attention rollout [1] to track down the information propagated from the input layer to the embeddings in the higher layers, and propose a non-target suppression loss to constrain the attention to be as small as possible in the artificially occluded region.

The main contributions of the paper are summarized as follows:

- A straightforward yet highly effective data augmentation technique, Ped-Mix, is proposed to tackle the problem of insufficient diversity of occlusion samples for occluded person ReID.
- A non-target suppression loss is proposed to reduce the information flow from the occluded region.
- Extensive experiments on two publicly occluded benchmarks demonstrate the superiority of our method.

2 Related Works

2.1 Occluded Person Re-identification

The occlusion problem has two main difficulties. Firstly, the features of the occluded region will interfere with the target pedestrian features and destroy the complete pedestrian features. Secondly, occlusion will cause the deficiency of local features of the target pedestrian, which will lead to misalignment in matching. To address the aforementioned problems, part-to-part matching strategy [7,18,19,21,33] tackle occluded scenarios by exploiting alignment relations among body parts according to the similarity of local spatial features across query and gallery images. These methods can be roughly summarized into two streams: 1) external cues like human parsing [2,19] or pose estimation [7,18,21] are leveraged to align parts of bodies and determine whether the local part are visible. Somers *et al.* [19] propose to learn multiple features for particular body parts with a global-identity local-triplet loss. Wang *et al.* [21] combine patch and key-point information to enhance local patch features. 2) identity prior is leveraged to generate the part region and to be used as pseudo-label to train the ReID network [33]. Other approaches [14,27] are based on feature reconstruction, which generate the feature of the occluded region from the other parts or from neighboring samples. Hou *et al.* [14] locate occluded human parts by key-points and propose Region Feature Completion (RFC) to recover the semantics of occluded regions in feature space. The above methods are limited by the fact that the current datasets for person re-identification usually exhibit limited variance in occlusion, which reduces the robustness of the trained networks when encountering new occlusion types.

2.2 Data Augmentation and Training Loss

Many methods [24,26,29] adopt data augmentation strategies to expand the diversity of occlusion, as well as to guide the attention network to focus more on the pure target person than the occlusion. Zhuo *et al.* [34] design an occlusion simulator to use the random patch from the background as the artificial occlusion to cover the full-body person image. Zhao *et al.* [29] add occlusion on the input image with an easy-to-hard strategy together with an adversarial suppression loss, making the network more robust to occlusion by gradually learning harder occlusion instead of hardest occlusion or random occlusion directly. Xia *et al.* [26] swaps the original region with background regions together with a learnable attention disturbance mask to divert attention away from actual occlusions during testing. Wang *et al.* [24] simulates multi-pedestrian images by diffusing characteristics of non-target pedestrian to the original features.

3 Proposed Method

In this section, we introduce the proposed data augmentation method Ped-Mix and non-target suppression loss in detail. The architecture is depicted in Fig. 2. We build our feature extractor based on the transformer-based image classification model ViT [6]. We first describe the detailed procedure of Ped-Mix. Following this, we introduce non-target suppression loss based on the augmentation region. Finally, the implementation of our training procedure is introduced.

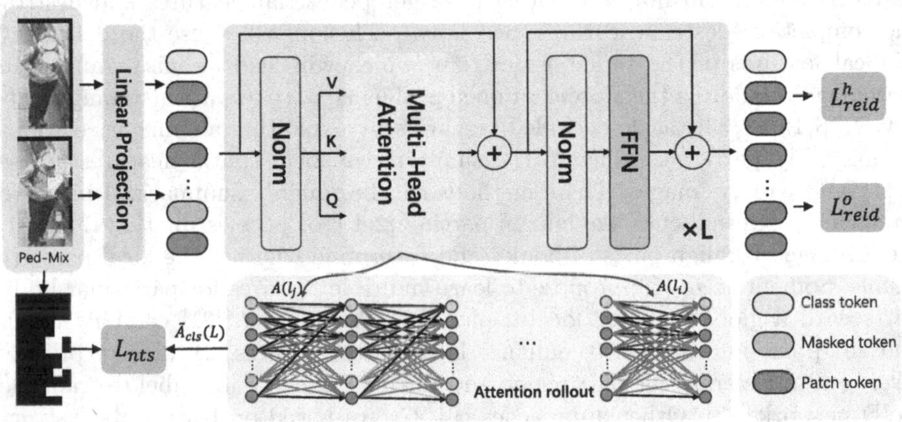

Fig. 2. Overview of the training procedure.

In the occluded person re-identification task, the pedestrian closest to the center of the bounding box is usually considered as the target pedestrian, while other pedestrians appearing in the same image can be considered as interference. Therefore, Ped-Mix is designed. The overall process is shown in Fig. 3.

3.1 Ped-Mix

Let $x \in \mathbb{R}^{W\times H\times C}$ and y denote a training image and its label, respectively. The goal of Ped-Mix is to generate a new training sample $\tilde{x}$ by combining two

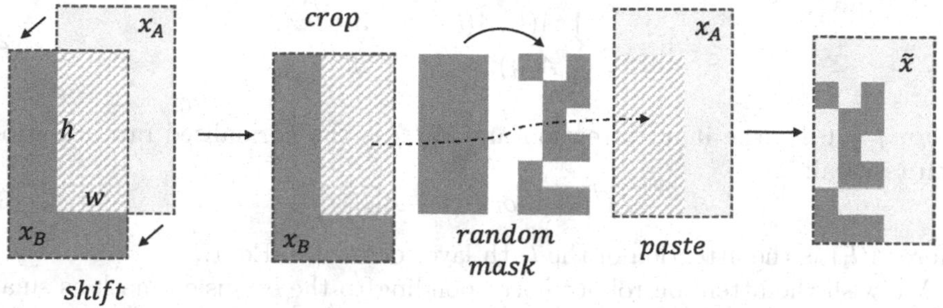

Fig. 3. Flowchart of the proposed Ped-Mix.

training samples x_A and x_B, where x_B need to shift away from the center of the bounding box. Firstly, we need to determine the amount and direction of the shift, where the amount of the offset corresponds to the size of the occlusion area O_c, and the direction determines the position of the occlusion. We initialize r_o which is the ratio of the occlusion size to the original image size. $r_w = \frac{w}{W}$ can be sampled from the distribution $\mathcal{U}(r_o, 1.0)$, where w is the width of the occlusion region. Then, we have $r_h = \frac{r_o}{r_w}$ and we can get $h = H \times r_h$ and $w = W \times r_w$, where h is the height of the selected region. In situations where there is occlusion between pedestrians, the individual located at the upper portion of the image is identified as the target pedestrian. Thus, we have two options for the shift directions, which are the bottom left and bottom right of x_A. After randomly choosing a direction, we can get the exact position of the overlapping region on x_A and x_B, respectively. The augmented sample $\tilde{x}$ is obtained by cropping the overlapping patch of x_B and pasting it to the corresponding region of x_A. Through experiments we found that using this full patch occlusion method may lead the model to only extract the most obvious visible parts, thus reducing its generalization ability. We would show it in Sect. 4.6. To address this issue, we randomly occlude certain patches as in [9]. Finally, the model can be trained with its original loss function using the generated training sample $(\tilde{x}, y)$.

3.2 Non-target Suppression Loss

The attention of each Transformer block reflects how information interaction between each feature. It can be formulated as:

$$A = softmax(\frac{QK^T}{\sqrt{c/N_h}}), \tag{1}$$

where Q and K are the queries and keys of the tokens, N_h is the number of heads in multi-head attention. The mixing of information gets increasing across layers. Attention rollout [1] is proposed to approximatively track the flow of information. The attention rollout from the l_j-th layer to the l_i-th layer can be obtained by recursively multiply the attention weights as:

$$\tilde{A}(l_i) = \begin{cases} \bar{A}(l_i)\tilde{A}(l_{i-1}), & \text{if } i > j; \\ \bar{A}(l_i), & \text{if } i = j. \end{cases} \tag{2}$$

where $j = 0$ for the input attention and $\bar{A(l_i)}$ is the normalized raw attention with residual:

$$\bar{A}(l_i) = norm(I + A(l_i)), \tag{3}$$

where $A(l_i)$ is the attention of the l_i-th layer defined in Eq. 1.

We wish the attention rollout corresponding to the occlusion mask as small as possible. To this end, a non-target suppression loss is designed, which can be formulated as:

$$L_{nts} = \|\tilde{A}_{cls}(L) \cdot M\|_2, \tag{4}$$

where M is the binary mask of the exchanged region of the augmented input, and $\tilde{A}_{cls}(L)$ indicates the attention rollout from each location to the class token.

3.3 Training Procedure

In this section, we utilize the double batch setting as in [29], and train the holistic batch and occluded batch separately as [26]. The architecture is depicted in Fig. 2. Given a batch of images $X \in \mathbb{R}^{B \times C \times H \times W}$, where B is the batch size and C is the number of the input channels, we first augment each sample within the batch by Ped-Mix and sent them into ViT backbone. Finally, the class token feature $F^h = [f_1^h, f_2^h, ..., f_B^h]$ and $F^o = [f_1^o, f_2^o, ..., f_B^o]$ of the last layer is chosen as the embedding for each sample, F^h and F^o are the feature of holistic samples and occluded samples, respectively. Besides, the attention matrices of each layer are also output from ViT. We optimize the network by adopting cross-entropy loss as identity loss $\mathcal{L}_{id}$ and soft margin triplet loss [13] $\mathcal{L}_{tri}$ as the metric loss:

$$\mathcal{L}_{id} = -\frac{1}{B}\sum_{i=1}^{B} log\frac{e^{(W^{y_i})^T f_i}}{\sum_{j=1}^{C} e^{(W^j)^T f_i}}, \tag{5}$$

$$\mathcal{L}_{tri} = \frac{1}{B}\sum_{i=1}^{B} ln(1 + e^{(\|f_i - f_i^p\|_2 - \|f_i - f_i^n\|_2)}), \tag{6}$$

where W represents the weight of the classifier, C is the number of classes, and y_i is the identity label for the $i-$th sample. The f^p and f^n in triplet loss refer to the hardest positive and negative features within the batch. As only the class token is used as the embedding, we only focus on the information flow towards the class token. The final loss is formulated as:

$$\mathcal{L}_{reid} = \mathcal{L}_{id} + \mathcal{L}_{tri}, \ \mathcal{L} = \mathcal{L}_{reid} + \mathcal{L}_{nts}. \tag{7}$$

4 Experiment

4.1 Datasets and Evaluation Measures

To evaluate the effectiveness of the proposed Ped-Mix, we conduct experiments on four benchmarks including Market-1501 [30], DukeMTMC-reID [31], Occluded-Duke [18], and Occluded-REID [34].

Market-1501 [30] contains 32,668 labeled images of 1,501 identities observed from 6 cameras. The dataset is split into training set with 12,936 images of 751 identities. Few images in this dataset are occluded.

DukeMTMC-reID [31] consists of 16,522 training images of 702 persons, 2,228 queries of 702 persons, and 17,661 gallery images of 702 persons from 8 cameras.

Occluded-Duke [18] is a large-scale dataset collected from the DukeMTMC-reID [31] for occluded person re-identification, which is by far the largest occluded ReID datasets. The training set consists of 15,618 images of 702 persons. The testing set contains 2,210 images of 519 persons as the query and 17,661 images of 1,110 persons as the gallery.

Occluded-REID [34] are captured by mobile cameras equipped on campus, including 2,000 annotated images belonging to 200 identities. Among the dataset, each person consists of 5 full-body person images and 5 occluded person images with various occlusions. Due to the absence of the same prescribed split of training and test set, the model is trained on Market-1501 [30], and all the images are adopted for testing.

Evaluation Protocols. We report the Cumulated Matching Characteristics (CMC) [8] and mean Average Precision (mAP) [30] for the proposed approach. All experiments are conducted in the single query mode.

4.2 Implementation Details

Unless otherwise specified, all images are resized to 256×128. The batch size is set to 64 with 4 images per ID. We adopt the ViT-B [6] pre-trained on ImageNet [3] as our backbone. SGD optimizer is employed with a momentum of 0.9 and the weight decay of 1e−4. The learning rate is initialized as 0.004 with cosine learning rate decay. The training images are augmented with random horizontal flipping, padding and random cropping for all experiments. Random erasing [32] is used for baseline, and is removed when using Ped-Mix. r_o is set to 0.5. The patch size and mask ratio of the random masking are set to 12 and 0.5, respectively. See the material for the analysis of the relevant parameters (Table 1).

4.3 Ablation Studies

In this section, we implement the ablation studies based on the Occluded-Duke dataset to analyze the influence of each module of the proposed method. In our

Table 1. Ablation study of each proposed module on Occluded-Duke dataset.

Methods	Occluded-Duke			
	Rank-1	Rank-5	Rank-10	mAP
baseline	59.7	75.3	80.7	49.8
RE [32]	61.3	77.7	81.8	53.5
NPO [24]	63.9	79.4	84.7	54.2
APD [26]	66.2	81.6	86.3	57.7
Ped-Mix	65.3	80.9	85.3	56.4
Ped-Mix + L_{nts}	66.3	81.7	86.4	57.2
Ped-Mix + NPO [24]	67.0	82.4	86.2	57.3
Ped-Mix + NPO [24] + L_{nts}	68.8	83.1	87.1	58.1

study, the baseline method adopts ViT as the backbone, which is trained based on the original softmax loss and triplet loss without any artificial occlusion. From the result, we can observe that training with images occluded by the Ped-Mix can significantly improve the model performance, the performance can be increased by +5.6% in Rank-1 and 6.6% in mAP, respectively, over the baseline. Besides, with the assistance of L_{nts}, the performance of the model can further increase from 65.3% to 66.3% in Rank-1 and 56.4% to 57.2% in mAP. Furthermore, as Ped-Mix aims to simulate non-target pedestrian occlusion, we combine it with NPO [24] and further improve Rank-1 and mAP from 65.3% to 67.0% and from 56.4% to 57.3%, respectively. Finally, by combining all the components, our model achieves 68.8% in Rank-1 and 58.1% in mAP. It is noteworthy that APD [26] is the latest augmentation method for occluded person ReID. We compare our method with it under the same setting and find that our method achieves comparable results with it.

4.4 Comparison with State-of-the-Art Methods

We compare Ped-Mix with existing state-of-the-art (SOTA) methods on two occluded datasets, and the results are shown in Table 2. Here we replace identity loss with Arcface loss [4] as in [20,26]. The compared methods can be divided into CNN-based and Transformer-based methods. As can be seen from Table 2, transformer-based methods outperform the CNN-based methods by a large margin. It can be seen, in the most challenging Occluded-Duke dataset, our proposed method Ped-Mix can achieve 70.8% in Rank-1 and 59.2% in mAP, respectively, significantly improving the Rank-1 accuracy by 4.0% and mAP by 6.6% over the CNN-based SOTA method BPBreID [19], which is a part-based method, with high-dimensional features when testing. Compared with the transformer-based SOTA method FED [24] which also considers the non-target pedestrian occlusion but with complex structures, our method achieves +2.7% in Rank-1 and +2.8% in mAP. PFD [23] achieves best result on mAP, which mainly benefit from the

Table 2. Comparison with state-of-the-art methods on Occluded-Duke and Occluded-REID. ∗ indicates that the backbone has a sliding-window setting and a smaller stride. † indicates that the backbone is HrNet [22].

Methods	Occluded-Duke		Occluded-REID	
	Rank-1	mAP	Rank-1	mAP
DSR [11]	40.8	30.4	72.8	62.8
Ad-Occluded [15]	44.5	32.2	–	–
PVPM [7]	47.0	37.7	66.8	59.5
HOReID [21]	55.1	43.8	80.3	70.2
ISP [33]†	62.8	52.3	–	–
BPBreID [19]	66.8	52.6	76.9	68.6
PAT [17]	64.5	53.6	81.6	72.1
TransReID [12]	64.2	55.7	–	–
PFD [23]	67.7	**60.1**	79.8	81.3
FED [24]	68.1	56.4	86.3	79.3
Ours	**70.8**	59.2	**87.1**	**81.4**
TransReID∗ [12]	66.4	59.2	–	–
PFD∗ [23]	69.5	61.8	81.5	83.0
DPM∗ [20]	71.4	61.8	85.5	79.7
Ours∗	**73.1**	**61.9**	**88.5**	**83.5**

part based model and matching with the visibility score. In contrast, our method achieves comparable results with single global feature. Furthermore, with a small step sliding-window setting, the proposed method can further achieve a higher performance of 73.1% in Rank-1 and 61.9% in mAP, respectively.

On the Occluded-REID dataset, our method also consistently outperform current SOTAs. Specifically, it achieves 87.1% in Rank-1 and 81.4% in mAP. Note that, in the transfer setting, the methods using data augmentation significantly outperform the other methods when facing unseen occlusions, which proves our point.

4.5 Visualization

We demonstrate how our method overcomes the occlusion constraint by providing several samples of person image ranking, Fig. 4 shows experimental results. We can observe that our method can overcome the non-pedestrian occlusions as well as the non-target pedestrian occlusion and identify images of the same pedestrian correctly (highlighted by green-color boxes). As a comparison, the baseline network is very sensitive to occlusions obviously.

Fig. 4. Ranking list for baseline and our proposed method. The green and red boxes highlight positive and negative matching. The image without a bound box is the query. (Color figure online)

4.6 Why Random Masking

Full patch occlusion method may lead the model to only extract the most obvious visible parts, thus reducing its generalization ability. We demonstrated this by comparing our method with the Cut-Mix method on the Occluded-REID dataset, where Cut-Mix means conducting Ped-Mix without random masking. The results are shown in Table 3. It can be seen that our method outperformed Cut-Mix in this transfer setting, demonstrating that our method has better generalization performance.

Table 3. Comparison with Cut-Mix

Method	Occluded-REID	
	Rank-1	mAP
Ours	88.5	83.5
Cut-Mix	87.5	81.3

Table 4. Results on holistic datasets.

Method	Market1501		DukeReID	
	Rank-1	mAP	Rank-1	mAP
Baseline	94.7	87.4	89.2	79.7
Ours	94.8	87.5	89.7	79.7

4.7 Results on Holistic Datasets

Though our method is specifically designed for occlusion situations, to verify its robustness, we also conducted experiments on holistic ReID datasets. The results are shown in Table 4, our method did not weaken the performance on the holistic dataset, demonstrating its robustness to various data.

5 Conclusion

In this paper, a straightforward yet highly effective data augmentation technique, Ped-Mix, is introduced to tackle the problem of insufficient diversity of occlusion samples for occluded person re-identification task, especially for non-target pedestrian occlusion. In addition, an non-target suppression loss is proposed to reduce the activation of the occluded region. Extensive experiments on two publicly occluded benchmarks demonstrate the superiority.

References

1. Abnar, S., Zuidema, W.: Quantifying attention flow in transformers. In: Proceedings of the 58th Annual Meeting of the Association for Computational Linguistics, January 2020. https://doi.org/10.18653/v1/2020.acl-main.385
2. Cheng, X., Jia, M., Wang, Q., Zhang, J.: More is better: multi-source dynamic parsing attention for occluded person re-identification. In: Proceedings of the 30th ACM International Conference on Multimedia, pp. 6840–6849 (2022)
3. Deng, J., Dong, W., Socher, R., Li, L.J., Li, K., Fei-Fei, L.: Imagenet: a large-scale hierarchical image database. In: 2009 IEEE Conference on Computer Vision and Pattern Recognition, pp. 248–255. IEEE (2009)
4. Deng, J., Guo, J., Xue, N., Zafeiriou, S.: Arcface: additive angular margin loss for deep face recognition. In: Proceedings of the IEEE/CVF Conference on Computer Vision and Pattern Recognition, pp. 4690–4699 (2019)
5. DeVries, T., Taylor, G.W.: Improved regularization of convolutional neural networks with cutout. arXiv preprint arXiv:1708.04552 (2017)
6. Dosovitskiy, A., et al.: An image is worth 16x16 words: transformers for image recognition at scale. arxiv 2020. arXiv preprint arXiv:2010.11929 (2010)
7. Gao, S., Wang, J., Lu, H., Liu, Z.: Pose-guided visible part matching for occluded person ReID. In: Proceedings of the IEEE/CVF Conference on Computer Vision and Pattern Recognition, pp. 11744–11752 (2020)
8. Gray, D., Brennan, S., Tao, H.: Evaluating appearance models for recognition, reacquisition, and tracking. In: Proc. IEEE International Workshop on Performance Evaluation for Tracking and Surveillance (PETS), vol. 3, pp. 1–7 (2007)
9. He, K., Chen, X., Xie, S., Li, Y., Dollar, P., Girshick, R.: Masked autoencoders are scalable vision learners. In: 2022 IEEE/CVF Conference on Computer Vision and Pattern Recognition (CVPR), June 2022. https://doi.org/10.1109/cvpr52688.2022.01553
10. He, K., Zhang, X., Ren, S., Sun, J.: Deep residual learning for image recognition. In: Proceedings of the IEEE Conference on Computer Vision and Pattern Recognition, pp. 770–778 (2016)
11. He, L., Liang, J., Li, H., Sun, Z.: Deep spatial feature reconstruction for partial person re-identification: alignment-free approach. In: Proceedings of the IEEE Conference on Computer Vision and Pattern Recognition, pp. 7073–7082 (2018)
12. He, S., Luo, H., Wang, P., Wang, F., Li, H., Jiang, W.: TransReID: transformer-based object re-identification. In: Proceedings of the IEEE/CVF International Conference on Computer Vision, pp. 15013–15022 (2021)
13. Hermans, A., Beyer, L., Leibe, B.: In defense of the triplet loss for person re-identification. arXiv preprint arXiv:1703.07737 (2017)
14. Hou, R., Ma, B., Chang, H., Gu, X., Shan, S., Chen, X.: Feature completion for occluded person re-identification. IEEE Trans. Pattern Anal. Mach. Intell. **44**(9), 4894–4912 (2021)
15. Huang, H., Li, D., Zhang, Z., Chen, X., Huang, K.: Adversarially occluded samples for person re-identification. In: Proceedings of the IEEE Conference on Computer Vision and Pattern Recognition, pp. 5098–5107 (2018)

16. Jia, M., Cheng, X., Lu, S., Zhang, J.: Learning disentangled representation implicitly via transformer for occluded person re-identification. IEEE Trans. Multimedia (2022)
17. Li, Y., He, J., Zhang, T., Liu, X., Zhang, Y., Wu, F.: Diverse part discovery: occluded person re-identification with part-aware transformer. In: Proceedings of the IEEE/CVF Conference on Computer Vision and Pattern Recognition, pp. 2898–2907 (2021)
18. Miao, J., Wu, Y., Liu, P., Ding, Y., Yang, Y.: Pose-guided feature alignment for occluded person re-identification. In: Proceedings of the IEEE/CVF International Conference on Computer Vision, pp. 542–551 (2019)
19. Somers, V., De Vleeschouwer, C., Alahi, A.: Body part-based representation learning for occluded person re-identification. In: Proceedings of the IEEE/CVF Winter Conference on Applications of Computer Vision, pp. 1613–1623 (2023)
20. Tan, L., Dai, P., Ji, R., Wu, Y.: Dynamic prototype mask for occluded person re-identification. In: Proceedings of the 30th ACM International Conference on Multimedia, pp. 531–540 (2022)
21. Wang, G., et al.: High-order information matters: learning relation and topology for occluded person re-identification. In: Proceedings of the IEEE/CVF Conference on Computer Vision and Pattern Recognition, pp. 6449–6458 (2020)
22. Wang, J., et al.: Deep high-resolution representation learning for visual recognition. IEEE Trans. Pattern Anal. Mach. Intell. **43**(10), 3349–3364 (2020)
23. Wang, T., Liu, H., Song, P., Guo, T., Shi, W.: Pose-guided feature disentangling for occluded person re-identification based on transformer. In: Proceedings of the AAAI Conference on Artificial Intelligence, vol. 36, pp. 2540–2549 (2022)
24. Wang, Z., Zhu, F., Tang, S., Zhao, R., He, L., Song, J.: Feature erasing and diffusion network for occluded person re-identification. In: Proceedings of the IEEE/CVF Conference on Computer Vision and Pattern Recognition, pp. 4754–4763 (2022)
25. Wei, L., Zhang, S., Gao, W., Tian, Q.: Person transfer GAN to bridge domain gap for person re-identification. In: Proceedings of the IEEE Conference on Computer Vision and Pattern Recognition, pp. 79–88 (2018)
26. Xia, J., Tan, L., Dai, P., Zhao, M., Wu, Y., Ji, R.: Attention disturbance and dual-path constraint network for occluded person re-identification. arXiv preprint arXiv:2303.10976 (2023)
27. Ye, Y., Zhou, H., Yu, J., Hu, Q., Yang, W.: Dynamic feature pruning and consolidation for occluded person re-identification. arXiv preprint arXiv:2211.14742 (2022)
28. Yun, S., Han, D., Oh, S.J., Chun, S., Choe, J., Yoo, Y.: Cutmix: regularization strategy to train strong classifiers with localizable features. In: Proceedings of the IEEE/CVF International Conference on Computer Vision, pp. 6023–6032 (2019)
29. Zhao, C., Lv, X., Dou, S., Zhang, S., Wu, J., Wang, L.: Incremental generative occlusion adversarial suppression network for person ReID. IEEE Trans. Image Process. **30**, 4212–4224 (2021)
30. Zheng, L., Shen, L., Tian, L., Wang, S., Wang, J., Tian, Q.: Scalable person re-identification: a benchmark. In: Proceedings of the IEEE International Conference on Computer Vision, pp. 1116–1124 (2015)
31. Zheng, Z., Zheng, L., Yang, Y.: Unlabeled samples generated by GAN improve the person re-identification baseline in vitro. In: Proceedings of the IEEE International Conference on Computer Vision, pp. 3754–3762 (2017)
32. Zhong, Z., Zheng, L., Kang, G., Li, S., Yang, Y.: Random erasing data augmentation. In: Proceedings of the AAAI Conference on Artificial Intelligence, vol. 34, pp. 13001–13008 (2020)

33. Zhu, K., Guo, H., Liu, Z., Tang, M., Wang, J.: Identity-guided human semantic parsing for person re-identification. In: Vedaldi, A., Bischof, H., Brox, T., Frahm, J.-M. (eds.) ECCV 2020, Part III. LNCS, vol. 12348, pp. 346–363. Springer, Cham (2020). https://doi.org/10.1007/978-3-030-58580-8_21
34. Zhuo, J., Chen, Z., Lai, J., Wang, G.: Occluded person re-identification. In: 2018 IEEE International Conference on Multimedia and Expo (ICME), pp. 1–6. IEEE (2018)

Object-Aware Transfer-Based Black-Box Adversarial Attack on Object Detector

Zhuo Leng[1], Zesen Cheng[1], Pengxu Wei[2], and Jie Chen[1,3(✉)]

[1] School of Electronic and Computer Engineering, Peking University, Shenzhen, China
jiechen2019@pku.edu.cn

[2] Sun Yat-Sen University, Guangzhou, China

[3] Peng Cheng Laboratory, Shenzhen, China

Abstract. Deep neural networks have been demonstrated to be vulnerable to adversarial noise from attacks. Compared with white-box attacks, black-box attacks fool deep neural networks to yield erroneous predictions without knowing the model parameters. Black-box attacks include query-based attacks and transfer-based attacks; the former rely on querying the model while the latter just rely on the transferability of adversarial examples, thus challenging. Existing transfer-based black-box adversarial attack methods focus on the image classification task. Especially, we empirically verify that those methods struggle to balance the attack on objects with different classes and sizes, and thus they perform poorly in the attack on object detectors. In this work, we propose an **Object-Aware mechanism** to address this issue. It includes **O**bject-**W**ise **G**radient (OWG) calculation to balance the attack on multiple objects and a **D**omain-**D**ivision **M**ap (DDM) to weigh the attack in size. Incorporating our method with seminal baselines (e.g., I-FGSM, MI-FGSM), we achieve superior attack performance on multiple object detectors (e.g., Faster R-CNN, DETR, SSD), which justifies the effectiveness and generality of our method.

Keywords: Black-box attack · Adversarial Attack · Object Detection

1 Introduction

Deep neural networks (DNN) are challenged by their vulnerability to adversarial examples [8,21]. By adding small and human-imperceptible noises to legitimate examples, the adversarial examples can make a model output attacker desire inaccurate predictions. Numerous attack methods have been proposed in recent years [6,8,11,14].

The adversarial examples have an intriguing property of transferability, where adversarial examples crafted by the current model can also fool other unknown models [23]. This character makes the adversarial samples feasible to perform transfer-based black-box attacks without any knowledge of the targeted model.

© The Author(s), under exclusive license to Springer Nature Singapore Pte Ltd. 2024
Q. Liu et al. (Eds.): PRCV 2023, LNCS 14436, pp. 278–289, 2024.
https://doi.org/10.1007/978-981-99-8555-5_22

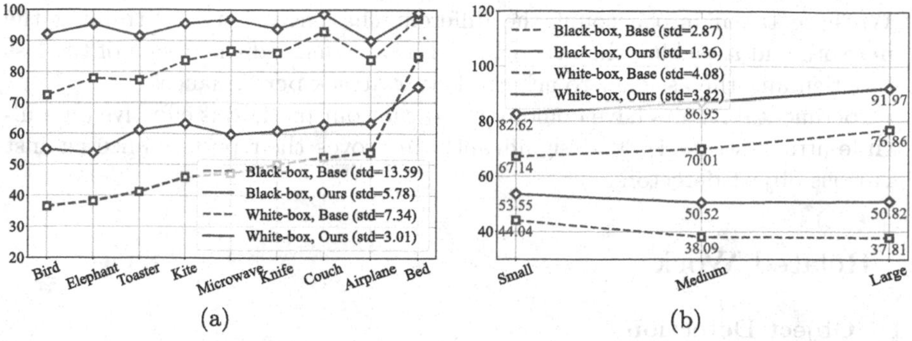

Fig. 1. Motivation and Contribution of our work. The vertical axis represents mAP drop rate (%), which represents the performance of the attack. (a) shows the imbalanced attack of the baseline (dash lines). Our improvements make the attack more balanced on different classes (solid lines). (b) shows that, with our improvements, attack performances are relatively more balanced on objects of different sizes. Both (a) and (b) show that we achieve a higher mAP drop rate than the baseline.

Recently, various methods have been proposed to enhance the transferability for black-box attacks [6,7,10,14,20,22,30]. However, existing methods are mostly designed for the image classification task, neglecting another significant task: object detection. Thus, we raise a question: *can we build a Transfer-based Black-box attack for the object detection task?*

We empirically find that existing image-classification attackers perform unsatisfactorily when directly applied to object detectors because they cannot balance the attacks on multiple objects of different classes and sizes. Shown in Fig. 1(a), objects with different analogies vary in vulnerability to the same attack: some categories are more vulnerable, *e.g.*, bed, couch; the others are more robust, *e.g.*, bird, toaster. Figure 1(b) shows that, in the black-box setting, smaller objects are more likely to be attacked.

To address the aforementioned problems, we propose an Object-Aware attack mechanism that can balance the attack on multiple targets. Specifically, we utilize an Object-Wise Gradient (OWG) calculation to balance the attack on multiple objects and a Domain-Division Map (DDM) to weigh the attack in size. Moreover, we improve the design of the loss function and assigner to accommodate the different requirements between training the network and conducting the attack. It's worth noting that our method does not require any knowledge of the model parameters, depending on querying [1,5,18], or producing noticeable adversarial patches [10,28] on the image. Our work has achieved significant improvements in both black-box and white-box attacks. Our main contributions can be summarized as follows:

- We have proposed an Object-Aware adversarial attack method for detectors to balance the attack on multiple objects with different categories and sizes.

- We have taken into account the different challenges between training the network and attacking the network, and have enhanced the design of the loss function and the assigner to achieve better attack performance.
- Experimental results have demonstrated that our method is effective on multiple attack strategies and significantly improves their performance against various object detectors.

2 Related Work

2.1 Object Detection

Object Detection is a fundamental computer vision task that localizes and classifies objects in images, and it's widely used in many fields, e.g., automatic driving, and security systems. In the past decade, seminal deep learning-based object detection models have been proposed, e.g., single-stage methods like SSD [19], RetinaNet [15], two-stage methods like Faster-RCNN [24], and Transformer [26] based models, e.g., DETR [2]. These methods greatly improve the performance of detectors. Recently, many multi-modal models[12, 13, 17] are proposed to handle various vision-language problems, including open-set or open-vocabulary object detection, e.g., GLIP [13], Grounding DINO [17]. In this work, we perform the attacks only on pure vision models.

2.2 Adverserial Attack

Adversarial Attack was first introduced in the image classification task at [8]. For a classifier f, let x be the original input tensor of the image, y and $f(x;\theta)$ be the ground truth label and the predicted label with the parameter θ, and $y = f(x;\theta)$ as the image is correctly predicted. Let $J(x, y;\theta)$ denote the loss function and x^{adv} be the adversarial image that the attacker should find. It should satisfy that $f(x;\theta) \neq f(x^{adv};\theta)$ with the constrain of p-norm distance within ϵ, that is, $\| (x - x^{adv}) \|_p < \epsilon$, and here we focus on $p = \infty$ like the previous works did.

Fast Gradient Sign Method (FGSM) is the most basic method to generate adversarial examples by adding perturbation to the image in one step towards the gradient direction:

$$x^{adv} = x + \epsilon \cdot \mathtt{sign}(\nabla_x J(x, y;\theta)). \quad (1)$$

The perturbation would not surpass the L_∞ norm distance as the $\mathtt{sign}(\cdot)$ function can constrain it within ϵ.

Iterative Fast Gradient Sign Method (I-FGSM) breaks the one-step of FGSM into multiple smaller ones to better maximize the loss function and it achieves better performance in white-box attack. Let α be the step size of each iteration, and I-FGSM runs T iterations of FGSM with a $\texttt{Clamp}(\cdot)$ function to constrain the perturbation in the ϵ-spherical space:

$$x_{t+1}^{adv} = \texttt{Clamp}_{x-\epsilon}^{x+\epsilon}\{x_t^{adv} + \alpha \cdot \texttt{sign}(\nabla_x J(x_t^{adv}, y; \theta))\}. \tag{2}$$

Similar to I-FGSM, PGD also performs iterative FGSM, but starts with random points in a uniform distribution.

Momentum Iterative Fast Gradient Sign Method (MI-FGSM) integrates the momentum into I-FGSM and it achieves higher attack performance both in white-box and black-box settings. Initialize x_0 by x, and g_0 by 0, let μ be the decay factor of g_t, and the g_{t+1} is calculated with a momentum:

$$g_{t+1} = \mu \cdot g_t + \frac{\nabla_x J(x_t^{adv}, y; \theta)}{\| \nabla_x J(x_t^{adv}, y; \theta) \|_1}. \tag{3}$$

Neterov Iterative Fast Gradient Sign Method (NI-FGSM) adopts Nesterov accelerated gradient method into MI-FGSM. It accumulates the gradient after adding momentum to the current data point for faster convergence and higher transferability. Its only difference with MI-FGSM is that NI-FGSM substitutes x^{nes} for x^{adv} to calculate the gradient:

$$x_t^{nes} = x_t^{adv} + \alpha \cdot \mu \cdot g_t. \tag{4}$$

Other Methods . Related studies can be divided into three categories: (a) more advanced optimization techniques for gradient-based methods [6,14], (b) ensemble-model attack [6,10,20], (c) input transformations [7,14,22,30] etc. We focus on the first one and have introduced the most important works above. The other gradient-based methods like VT [27] are mainly improved based on these methods. In addition, model ensemble methods and input transformation such as DIM [30], TIM [7], SIM [14], etc. can also improve the transferability of black-box attacks. In this work, we focus on an optimization-based method.

3 Method

In this section, we will describe our approach in detail. In Sect. 3.1, we'll introduce our Object-Aware mechanism for adversarial attacks and explain our validity. In Subsects. 3.2 and 3.3, we'll discuss the designs of assigner and loss functions of detectors when generating adversarial examples and explain their differences with the designs of training the networks.

3.1 Object-Aware Mechanism

Our Object-Aware mechanism focused on solving the problem of multi-target attacks. We emphasize that the biggest challenge is balancing the attacks of different objects with various categories and sizes. Correspondingly, our method consists of two components; one is the Object-Wise Gradient (OWG) calculation, and the next is the Domain-Division Map (DDM) for specific perturbations.

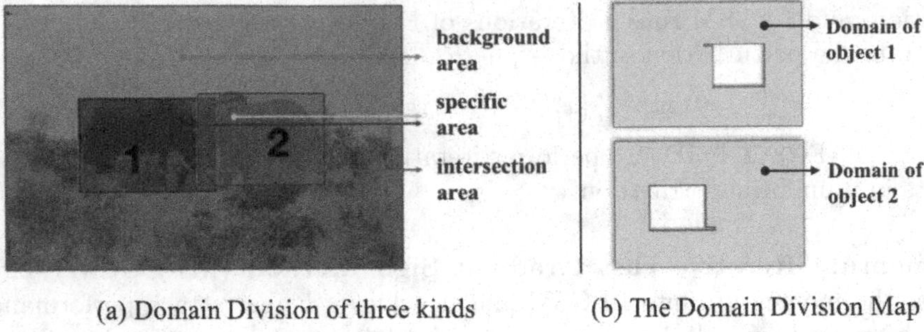

(a) Domain Division of three kinds (b) The Domain Division Map

Fig. 2. The mechanism of Domain-Division Map. White pixels indicate 1 and are to be optimized to attack the special object. Black pixels are masked.

Object-Wise Gradient (OWG) Calculation. We calculate the gradient for each object in the image using object-wise loss. And then, unlike [4,29], we do not sum the total loss weighted by the number of the corresponding classes, but directly obtain the gradient by the backward propagation for each object. In detail, given an image x with N objects$\{e_1, e_2, ..., e_N\}$, the goal is to obtain the adversarial image x^{adv} within a ϵ-spherical space. For each object e_i, the assigner will arrange m_i (one or more) proposals to predict it, denoted as $\{p_{i,1}, p_{i,2}, ..., p_{i,m_i}\}$, and the loss for e_i can be formalized as:

$$\mathcal{L}_i = \frac{1}{m_i} \cdot \sum_{j=0}^{m_i} \mathcal{L}(e_i, p_{i,j}), \tag{5}$$

where $\mathcal{L}(\cdot)$ is the loss function of the detector including classification loss and regression loss. For each object e_i, we obtain the corresponding $\mathcal{L}_i$ with the above equation. And then, instead of calculating the sum of losses, we perform a backward propagation algorithm for $\mathcal{L}_i$ to calculate the Object-Wise Gradient of the image ∇_{e_i}. The gradients are not necessarily normalized by 1-norm distance since the $\mathtt{sign}(\cdot)$ function, which will be executed in the optimization step, can project it to a 1−spherical space.

Domain-Division Map (DDM). We further propose a Domain-Division Map to solve the imbalanced attack on different sizes and achieve better performance. DDM segments the image into parts only using the ground truth bounding box of each object. To attack each object, we update the pixels with the Domain-Division Map. Shown in Fig. 2, the boxes divide the pixels into three kinds: background (no object covering), independent area (one specific object covering), and intersection area (at least two objects covering). For each object, we don't update the pixels of independent areas of other objects. DDM increases the weight of attacks on large objects, but interestingly, this does not exacerbate the imbalance of white-box attacks on objects of different sizes. The DDM is denoted as $M \in \{0,1\}^{N\times H\times W}$, where N denotes the number of objects in the

Algorithm 1: Object-Aware Attack on Detectors with I-FGSM

```
Input: Image x, N objects {e_1, e_2, ..., e_N}. Corresponding predictions of each
       object e_i ({p_{i,1}, p_{i,2}, ..., p_{i,m_i}}). Domain-Division Map M. The iteration
       number is T, the step size is α, and the constrain factor of I-FGSM is ε.
Output: Image after adversarial attack x^{adv}
Initialize x^{adv} = x
while iter = 0; iter! = T; iter + + do
    perturb = 0;
    for e_i in {e_1, e_2, ..., e_N} do
        Calculate loss in Eq. 5;
        Clean the existing gradient;
        Backward the gradient of L_i;
        Compute the gradient of each object ∇_{e_i};
        // Update perturb as I-FGSM:
        perturb = perturb + M_i · α · sign(∇_{e_i});
    end
    x^{adv} = Clamp_{x-ε}^{x+ε}{x^{adv} + perturb}
end
```

image, H denotes the height of the image, W denotes the width of the image. First, we annotate the background pixels that are not covered by any object bounding box. Then, for each object, we set $M_{i,j,k} = 1$ if pixels at location (j, k) is in the box area of object e_i or background area which is marked in the last step. Otherwise, $M_{i,j,k} = 0$ is set. The process can be formalized as follows:

$$M_{i,x,y} = \begin{cases} 1, & (j,k) \in bg \mid (j,k) \in e_i, \\ 0, & otherwise, \end{cases} \tag{6}$$

where "bg" denotes the background area. With the Object-Wise Gradient and Domain-Division Map, we can use any gradient-based attack method including I-FGSM and MI-FGSM to generate adversarial examples. We propose the Complete Object-Aware Attack using I-FGSM in Algorithm 1.

3.2 Assigners and Samplers of Detectors for Attack

In this work, we assign more positive boxes with a low threshold of IoU and meanwhile, we use a sampler with a higher ratio of positive boxes to get more positive examples. This may be counter-intuitive, because it will assign many background regions as positive predictions, thus leading to low precision. However, since the purpose of the attack is to maximize the loss, the false positive samples will not lead to much degradation of the attack performance. Figure 3 visually shows the motivation.

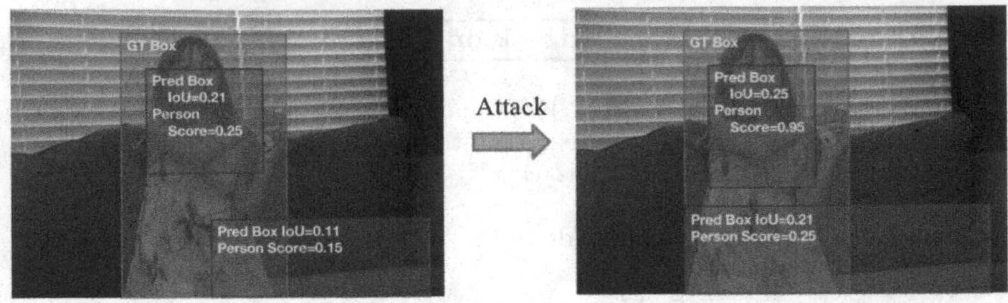

Fig. 3. Although some prediction boxes have a small IoU with ground truth, they should be regarded as positive samples to benefit the attack, just like the partially framed child in the picture who retains the strong feature of "person".

3.3 Loss Functions of Detectors for Attack

Different from the training process, attackers should pay more attention to easier-predicted objects, as the easier it is to be predicted, the harder to be attacked. In this work, we replace the focal loss function of RetianNet [15] with cross-entropy loss. We confirmed that, when attacking the whole image without our proposed Object-Aware mechanism, the cross-entropy loss is much better than focal loss, as we can show in Table 4(c). We attribute this to the fact that focal loss decreases the loss-weight of easily predicted samples and increases the loss-weight of difficultly predicted samples. It works when training the network for better convergence, but is the exact opposite of what is needed when attacking the network. But the difference is much smaller when equipped with our Object-Aware mechanism, which is also presented in Table 4(c).

4 Experiments

4.1 Implementation Details

For detectors, we implement our approach based on mmdetection [3]. We chose 5 models for our experiments: SSD300, SSD512, RetinaNet, Faster-RCNN, and DETR. SSD300 and SSD512 share the same architecture and the backbone of vgg16 [25], but they have the different size of neck and head, and input sizes of the image (The height and width of the SSD300 input image are 300, and the latter is 512). We choose RetinaNet [9] because it uses focal loss to train the detector, and it's also a single-stage detector, just like SSD. We choose Faster-RCNN and DETR as they are the most classic two-stage detectors and transformer-based detectors. The RetinaNet, Fatser-RCNN, and DETR share the same backbone, Resnet50, and to align with SSD512, we resize the input image to 512 pixels both in height and width for the three models. All models are trained on COCO [16] in 12 epochs except DETR, trained for 150 epochs.

For attackers, we select five optimization-based methods: FGSM, I-FGSM, PGD, MI-FGSM, and NI-FGSM, which have been formulated in Sect. 2.2. We set

$\epsilon = 20$ to constrain the pertubation. For iterative methods, we set the iteration number $T = 20$ and the step size $\alpha = 5$, and we reset $\alpha = 1$ after 15 iterations. For MI-FGSM and NI-FGSM, we set the decay factor at $\mu = 1$.

Table 1. mAP drop rate compares with the baseline of models and attackers. "RCNN" denotes "Faster R-CNN".

Source	Attackers	DETR	RCNN	RetinaNet	SSD300	SSD512
–	(clean)	32.4	30.5	28.9	25.5	29.5
–	(noisy)	31.6	29.7	28.2	25.0	28.4
DETR	I-FGSM	17.0	25.4	23.4	23.7	26.4
	I-FGSM+ours	16.3 ↓ **2.2%**	25.3 ↓ **0.3%**	23.4 ↑ 0.0%	23.7 ↑ 0.0%	26.4 ↑ 0.0%
RCNN	I-FGSM	24.3	14.9	19.3	23.2	25.6
	I-FGSM+ours	21.0 ↓ **10.2%**	9.1 ↓ **19.0%**	14.5 ↓ **16.6%**	22.2 ↓ **3.9%**	24.4 ↓ **4.1%**
RetinaNet	I-FGSM	23.3	20.0	13.7	23.2	25.6
	I-FGSM+ours	23.1 ↓ **0.6%**	20.1 ↑ 0.3%	12.9 ↓ **2.8%**	22.8 ↓ **1.6%**	25.3 ↓ **1.0%**
SSD300	I-FGSM	27.5	26.1	25.0	9.0	21.9
	I-FGSM+ours	24.9 ↓ **8.0%**	23.4 ↓ **8.9%**	22.3 ↓ **9.3%**	2.9 ↓ **23.9%**	16.9 ↓ **16.9%**
SSD512	I-FGSM	27.7	26.2	25.2	20.9	7.7
	I-FGSM+ours	26.0 ↓ **5.2%**	24.4 ↓ **5.9%**	23.0 ↓ **7.6%**	18.8 ↓ **8.2%**	2.9 ↓ **16.3%**
DETR	MI-FGSM	5.9	15.8	14.2	18.2	18.6
	MI-FGSM+ours	4.1 ↓ **5.6%**	14.4 ↓ **4.6%**	13.0 ↓ **4.2%**	17.6 ↓ **2.4%**	17.6 ↓ **3.4%**
RCNN	MI-FGSM	14.7	8.0	11.4	18.1	18.6
	MI-FGSM+ours	8.2 ↓ **20.1%**	3.4 ↓ **15.1%**	5.5 ↓ **20.4%**	15.0 ↓ **12.2%**	14.8 ↓ **12.9%**
RetinaNet	MI-FGSM	15.3	13.4	8.7	18.4	19.0
	MI-FGSM+ours	9.8 ↓ **17.0%**	8.0 ↓ **17.7%**	3.3 ↓ **18.7%**	15.9 ↓ **9.8%**	15.9 ↓ **10.5%**
SSD300	MI-FGSM	21.7	21.2	20.5	7.4	17.2
	MI-FGSM+ours	17.2 ↓ **13.9%**	17.1 ↓ **13.4%**	16.3 ↓ **14.5%**	2.1 ↓ **20.8%**	11.1 ↓ **20.7%**
SSD512	MI-FGSM	21.8	20.9	20.2	16.4	4.9
	MI-FGSM+ours	18.5 ↓ **10.2%**	17.8 ↓ **10.2%**	16.9 ↓ **11.4%**	12.3 ↓ **16.1%**	1.2 ↓ **12.5%**

For evaluation, mean Average Precision (mAP) is commonly used to measure the performance of detectors, so we use **mAP drop rate** (%) to assess the attack performance. We generate 25 sets of adversarial examples across the combination of the five models and five attackers on the COCO val2017 set and conduct the evaluation of each set of adversaries with each model. The mAP drop rate is a statistic, and it can be formalized as:

$$mAP_{drop\ rate}(\%) = (1 - \frac{mAP_{x^{adv}}}{mAP_{clean}}) \cdot 100\% \tag{7}$$

where $mAP_{x^{adv}}$ is the mean Average Precision on adversarial images, and mAP_{clean} is the mean Average Precision on clean images.

4.2 Main Results

Part of the main results of attack performance are shown in Table 1. The logits indicate mAP on adversarials, and the percentages aside represent the differ-

Table 2. mAP drop rate (%) of different attackers averaged on the five source models (DETR, Faster R-CNN, RetinaNet, SSD300, SSD512).

Settings	Attackers	FGSM	I-FGSM	I-PGD	MI-FGSM	**NI-FGSM**
White-box	baseline	44.08	57.98	53.61	75.97	76.01
White-box	baseline+ours	**53.88**	**70.80**	**66.49**	**90.50**	**90.56**
Black-box	baseline	37.00	17.34	19.39	39.12	39.03
Black-box	baseline+ours	**41.78**	**22.75**	**24.02**	**51.39**	**51.38**

Table 3. mAP drop rate (%) of different source models averaged on the five attackers (FGSM, I-FGSM, PGD, MI-FGSM, NI-FGSM). "RCNN" denotes Faster R-CNN.

Settings	Source	DETR	RCNN	Retina Net	SSD300	SSD512
White-box	baseline	60.06	56.30	55.60	60.72	69.86
White-box	baseline+ours	**64.18**	**72.59**	**66.74**	**81.76**	**83.59**
Black-box	baseline	28.27	31.91	31.85	26.55	23.87
Black-box	baseline+ours	**30.28**	**43.57**	**39.29**	**37.84**	**31.98**

ence of mAP drop rate between our method and the baseline. As we can see, our approach is effective in the vast majority of cases, both in the white-box and black-box settings. Further, we analyze the effect of different optimization methods on black-box and white-box attacks and our improvement to them. The average results are calculated in Table 2. As is shown in the table, our method can reach a higher average mAP drop rate by 9–15% in white-box attack and 5–12% in black-box attack than the baseline, regardless of the method of the attacker. Our method has great improvements for MI-FGSM and NI-FGSM, and achieves state-of-the-art attack performance when using them as attackers both in white-box and black-box settings. Similarly, we analyzed the improvement of our attack on different source models. The average results are calculated in Table 3. Again, except DETR, our method benefits most source models greatly with 10–21% additional mAP drop rate in the white-box setting and $8 \sim 17\%$ in the black-box setting compared with the baseline.

4.3 Ablation Study

Object-Wise Gradient and Domain-Division Map. We have analyzed the results per class and justified that our improvements could balance the attack on different categories. As Fig. 1(a) shows, attackers equipped with our Object-Aware mechanism exhibit more balanced damage in different classes. Figure 1(b) shows that we further improve the balanced attack performance in sizes, with a higher increase of mAP drop rate on larger objects. And Table 4(a) shows the ablation results on OWG and DDM to justify the respective validity.

Assigner for More Positive Example. The result in Table 4(b) shows the ablation study of assigners. As we can see, appropriately decreasing the IoU threshold and increasing the proportion of positive samples are beneficial. There is a possible explanation given in Fig. 3.

Focal Loss vs CE Loss. We have studied the impact of focal loss and cross-entropy (CE) loss on attack. The experiments are carried out on RetinaNet. And we find that CE loss performs much better, as the results in Table 4(c) can show. We have proposed our assumption in Sect. 3.3.

Table 4. Ablation Studies of our design choices. The mAP drop rates (%) of (a), (b) are calculated by averaging attack performance of five detectors, i.e., DETR, Faster R-CNN, RetinaNet, SSD300, SSD512.

(a) Ablations of OWG and DDM.

Attacker	**OWG**	**DDM**	**White-box**	**Black-box**
MI-FGSM			75.97	39.12
	✓		80.47	44.89
	✓	✓	**90.50**	**51.39**
NI-FGSM			76.01	39.03
	✓		80.44	45.02
	✓	✓	**90.56**	**52.38**

(b) Ablations of the assigner and sampler.

Attacker	**IoU thre**	**Positive ratio**	**White-box**	**Black-box**
MI-FGSM	0.5	0.25	83.46	44.23
	0.2	0.25	89.05	48.33
	0.2	0.5	**90.50**	**51.39**

(c) Ablations of CE-loss and focal loss on RetinaNet.

Detector	**Attacker**	**Loss**	**White-box**	**Black-box**
RetinaNet	MI-FGSM	Focal loss	69.90	43.07
	MI-FGSM	CE loss	82.33	48.79
	MI-FGSM+ours	Focal loss	85.67	54.31
	MI-FGSM+ours	CE loss	**88.58**	**56.82**

5 Conclusion

In this paper, we propose an Object-Aware mechanism to balance the adversarial attack on multiple objects of different categories and sizes and perform transfer-based black-box attack on detectors. Our method decouples the attack of each target in the image by Object-Wise Gradient and weighs the attack for objects of different sizes with a Domain-Division Map. We also improve the design of assigners and losses to perform attack better. Plenty of experiments on different models and attack strategies show, our method achieves a huge improvement both in white-box attack and black-box attack settings.

Acknowledgements. This work was supported in part by the National Key R&D Program of China (No. 2022ZD0118201), Natural Science Foundation of China (No. 61972217, 32071459, 62176249, 62006133, 62271465).

References

1. Brendel, W., Rauber, J., Bethge, M.: Decision-based adversarial attacks: Reliable attacks against black-box machine learning models. arXiv preprint arXiv:1712.04248 (2017)
2. Carion, N., Massa, F., Synnaeve, G., Usunier, N., Kirillov, A., Zagoruyko, S.: End-to-end object detection with transformers. In: Vedaldi, A., Bischof, H., Brox, T., Frahm, J.-M. (eds.) ECCV 2020. LNCS, vol. 12346, pp. 213–229. Springer, Cham (2020). https://doi.org/10.1007/978-3-030-58452-8_13
3. Chen, K., et al.: MMDetection: open mmlab detection toolbox and benchmark. arXiv preprint arXiv:1906.07155 (2019)
4. Chen, P.C., Kung, B.H., Chen, J.C.: Class-aware robust adversarial training for object detection. In: Proceedings of the IEEE/CVF Conference on Computer Vision and Pattern Recognition, pp. 10420–10429 (2021)
5. Dong, Y., et al.: Benchmarking adversarial robustness on image classification. In: Proceedings of the IEEE/CVF Conference on Computer Vision and Pattern Recognition, pp. 321–331 (2020)
6. Dong, Y., Liao, F., Pang, T., Su, H., Zhu, J., Hu, X., Li, J.: Boosting adversarial attacks with momentum. In: Proceedings of the IEEE/CVF Conference on Computer Vision and Pattern Recognition, pp. 9185–9193 (2018)
7. Dong, Y., Pang, T., Su, H., Zhu, J.: Evading defenses to transferable adversarial examples by translation-invariant attacks. In: Proceedings of the IEEE/CVF Conference on Computer Vision and Pattern Recognition, pp. 4312–4321 (2019)
8. Goodfellow, I.J., Shlens, J., Szegedy, C.: Explaining and harnessing adversarial examples. arXiv preprint arXiv:1412.6572 (2014)
9. He, K., Zhang, X., Ren, S., Sun, J.: Deep residual learning for image recognition. In: Proceedings of the IEEE/CVF Conference on Computer Vision and Pattern Recognition, pp. 770–778 (2016)
10. Huang, H., Wang, Y., Chen, Z., Tang, Z., Zhang, W., Ma, K.K.: RPattack: refined patch attack on general object detectors. In: IEEE International Conference on Multimedia and Expo, pp. 1–6. IEEE (2021)
11. Kurakin, A., Goodfellow, I.J., Bengio, S.: Adversarial examples in the physical world. In: Artificial Intelligence Safety and Security, pp. 99–112. Chapman and Hall/CRC (2018)
12. Li, H., et al.: TG-VQA: ternary game of video question answering. arXiv preprint arXiv:2305.10049 (2023)
13. Li, L.H., et al.: Grounded language-image pre-training. In: Proceedings of the IEEE/CVF Conference on Computer Vision and Pattern Recognition, pp. 10965–10975 (2022)
14. Lin, J., Song, C., He, K., Wang, L., Hopcroft, J.E.: Nesterov accelerated gradient and scale invariance for adversarial attacks. arXiv preprint arXiv:1908.06281 (2019)
15. Lin, T.Y., Goyal, P., Girshick, R., He, K., Dollár, P.: Focal loss for dense object detection. In: Proceedings of the IEEE/CVF International Conference on Computer Vision, pp. 2980–2988 (2017)

16. Lin, T.-Y., et al.: Microsoft COCO: common objects in context. In: Fleet, D., Pajdla, T., Schiele, B., Tuytelaars, T. (eds.) ECCV 2014. LNCS, vol. 8693, pp. 740–755. Springer, Cham (2014). https://doi.org/10.1007/978-3-319-10602-1_48
17. Liu, S., et al.: Grounding dino: marrying dino with grounded pre-training for open-set object detection. arXiv preprint arXiv:2303.05499 (2023)
18. Liu, S., et al.: Efficient universal shuffle attack for visual object tracking. In: IEEE International Conference on Acoustics, Speech and Signal Processing, pp. 2739–2743. IEEE (2022)
19. Liu, W., et al.: SSD: single shot multibox detector. In: Leibe, B., Matas, J., Sebe, N., Welling, M. (eds.) ECCV 2016. LNCS, vol. 9905, pp. 21–37. Springer, Cham (2016). https://doi.org/10.1007/978-3-319-46448-0_2
20. Liu, Y., Chen, X., Liu, C., Song, D.: Delving into transferable adversarial examples and black-box attacks. arXiv preprint arXiv:1611.02770 (2016)
21. Madry, A., Makelov, A., Schmidt, L., Tsipras, D., Vladu, A.: Towards deep learning models resistant to adversarial attacks. arXiv preprint arXiv:1706.06083 (2017)
22. Naseer, M., Khan, S., Hayat, M., Khan, F.S., Porikli, F.: A self-supervised approach for adversarial robustness. In: Proceedings of the IEEE/CVF Conference on Computer Vision and Pattern Recognition, pp. 262–271 (2020)
23. Papernot, N., McDaniel, P., Goodfellow, I., Jha, S., Celik, Z.B., Swami, A.: Practical black-box attacks against machine learning. In: Proceedings of the 2017 ACM on Asia Conference on Computer and Communications Security, pp. 506–519 (2017)
24. Ren, S., He, K., Girshick, R., Sun, J.: Faster R-CNN: towards real-time object detection with region proposal networks. In: Advances in Neural Information Processing Systems, vol. 28 (2015)
25. Simonyan, K., Zisserman, A.: Very deep convolutional networks for large-scale image recognition. arXiv preprint arXiv:1409.1556 (2014)
26. Vaswani, A., et al.: Attention is all you need. In: Advances in Neural Information Processing Systems, vol. 30 (2017)
27. Wang, X., He, K.: Enhancing the transferability of adversarial attacks through variance tuning. In: Proceedings of the IEEE/CVF Conference on Computer Vision and Pattern Recognition, pp. 1924–1933 (2021)
28. Wu, S., Dai, T., Xia, S.T.: Dpattack: diffused patch attacks against universal object detection. arXiv preprint arXiv:2010.11679 (2020)
29. Xie, C., Wang, J., Zhang, Z., Zhou, Y., Xie, L., Yuille, A.: Adversarial examples for semantic segmentation and object detection. In: Proceedings of the IEEE/CVF International Conference on Computer Vision, pp. 1369–1378 (2017)
30. Xie, C., et al.: Improving transferability of adversarial examples with input diversity. In: Proceedings of the IEEE/CVF Conference on Computer Vision and Pattern Recognition, pp. 2730–2739 (2019)

HTNet: A Hybrid Model Boosted by Triple Self-attention for Crowd Counting

Yang Li and Baoqun Yin(✉)

University of Science and Technology of China, Hefei, China
liyang99@mail.ustc.edu.cn, bqyin@ustc.edu.cn

Abstract. The swift development of convolutional neural network (CNN) has enabled significant headway in crowd counting research. However, the fixed-size convolutional kernels of traditional methods make it difficult to handle problems such as drastic scale change and complex background interference. In this regard, we propose a hybrid crowd counting model to tackle existing challenges. Firstly, we leverage a global self-attention module (GAM) after CNN backbone to capture wider contextual information. Secondly, due to the gradual recovery of the feature map size in the decoding stage, the local self-attention module (LAM) is employed to reduce computational complexity. With this design, the model can fuse features from global and local perspectives to better cope with scale change. Additionally, to establish the interdependence between spatial and channel dimensions, we further design a novel channel self-attention module (CAM) and combine it with LAM. Finally, we construct a simple yet useful double head module that outputs a foreground segmentation map in addition to the intermediate density map, which are then multiplied together in a pixel-wise style to suppress background interference. The experimental results on several benchmark datasets demonstrate that our method achieves remarkable improvement.

Keywords: Crowd Counting · Deep Learning · Self-Attention · Hybrid Model

1 Introduction

Crowd counting is a typical computer vision task with numerous real-world applications across various domains, including public safety, event management, and regional planning. Along with its prevalence, some related domains, such as cell counting, traffic flow analysis, and wildlife monitoring, can also gain valuable reference and inspiration from it.

This work was supported in part by the National Natural Science Foundation of China under Grant 62133013 and in part by the Chinese Association for Artificial Intelligence (CAAI)-Huawei MindSpore Open Fund.

© The Author(s), under exclusive license to Springer Nature Singapore Pte Ltd. 2024
Q. Liu et al. (Eds.): PRCV 2023, LNCS 14436, pp. 290–301, 2024.
https://doi.org/10.1007/978-981-99-8555-5_23

Recently, a mainstream process for crowd counting involves developing a network architecture that regresses a density map from an input image and obtains the corresponding number of people by integrating the entire density map. As a classic neural network architecture, CNN has been widely applied in computer vision tasks due to its simplicity and effectiveness. Consequently, most previous works on crowd counting chose CNN to build models [1–4]. However, CNN is designed based on fixed-size convolutional kernels, which lead to challenges for such counting models in dealing with two scenarios: 1) Difficulty in adapting to drastic variations in crowd scale caused by perspective distortion; 2) Confusion between dense crowds and complex backgrounds with similar textures. Especially for the latter, it is hard for even human experts to accurately distinguish foreground from background if only local image information is concerned. Therefore, the key to solving the above problems lies in breaking through the limitations of the local receptive field and enabling the network to understand the input image from a global perspective.

The self-attention mechanism [5] is capable of modeling long-range dependencies and has been used to handle a variety of tasks, such as image classification [6,7], object detection [8], and semantic segmentation [9]. Currently, some crowd counting methods also employ self-attention mechanism to construct network modules.

Building upon these cutting-edge works, we further explore the specific application of the self-attention mechanism to address the problems mentioned before. Considering that CNN excels at capturing local details and that injecting convolutional inductive bias into the early stage of ViT can improve convergence speed and training stability [10], we propose a hybrid architecture that combines the strengths of CNN and self-attention mechanisms, which enables the model to comprehensively understand both the local and global information in crowd images. To capture the interdependence between spatial and channel dimensions, we specifically introduce a novel module based on the channel self-attention mechanism and integrate it with the spatial part. Additionally, we devise a simple but reliable double head module that outputs an intermediate density map and a head segmentation map, where the latter serves as a guide for the former to focus on the regions of interest.

In short, the main contributions of this paper are summarized as follows:

1. We propose a hybrid crowd counting model boosted by global, local, and channel self-attention mechanisms, employed to acquire wider context and establish dependencies between spatial and channel dimensions, respectively.
2. Extensive experiments on four crowd datasets are conducted to validate the solid advancements achieved by our method, including ShanghaiTech A, UCF-QNRF, JHU-Crowd++, and NWPU-Crowd.

2 Related Work

The primary objective of crowd counting is to estimate the number of people in a given input image, while the density estimation task further provides additional information about the spatial distribution of the crowd, which is currently gaining prominence in crowd counting research. The total number of people can be obtained by summing up all the density values in the map.

2.1 CNN in Crowd Counting

Thanks to the rapid development of CNN, diverse network architectures have been proposed to address various problems in the field of crowd counting. MCNN [1] and Switch-CNN [11] employ branches with different kernel sizes to handle crowds at varying scales. CSRNet [2] and SPN [12]utilize dilated convolutions to enlarge the receptive field, capturing contextual information over a broader range. MBTTBF [3] and SASNet [4] leverage the hierarchical structure of backbone network to acquire and fuse multi-scale information. To suppress background noise, ASNet [13] and CFANet [14] introduce auxiliary segmentation tasks through attention mechanisms. In addition, PGCNet [15] and PFDNet [16] integrate perspective analysis into methods, expanding the research scope of crowd counting.

2.2 Self Attention in Crowd Counting

Despite the notable achievements of CNN in processing vision tasks, their performance is intrinsically limited due to the difficulty in capturing long-range dependencies. In contrast, Transformers, based on self-attention mechanisms, excel in modeling global context information. As a result, researchers have migrated the self-attention mechanism to the domain of computer vision. For instance, DETR [8] uses a Transformer-based encoder-decoder structure for object detection. ViT [6] divides an image into a sequence of patches and feeds them as tokens into a standard Transformer encoder for image classification tasks. Swin Transformer [7] improves upon ViT [6] by introducing a shifted windowing scheme, which performs self-attention computation within local window regions and allows cross-window connection. SETR [9] replaces the CNN encoder with a Transformer in semantic segmentation task to carry out sequence-to-sequence prediction task. Analogously, researchers start to incorporate self-attention mechanisms into counting networks. CrowdFormer [17] develops an *Overlap Patching Transformer Block* to explore global dependencies. MAN [18] proposes a *Learnable Region Attention* to dynamically assign exclusive attention region for each feature position. CUT [19] adopts Twins-PCPVT [20] as the backbone and constructs a *Segmentation As Attention Module* to obtain fine-grained features with rich semantic information.

Inspired by these seminal works, this paper further extends the application of self-attention mechanisms in the field of crowd counting. Briefly, we design triple self-attention modules from global, local, and channel perspectives to boost the accuracy of our counting network.

3 Proposed Method

In this section, we first introduce a hybrid model based on triple self-attention mechanisms, which consist of global, local and channel attention modules. Then we present a simple double head module and corresponding loss function. The architecture of HTNet is shown in Fig. 1.

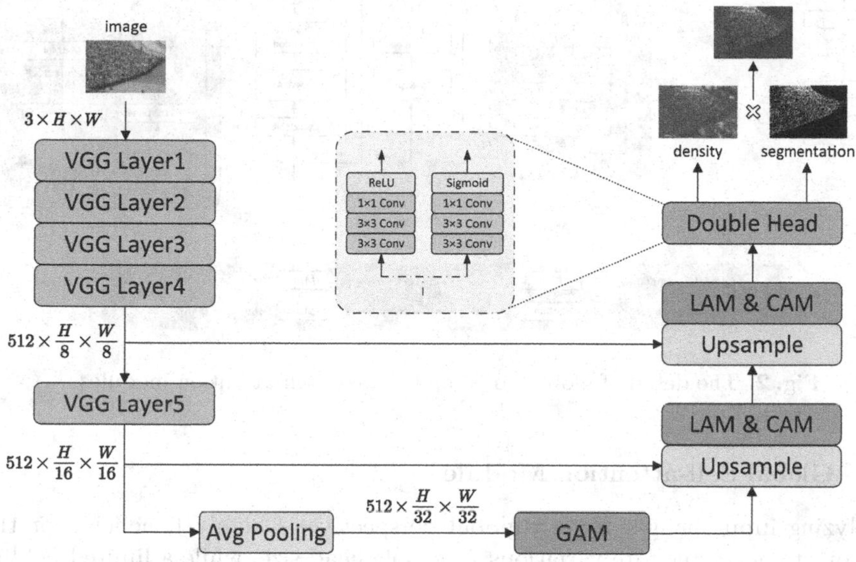

Fig. 1. The architecture of HTNet that is made up of CNN backbone, triple self-attention, and a simple but effective double head module.

To begin with, we utilize the first 13 layers in VGG-16BN [21] as backbone to extract multi-levels local features from the input image. Then a global self-attention module (GAM) follows the features to gain a larger range of context. During the decoding stage, as the feature map resolution is gradually recovered, a local self-attention module (LAM) is adopted to reduce the computational cost. Meanwhile, we also integrate a channel self-attention module (CAM) into the LAM to capture the interaction between spatial and channel dimensions. Finally, a double head module feature map simultaneously outputs the intermediate density estimation and the foreground segmentation map, which are multiplied together to obtain the final crowd density map.

3.1 CNN-Based Backbone

According to [10], the introduction of convolutional inductive biases in the early stages can improve optimization stability and converge faster. Thus we utilize the first 13 layers in VGG-16BN [21] as stem to extract local detail features. Two feature maps with sizes of $\frac{1}{8}$ and $\frac{1}{16}$ are used to construct the FPN [22] structure, which allows better utilization of multi-scale information.

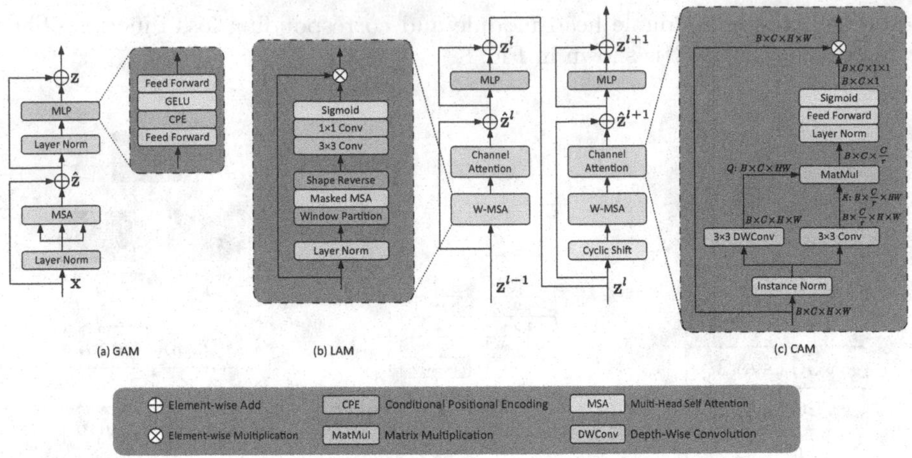

Fig. 2. The detail of global, local and channel self-attention modules.

3.2 Global Self-attention Module

Analyzing input images from a global perspective is highly beneficial for the network to accommodate variations in people head size, while a limited field of view may even make it challenging for human experts to accurately distinguish dense crowds and complex backgrounds. Hence, after extracting local features from the convolutional stem, we place a global self-attention module (GAM), which expands the receptive field from local to global, as depicted in Fig. 2(a).

The GAM is composed of a standard multi-head self-attention (MSA) and a multi-layer perceptron (MLP). A conditional positional encoding scheme [23] is implemented using depth-wise separable convolutions with zero-padding. To reduce computational complexity, we adopt an average pooling layer with a stride of 2 to perform the patch embedding projection. Then the patches are flattened into a 2D sequence x and fed into GAM, in which the calculation process can be formulated as follows:

$$\begin{aligned} \hat{\mathrm{z}} &= \mathrm{MSA}\left(\mathrm{LN}\left(\mathrm{x}\right)\right) + \mathrm{x}, \\ \mathrm{z} &= \mathrm{MLP}\left(\mathrm{LN}\left(\hat{\mathrm{z}}\right)\right) + \hat{\mathrm{z}} \end{aligned} \quad (1)$$

where $\hat{\mathrm{z}}$ and z denote the output features of the MSA and MLP, respectively; LN stands for Layer Norm.

3.3 Local and Channel Self-attention Module

As HTNet is based on the FPN architecture, we upsample the deep-level feature maps to restore resolution and add them to the shallow feature maps. The upsampling operation is accomplished through the Warp function introduced from FAM [24]. Then self-attention computation is performed on the fused feature maps. To establish dependencies between spatial and channel dimensions, we construct spatial and channel self-attention modules separately, which are combined in series.

Local Self-attention Module. With the gradual restoration of the feature map resolution, we consider a local self-attention module (LAM) to lower the computation cost, as shown in Fig. 2(b). Firstly, we utilize a Window Partition operation to divide the feature maps into several windows. Subsequently, a Masked MSA is applied to restrict the self-attention computation within each window. After that, we use convolutional layers and Sigmoid function to build an attention gate, which can selectively emphasize informative features while suppress less useful ones. Lastly, the sequence is reshaped back into a feature map through the Shape Reverse. In addition, we also follow Swin Transformer [7] and apply Cyclic Shift to the feature maps before feeding into the next Masked MSA layer, facilitating interactions between adjacent windows. It is noteworthy that the succeeding MLP excludes LN, as it performs point-wise normalization and weakens the enhancement or suppression effect of LAM on features.

Channel Self-attention Module. The structure of the channel self-attention module (CAM) is depicted in Fig. 2(c). CAM first normalizes the feature maps at channel dimension and then projects them into queries and keys independently. Before mapping the feature map into keys, we conduct dimension reduction to make representations of the keys more compact and reduce computational complexity. Then, for a given query and all N keys, we can obtain N dot product similarities, which will serve as input to a fully connected layer to generate an attention score. Finally, the original channel feature corresponding to the query is multiplied by this score, which plays a dynamic emphasis or suppression role. Notably, since CAM operates on channel-wise feature maps, Instance Normalization (IN) is employed instead of the regular LN, and the mapping process is implemented using depth-wise separable convolutions.

3.4 Double Head Module

We construct a concise double head module (DHM) to mitigate background interference, which generates an intermediate feature map and a foreground segmentation map. Then an element-wise product is applied on these two maps to output the final density estimation map. We show this module structure in Fig. 1. Each branch in DHM independently comprises two consecutive 3×3 and a subsequent 1×1 convolutional layers. The last layer of the density regression branch employs ReLU as activation function, whereas the segmentation task incorporates a Sigmoid. The remaining activation functions within DHM are GELU.

3.5 Loss Function

Corresponding to the regression and segmentation task in DHM, we implement a loss function comprising two components, each of which serves to supervise the above-mentioned maps individually.

As for the density regression head, we utilize Euclidean loss to measure pixel-wise prediction error. Let X_i be the *i-th* image in the training batch, $D(X_i)$ stand for the ground truth density map corresponding to X_i, and $\hat{D}(X_i;\Theta)$ indicate the crowd density map estimated by the model with parameters represented as Θ. Then the density map estimation loss function L_{den} is given by:

$$L_{den} = \frac{1}{B}\sum_{i=1}^{B}\left\|\hat{D}(X_i;\Theta) - D(X_i)\right\|_2^2 \tag{2}$$

where B is the size of the training batch.

With regard to the segmentation task, we employ Focal Loss [25], named FL. Let $M(X_i)$ denote the ground truth foreground-background mask for the input image X_i, and $\hat{M}(X_i;\Theta)$ represent the classification probability map output by the network. The expression for FL can be written as:

$$\mathrm{FL}(\hat{y}, y) = -y(1-\hat{y})^{\gamma}\log\hat{y} - (1-y)\hat{y}^{\gamma}log(1-\hat{y}) \tag{3}$$

$$L_{seg} = \frac{1}{B}\sum_{i=1}^{B}\mathrm{FL}\left(\hat{M}(X_i;\Theta), M(X_i)\right) \tag{4}$$

where y stands for the actual label (with a value of 1 for the foreground and 0 for the background), while $\hat{y}$ is the predicted probability of the corresponding pixel belonging to the foreground. γ is a hyperparameter that is set to 2 in the experiments.

The final loss function is a weighted combination of the above two terms, as shown below:

$$L_{final} = L_{den} + \lambda L_{seg} \tag{5}$$

where λ is a hyperparameter to balance the contributions of L_{den} and L_{seg}.

4 Experiments

In the experimental section, we first introduce the ground truth generation procedures and implementation details. Subsequently, we report the experimental results of HTNet on ShanghaiTech A [1], UCF-QNRF [26], JHU-Crowd++ [27], and NWPU-Crowd [28]. Finally, we conduct complete ablation studies to analyze the contributions of each proposed module.

4.1 Ground Truth Generation

The ground truth density map $D(x)$ is generated by convolving head annotation map with a normalized Gaussian kernel with a fixed standard deviation of 4. As for the segmentation task, given its auxiliary role in our method, we simply regard the non-zero regions in the $D(x)$ as foreground, otherwise as background, thus generating the ground truth foreground-background mask.

4.2 Implementation Details

Our HTNet employs VGG-16BN [21] that has been pretrained on ImageNet [29] as the backbone, and the remaining convolutional and attention layers are initialized using a Gaussian distribution with a mean of 0 and a standard deviation of 0.01. During training, the crop size is 256 × 256 for all datasets. The model is optimized by AdamW [30] with a learning rate of 1e−5. Horizontal Flipping, Gaussian noise, and Color jitter are adopted for data augmentation.

4.3 Evaluations and Comparisons

To quantitatively evaluate HTNet, we conduct extensive experiments on four benchmark datasets and compare the performance with state-of-the-art methods. Mean Absolute Error (MAE) and Mean Square Error (MSE) are uesd as evaluation metrics. The experimental results are summarized in Table 1 and 2, where the optimal performances are indicated by **bold** numbers.

Table 1. Comparison with mainstream methods on ShanghaiTech A and UCF-QNRF.

Method	Venue	ShanghaiTech A		UCF-QNRF	
		MAE	MSE	MAE	MSE
MCNN [1]	CVPR 16	110.2	173.2	277.0	426.0
CSRNet [2]	CVPR 18	68.2	115.0	–	–
BL [31]	ICCV 19	62.8	101.8	88.7	154.8
DM-Count [32]	NIPS 20	59.7	95.7	85.6	148.3
ASNet [13]	CVPR 20	57.8	90.1	91.6	159.7
BM-Count [33]	IJCAI 21	57.3	90.7	81.2	138.6
S3 [34]	IJCAI 21	57.0	96.0	80.6	139.8
CLTR [35]	ECCV 22	56.9	95.2	85.8	141.3
ChfL [36]	CVPR 22	57.5	94.3	80.3	137.6
GauNet [37]	CVPR 22	54.8	89.1	81.6	153.7
MAN [18]	CVPR 22	56.8	90.3	77.3	**131.5**
HTNet	–	**52.5**	**87.9**	**77.1**	137.4

ShanghaiTech Part A [1] is comprised of 482 images sourced from the internet, which are divided into 300 training images and 182 testing images. The dataset contains a total of 241,677 annotated points, with the number of annotations per image varying between 33 and 3139. We report the best performance achieved by HTNet on ShanghaiTech A in Table 1, which reduces the MAE and MSE of the second-ranked GauNet [37] from 54.8 to 52.5 and from 89.1 to 87.9.

UCF-QNRF [26] contains 1,535 high-resolution images with approximately 1.25 million annotated points. The dataset features an exceptional variety of crowd densities, with the lowest and highest counts being 49 and 12,865, respectively. The training set is composed of 1,201 images, while the remaining 334 images are reserved for testing. As shown in Table 1, we achieve the best MAE and the second best MSE performance on UCF-QNRF.

Table 2. Comparison with mainstream methods on JHU-Crowd++ and NWPU-Crowd.

Method	Venue	JHU-Crowd++		NWPU-Crowd	
		MAE	MSE	MAE	MSE
MCNN [1]	CVPR 16	188.9	483.4	232.5	714.6
CSRNet [2]	CVPR 18	85.9	309.2	121.3	387.8
BL [31]	ICCV 19	75.0	299.9	105.4	454.2
DM-Count [32]	NIPS 20	–	–	88.6	388.6
BM-Count [33]	IJCAI 21	61.5	263	83.4	358.4
P2PNet [38]	ICCV 21	–	–	77.44	362
S3 [34]	IJCAI 21	59.4	244.0	83.5	346.9
CLTR [35]	ECCV 22	59.5	240.6	74.3	333.8
GauNet [37]	CVPR 22	58.2	245.1	–	–
ChfL [36]	CVPR 22	57.0	235.7	76.8	343.0
HTNet	–	61.6	261.5	**69.3**	**295.0**

JHU-Crowd++ [27] is a massive dataset with 4,372 images encompassing varied weather conditions and density levels. The head annotations for each image range from 0 to 25,791. Among the images, 2,772 are designated for training, 500 are in the validation set, and the remaining 1,600 are for testing. According to Table 2, the performance of HTNet on JHU is slightly worse than that of state-of-the-art methods, but still acceptable.

NWPU-Crowd [28] consists of 5,109 high-resolution images with over 2.13 million annotated points, of which 3,109 are assigned to the training set, 500 are used for validation, and the remaining 1,500 are divided into testing set. From Table 2, it is evident that HTNet performs the best in terms of MAE and MSE, with performance gains of 6.7% and 11.6% respectively.

4.4 Ablation Studies

We perform comprehensive ablation experiments on ShanghaiTech A to validate the effectiveness of each proposed module, including GAM, LAM, CAM, and DHM. The quantitative results are provided in Table 3. Row 2 represents a baseline model, which eliminates all the aforementioned modules. Rows 3 to 9 show the performance when different combinations of modules are included.

Table 3. Ablation studies for each module in HTNet on ShanghaiTech A

GAM	LAM	CAM	DHM	MAE	MSE
				63.32	104.05
			√	57.51	91.40
√			√	56.93	90.89
	√		√	55.34	91.90
		√	√	55.39	93.06
	√	√	√	55.17	95.78
√	√	√		55.56	93.38
√	√		√	53.42	92.17
√		√	√	53.94	93.36
√	√	√	√	52.47	87.90

Ablation for GAM. The removal of GAM leads to severe decrease in MSE, highlighting the critical role of global priors in guiding the model to comprehend the input image.

Ablation for LAM. Eliminating LAM refers to replacing the W-MSA in Fig. 2(b) with an identity connection. In this case, both MAE and MSE show a certain degree of deterioration, suggesting that the local information also contributes to boosting the model.

Ablation for CAM. When we ablate CAM, the remaining modules lose the ability to capture the spatial-channel dependency. Hence, the model exhibits suboptimal performance.

Ablation for DHM. In the scenario where we discard the segmentation task of DHM, resulting in a plain density regression module, we observe the worst MAE. This indicates that relying solely on a density regression task is insufficient to leverage the extracted rich features.

5 Conclusion

In this paper, we propose a crowd counting model, HTNet, which consists of a CNN stem for capturing detailed information and three modules based on global, local, and channel self-attention mechanisms respectively. With this design, the model can overcome the restriction of a limited receptive field, analyze the input image from a global perspective, and capture the interaction between spatial and channel dimensions. Additionally, we introduce a double head module to further alleviate confusion caused by complex backgrounds with textures similar to those of dense crowds. Experimental results on benchmark datasets demonstrate that our proposed approach achieves superior performance.

References

1. Zhang, Y., Zhou, D., Chen, S., Gao, S., Ma, Y.: Single-image crowd counting via multi-column convolutional neural network. In: CVPR, pp. 589–597 (2016)
2. Li, Y., Zhang, X., Chen, D.: CSRnet: dilated convolutional neural networks for understanding the highly congested scenes. In: CVPR, pp. 1091–1100 (2018)
3. Sindagi, V., Patel, V.: Multi-level bottom-top and top-bottom feature fusion for crowd counting. In: ICCV, pp. 1002–1012 (2019)
4. Song, Q., et al.: To choose or to fuse? Scale selection for crowd counting. In: AAAI, pp. 2576–2583 (2021)
5. Vaswani, A., et al.: Attention is all you need. In: NeurIPS (2017)
6. Dosovitskiy, A., et al.: An image is worth 16x16 words: transformers for image recognition at scale. In: ICLR (2021)
7. Liu, Z., et al.: Swin transformer: hierarchical vision transformer using shifted windows. In: ICCV, pp. 9992–10002 (2021)
8. Carion, N., Massa, F., Synnaeve, G., Usunier, N., Kirillov, A., Zagoruyko, S.: End-to-end object detection with transformers. In: Vedaldi, A., Bischof, H., Brox, T., Frahm, J.-M. (eds.) ECCV 2020. LNCS, vol. 12346, pp. 213–229. Springer, Cham (2020). https://doi.org/10.1007/978-3-030-58452-8_13
9. Zheng, S., et al.: Rethinking semantic segmentation from a sequence-to-sequence perspective with transformers. In: CVPR, pp. 6877–6886 (2021)
10. Xiao, T., Singh, M., Mintun, E., Darrell, T., Dollar, P., Girshick, R.: Early convolutions help transformers see better. In: NeurIPS, pp. 30392–30400 (2021)
11. Sam, D.B., Surya, S., Babu, R.V.: Switching convolutional neural network for crowd counting. In: CVPR, pp. 4031–4039 (2017)
12. Chen, X., Bin, Y., Sang, N., Gao, C.: Scale pyramid network for crowd counting. In: WACV, pp. 1941–1950 (2019)
13. Jiang, X., et al.: Attention scaling for crowd counting. In: CVPR, pp. 4705–4714 (2020)
14. Rong, L., Li, C.: Coarse- and fine-grained attention network with background-aware loss for crowd density map estimation. In: WACV, pp. 3674–3683 (2021)
15. Yan, Z., et al.: Perspective-guided convolution networks for crowd counting. In: ICCV, pp. 952–961 (2019)
16. Yan, Z., Zhang, R., Zhang, H., Zhang, Q., Zuo, W.: Crowd counting via perspective-guided fractional-dilation convolution. IEEE Trans. Multimedia, pp. 2633–2647 (2022)
17. Yang, S., Guo, W., Ren, Y.: Crowdformer: an overlap patching vision transformer for top-down crowd counting. In: IJCAI, pp. 1545–1551 (2022)
18. Lin, H., Ma, Z., Ji, R., Wang, Y., Hong, X.: Boosting crowd counting via multi-faceted attention. In: CVPR, pp. 19596–19605 (2022)

19. Qian, Y., Zhang, L., Hong, X., Donovan, C., Arandjelovic, O.: Segmentation assisted u-shaped multi-scale transformer for crowd counting. In: BMVC (2022)
20. Chu, X., et al.: Twins: Revisiting the design of spatial attention in vision transformers. In: NeurIPS, pp. 9355–9366 (2021)
21. Simonyan, K., Zisserman, A.: Very deep convolutional networks for large-scale image recognition. arXiv preprint arXiv:1409.1556 (2014)
22. Lin, T.Y., Dollar, P., Girshick, R., He, K., Hariharan, B., Belongie, S.: Feature pyramid networks for object detection. In: CVPR, pp. 936–944 (2017)
23. Chu, X., et al.: Conditional positional encodings for vision transformers. arXiv preprint arXiv:2102.10882 (2021)
24. Li, X., et al.: Semantic flow for fast and accurate scene parsing. In: Vedaldi, A., Bischof, H., Brox, T., Frahm, J.-M. (eds.) ECCV 2020. LNCS, vol. 12346, pp. 775–793. Springer, Cham (2020). https://doi.org/10.1007/978-3-030-58452-8_45
25. Lin, T.Y., Goyal, P., Girshick, R., He, K., Dollár, P.: Focal loss for dense object detection. In: ICCV, pp. 2999–3007 (2017)
26. Idrees, H., et al.: Composition loss for counting, density map estimation and localization in dense crowds. In: Ferrari, V., Hebert, M., Sminchisescu, C., Weiss, Y. (eds.) ECCV 2018. LNCS, vol. 11206, pp. 544–559. Springer, Cham (2018). https://doi.org/10.1007/978-3-030-01216-8_33
27. Sindagi, V.A., Yasarla, R., Patel, V.M.: Jhu-crowd++: large-scale crowd counting dataset and a benchmark method. Technical report (2020)
28. Wang, Q., Gao, J., Lin, W., Li, X.: NWPU-crowd: a large-scale benchmark for crowd counting and localization. IEEE Trans. Pattern Anal. Mach. Intell. 2141–2149 (2021)
29. Deng, J., Dong, W., Socher, R., Li, L.J., Li, K., Fei-Fei, L.: Imagenet: a large-scale hierarchical image database. In: CVPR, pp. 248–255 (2009)
30. Loshchilov, I., Hutter, F.: Decoupled weight decay regularization. In: ICLR (2019)
31. Ma, Z., Wei, X., Hong, X., Gong, Y.: Bayesian loss for crowd count estimation with point supervision. In: ICCV, pp. 6141–6150 (2019)
32. Wang, B., Liu, H., Samaras, D., Nguyen, M.H.: Distribution matching for crowd counting. In: NeurIPS, pp. 1595–1607 (2020)
33. Liu, H., Zhao, Q., Ma, Y., Dai, F.: Bipartite matching for crowd counting with point supervision. In: IJCAI, pp. 860–866 (2021)
34. Lin, H., et al.: Direct measure matching for crowd counting. In: IJCAI, pp. 837–844 (2021)
35. Liang, D., Xu, W., Bai, X.: An end-to-end transformer model for crowd localization. In: Avidan, S., Brostow, G., Cissé, M., Farinella, G.M., Hassner, T. (eds.) ECCV 2022. LNCS, vol. 13661, pp. 38–54. Springer, Cham (2022). https://doi.org/10.1007/978-3-031-19769-7_3
36. Shu, W., Wan, J., Tan, K.C., Kwong, S., Chan, A.B.: Crowd counting in the frequency domain. In: CVPR, pp. 19586–19595 (2022)
37. Cheng, Z.Q., Dai, Q., Li, H., Song, J., Wu, X., Hauptmann, A.G.: Rethinking spatial invariance of convolutional networks for object counting. In: CVPR, pp. 19606–19616 (2022)
38. Song, Q., et al.: Rethinking counting and localization in crowds: a purely point-based framework. In: ICCV, pp. 3345–3354 (2021)

Reliable Boundary Samples-Based Proxy Pairs for Unsupervised Person Re-identification

Chang Zou[1], Zeqi Chen[2], Yuehu Liu[2(✉)], and Chi Zhang[2]

[1] School of Software Engineering, Xi'an Jiaotong University, Xi'an 710049,, China
zouchang@stu.xjtu.edu.cn

[2] Institute of Artificial Intelligence and Robotics, Xi'an Jiaotong University, Xi'an 710049,, China
chenzeqi@stu.xjtu.edu.cn, liuyh@mail.xjtu.edu.cn, chizhang@xjtu.edu.cn

Abstract. Contrastive learning methods based on the memory bank have shown promising results for unsupervised person re-identification. However, most methods maintain a uni-proxy for each cluster in the memory that only describes the average information but can not represent the intra-class variation. As a result, contrastive learning based on the uni-proxy cannot effectively guide the model to reduce the variation. To address this issue, we maintain a proxy pair for each cluster updated by the least similar boundary sample pair since they concretely reveal the intra-class variation of the cluster. Through contrastive loss, the proxy updated based on one boundary sample generates strong pulls to another one and its surrounding samples due to the low similarity, and two proxies collaboratively form bidirectional strong pulls to effectively reduce intra-class variance. To mitigate the impact of boundary samples being noisy, we further propose local-global consistency guided label refinement, which utilizes local fine-grained cues to select reliable samples with high overlap in the local and global feature neighborhoods. Comprehensive experiments on Market-1501 and MSMT17 demonstrate that the proposed method surpasses state-of-the-art approaches.

Keywords: Person re-identification · Contrastive learning · Proxy pairs

1 Introduction

Unsupervised person re-identification (Re-ID) searches for images of a target person across non-overlapping cameras without annotations [23]. Currently, most state-of-the-art methods [1,6,16,17] utilize a memory bank to maintain cluster proxies according to the pseudo labels generated by clustering algorithms, then employ contrastive learning [12] based on the proxies to train the model. The

This work is supported by the Key R&D Plan of Shaanxi Province (Program No. 2023-YBGY-029).

© The Author(s), under exclusive license to Springer Nature Singapore Pte Ltd. 2024
Q. Liu et al. (Eds.): PRCV 2023, LNCS 14436, pp. 302–314, 2024.
https://doi.org/10.1007/978-981-99-8555-5_24

contrastive loss is designed to pull a sample closer to the proxy of its cluster while pushing it away from proxies of other clusters. The effectiveness of the loss heavily relies on the quality of the proxies. Therefore, several works have been conducted to study this issue. SpCL [8] maintains instance features in the memory and uses the cluster centroids as proxies. CCL [6] directly stores cluster-level proxies to keep the updating consistency. HCM [16] integrates both cluster features and instance features as proxies. The commonality of these methods is using a uni-proxy as the cluster-level representation for each cluster. However, due to various factors such as pose, viewpoint, illumination, occlusion, etc., large intra-class variation usually exists. The conventional uni-proxy only describes the average information and fails to capture the intra-class variation. Consequently, as shown in Fig. 1a, a uni-proxy can only generate a unidirectional pull to the samples in the cluster through contrastive loss, but not a targeted pull that brings samples with large distances closer to each other and therefore cannot effectively reduce intra-class variation. There are also some methods [1,5,17] maintaining multiple camera-aware proxies at the cluster level to reduce intra-class variation caused by changing viewpoints, but they rely on camera labels and cannot mitigate the variation caused by other factors. To solve the problem, we propose a

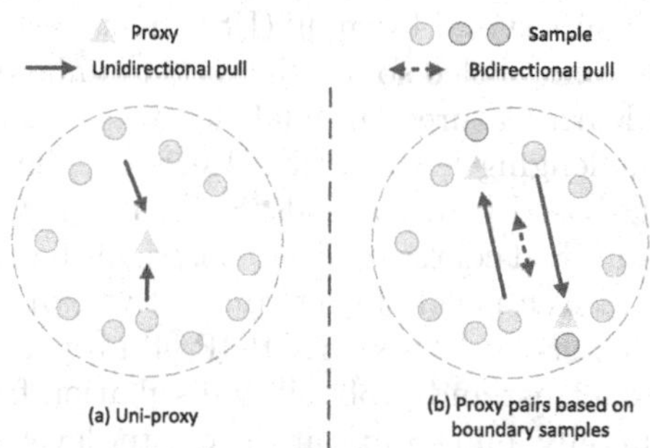

Fig. 1. An illustration showing that a uni-proxy only generates weak unidirectional pulls to samples, while discrepant proxy pairs based on boundary samples generate strong bidirectional pulls to make the samples close to each other.

contrastive learning method that maintains a proxy pair for each cluster based on reliable boundary samples. The boundary sample pair with the lowest similarity reveals the current largest intra-class variation, thus a proxy pair based on them can accurately represent the existing variance. Moreover, each proxy can produce a strong pull on samples around the other boundary sample due to their low similarity, thus strong bidirectional pulls are collaboratively formed to decrease intra-class variation like Fig. 1b. Nevertheless, considering that the generated pseudo labels are not all correct while boundary samples are most likely to be noisy and to significantly degrade performance [5], we propose to select reliable samples in clusters based on the neighborhood consistency of local and global features. Our key idea is that the label noise caused by the global appearance similarity can be corrected by the fine-grained visual cues included in the local features. We calculate the consistency of a sample by the overlap between the k-nearest neighbors of its local feature and the samples sharing its global label. Based on the consistency of each sample, we introduce a dynamic changing threshold to filter reliable samples with high consistency, which effectively mitigates the impact of label noise. The main contributions of our work are as follows:

- We propose boundary samples-based proxy pairs that can explicitly represent the intra-class variation and effectively reduce the variation through the generated strong bidirectional pulls in a cluster.
- We propose a local-global consistency guided pseudo label refinement method, which utilizes the finer-grained information of local features to filter reliable samples with high-confidence global pseudo labels.

2 Related Work

Unsupervised Person Re-identification. Current unsupervised approaches can be generally categorized into unsupervised domain adaptive (UDA) methods [7,19] and purely unsupervised learning (USL) methods [1,5,6]. UDA methods transfer knowledge from labeled source datasets to unlabeled target datasets. USL methods directly learn features on unlabeled target datasets. Our method belongs to the more challenging USL setting. Recently, significant progress has been made in clustering-based USL methods. SpCL [8] gradually selects reliable clusters and proposes a unified contrastive loss based on the cluster centroids and outliers. CCL [6] introduces a novel cluster contrastive learning framework that stores and updates a cluster-level proxy. HCPDP [4] proposes a hybrid dynamic local-global clustering contrast and probability distillation framework. However, these methods maintain only a uni-proxy at the cluster level and fail to represent intra-class variation. Therefore, CAP [17] forms multiple camera-aware proxies within a cluster to alleviate the camera domain gaps. MRCN [19] utilizes multiple instance proxies to mitigate the influence of noisy samples. Unlike these methods, we propose to maintain a discrepant proxy pair that explicitly represents the intra-class variation based on reliable boundary samples and reduce the variation by strong bidirectional pulls.

Noise Reduction of Pseudo Labels. Recently, reducing label noise has become a research focus in unsupervised person Re-ID. MMT [7] utilizes mutual learning by leveraging predictions from a mean teacher network to refine pseudo labels. RLCC [21] estimates pseudo-label similarity over consecutive training generations and proposes temporal propagation and ensemble. DCCT [6] proposes parameter-differentiated dual clustering to select consistent samples. PPLR [5] considers the complementary relationship between global and local features to generate soft pseudo-labels. SECRET [10] utilizes the consistency between local and global label spaces for reliable sample selection. In contrast to these methods, we utilize fine-grained clues from local features to exclude unreliable samples in the global label space based on the consistency between local and global feature neighborhoods.

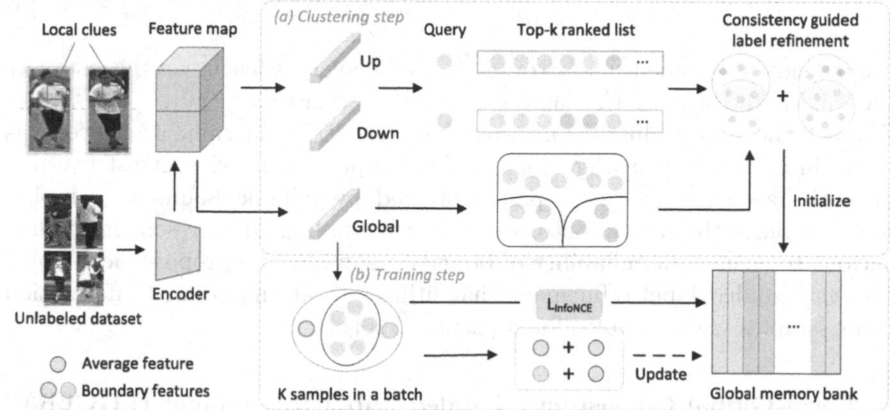

Fig. 2. The illustration of our framework. In the clustering step, we calculate local-global consistency to select reliable samples. In the training step, we update the proxy pair of the identity based on the two least similar boundary samples.

3 Method

3.1 Overview

Let $\mathcal{D} = \{x_i\}_{i=1}^{N_\mathcal{D}}$ denote the unlabeled dataset, where x_i is the i-th image and N_D is the number of images. $\mathcal{F}_m = \{F_i\}_{i=1}^{N_\mathcal{D}}$ denotes the feature maps extracted by encoder f_θ. $\mathcal{F}_v = \{f_i\}_{i=1}^{N_\mathcal{D}}$ denotes the features after the pooling on feature maps. f_q denotes the feature vector of a query instance.

Currently, most USL methods [1,6,8] adopt a two-step alternating strategy of clustering and training. They fist obtain a labeled dataset $\mathcal{D}' = \{(x_i, y_i)\}_{i=1}^{N'_\mathcal{D}}$ through DBSCAN clustering [15], where $y_i \in \{1, 2, \ldots, c\}$ denotes the pseudo label of the i-th image, c denotes the number of clusters and N'_D is the number of images after discarding outliers. Then the memory mechanism and contrastive learning are applied to $\mathcal{D}'$. Most methods [3,6] initialize the memory bank $\mathcal{M}$ with the average feature of samples in each cluster as its cluster-level uni-proxy, which plays the role of a non-parametric classifier in the InfoNCE loss function [12]. The formulation of the InfoNCE loss can be summarized as follows:

$$\mathcal{L} = -\log \frac{\exp\left(f_q \cdot p^+/\tau\right)}{\sum_{i=1}^{c} \exp\left(f_q \cdot p_i/\tau\right)}, \tag{1}$$

where f_q is the query instance feature. p_i is the proxy of the i-th cluster in the memory bank $\mathcal{M}$. p^+ is the proxy of the cluster to which the query belongs. τ is a temperature factor. c is the number of clusters. Since both f_q and p_i are L_2-normalized, the cosine similarity $f_q \cdot p_i$ is specified as their similarity score. During the back-propagation, the memory bank is updated as follows:

$$p^+ \leftarrow \mu \cdot p^+ + (1-\mu) \cdot q_{mean}, \tag{2}$$

where μ denotes a momentum factor. q_{mean} is the average feature of the instances in the mini-batch sharing the same pseudo label with p^+.

We set the above cluster uni-proxy-based contrast learning framework as our baseline. As shown in Fig. 2, this paper proposes a novel contrast learning framework based on the proxy pairs obtained by reliable boundary samples, which overcomes the limitation of the uni-proxy failing to represent intra-class variation. To ensure the reliability of boundary samples, we propose local-global consistency guided label refinement that utilizes local fine-grained information to filter samples with reliable global pseudo labels.

3.2 Local-Global Consistency Guided Label Refinement (LGCLR)

The global features of different persons often show high similarity due to under the same camera view or wearing similar clothes. In this case, local visual cues such as bags, patterns on clothes, shoes, etc. are the key factors to distinguish similar persons. Therefore, we propose local-global consistency guided label refinement which defines the overlap between the k -nearest neighbors of a local feature and the samples sharing its global label as local-global consistency and selects samples with high consistency.

We extract the feature map $F_i \in \mathbb{R}^{C\times H\times W}$ of an image x_i with the encoder f_θ, where C, H, W denote the sizes of the channel, height, and width. Then the global feature f_i^g is obtained by Generalized-Mean (GEM) pooling [14] over F_i, while two local features $\left(f_i^{up}, f_i^{dw}\right)$ are obtained by dividing the feature map horizontally into two uniform regions $\mathbb{R}^{C\times\frac{H}{2}\times W}$. We calculate the k-reciprocal Jaccard Distance [24] matrices D^g, D^{up} and D^{dw} of the train set based on the global and local feature sets f^g, f^{up} and f^{dw} respectively, where k is set to 30 and the re-ranking technique is applied. Then the pseudo labels are generated by applying DBSCAN on the global distance matrix D^g only. For a global feature f_i^g of the sample x_i, we get the cluster it belongs to and its global label y_i. We denote the sample set of the cluster as $\mathcal{I}\left(f_i^g\right)$ and the number of samples as n_i. The top-n_i nearest neighbors of x_i according to distance matrices D^{up} and D^{dw} can be denoted as $\mathcal{I}\left(f_i^{up}\right)$ and $\mathcal{I}\left(f_i^{dw}\right)$. We propose the following metrics to measure the consistency of the global and two local feature neighborhoods,

$$\begin{aligned} \mathcal{C}_{\text{up}}\left(f_i^g\right) &= \frac{\left|\mathcal{I}\left(f_i^g\right) \cap \mathcal{I}\left(f_i^{up}\right)\right|}{\left|\mathcal{I}\left(f_i^g\right)\right|} \in [0,1], \\ \mathcal{C}_{\text{dw}}\left(f_i^g\right) &= \frac{\left|\mathcal{I}\left(f_i^g\right) \cap \mathcal{I}\left(f_i^{dw}\right)\right|}{\left|\mathcal{I}\left(f_i^g\right)\right|} \in [0,1], \end{aligned} \tag{3}$$

where $|\cdot|$ denotes the number of samples in the set. Larger $\mathcal{C}_{\text{up}}(f_i^g)$ and $\mathcal{C}_{\text{dw}}(f_i^g)$ indicates a higher consistency and reliability for f_i^g, *i.e.*, even from a local perspective that contains more detailed information, samples with the same global pseudo label are still most likely to be clustered into the same category. We calculate the final consistency $\mathcal{C}(f_i^g)$ as the mean value of $\mathcal{C}_{\text{up}}(f_i^g)$ and $\mathcal{C}_{\text{dw}}(f_i^g)$.

After obtaining the local-global consistency of each sample, we introduce a threshold $\alpha \in [0, 1]$ to select reliable samples. Specifically, we retain samples with consistency $\mathcal{C} > \alpha$, while the remaining samples are regarded as outliers. The consistency of samples will increase with training, and more and more samples will be considered reliable ones to enhance feature learning.

3.3 Reliable Boundary Samples Based Proxy Pairs (RBSPP)

Considering that the uni-proxy merely represents the average information within a cluster, failing to capture the intra-class variations whereas the least similar boundary samples accurately reflect the largest intra-class variance, we form discrepant proxy pairs based on reliable boundary samples after LGCLR to effectively represent and reduce the intra-class variance.

Memory Initialization. Assume that there remain c' clusters after LGCLR, we initialize the proxy pair (p_j^1, p_j^2) of the j-th cluster with the cluster centroid c_j at the beginning of each epoch as follows,

$$c_j = \frac{1}{|\mathcal{H}_j|} \sum_{f_i^g \in \mathcal{H}_j} f_i^g, \tag{4}$$

where H_j denote the set of global features in the j-th cluster. Note that part features are only used for label refinement. Hence, the memory bank $\mathcal{M}$ has a size of $\mathbb{R}^{2c' \times d}$, with d representing the feature dimension.

Memory Update. Following previous works, we randomly sample K instances for each of P person identities to form a mini-batch. For the i-th identity, we select the boundary sample pair $\left(f_{b_1}^g, f_{b_2}^g\right)$ with the lowest cosine similarity among the K instances. Considering that utilizing only the boundary sample pairs to update the proxies ignores the information of other samples in the clusters, which may lead to biased training, we weighted fuse the boundary sample pairs with the average features of the remaining $K-2$ samples before updating,

$$f_{b_t}^w = \beta f_{b_t}^g + (1-\beta) \frac{1}{K-2} \sum_{j=1}^{K-2} f_j^g, \tag{5}$$

where $t = \{1, 2\}$ and β is the fusion weight. Then $\left(f_{b_1}^w, f_{b_2}^w\right)$ are used to update the proxy pair (p_i^1, p_i^2) of the i-th identity. To avoid the drastic changes in optimization direction caused by significant updates of proxies, we update each proxy as Eq. 2 by replacing q_{mean} with corresponding $f_{b_t}^w$ with higher similarity to the proxy.

Contrastive Loss. With the proxy pair of each cluster, we form the contrastive loss for any query instance as follows:

$$\mathcal{L}_{RBSPP} = -\frac{1}{2}\sum_{j=1}^{2}\log\frac{\exp\left(f_q^g \cdot p_j^+/\tau\right)}{\sum_{i=1}^{c'}\exp\left(f_q^g \cdot p_i^j/\tau\right)}, \tag{6}$$

where p_i^j is the j-th proxy of the i-th cluster. p_j^+ shares the same label with the global query feature f_q^g and is the j-th proxy for that cluster. Updating a proxy based on a boundary sample brings a decrease in the similarity between a proxy and the samples around another boundary sample. Therefore, the contrastive loss can produce larger gradients, resulting in a stronger bidirectional pull to effectively reduce intra-class variation.

4 Experiments

4.1 Datasets and Evaluation Protocols

The evaluations are conducted on the Market-1501 [22] dataset and the MSMT17 [18] dataset. Market-1501 is gathered from the Tsinghua University campus using 6 cameras and has 32,668 images belonging to 1,501 unique person identities. A training set of 12,936 images from 751 identities and a test set of 19,732 images from 750 identities make up the dataset. MSMT17 is a larger, more difficult dataset with 126,441 images and 4,101 person identities captured by 15 cameras. The MSMT17 training set comprises 32,621 images from 1,041 identities, while the test set comprises 93,820 images from 3,060 identities. In our experiments, we assessed performance using the top-1, top-5, and top-10 accuracies of the Cumulative Matching Characteristics (CMC), as well as the mean Average Precision (mAP).

4.2 Implementation Details

We adopt ResNet50 [9] as the backbone. All layers after layer-4 are removed and the GEM pooling [14] layer is added. The input image size is 320×128. The maximum distance in DBSCAN is set to 0.45 for Market-1501 and 0.7 for MSMT17. The consistency threshold α is set to the value that ranks the 1% position. Considering that label noise is most severe during the early training stage, we only perform label refinement in the first 20 epochs to ensure an adequate number of training samples. The weight β in Eq. 5 is set to 0.7 and 0.6 for Market-1501 and MSMT17, respectively. A mini-batch has 256 samples comprising 16 IDs and 16 images for each ID. The temperature hyper-parameter τ in the contrastive loss (Eq. 6) is set to 0.05. We use an Adam optimizer with weight decay of 5×10^{-4}. The initial learning rate is set to 3.5×10^{-5} and divided by 10 every 20 epochs. We train the model for 60 epochs on both datasets.

4.3 Comparison with State-of-the-Arts

We initially compare RBSPP to state-of-the-art methods on Market-1501 and MSMT17. From Table 1, We can see that our approach performs substantially better than the prior approaches without any labels. We achieve mAP/top-1 of 85.4%/94.0% and 38.6%/67.3% on Market-1501 and MSMT17. The mAP of our method substantially surpasses the state-of-the-art method CCL [6] by 1.3% and 5.0% on Market-1501 and MSMT17. In comparison to the methods using camera labels, our method without any camera information outperforms the mAP of IICS [20], CAP [17], ICE [1], and PPLR [5] on Market1501, and surpasses the mAP of IICS [20] and CAP [17] on MSMT17. Moreover, under the supervised condition, our method greatly exceeds the mAP/top-1 of ICE [1] by 11.1%/6.8% on MSMT17, which shows our potential on large datasets. It is also worth noting that our method demonstrates comparable performance to the widely recognized supervised method ADBNet [2].

Table 1. Comparison with state-of-the-art methods on Market-1501 and MSMT17. **Bold** is used to mark the best results of unsupervised methods without any labels.

Method		Market-1501				MSMT17			
		mAP	top-1	top-5	top-10	mAP	top-1	top-5	top-10
Unsupervised methods with camera labels									
IICS [20]	CVPR'21	72.9	89.5	95.2	97.0	26.9	56.4	68.8	73.4
AP [17]	AAAI'21	79.2	91.4	96.3	97.7	36.9	67.4	78.0	81.4
ICE [1]	ICCV'21	82.3	93.8	97.6	98.4	38.9	70.2	80.5	84.4
PPLR [5]	CVPR'22	84.4	94.3	97.8	98.6	42.2	73.3	83.5	86.5
Unsupervised methods without any labels									
SpCL [8]	NeurIPS'20	73.1	88.1	95.1	97.0	19.1	42.3	55.6	61.2
ICE [1]	ICCV'21	79.5	92.0	97.0	98.1	29.8	59.0	71.7	77.0
MCRN [19]	AAAI'22	80.8	92.5	-	-	31.2	63.6	-	-
SECRET [10]	AAAI'22	81.0	92.6	-	-	31.3	60.4	-	-
PPLR [5]	CVPR'22	81.5	92.8	97.1	98.1	31.4	61.1	73.4	77.8
HDCPD [4]	TIP'22	81.7	92.4	97.4	98.1	24.6	50.2	61.4	65.7
CCL [6]	ACCV'22	84.2	93.4	**97.6**	98.3	33.6	63.3	73.3	78.0
MPC [11]	CVIU'23	79.9	91.1	96.1	97.4	-	-	-	-
RMCL [13]	KBS'23	81.7	93.0	**97.6**	**98.4**	32.5	62.3	73.6	78.0
RBSPP	This paper	**85.4**	**94.0**	**97.6**	98.3	**38.6**	**67.3**	**77.2**	**81.2**
Supervised methods									
ABD-Net [2]	ICCV'19	88.3	95.6	-	-	60.8	82.3	90.6	-
ICE (w/ ground truth) [1]	ICCV'21	86.6	95.1	98.3	98.9	50.4	76.4	86.6	90.0
RBSPP (w/ ground truth)	This paper	88.9	95.5	98.4	99.0	61.5	83.2	91.5	93.7

4.4 Ablation Study

To evaluate the effectiveness of each proposed component, we undertake extensive experiments on Market-1501 and MSMT17 in this subsection. We define the framework introduced in Sect. 3.1 that uses the uni-proxy as our *baseline*.

Table 2. Ablation studies on proposed components of our method.

Method	Market-1501		MSMT17	
	mAP	top-1	mAP	top-1
Baseline	81.1	92.5	35.4	65.6
Baseline+LGCLR	83.1	93.0	35.9	66.4
Baseline+BSPP	84.7	93.5	36.9	65.8
RBSPP	**85.4**	**94.0**	**38.6**	**67.3**

Table 3. Comparison with different strategies for LGCLR and RBSPP.

Method	Market-1501		MSMT17	
	mAP	top-1	mAP	top-1
RBSPP	**85.4**	**94.0**	**38.6**	**67.3**
RBSPP-up	84.7	93.4	37.0	66.1
RBSPP-down	84.2	93.0	35.6	63.8
RBSPP-sep	82.5	92.0	37.7	66.2

(a) Alternative definitions of consistency

Method	Market-1501		MSMT17	
	mAP	top-1	mAP	top-1
RBSPP	**85.4**	**94.0**	**38.6**	**67.3**
Baseline	81.1	92.5	35.4	65.6
RBSPP-BF	85.0	93.8	27.4	52.7
RBSPP-RF	82.5	92.6	35.9	66.6

(b) Different features for proxy updating

Effectiveness of LGCLR. We add local-global consistency guided label refinement (LGCLR) to the baseline and BSPP (*i.e.* RBSPP without label refinement), respectively. As shown in Table 2, LGCLR boosts the mAP/top-1 of baseline by 2.0%/0.5% and 0.5%/0.8% on Market-1501 and MSMT17. In addition, compared to BSPP, RBSPP (BSPP + LGCLR) improves mAP/top-1 by 0.7%/0.5% and 1.7%/1.5% on the two datasets, respectively. This demonstrates that the proposed LGCLR can effectively mitigate the impact of label noise in the early training stage by utilizing local detail cues as complementation for global information to select reliable samples. We also explore alternative consistency definitions to validate our assumption on it. In Table 3(a), RBSPP-up/RBSPP-down regards the consistency of the up/down part as the consistency of a sample. RBSPP-sep denotes that two thresholds are set for up-global and down-global consistency separately. According to the experimental results, RBSPP aggregating the two local-global consistency achieves the optimum, presumably because an individual part loses information contained in the other part and instead easily confuses two samples that are not similar from the global view.

Effectiveness of BSPP. As shown in Table 2, the proposed boundary samples-based proxy pairs (BSPP) improve mAP/top-1 by 3.6%/1.0% and 1.5%/0.2% on Market-1501 and MSMT17 compared to the uni-proxy-based baseline. Moreover, the RBSPP including LGCLR significantly outperforms the baseline by 4.3%/1.5% and 3.2%/1.7% on the two datasets, which shows the effectiveness of BSPP. We also study different strategies to get the proxy pair in Table 3(b). RBSPP-BF directly updates the proxy pair by the boundary sample pair without fusing the remaining features. RBSPP-RF updates the proxy with a randomly selected sample pair. We can see that RBSPP outperforms RBSPP-BF on both datasets by 0.4% and 9.2% in mAP, which demonstrates that ignoring the remaining samples does result in biased training, especially on MSMT17. Compared to RBSPP-RF, RBSPP improved mAP/top-1 by 3.1%/1.4% and 2.7%/0.7% on the two datasets, which proves the boundary samples and remaining samples are both important for cluster representation.

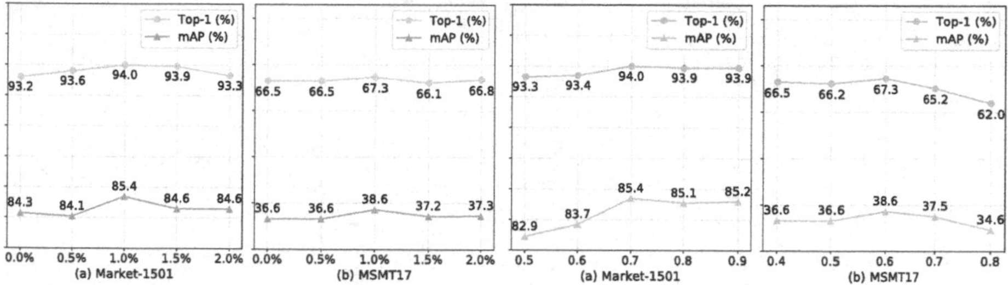

Fig. 3. Analysis of threshold proportion.

Fig. 4. Analysis of fusion weight β.

The Consistency Threshold Proportion for LGCLR. Considering that the consistency of samples increases as training and a fixed threshold is not appropriate, we first calculate the consistency of each sample and average it within each cluster. Then the mean values are ranked and we identify the value at a certain proportion position as the threshold. We analyze the effect of different proportions on LGCLR. As shown in Fig. 3, our method is sensitive to proportion. A smaller proportion causes the filtered samples still contain a large number of noisy samples and leads to performance degradation, while a larger one causes many correctly labeled samples to be also regarded as outliers, resulting in fewer data available for training. Therefore, we set the scale value to 1.0%.

The Fusion Weight of Boundary Samples for BSPP. Fig. 4 shows the effect of the fusion weight β on BSPP. We can see that $\beta = 0.6$ and $\beta = 0.7$ achieve optimal accuracy for Market-1501 and MSMT17. We speculate that larger β may lead to noisy samples dominating the updating of the proxies and missing the information contained in other reliable samples, which further causes the wrong training direction. In contrast, smaller β allows the mean feature of internal similar samples to play a leading role and decrease the discrepancy of the proxy pair. Consequently, the proxy pair cannot accurately represent the intra-class variation and cannot generate strong bidirectional pulls to reduce the variation.

4.5 Clustering Quality

As shown in Fig. 5, the intra-variation of all classes is significantly reduced in our method compared to the baseline. For several classes difficult to distinguish or already mixed in the baseline, our method succeeds in separating them, which demonstrates that our method learns more discriminative features.

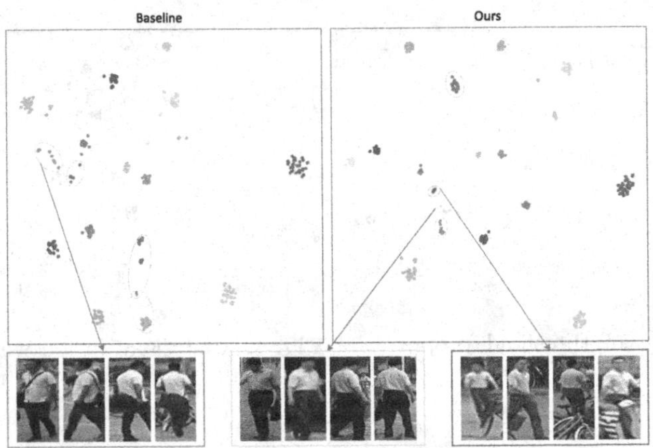

Fig. 5. The t-SNE visualization of 20 random classes in Market-1501 between baseline and our method. Different colors represent different IDs.

5 Conclusion

In this paper, we propose a contrastive learning framework maintaining discrepant proxy pairs for each cluster based on reliable boundary samples, which accurately represent the intra-class variation and generate strong bidirectional pulls to effectively reduce intra-class variance. To ensure the reliability of boundary samples, we also propose local-global consistency guided label refinement that utilizes local visual cues to exclude unreliable noisy samples caused by global similarity. Comprehensive experiments have shown that our framework outperforms prior state-of-art methods on Market-1501 and MSMT17 datasets. In further work, we will investigate using inter-class relationships to further improve cluster representation.

References

1. Chen, H., Lagadec, B., Bremond, F.: Ice: Inter-instance contrastive encoding for unsupervised person re-identification. In: Proceedings of the IEEE/CVF International Conference on Computer Vision, pp. 14960–14969 (2021)
2. Chen, T., et al.: Abd-net: attentive but diverse person re-identification. In: Proceedings of the IEEE/CVF International Conference on Computer Vision, pp. 8351–8361 (2019)
3. Chen, Z., Cui, Z., Zhang, C., Zhou, J., Liu, Y.: Dual clustering co-teaching with consistent sample mining for unsupervised person re-identification. IEEE Trans. Circuits Syst. Video Technol. (2023)
4. Cheng, D., Zhou, J., Wang, N., Gao, X.: Hybrid dynamic contrast and probability distillation for unsupervised person re-id. IEEE Trans. Image Process. **31**, 3334–3346 (2022). https://doi.org/10.1109/TIP.2022.3169693
5. Cho, Y., Kim, W.J., Hong, S., Yoon, S.E.: Part-based pseudo label refinement for unsupervised person re-identification. In: Proceedings of the IEEE/CVF Conference on Computer Vision and Pattern Recognition, pp. 7308–7318 (2022)
6. Dai, Z., Wang, G., Yuan, W., Zhu, S., Tan, P.: Cluster contrast for unsupervised person re-identification, pp. 1142–1160 (2022)
7. Ge, Y., Chen, D., Li, H.: Mutual mean-teaching: pseudo label refinery for unsupervised domain adaptation on person re-identification. arXiv preprint arXiv:2001.01526 (2020)
8. Ge, Y., Zhu, F., Chen, D., Zhao, R.: Self-paced contrastive learning with hybrid memory for domain adaptive object re-id. Adv. Neural. Inf. Process. Syst. **33**, 11309–11321 (2020)
9. He, K., Zhang, X., Ren, S., Sun, J.: Deep residual learning for image recognition. In: Proceedings of the IEEE Conference on Computer Vision and Pattern Recognition, pp. 770–778 (2016)
10. He, T., Shen, L., Guo, Y., Ding, G., Guo, Z.: Secret: Self-consistent pseudo label refinement for unsupervised domain adaptive person re-identification. In: Proceedings of the AAAI Conference on Artificial Intelligence, vol. 36, pp. 879–887 (2022)
11. Li, X., Li, Q., Liang, F., Wang, W.: Multi-granularity pseudo-label collaboration for unsupervised person re-identification. Comput. Vis. Image Underst. **227**, 103616, Jananuary 2023. https://doi.org/10.1016/j.cviu.2022.103616
12. van den Oord, A., Li, Y., Vinyals, O.: Representation learning with contrastive predictive coding. arXiv preprint arXiv:1807.03748 (2018)
13. Pang, Z., Wang, C., Wang, J., Zhao, L.: Reliability modeling and contrastive learning for unsupervised person re-identification. Knowl.-Based Syst. **263**, 110263 (2023). https://doi.org/10.1016/j.knosys.2023.110263
14. Radenović, F., Tolias, G., Chum, O.: Fine-tuning CNN image retrieval with no human annotation. IEEE Trans. Pattern Anal. Mach. Intell. **41**(7), 1655–1668 (2018)
15. Ram, A., Jalal, S., Jalal, A.S., Kumar, M.: A density based algorithm for discovering density varied clusters in large spatial databases. Int. J. Comput. Appl. **3**(6), 1–4 (2010)
16. Si, T., He, F., Zhang, Z., Duan, Y.: Hybrid contrastive learning for unsupervised person re-identification. IEEE Trans. Multimed., 1 (2022). https://doi.org/10.1109/TMM.2022.3174414
17. Wang, M., Lai, B., Huang, J., Gong, X., Hua, X.S.: Camera-aware proxies for unsupervised person re-identification. In: AAAI, vol. 2, p. 4 (2021)

18. Wei, L., Zhang, S., Gao, W., Tian, Q.: Person transfer gan to bridge domain gap for person re-identification. In: 2018 IEEE/CVF Conference on Computer Vision and Pattern Recognition. pp. 79–88. IEEE, Salt Lake City, UT, USA, June 2018. https://doi.org/10.1109/CVPR.2018.00016
19. Wu, Y., Huang, T., Yao, H., Zhang, C., Shao, Y., Han, C., Gao, C., Sang, N.: Multi-centroid representation network for domain adaptive person re-id. AAAI **36**(3), 2750–2758, June 2022. https://doi.org/10.1609/aaai.v36i3.20178
20. Xuan, S., Zhang, S.: Intra-inter camera similarity for unsupervised person re-identification. In: Proceedings of the IEEE/CVF Conference on Computer Vision and Pattern Recognition, pp. 11926–11935 (2021)
21. Zhang, X., Ge, Y., Qiao, Y., Li, H.: Refining pseudo labels with clustering consensus over generations for unsupervised object re-identification. In: Proceedings of the IEEE/CVF Conference on Computer Vision and Pattern Recognition, pp. 3436–3445 (2021)
22. Zheng, L., Shen, L., Tian, L., Wang, S., Wang, J., Tian, Q.: Scalable person re-identification: a benchmark. In: Proceedings of the IEEE International Conference on Computer Vision, pp. 1116–1124 (2015)
23. Zheng, L., Yang, Y., Hauptmann, A.G.: Person re-identification: past, present and future. arXiv preprint arXiv:1610.02984 (2016)
24. Zhong, Z., Zheng, L., Cao, D., Li, S.: Re-ranking person re-identification with k-reciprocal encoding. In: Proceedings of the IEEE Conference on Computer Vision and Pattern Recognition, pp. 1318–1327 (2017)

High-Resolution Feature Representation Driven Infrared Small-Dim Object Detection

Yuhang Dong[1], Yingying Wang[2], Linyu Fan[1], Xinghao Ding[1,2(✉)], and Yue Huang[1,2]

[1] School of Informatics, Xiamen University, Xiamen 361001, China
[2] Institute of Artificial Intelligence, Xiamen University, Xiamen 361001, China
dxh@xmu.edu.cn

Abstract. Infrared small-dim object detection is a challenging task due to the small size, weak features, lack of prominent structural information, and vulnerability to background interference. During the process of deep learning-based feature extraction, as the number of layers increases, the size of the feature map decreases, resulting in a reduction in resolution for small object features. This reduction negatively affects the network's ability to capture fine-grained details and compromises the detection efficiency. Besides, the infrared objects can be easily overwhelmed by strong background interference, which further diminishes the original faint representations. To solve these issues, we proposes a high-resolution feature representation driven network for infrared small-dim object detection (HRFRD-Net). This network comprises three key components: High-Resolution Feature Representation Branch (HRFR), Infrared Small-Dim Object Detection Branch (ISDOD), and Spatial-Frequency Interaction Feature Enhancement Module (SFIFE). The HRFR branch employs implicit neural representation to super-resolve the infrared small objects in a self-supervised learning scheme. To effectively detect the small-scale objects, ISDOD leverages the shared encoder from HRFR to construct high-resolution and high-quality representation of infrared small objects in a resolution-free manner. To address the issue of dim objects, SFIFE incorporates a global-local mixed receptive field via the features interaction in spatial-frequency dual domains, which significantly improves the accuracy of infrared dim object detection. Experiments conducted on the MSISTD and MDvsFA datasets demonstrate the effectiveness of our approach, especially in complex scenarios where the objects are heavily obscured by the background and background interference closely resembles the objects.

Y. Dong and Y. Wang—Contribute equally to this work.

Supplementary Information The online version contains supplementary material available at https://doi.org/10.1007/978-981-99-8555-5_25.

© The Author(s), under exclusive license to Springer Nature Singapore Pte Ltd. 2024
Q. Liu et al. (Eds.): PRCV 2023, LNCS 14436, pp. 315–327, 2024.
https://doi.org/10.1007/978-981-99-8555-5_25

Keywords: Infrared small-dim object detection · High-Resolution Feature Representation · Implicit neural representation · Spatial-frequency interaction feature enhancement

1 Introduction

Infrared sensors are widely utilized in military applications such as missile tracking [7], maritime surveillance, and early-warning systems, benefiting from their notable features of strong cloud penetration, anti-interference, and blind spot detection. However, due to the long imaging distance, the infrared objects generally appear to be very small, sometimes even as a single pixel in infrared images. Moreover, the infrared radiation energy diminishes significantly over distance, leading to extremely dim objects that are frequently overwhelmed by heavy noise and cluttered backgrounds. In addition, the lack of shape, color, and texture information further exacerbates the difficulty in detection of infrared objects [16], leading to a higher risk of false alarms or missed detections. High quality infrared small-dim object detection, also known as infrared small-dim object segmentation, plays a crucial role in various scenarios.

Traditional methods for infrared small-dim object detection can be broadly classified into three categories: filter-based methods, human vision system-based methods, and low-rank-based methods. Filter-based methods are suitable only for single and uniform scenes [1], while human vision system-based methods work well for objects with relatively high brightness and distinct differences from the surrounding background [17]. The low-rank-based methods are applicable to complex and rapidly changing backgrounds, but suffer from slower computational speed in practice [21]. Although traditional model-driven methods have already achieved relatively good results, they still face challenges such as complex parameters setting, low accuracy, low robustness, and heavy reliance on prior knowledge of specific scenarios and manual hyperparameters tuning.

With the rapid development of deep learning techniques, deep learning-based methods have received considerable attention in recent years. In contrast to traditional methods, deep learning approaches are able to automatically extract object features, leading to superior representation of the infrared small-dim objects and enabling more accurate detection with less false alarms. Dai et al. [5] introduced a novel approach that combines discriminative networks with conventional model-driven methods and utilizes the local contrast prior to improve the detection of infrared small objects. MDvsFA-cGan was introduced to combine the generative adversarial network with a generator divided into miss detection and false alarm subtasks [15], successfully balancing miss detection (MD) and false alarm (FA) in infrared small object detection. DNA-Net was put forward to achieve the progressive interaction between high-level and low-level features [9], avoiding feature loss in small infrared objects to a certain extent. Additionally, attention-guided pyramid context networks (AGPC) was introduced by Zhang et al. [20] to establish global associations between semantics and enhance the feature representation of the infrared objects. Although existing deep learning methods perform well in some cases, they still share a limitation. During the feature extraction process, as the network depth increases, the feature map size decreases, resulting in a loss of resolution. The loss of resolution adversely affects the capture of fine-grained details. This issue is particularly noticeable in infrared small-dim object detection tasks, where the structure of the object is poorly represented and easily overwhelmed by the background, resulting in challenging detection of the infrared objects.

To address the above issues, we propose a high-resolution feature representation driven network for infrared small-dim object detection (HRFRD-Net). This network comprises three key components: (a) High-Resolution Feature Representation Branch (HRFR), (b) Infrared Small-Dim Object Detection Branch (ISDOD), and (c) Spatial-Frequency Interaction Feature Enhancement Module (SFIFE). In recent years, Implicit Neural Representation (INR) has demonstrated its efficacy in representing signals in the continuous domain, delivering favorable outcomes in various tasks such as image super-resolution [2] and 3D reconstruction [8,11]. To effectively tackle small-scale objects, we integrate INR into our HRFR branch and super-resolve the infrared small objects in a self-supervised learning strategy. By leveraging the super-resolution guidance from HRFR to ISDOD branch through a shared encoder, the ISDOD branch can extract high-resolution and high-quality representations of infrared small objects, thereby achieving more accurate object detection. To address the issues of object dimness and vulnerability to background interference, we utilize the SFIFE module. This module improves the detection of infrared small-dim objects by facilitating a global-local mixed receptive field approach through the interaction of spatial-frequency dual domains [12]. The interaction facilitates feature enhancement for the dim objects while simultaneously enhancing the contrast information between the objects and their corresponding backgrounds.

In summary, our contributions in this paper are as follows:

- We propose a high-resolution feature representation driven network for infrared small-dim object detection (HRFRD-Net), which enables to construct high-resolution and high-quality representation of infrared small objects via implicit neural representation in a self-supervised learning scheme.
- To address the issues of object dimness and vulnerability to background interference, we introduce a feature enhancement module for infrared small-dim object detection. This module integrates global and local mixed receptive fields through spatial-frequency dual domain interaction, leading to significant improvements in detection accuracy.
- Extensive experiments conducted on two infrared datasets demonstrate the superiority of our proposed method over state-of-the-art detection methods.

2 Methods

The overall framework of our proposed network HRFRD-Net is presented in Fig. 1, which comprises three key components: High-Resolution Feature Representation Branch (HRFR), Infrared Small-Dim Object Detection Branch

(ISDOD), and Spatial-Frequency Interaction Feature Enhancement Module (SFIFE). Our training strategy involves two phases. In the first phase, we train the HRFR branch via self-supervised learning to acquire high-resolution feature representations for infrared small-dim objects. In the second phase, we freeze the HRFR encoder and share it with the ISDOD branch. By leveraging the high-resolution feature representation and incorporating the SFIFE feature enhancement module, the accuracy of infrared object detection can be significantly improved.

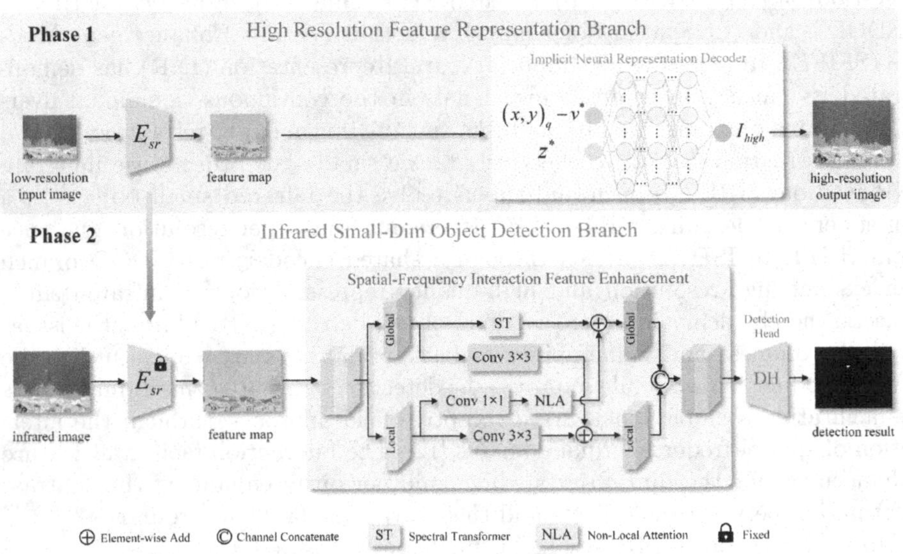

Fig. 1. The overall framework of our proposed network.

2.1 High-Resolution Feature Representation (HRFR)

Infrared objects are generally small in size and lack prominent structural information, posing challenges in objects detection. To overcome this limitation, we introduce implicit neural representations into the HRFR branch during the initial training phase to achieve high-resolution feature representation of infrared small objects. We generate pairs of infrared samples by applying four times downsampling on the original infrared image using bicubic interpolation, resulting in the creation of low-resolution and high-resolution pairs. Then, we train the HRFR branch in a self-supervised learning approach. The super-resolution encoder E_{sr} is utilized to extract the deep features of the low-resolution infrared image I_{low} and obtain the latent codes z^*. With the arbitrary query coordinates $(x,y)_q$ in the continuous domain, the nearest latent codes z^* from $(x,y)_q$, and the distance between $(x,y)_q$ and coordinates v^* of the latent codes z^* are fed into the implicit neural representation MLP decoder D_θ together to generate the high-resolution result I_{high}. This approach enables the representation of small objects in a high-resolution and high-quality manner without being restricted by grid resolution [13]. The formulas are expressed as below

$$
\begin{aligned}
z^* &= E_{sr}(I_{low}), \\
I_{high} &= \mathcal{D}_\theta\left(z^*, (x,y)_q - v^*\right).
\end{aligned} \tag{1}
$$

The high-resolution result I_{high} is represented on a 2D continuous feature map $M \in R^{H \times W \times C}$. To enrich the information on feature map M, we employ feature unfolding to expand it

$$
M_{unfold} = \text{Concat}\left(\{M_{l,m}\}_{l,m \in \{-1,0,1\}}\right), \tag{2}
$$

where each latent code in M_{fold} is obtained by the concatenation of the 3×3 neighboring latent codes $M_{l,m}$.

2.2 Infrared Small-Dim Object Detection (ISDOD)

In the second training phase, we freeze the super-resolution encoder E_{sr} in HRFR branch and share it with the ISDOD branch, which enables the ISDOD branch to extract high-resolution and high-quality representations of infrared small objects

$$
M^* = E_{sr}(I), \tag{3}
$$

where I denotes the input infrared image, E_{sr} represents the super-resolution encoder in the HRFR branch, which is shared with the ISDOD branch. This sharing allows for the generation of high-resolution feature representation M^*. To further improve the accuracy of objects detection, we introduce SFIFE module into the ISDOD branch. The details will be illustrated as follows.

2.3 Spatial-Frequency Interaction Feature Enhancement (SFIFE)

While the high-resolution feature representation can mitigate small objects issue and improve the detection accuracy, the inherent weaknesses and susceptibility to background interference still hinder the performance of object detection. To solve these issues, we introduce SFIFE module to enhance the infrared dim object representation. As well known, the spectral convolution theorem in Fourier theory demonstrates that modifying a point in the spectral domain exerts a global influence on all the input features [3]. Inspired by this, we employ a SFIFE module based on fast Fourier convolution which incorporates an image-wide receptive field and integrates global and local mixed receptive fields through spatial-frequency dual domain interaction [12]. This interaction can not only facilitate the feature enhancement for the dim objects but also simultaneously enhance the contrast information between the objects and their corresponding

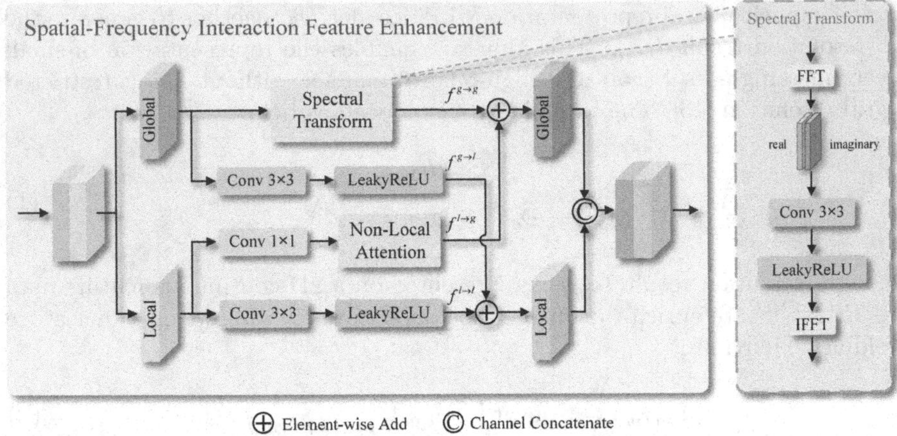

Fig. 2. Spatial-Frequency Interaction Feature Enhancement Module.

background. We first split the high-resolution feature representation M^* into the global M_g and local M_l parts along the feature channel dimension

$$M_g, M_l = \text{split}(M^*). \tag{4}$$

As shown in Fig. 2, the feature enhancement module consists of four branches: local-to-local, local-to-global, global-to-local and global-to-global to achieve the mixed receptive fields through the interaction on spatial-frequency dual domain. Both local-to-local ($f^{l\to l}$) and global-to-local branches ($f^{g\to l}$) capture the local features via a 3×3 convolution, the local-to-global branch ($f^{l\to g}$) leverages a non-local attention mechanism to explore the global dependency for each query pixel with its surrounding parts, and the global-to-global branch ($f^{g\to g}$) utilizes Fourier transform to widen the receptive field and capture long-range context. The procedures can be described as follows

$$\begin{aligned} f^{g\to g}(M_g) &= \mathcal{ST}(M_g), \\ f^{g\to l}(M_g) &= \text{Conv}_{3\times 3}^{g\to l}(M_g), \\ f^{l\to g}(M_l) &= \mathcal{NL}\left(\text{Conv}_{1\times 1}^{l\to g}(M_l)\right), \\ f^{l\to l}(M_l) &= \text{Conv}_{3\times 3}^{l\to l}(M_l), \end{aligned} \tag{5}$$

where $\mathcal{NL}$ denotes non-local attention mechanism, $\mathcal{ST}$ is the spectral transformation which utilize Fourier transform to enlarge the receptive field to the full resolution of input feature map in an efficient way. We first adopt 2D Fast Fourier Transform (FFT) to transform the spatial features into frequency domain with both real and imaginary parts of the signal. Subsequently, a 3×3 convolution and leakyrelu [19] activation operation $\mathcal{O}$ is applied before converting it back to the spatial domain via the inverse Fourier transform

$$\begin{aligned} R\left(M_g\right), I\left(M_g\right) &= \mathcal{F}\left(M_g\right), \\ R\left(M_g\right) &= \mathcal{O}\left(R\left(M_g\right)\right), \\ I\left(M_g\right) &= \mathcal{O}\left(I\left(M_g\right)\right), \\ M_{st} &= \mathcal{F}^{-1}(R(M_g), I(M_g)), \end{aligned} \tag{6}$$

where $\mathcal{F}$ and $\mathcal{F}^{-1}$ denotes the Fourier transformation and inverse Fourier transformation respectively. R and I means the real and imaginary components of the signal. By summing up the global-to-global and local-to-global branches, we obtain the global feature, and in the same way we obtain the local features from the other two branches

$$\begin{aligned} F_g &= f^{g\rightarrow g}\left(M_g\right) + f^{l\rightarrow g}\left(M_l\right) = M_{st} + f^{l\rightarrow g}\left(M_l\right), \\ F_l &= f^{g\rightarrow l}\left(M_g\right) + f^{l\rightarrow l}\left(M_l\right), \end{aligned} \tag{7}$$

where F_g and F_l denote global and local features respectively. Afterwards, they are concatenated to generate the enhanced feature F

$$F = \text{Cat}\left(F_g, F_l\right). \tag{8}$$

After that, the enhanced feature F is passed through the detection head DH to generate the final detection result R

$$R = DH(F). \tag{9}$$

2.4 Loss Function

There are three loss functions in the proposed approach.

Phase 1. During the first phase, we train the HRFR branch to super-resolve the infrared small objects via self-supervised learning.

Super-Resolution Loss. We adopt L1 loss, which measures the average absolute difference between the object y and predicted values $f(x)$, where n denotes the total pixels. The loss is computed as below

$$\mathcal{L}_{sr} = \frac{\sum_{i=1}^{n}|f(x) - y|}{n}. \tag{10}$$

Phase 2. During the second phase, we share HRFR branch fixed encoder to the ISDOD branch to obtain the high-resolution feature representation. Another two loss functions are utilized in this phase to ensure the accurate detection.

BCE Loss. The detection loss is calculated using Binary CrossEntropy (BCE) due to its sharper performance and faster convergence compared to the blurry results obtained from Mean Squared Error. The loss is defined as follows

$$\mathcal{L}_{bce} = -(ylog(p(x)) + (1-y)log(1-p(x))). \tag{11}$$

Dice Loss. Since the BCE Loss does not take into account the similarity between the predicted image and the ground truth, we introduce a similarity loss using the Dice coefficient. A higher Dice coefficient indicates a higher level of similarity. The loss is defined as follows

$$\mathcal{L}_{dice} = 1 - \frac{2 \cdot \sum_{i=1}^{N} y_i \hat{y}_i}{\sum_{i=1}^{N} y_i + \sum_{i=1}^{N} \hat{y}_i}. \tag{12}$$

Total Loss. The overall loss in phase 2 is computed by

$$\mathcal{L}_{detect} = \lambda_{bce}\mathcal{L}_{bce} + \lambda_{dice}\mathcal{L}_{dice}, \tag{13}$$

where λ_{bce} and λ_{dice} are set as 0.2 and 0.8 respectively.

3 Experiments

3.1 Baseline Methods

To evaluate the effectiveness of our method, we compare it with three traditional methods: ADMD [10], MPCM [6], RLCM [6], and seven deep learning-based methods, including MDvsFA-cGAN [15], YOLOv5 [22], DNANet [9], ACM [4], ALCNet [5], AGPCNet [20] and UIUNet [18].

3.2 Datasets and Implementation Details

Datasets. We choose MSISTD [14] (multiscene infrared small object dataset) and MDvsFA [15] (miss detection vs. false alarm) as our experimental datasets. MSISTD consists of 1077 images with 1343 instances, while MDvsFA contains 10000 training images and 100 test images.

Implementation Details. We implement our network on a PC equipped with a single NVIDIA TITAN GPU using the PyTorch framework. Our training strategy can be divided into two phases. In the first phase, the HRFR branch is trained to obtain the high-resolution feature representation of infrared small objects, which utilizes the $\mathcal{L}_{sr}$ as loss function. In the second phase, the encoder E_{sr} in HRFR branch encoder is frozen and shared to the ISDOD branch. The ISDOD branch is responsible for accurate detection of infrared small-dim objects, which adopts $\mathcal{L}_{bce}$ and $\mathcal{L}_{dice}$ as the loss functions. We try several experimental combinations and finally set the $\mathcal{L}_{bce}$=0.2 and $\mathcal{L}_{dice}$=0.8 empirically. During training, we optimize the model parameters using the Adam optimizer for 300 epochs. The initial learning rate is set to 1×10^{-4}, and the batch size is set to 10. We apply a learning rate decay by multiplying 0.5 when reaching 50 epochs.

Metrics. To evaluate the performance, we adopt the following evaluation metrics: IOU, nIOU [4], F1-score and AUC. IOU serves as a crucial indicator for object detection, while the normalized Intersection over Union (nIOU) play an complementary role. F1-score provides a comprehensive measurement by combining precision and recall. Moreover, AUC serves as a quantitative metric for evaluating the ROC curve by measuring the area under the ROC curve.

3.3 Comparison with State-of-the-art Methods

Quantitative Comparison. Table 1 compares the performance of our method with several methods on MSISTD and MDvsFA datasets. The optimal results for each respective column are highlighted in red. Our method achieves the best results across almost all metrics on both datastes, with the exception of the F1-score which is only 0.01 lower than the ACM [4] method on the MSISTD dataset. Moreover, our approach demonstrates significant improvements in IOU and nIOU, indicating its outstanding advantages in predicting object shape integrity and edge details. Furthermore, our method outperforms other approaches in AUC, suggesting its general validity across various cases. The HRFR branch (phase 1) comprises 1.6M parameters, while the ISDOD branch (phase 2) contains 0.5M parameters. The inference speed is 0.31 s per image.

Visual Comparison. We provide detection visualization results in Fig. 3, Fig. 4 and Fig. 5. These results depict three typical scenarios: small objects with ambiguous features in the presence of significant background interference, the

Table 1. Quantitative comparison of reference metrics on MSISTD and MDvsFA datasets. Best results are highlighted by red. ↑ indicates that the larger the value, the better the performance, and ↓ indicates that the smaller the value, the better the performance.

Method	MSISTD				MDvsFA			
	IOU↑	nIOU↑	F1-score↑	AUC↑	IOU↑	nIOU↑	F1-score↑	AUC↑
ADMD	0.133	0.133	0.108	0.681	0.116	0.116	0.238	0.740
MPCM	0.174	0.174	0.141	0.725	0.219	0.219	0.303	0.780
RLCM	0.262	0.262	0.216	0.757	0.214	0.214	0.160	0.808
MDvsFA-cGAN	0.397	0.490	0.551	0.993	0.353	0.373	0.523	0.921
YOLOv5	0.424	0.424	0.698	0.901	0.377	0.377	0.470	0.924
AGPCNet	0.642	0.646	0.819	0.956	0.423	0.434	0.578	0.876
ALCNet	0.646	0.646	0.801	0.920	0.362	0.363	0.545	0.903
ACM	0.637	0.638	0.828	0.988	0.426	0.427	0.582	0.890
DNANet	0.619	0.624	0.799	0.923	0.393	0.390	0.576	0.756
UIUNet	0.636	0.635	0.733	0.928	0.314	0.322	0.415	0.919
HRFRD-Net(Ours)	0.668	0.693	0.812	0.995	0.427	0.440	0.593	0.941

scenario where numerous objects in the background possess similar characteristics to the small objects, and the scenario which requires not only simple object detection but also precise segmentation of the small object's edges. The results show that our method outperforms others in scenes with blurred features and heavy background interference. For instance, as shown in Fig. 3, the object is completely hidden under the cloud layer, which poses a great challenge for feature extraction. Only our method can successfully detect all the objects, which proves the effectiveness of our approach with strong anti-interference capability for small-dim objects detection.

3.4 Ablation Study

To evaluate the effectiveness of each component, we conduct ablation studies on both MSISTD and MDvsFA dataset. Initially, we remove the SFIFE module to assess its impact. Subsequently, we eliminate both the HRFR branch and the shared encoder from it in the ISDOD branch. The performance comparison is presented in Table 2.

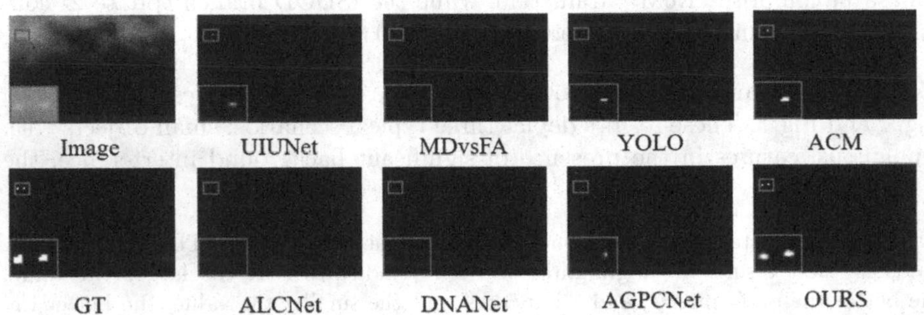

Fig. 3. Small objects with ambiguous features in the presence of significant background interference on MSISTD dataset, red boxes indicate correct detection. (Color figure online)

Table 2. Ablation study about two key components on MSISTD and MDvsFA dataset.

Config	HRFR	SFTFE	MSISTD				MDvsFA			
			IOU↑	nIOU↑	F1-score↑	AUC↑	IOU↑	nIOU↑	F1-score↑	AUC↑
(I)	✓		0.627	0.651	0.745	0.949	0.397	0.414	0.579	0.939
(II)		✓	0.625	0.653	0.792	0.989	0.374	0.403	0.539	0.900
Ours	✓	✓	0.668	0.693	0.812	0.995	0.427	0.440	0.593	0.941

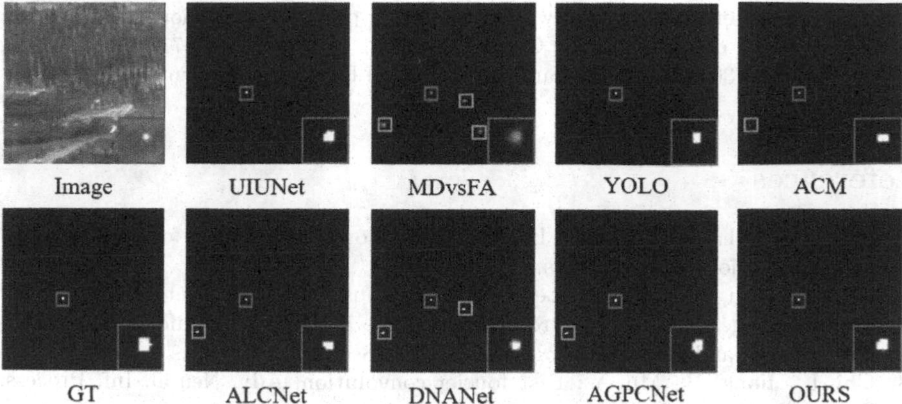

Fig. 4. Numerous objects in the background possess similar characteristics to the small objects on MSISTD dataset, red boxes indicate correct detection, green boxes indicate incorrect detection. (Color figure online)

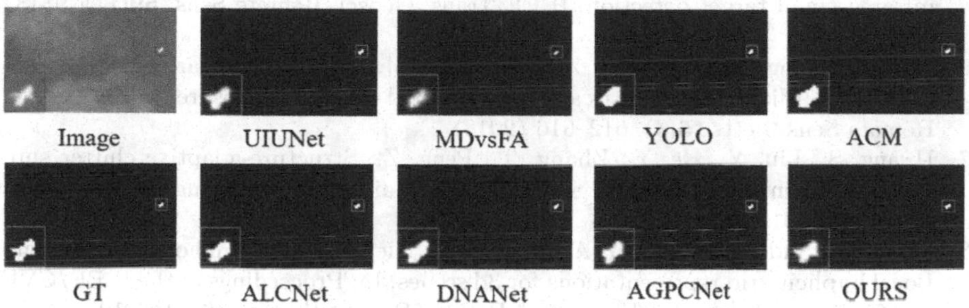

Fig. 5. The scenario which requires not only simple object detection but also precise segmentation of the infrared small-dim object's edges on MSISTD dataset, red boxes (Color figure online) indicate correct detection.

4 Conclusion

In this paper, we propose a novel high-resolution feature representation driven network for infrared small-dim object detection (HRFRD-Net). Our framework is able to preserve high-resolution and high-quality representations of the infrared small objects in a resolution-free manner, thereby achieving accurate objects detection. Additionally, the SFIFE module integrates global and local mixed receptive fields, enhancing the representation of dim objects and improving the contrast between objects and their background. Extensive experiments conducted on the MSISTD and MDvsFA datasets demonstrate the effectiveness of our approach, especially in challenging scenarios where objects are heavily obscured by the background, numerous background objects share similar characteristics with small objects, and the need for both accurate object detection and precise segmentation.

Acknowledgements. The work was supported in part by the National Natural Science Foundation of China under Grant 82172033, U19B2031, 61971369, 52105126, 82272071, 62271430, and the Fundamental Research Funds for the Central Universities 20720230104.

References

1. Bae, T.W., Zhang, F., Kweon, I.S.: Edge directional 2d lms filter for infrared small target detection. Infrared Phys. Technol. **55**(1), 137–145 (2012)
2. Chen, Y., Liu, S., Wang, X.: Learning continuous image representation with local implicit image function. In: Proceedings of the IEEE/CVF Conference on Computer Vision and Pattern Recognition, pp. 8628–8638 (2021)
3. Chi, L., Jiang, B., Mu, Y.: Fast fourier convolution. Adv. Neural. Inf. Process. Syst. **33**, 4479–4488 (2020)
4. Dai, Y., Wu, Y., Zhou, F., Barnard, K.: Asymmetric contextual modulation for infrared small target detection. In: Proceedings of the IEEE/CVF Winter Conference on Applications of Computer Vision, pp. 950–959 (2021)
5. Dai, Y., Wu, Y., Zhou, F., Barnard, K.: Attentional local contrast networks for infrared small target detection. IEEE Trans. Geosci. Remote Sens. **59**(11), 9813–9824 (2021)
6. Han, J., Liang, K., Zhou, B., Zhu, X., Zhao, J., Zhao, L.: Infrared small target detection utilizing the multiscale relative local contrast measure. IEEE Geosci. Remote Sens. Lett. **15**(4), 612–616 (2018)
7. Huang, S., Liu, Y., He, Y., Zhang, T., Peng, Z.: Structure-adaptive clutter suppression for infrared small target detection: chain-growth filtering. Remote Sens. **12**(1), 47 (2019)
8. Jiang, C., Sud, A., Makadia, A., Huang, J., Nießner, M., Funkhouser, T., et al.: Local implicit grid representations for 3d scenes. In: Proceedings of the IEEE/CVF Conference on Computer Vision and Pattern Recognition, pp. 6001–6010 (2020)
9. Li, B., et al.: Dense nested attention network for infrared small target detection. IEEE Trans. Image Process. (2022)
10. Moradi, S., Moallem, P., Sabahi, M.F.: Fast and robust small infrared target detection using absolute directional mean difference algorithm. Sig. Process. **177**, 107727 (2020)
11. Park, J.J., Florence, P., Straub, J., Newcombe, R., Lovegrove, S.: Deepsdf: learning continuous signed distance functions for shape representation. In: Proceedings of the IEEE/CVF Conference on Computer Vision and Pattern Recognition, pp. 165–174 (2019)
12. Sinha, A.K., Moorthi, S.M., Dhar, D.: Nl-ffc: non-local fast fourier convolution for image super resolution. In: Proceedings of the IEEE/CVF Conference on Computer Vision and Pattern Recognition, pp. 467–476 (2022)
13. Sitzmann, V., Martel, J., Bergman, A., Lindell, D., Wetzstein, G.: Implicit neural representations with periodic activation functions. Adv. Neural. Inf. Process. Syst. **33**, 7462–7473 (2020)
14. Wang, A., Li, W., Huang, Z., Wu, X., Jie, F., Tao, R.: Prior-guided data augmentation for infrared small target detection. IEEE J. Sel. Top. Appl. Earth Observations Remote Sens. **15**, 10027–10040 (2022)
15. Wang, H., Zhou, L., Wang, L.: Miss detection vs. false alarm: adversarial learning for small object segmentation in infrared images. In: Proceedings of the IEEE/CVF International Conference on Computer Vision, pp. 8509–8518 (2019)

16. Wang, X., Peng, Z., Zhang, P., He, Y.: Infrared small target detection via nonnegativity-constrained variational mode decomposition. IEEE Geosci. Remote Sens. Lett. **14**(10), 1700–1704 (2017)
17. Wei, Y., You, X., Li, H.: Multiscale patch-based contrast measure for small infrared target detection. Pattern Recogn. **58**, 216–226 (2016)
18. Wu, X., Hong, D., Chanussot, J.: Uiu-net: U-net in u-net for infrared small object detection. IEEE Trans. Image Process. **32**, 364–376 (2022)
19. Xu, J., Li, Z., Du, B., Zhang, M., Liu, J.: Reluplex made more practical: Leaky relu. In: 2020 IEEE Symposium on Computers and communications (ISCC), pp. 1–7. IEEE (2020)
20. Zhang, T., Cao, S., Pu, T., Peng, Z.: Agpcnet: attention-guided pyramid context networks for infrared small target detection. arXiv preprint arXiv:2111.03580 (2021)
21. Zhu, H., Ni, H., Liu, S., Xu, G., Deng, L.: Tnlrs: target-aware non-local low-rank modeling with saliency filtering regularization for infrared small target detection. IEEE Trans. Image Process. **29**, 9546–9558 (2020)
22. Zhu, X., Lyu, S., Wang, X., Zhao, Q.: Tph-yolov5: Improved yolov5 based on transformer prediction head for object detection on drone-captured scenarios. In: Proceedings of the IEEE/CVF International Conference on Computer Vision, pp. 2778–2788 (2021)

Few-Shot Object Detection Algorithm Based on Adaptive Relation Distillation

Danting Duan[1], Wei Zhong[2], Liang Peng[1], Shuang Ran[1], and Fei Hu[2](✉)

[1] Key Laboratory of Media Audio and Video (Communication University of China), Ministry of Education, Beijing 100024, China
{dantingduan,rans}@cuc.edu.cn

[2] State Key Laboratory of Media Convergence and Communication, Communication University of China, Beijing 100024, China
{wzhong,hufei}@cuc.edu.cn

Abstract. Deep learning methods have advanced the accuracy and speed of object detection. However, acquiring labeled data is especially challenging in real-world scenarios. As a result, the general-purpose object detection algorithms based on deep learning experience significant performance degradation with limited samples. To tackle the issue of declining detection accuracy in the presence of few-shot data, this paper introduces a few-shot object detection algorithm based on adaptive relation distillation, which improves upon existing algorithms by enhancing the fusion of query and support features. In the proposed method, the adaptive relational distillation module discards the hand-designed and inefficient query information utilization strategy employed in previous algorithms and adaptively fuses the features of support and query images using convolutional networks. To augment the learning capability of the adaptive relational distillation module, we utilize a hybrid attention module in the support branch to emphasize the regions crucial for detecting specific classes of objects. The experimental results demonstrate our proposed algorithm achieves an average accuracy of 47.6% on three data divisions and five sample size settings for the PASCAL VOC dataset, which marks an improvement of 8.3% over DCNet.

Keywords: Object detection · Few-shot learning · Meta learning · Attention mechanism

1 Introduction

Object detection is one of the most challenging fundamental tasks in the field of computer vision. It consists of two subtasks, i.e., classification and localization.

This work is supported by National Key R&D Program of China under Grant No. 2022YFF0902401, the National Natural Science Foundation of China under Grant 62271455 and the Fundamental Research Funds for the Central Universities under Grant Nos. CUC210C013 and CUC18LG024.

© The Author(s), under exclusive license to Springer Nature Singapore Pte Ltd. 2024
Q. Liu et al. (Eds.): PRCV 2023, LNCS 14436, pp. 328–339, 2024.
https://doi.org/10.1007/978-981-99-8555-5_26

Specifically, classification refers to the identification of target categories in an image, while localization is the accurate search of the location of the target category. Currently, object detection has become a research hotspot in security, military, transportation and medical areas [1–3].

In recent years, artificial intelligence technology has developed rapidly, but it is undeniable that today's artificial intelligence is far from the true human intelligence. Indeed, the success of artificial intelligence relies on the creation of large-scale data sets. However, humans need to see an object only once or very few times to complete the perception of an object, even for a child. Therefore, the few-shot learning problem is likewise an important bottleneck limiting the development of artificial intelligence. Most of the current few-shot learning methods target few-shot image classification tasks. In the few-shot image classification task, a network model is trained on the base class samples containing enough training data for good generalization ability and thus perform well on a new class with only few samples. To train such a network model, the meta-learning methods are widely used [4–6]. The meta-learning based few-shot objective detection network has two inputs, the query image and the support set. The support set needs to be referenced when detecting objects in the query image. Since meta-learning was proposed relatively late, there is a great potential for improvement of few-shot object detection methods based on meta-learning. In the existing work, there are relatively few studies on the relationship between support features and query features. Researchers usually directly perform global pooling operations on the support features to adjust the query branches. For example, Meta YOLO [7] uses a reweighting module to transform support images into weighted vectors that perform channel weighting operations on meta-features of query images. Meta R-CNN [8] transforms support images into attention-like vectors in a prediction head reshaping network that performs channel soft attention operations on RoI features. However, the global pooling operation used in the above work tends to lose the local detail information in the support set.

As a matter of fact, the local detail information is extremely important for solving the occlusion problem of object detection. The variations in the appearance of objects lead to misclassification between similar categories, while the object occlusion brings incomplete feature representation, leading to misclassification and missing detection. Without sufficient discriminative information, the model will not be able to learn the key features for classification and bounding box prediction. This problem is even more prominent in few-shot object detection where training examples are extremely sparse. At this point, using the local detail information in the support set becomes the key to detection accuracy improvement.

To make better use of the detailed information from the support set, Fan et al. [9] used a variety of relationship headers to model different relationships between the query image and the support image. Among them, the global relation head uses a global representation to match images, the local relation head captures pixel-level matching relations, and the block relation head models one-to-many pixel relations. In contrast, a dense relational distillation module is used in DCNet [10] to match query and support features at the pixel level, which effectively improves the accuracy of few-shot detection.

To this end, we introduce the adaptive relation distillation module (ARDM) to make full use of the support set, not only limited to the global information or local detail information of the support set. We implement the ARDM by mixed attention and apply it to few-shot object detection, and propose the adaptive relation distillation based detection network (ARDNet). In the ARDNet, given a query image and a few new classes of support images, the query features and support features are first extracted with a shared feature learner, and then fed into the ARDM for fusion of the two sets of features. Finally, the fused features are fed into the subsequent region proposal generation and detection process.

2 Related Work

Meta-learning addresses the shortcomings of traditional neural network models with insufficient generalization performance and poor adaptability to new kinds of tasks. Since the concept of meta-learning was introduced by Schmidhuber [11] and Hinton [12] in 1987, the meta-learning has become a common learning strategy in few-shot learning. The algorithms of meta-learning have also achieved great success in the field of few-shot object detection because of their powerful generalization ability. Since most current meta-learning methods for detection are conditioned on a set of supported examples, they can also be considered as example-based visual search.

Recently, Fu et al. [13] proposed Meta SSD, which attaches a meta-learner to the SSD and uses the meta-learner to optimize the initial values of network parameters and the learning rate, while the detector updates its weights under the guidance of the meta-learner. Kang et al. [7] proposed Meta YOLO, which uses a meta-learning approach to learn the meta-features of the base class and introduces a channel attention approach to reweight the importance of features to fit the new class target. Wang et al. [14] proposed MetaDet, which uses meta-level knowledge generated about the model parameters of a new class-specific component. Karlinsky et al. [15] proposed RepMet, which uses a classifier head based on distance metric learning to replace the RoI classifier head in a two-stage detection algorithm to extract representative embedding vectors by clustering and calculate the distance between the query and the supported instances for classification. Yan et al. [8] proposed Meta R-CNN, which reweighted RoI features based on the Faster R-CNN approach. Yang et al. [16] proposed NP-RepMet, which divides the class representation in RepMet into two modules to learn negative and positive representations, and replaces the embedding vector given a suggestion with a new positive and negative embedding. Fan et al. [9] proposed FSOD, which uses support information in attention RPN to filter out most background boxes and mismatched categories in background boxes. Hu et al. [10] proposed DCNet, containing a dense relation extraction module, which matches queries and support features at the pixel level, and a context-aware aggregation module, which uses three different pooling resolutions to capture richer contextual features. Li et al. [17] proposed CME to transform the small-sample detection problem into a small-sample classification problem using a fully connected layer, while introducing class margin loss to interfere with the features of new class instances in an adversarial min-max manner, thus achieving margin equalization.

In the meta-learning approach, how to use the support set to adjust the features of the query image is the key to improve the detection accuracy. Although the above works improves the accuracy of few-shot object detection, the relationship between support sets and query images has not been fully investigated. Neither the global features nor the detailed features of the support set can be used to fully exploit the query set. Therefore, in this paper, the query set is utilized adaptively with the help of deep networks to eliminate the hand-designed way of utilizing query features, and satisfactory accuracy is achieved.

3 Our Approach

To extract enough information from the support features, we propose the ARDM using a hybrid attention mechanism, whose structure is shown in Fig. 1. In the query branch of ARDM, a query encoder is used to halve its channel count. At the same time, in the support branch of ARDM, we first perform the averaging operation on the class dimension to change the support features of $N \times C \times H \times W$ into $1 \times C \times H \times W$, where N is the number of classes. The support features are then fed into a hybrid attention module that contains both channel and spatial attention. Later, the support encoder is used for dimensionality reduction. The support features are bilinearly interpolated and deflated so that their width and height are the same as the query features. Finally, the channel halved query features and adjusted support features are stitched together in the channel dimension. In this case, both the query encoder and the support encoder are implemented by a single 3×3 convolutional layer. The operation of the adaptive relational distillation module ensures that its output size is consistent with the size of the query features.

In order to enhance the learning ability of the ARDM, the convolutional block attention module (CBAM) [18] is used in the supporting branch. The CBAM is a simple and effective attention module for feedforward convolutional neural networks, whose structure is shown in Fig. 2. Given an intermediate feature map, the CBAM module sequentially infers the attention map along two independent dimensions (channel and space), and then multiplies the attention map with the input feature map for adaptive feature optimization. Since CBAM is an attention mechanism module that combines space and channel, it can achieve better results than attention mechanisms that focus only on channel or space.

We apply the ARDM to the few-shot object detection network DCNet to obtain the ARDNet proposed in this paper, whose structure is shown in Fig. 3. In the ARDNet network, the inputs are query images and support images, and the query features and support features are obtained separately using a shared feature extractor and fed into the adaptive relational distillation module. The

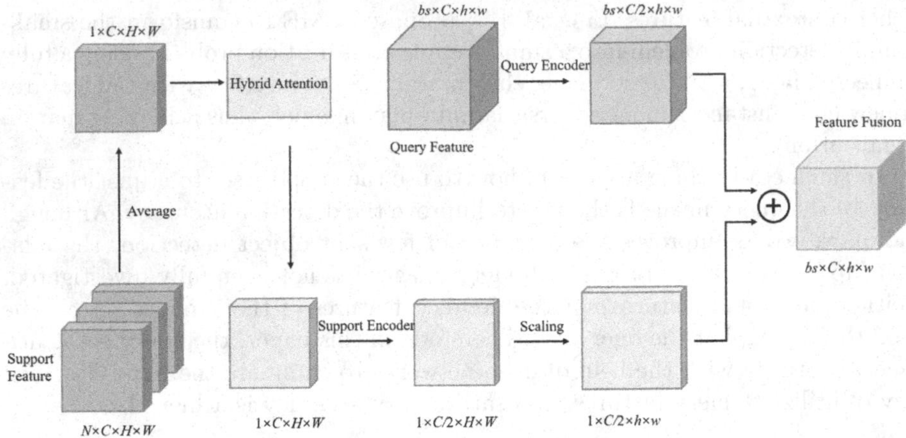

Fig. 1. Adaptive relation distillation module.

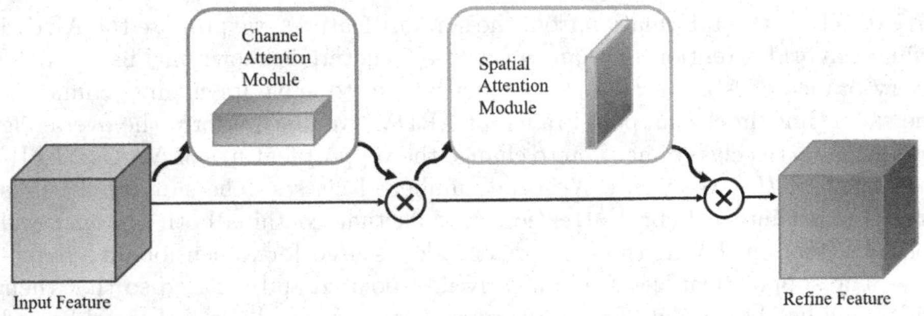

Fig. 2. The convolutional block attention module.

fused features containing both query information and rich support information are obtained by the adaptive relational distillation module. After that, the fused features are fed into the RPN network to generate candidate region proposals. RoI Align uses the region proposals, fused features, and three pooling resolutions to obtain RoI features, which are fed into the final classification and localization module.

4 Experiment

4.1 Dataset

We evaluate our model on the PASCAL VOC dataset [19], which is a common dataset for few-shot object detection. The VOC dataset provides the annotation information required for tasks such as image classification, object detection, and semantic segmentation. The dataset contains 20 class targets and 1 background class. Specifically, the training set contains 10582 images, the validation

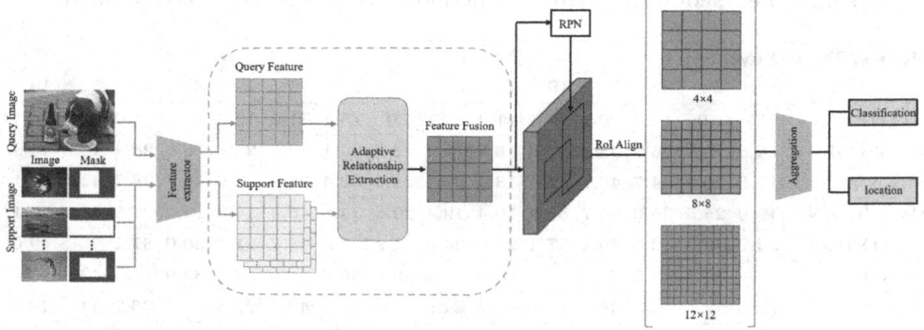

Fig. 3. The overall framework of our proposed ARDNet.

set contains 1449 images, and the test set contains 1456 images. For researchers, the VOC 2012 and VOC 2007 are the two most popular datasets, which have a total of 4 major classes (vehicle, household, animal and person) and 20 subclasses (aeroplane (aero), bicycle, bird, boat, bottle, bus, car, cat, chair, cow, diningtable, dog, horse, motorbike (mbike), person, pottedplant, sheep, sofa, train, tvmonitor) and 1 background class.

For the few-shot object detection task, the 20 classes of the VOC dataset are randomly divided into 15 base classes and 5 new classes. The number of instances of each new class is $K = 1, 2, 3, 5, 10$. Samples are taken from a combination of the training validation sets of VOC 2007 and VOC 2012, and the VOC 2007 test set is used for evaluation. The AP50 of the new class is used for the evaluation metrics (the IoU matching threshold is set to 0.5). Three random groupings need to be considered for the evaluation algorithm, and the three novel sets in the standard configuration are (bird, bus, cow, motorbike, sofa), (aeroplane, bottle, cow, horse, sofa), (boat, cat, motorbike, sheep, sofa).

4.2 Implementation Details

In the meta-learning based K-sample learning task, each training event consists of sampling: 1) a support set $S = \{x_i, y_i\}_{i=1}^{N}$ containing image mask pairs of different classes, where $x_i \in R^{h \times w \times 3}$ is the RGB image, $y_i \in R^{h \times w}$ is the binary mask of the object of class i in the support image generated by the bounding box annotation, N is the number of object classes in the training set; 2) the query image q and the annotations of the training classes used in the query image m. The inputs to the model are the support set and the query image, and the outputs are the detection predictions of the query image.

Both training and testing are performed on the fixed-scale images. The shorter edges of the query images are scaled to 800 pixels and the longer edges are scaled to less than 1333 pixels while maintaining the aspect ratio. The support images are adjusted to the square images of 256 × 256 pixels. We use ResNet-101 as the feature extractor and RoI Align as the RoI feature extractor. The weights

Table 1. Few-shot object detection performance on PASCAL VOC dataset.

Models/Shots	Novel Set 1					Novel Set 2					Novel Set 3				
	1	2	3	5	10	1	2	3	5	10	1	2	3	5	10
LSTD	8.2	1.0	12.4	29.1	38.5	11.4	3.8	5.0	15.7	31.0	12.6	8.5	15.0	27.3	36.3
Meta YOLO	14.8	15.5	26.7	33.9	47.2	15.7	15.2	22.7	30.1	40.5	21.3	25.6	28.4	42.8	45.9
FRCN	15.2	20.3	44.7	40.1	45.5	13.4	20.6	28.6	32.4	38.8	19.6	20.8	28.7	42.2	42.1
Meta RCNN	19.9	25.5	35.0	45.7	51.5	10.4	19.4	29.6	34.8	45.4	14.3	18.2	27.5	41.2	48.1
FsDetView	24.2	35.3	42.2	49.1	57.4	21.6	24.6	31.9	37.0	45.7	21.2	30.0	37.2	**43.8**	49.6
DCNet	**33.9**	37.4	43.7	*51.1*	*59.6*	*23.2*	*24.8*	30.6	36.7	*46.6*	**32.3**	34.9	*39.7*	42.6	**50.7**
DCNet*	22.6	*42.3*	*45.9*	48.3	55.7	15.4	24.5	*37.6*	*37.8*	42.4	22.3	*35.2*	36.7	43.4	49.1
ARDNet(ours)	*33.4*	**50.5**	**57.2**	**59.0**	**65.4**	**28.0**	**47.6**	**53.9**	**56.4**	**59.4**	*27.9*	**39.6**	**41.7**	*43.7*	*49.8*

of the backbone are pre-trained on ImageNet. After the model is trained on the base class, only the last fully connected layer used for classification is removed and replaced with a new layer initialized randomly. All parts of the model are involved in the learning process of the second meta-refinement phase without any freezing operation. The batch size during training is 8, using a Tesla M40 24GB GPU. We use an SGD optimizer with a momentum of 0.9 and a weight decay of 0.0001. For meta-training of PASCAL VOC, the model is trained for 12k, 4k and 2k iterations with learning rates of 0.01, 0.001 and 0.0001, respectively. For meta fine tuning of PASCAL VOC, the batch size is 4, and the model is trained for 1300, 400 and 300 iterations with learning rates of 0.005, 0.0005 and 0.00005.

4.3 Comparisons with State-of-the-Art Methods

In Table 1, we compare the method proposed in this paper with previous state-of-the-art methods, where the red bold represents the best results and the blue bold italic represents the suboptimal results. ARDNet represents the small-sample target detection algorithm based on adaptive relational distillation proposed in this paper. DCNet* is the result obtained by running the baseline model code directly, and DCNet represents the result given in the original paper. It can be seen from Table 1 that ARDNet achieves state-of-the-art results on 11 of the 15 data divisions and sample size settings, achieves suboptimal results on 4 data, and largely outperforms the baseline model and other previous methods. Specifically, the detection accuracy of the proposed method is 13.1% higher than DCNet in the 2-sample setting of the novel set 1. This provides a convincing evidence that our ARDNet is able to capture the local details of the support set to overcome the challenges posed by the large differences between the test and training samples.

Table 2. New class detection accuracy comparison.

Shots	Models	Novel Set 1					Novel Set 2					Novel Set 3				
		bird	bus	cow	mbike	sofa	aero	bottle	cow	horse	sofa	boat	cat	mbike	sheep	sofa
1	DCNet*	12.6	43.1	10.3	44.7	2.3	40.0	3.0	13.7	5.3	15.2	5.5	42.0	42.6	7.1	14.2
	ARDNet	17.4	69.7	26.1	45.4	8.4	46.1	25.3	23.6	27.6	17.0	10.5	50.3	50.1	16.7	11.8
2	DCNet*	26.9	60.7	20.0	54.5	49.3	47.7	9.1	11.8	20.6	33.2	14.8	40.2	59.7	27.5	33.6
	ARDNet	36.0	78.4	43.9	57.3	36.9	53.2	25.6	45.4	70.8	43.1	15.4	48.5	54.7	42.7	38.8
3	DCNet*	35.0	61.1	34.8	51.1	47.0	52.2	9.1	34.4	49.9	42.6	18.6	48.7	49.9	28.5	37.9
	ARDNet	51.0	77.9	55.0	57.3	44.9	59.7	43.2	40.7	77.2	48.9	19.1	55.3	48.2	37.1	48.6
5	DCNet*	35.2	56.7	37.4	57.3	54.8	48.1	10.5	48.6	30.4	51.5	22.6	56.1	55.1	35.6	47.7
	ARDNet	50.5	77.6	62.1	57.1	48.0	59.0	43.7	53.7	76.0	49.5	21.7	59.0	51.2	35.3	51.2
10	DCNet*	43.5	67.9	50.7	60.9	55.3	51.0	9.9	48.5	48.8	53.7	31.0	55.9	66.1	43.0	49.6
	ARDNet	63.4	83.9	66.3	62.4	51.0	62.5	47.9	58.0	75.5	53.0	27.9	39.6	41.7	43.7	49.8

Table 3. Comparison of base class detection accuracy in the basic training phase.

Models	Novel Set 1	Novel Set 2	Novel Set 3
DCNet*	82.3	81.8	82.7
ARDNet	81.0	81.1	81.6

4.4 Comparisons with DCNet

In Table 2, we compare the detection accuracy of DCNet and ARDNet on each novel set for the three PASCAL VOC data divisions, with the best results for each data division and sample size setting shown in red bold. ARDNet achieves the highest accuracy increment of 22.3% for the bottle category with novel set 2, and is only less effective than DCNet for the couch with novel set 3, which is 2.4% lower.

A comparison of the average base class detection accuracies for the basic training phase (meta-training phase) is shown in Table 3, and the best results are shown in red bold. The base training phase uses large-scale base class training data, and the base class detection accuracy of ARDNet is slightly lower than that of DCNet* at this time. This phenomenon also shows the difference between few-shot object detection and general target detection, where high detection accuracy on the base class data does not mean high detection accuracy on the new class data.

In Table 4, the average detection accuracies of the two models are compared for the new class fine-tuning phase (meta-testing phase) on all classes containing both new and base classes, with the best results shown in red bold. Our method not only achieves excellent detection accuracy on new classes, but also achieves the best results when detecting both new classes and base classes, which demonstrates the effectiveness of our method.

A comparison of the inference speed of the proposed model and DCNet is shown in Table 5. The experimental data are derived on the test set and are averaged over three times. It can be seen from Table 5 that the inferring time

Table 4. Average detection accuracy of the fine-tuning model over all classes.

Models	Novel Set 1					Novel Set 2					Novel Set 3				
	1	2	3	5	10	1	2	3	5	10	1	2	3	5	10
DCNet*	50.3	62.1	63.1	64.8	68.5	48.1	54.1	60.4	61.5	64.6	49.1	58.4	60.2	64.0	66.0
ARDNet	53.5	63.9	66.2	67.9	70.7	52.5	63.4	65.2	67.2	69.4	55.0	60.5	61.7	63.9	66.3

Table 5. Comparison of reasoning speed of DCNet and ARDNet.

Models	DCNet	ARDNet
Inferring Time	0.208	0.212
FPS	4.81	4.72

of ARDNet is comparable with that of DCNet (0.208 versus 0.212), so does the performance on the frame rate (4.81 versus 4.72).

4.5 Ablation Study

To demonstrate that the deep network has the ability to automatically utilize support information, we also design another simple version of the ARDM as shown in Fig. 4. Specifically, after obtaining the support features and query features, we first halve the number of channels of the query features directly using the query encoder. For the support features, after halving their number of channels using the support encoder, an averaging operation on the class dimension is performed to change the support features of $N \times C/2 \times H \times W$ into $1 \times C/2 \times H \times W$, where N is the number of classes. After that, the support features are bilinearly interpolated to make their width and height consistent with the query features. Finally, the query features of channel halving and the adjusted support features are stitched together in the channel dimension.

The average detection accuracies of the baseline model and the model using two adaptive relational distillation modules are compared on the three new class divisions of PASCAL VOC in Table 6. The best performance in each accuracy column are marked in red bold. ARDNet- denotes the small-sample target detection network using the simple adaptive relational distillation module. It can be seen from Table 6 that, the average accuracy of DCNet* is slightly lower than that of DCNet by 2% on AP metrics. In contrast, the accuracy of ARDNet-is approximately the same as DCNet, with only a 0.1% decrease in accuracy. This shows that the detection network can achieve good accuracy without using elaborate support feature utilization strategies. It should be noticed that, the proposed ARDNet obtains the best results on all three data divisions, with an average accuracy improvement of 8.3%. This indicates that the adaptive relational distillation module with the addition of mixed attention further reduces the learning difficulty of the support set information.

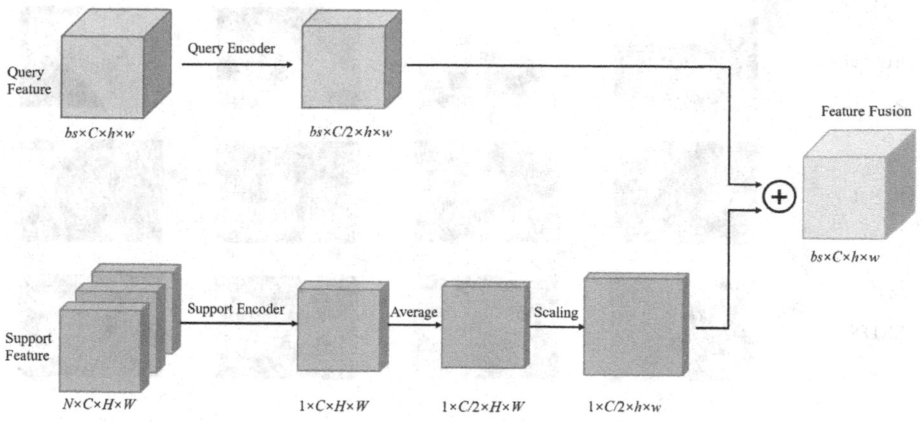

Fig. 4. Simple adaptive relation distillation module.

Table 6. The average accuracy AP and accuracy increment Δ for different models on three data divisions of PASCAL VOC.

Models	Novel Set 1		Novel Set 2		Novel Set 3		Average	
	AP	Δ	AP	Δ	AP	Δ	AP	Δ
DCNet	45.1	-	32.4	-	40.4	-	39.3	-
DCNet*	43.0	−2.1	31.5	−0.9	37.3	−3.1	37.3	−2.0
ARDNet-	46.6	+1.5	33.0	+0.6	38.0	−2.4	39.2	−0.1
ARDNet	53.1	+8.0	49.1	+16.7	40.5	+0.1	47.6	+8.3

The Fig. 5 shows the feature map visualization of the test images after going through two different relationship extraction modules in DCNet and ARDNet. It can be seen from Fig. 5 that the ARDM generates feature maps with less noise and greater activation values in the object region than the dense relational extraction module in DCNet. This indicates that our proposed ARDM is able to focus on the important object detail information in the query image guided by the support set images and is more effective in modeling the relationship between the query image and the support set.

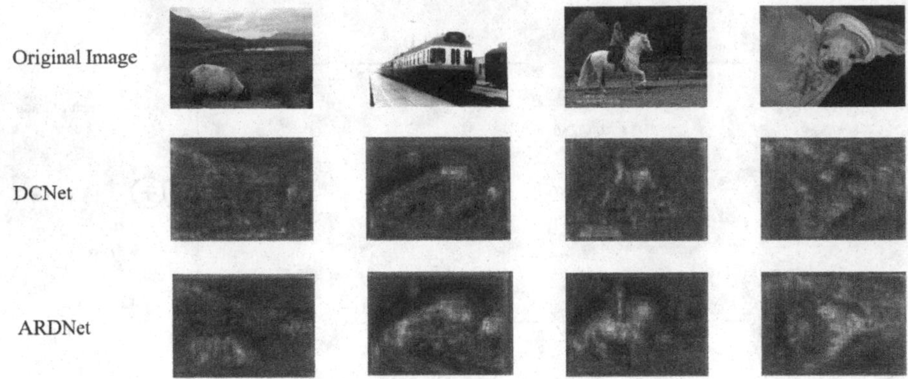

Fig. 5. The visualizations of the feature map.

5 Conclusion

In this paper, we improve the meta-learning based few-shot object detection algorithm. We first analyze the problems of current meta-learning algorithms that cannot fully utilize the detailed information from the support set and the need to manually design the support set utilization strategy. To address these problems, we propose ARDM, and apply it to few-shot object detection networks by proposing ARDNet. ARDM eliminates the manually designed and inefficient support feature utilization strategies of previous work, and effectively incorporates the rich information from the support set into the query features. Through extensive experiments, we demonstrate that ARDM has the ability to automatically mine effective information from support images, and the few-shot object detection network ARDNet based on adaptive relational distillation shows a significant improvement in detection accuracy compared to the baseline DCNet.

References

1. Diwan, T., Anirudh, G., Tembhurne, J.V.: Object detection using yolo: challenges, architectural successors, datasets and applications. Multimed. Tools Appl. **82**(6), 9243–9275 (2023)
2. Wang, C.Y., Bochkovskiy, A., Liao, H.Y.M.: Yolov7: trainable bag-of-freebies sets new state-of-the-art for real-time object detectors. In: Proceedings of the IEEE/CVF Conference on Computer Vision and Pattern Recognition, pp. 7464–7475 (2023)
3. Zeng, N., Wu, P., Wang, Z., Li, H., Liu, W., Liu, X.: A small-sized object detection oriented multi-scale feature fusion approach with application to defect detection. IEEE Trans. Instrum. Meas. **71**, 1–14 (2022)
4. Vs, V., Poster, D., You, S., Hu, S., Patel, V.M.: Meta-uda: Unsupervised domain adaptive thermal object detection using meta-learning. In: Proceedings of the IEEE/CVF Winter Conference on Applications of Computer Vision, pp. 1412–1423 (2022)

5. Su, H., Xiang, L., Hu, A., Xu, Y., Yang, X.: A novel method based on meta-learning for bearing fault diagnosis with small sample learning under different working conditions. Mech. Syst. Signal Process. **169**, 108765 (2022)
6. Huang, S.F., Lin, C.J., Liu, D.R., Chen, Y.C., Lee, H.y.: Meta-tts: meta-learning for few-shot speaker adaptive text-to-speech. IEEE/ACM Trans. Audio Speech Lang. Process. **30**, 1558–1571 (2022)
7. Kang, B., Liu, Z., Wang, X., Yu, F., Feng, J., Darrell, T.: Few-shot object detection via feature reweighting. In: Proceedings of the IEEE/CVF International Conference on Computer Vision. pp. 8420–8429 (2019)
8. Yan, X., Chen, Z., Xu, A., Wang, X., Liang, X., Lin, L.: Meta r-cnn: towards general solver for instance-level low-shot learning. In: Proceedings of the IEEE/CVF International Conference on Computer Vision, pp. 9577–9586 (2019)
9. Fan, Q., Zhuo, W., Tang, C.K., Tai, Y.W.: Few-shot object detection with attention-rpn and multi-relation detector. In: Proceedings of the IEEE/CVF Conference on Computer Vision and Pattern Recognition, pp. 4013–4022 (2020)
10. Hu, H., Bai, S., Li, A., Cui, J., Wang, L.: Dense relation distillation with context-aware aggregation for few-shot object detection. In: Proceedings of the IEEE/CVF Conference on Computer Vision and Pattern Recognition, pp. 10185–10194 (2021)
11. Schmidhuber, J.: Evolutionary principles in self-referential learning, or on learning how to learn: the meta-meta-... hook. Ph.D. thesis, Technische Universität München (1987)
12. Hinton, G.E., Plaut, D.C.: Using fast weights to deblur old memories. In: Proceedings of the 9th Annual Conference of the Cognitive Science Society, pp. 177–186 (1987)
13. Fu, K., Zhang, T., Zhang, Y., Yan, M., Chang, Z., Zhang, Z., Sun, X.: Meta-ssd: towards fast adaptation for few-shot object detection with meta-learning. IEEE Access **7**, 77597–77606 (2019)
14. Wang, Y.X., Ramanan, D., Hebert, M.: Meta-learning to detect rare objects. In: Proceedings of the IEEE/CVF International Conference on Computer Vision, pp. 9925–9934 (2019)
15. Karlinsky, L., et al.: Repmet: representative-based metric learning for classification and few-shot object detection. In: Proceedings of the IEEE/CVF Conference on Computer Vision and Pattern Recognition, pp. 5197–5206 (2019)
16. Yang, Y., Wei, F., Shi, M., Li, G.: Restoring negative information in few-shot object detection. Adv. Neural. Inf. Process. Syst. **33**, 3521–3532 (2020)
17. Li, B., Yang, B., Liu, C., Liu, F., Ji, R., Ye, Q.: Beyond max-margin: Class margin equilibrium for few-shot object detection. In: Proceedings of the IEEE/CVF Conference on Computer Vision and Pattern Recognition, pp. 7363–7372 (2021)
18. Woo, S., Park, J., Lee, J.Y., Kweon, I.S.: Cbam: convolutional block attention module. In: Proceedings of the European Conference on Computer Vision (ECCV), pp. 3–19 (2018)
19. Everingham, M., Van Gool, L., Williams, C.K., Winn, J., Zisserman, A.: The pascal visual object classes (voc) challenge. Int. J. Comput. Vision **88**(2), 303–338 (2010)

A Real-Time Safety Detector Based on Re-parameterization Multiscale Feature Fusion for Forklift Driving

Linhua Ye[1,2], Songhang Chen[2,3](✉), Zhiqing Lai[2], and Meng Guo[1]

[1] College of Computer and Cyber Security, Fujian Normal University, Fuzhou 350117, China

[2] Fujian Institute of Research on the Structure of Matter, Chinese Academy of Sciences, Fujian 350108, China

songhang.chen@fjirsm.ac.cn

[3] Fujian Provincial Key Laboratory of Intelligent Identification and Control of Complex Dynamic System, Quanzhou 362200, China

Abstract. The application of object detection in intelligent logistics has received considerable attention. However, existing detector models face challenges such as high computational costs, slow detection speed, and difficulty in deployment on edge devices with limited computational resources. This paper proposes a novel real-time safety detector model (RMFFDet) based on YOLOv8s for forklift driving. A hardware-friendly FasterNeXt module is designed to optimize feature extraction and reduce the computational costs in the Backbone. Inspired by the work on RepGhost, a re-parameterization multiscale feature fusion Neck (RMFFNeck) is proposed in this paper. Reconstructing the Neck based on RMFFNeck improves the capture of contextual logistics background feature information while reducing the model parameters. Finally, the Wise-IoU (WIoU) is introduced as a bounding box regression loss combined with a dynamic non-monotonic focusing mechanism to improve the model's overall performance. Experiments show that RMFFDet achieves a mean Average Precision (mAP) of 95.2% on the KITTI dataset and 92.8% on the self-built Forklift-3k dataset. Compared to YOLOv8s, the model parameters are reduced by 34.5%. On the Jetson Nano edge platform and 640×640 input size, RMFFDet requires only 100.2ms inference time. RMFFDet offers an excellent trade-off between inference speed and detection accuracy. It meets the industrial requirements of logistics scenarios.

Keywords: Object detection · Intelligent logistics · Feature fusion

1 Introduction

The logistics industry is experiencing rapid growth, and forklifts have become indispensable in various logistics scenarios. While forklifts offer convenience in

This work was supported by the Science and Technology Program of Quanzhou, China (Grant No. 2023C011R).

© The Author(s), under exclusive license to Springer Nature Singapore Pte Ltd. 2024

Q. Liu et al. (Eds.): PRCV 2023, LNCS 14436, pp. 340–351, 2024.
https://doi.org/10.1007/978-981-99-8555-5_27

industrial manufacturing, they pose significant safety risks. Excessive stacking of cargo obstructs the visibility of forklift operators, limiting their ability to observe pedestrians in front of them. Moreover, the elevated body of the forklift creates a blind spot in the rearview, further compromising safety. Additionally, forklift operators suffer from fatigue, inattention, and potential hazards and irregularities. To address these issues, machine vision-based safety warning systems have emerged as effective solutions. In such systems, the detector model assumes a critical role. Equipping forklifts with RGB-D depth cameras and integrating advanced detector models can enable accurate detection of target locations and distances, thereby effectively preventing accidents.

Deep learning-based object detector models, as an essential branch in the field of machine learning [5,22,26], have demonstrated superior performance compared to traditional methods. Object detector models can be classified into two main categories. Firstly, there are region proposal-based two-stage detector models, with Faster R-CNN [17] being the most typical representative. This approach generates candidate bounding boxes and performs classification and regression on these boxes. While two-stage detectors achieve impressive accuracy, their detection speed is often inadequate for real-time applications. In contrast, single-stage detector models, such as the YOLO series [1,14–16], take a regression-based approach to object detection, allowing direct prediction of object categories and precise localization. These detectors have faster detection speed and can meet the requirements of real-time detection.

Despite the remarkable performance of one-stage detector models on devices with abundant computational resources, they suffer from slow inference speed and compromised detection accuracy when deployed on edge devices with limited computing power. This issue is particularly critical in complex logistics scenarios where forklift drivers encounter problems such as narrow aisles and random cargo stacking. Thus, the trade-off between detection accuracy and speed has been a significant challenge [13,23,24]. Another challenge arises from the lack of suitable datasets capturing logistics scenes, posing difficulties for research endeavours. The logistics scenarios is characterized by its complexity and variability, including different cargo types, stacking methods, and lighting conditions, making data collection challenging. To meet the requirements of real-time detection and high detection accuracy in complex logistics scenarios, this study addresses the challenges of high computational costs and inferior detection accuracy on edge devices by utilizing the proposed RMFFDet model. The main contributions of this paper are as follows:

- A hardware-friendly FasterNeXt module is designed to optimize the Backbone in RMFFDet and reduce the computational costs.
- Reconstructing the Neck based on the proposed RMFFNeck makes the model better for obtaining complex logistics background feature information while reducing the model parameters.

- The WIoU is introduced as a bounding box regression loss to improve the computational procedure of the loss function in RMFFDet.

2 Related Works

Real-time detection and high detection accuracy are crucial in practical deployments. The YOLO series has an excellent balance between speed and accuracy, especially the current latest YOLOv8 [19]. Although YOLOv8 has excellent performance on the GPUs side, it cannot achieve real-time detection on edge devices due to excessive hardware requirements. Lightweight models have gained wide attention and applications [7,9,21]. GhostNet [8] is a lightweight convolutional neural network. It effectively reduces the computational costs by using cheap operations to generate more feature maps. Similarly, ShuffleNetV2 [12] improves efficiency by processing half of the feature channels and reserving the other half for concat operations. Although their concat operations are parametric-free, their computational costs on hardware devices is not negligible. To address this issue, structural re-parameterization techniques [4] to achieve feature reuse provide a new perspective. Inspired by lightweight models and structural re-parameterization techniques, this paper proposes the RMFFDet model. It reaches more efficient model design on mobile devices with limited computational resources by reducing the computational costs of the model while maintaining the detection performance.

3 Proposed Methods

3.1 FasterNeXt

FasterNet [3] is a novel neural network that achieves fast-running speeds and is well-suited for hardware implementations. It is built upon the FasterNet Block and utilizes 1×1 Conv as its basic module. The FasterNeXt module is designed based on the FasterNet Block to optimize the Backbone in RMFFDet. Figure 1 illustrates the FasterNeXt module, which efficiently uses inverted residual blocks and cross-layer connections to reuse input features. The FasterNet Block consists of a PConv layer and two 1×1 Conv layers. The PConv layer employs a straightforward structure to minimize computational redundancy and optimize memory access. It selectively applies regular Conv operations to a subset of the input channels while preserving the unchanged channels.

3.2 RepGhost

The RepGhost module [2] is an innovative lightweight network that utilizes structural re-parameterization techniques. This module allows efficient feature reuse by providing an alternative to the inefficient concatenation operators used in the Ghost module. Figure 2 visually illustrates the differences between the RepGhost bottleneck and the Ghost bottleneck. In particular, the middle channel of the

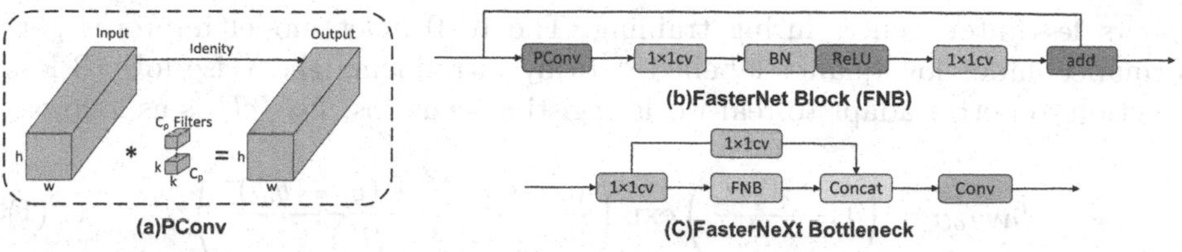

Fig. 1. FasterNeXt module. 1x1cv: 1x1 conv.

RepGhost Bottleneck is narrower, while the last channel is coarser compared to its Ghost counterpart. Furthermore, RepGhost Bottleneck replaces Ghost Bottleneck's Concat operation with the more efficient Add operation. During the inference process, RepGhost Bottleneck performs calculations using an equivalent simple convolution operation, improving performance.

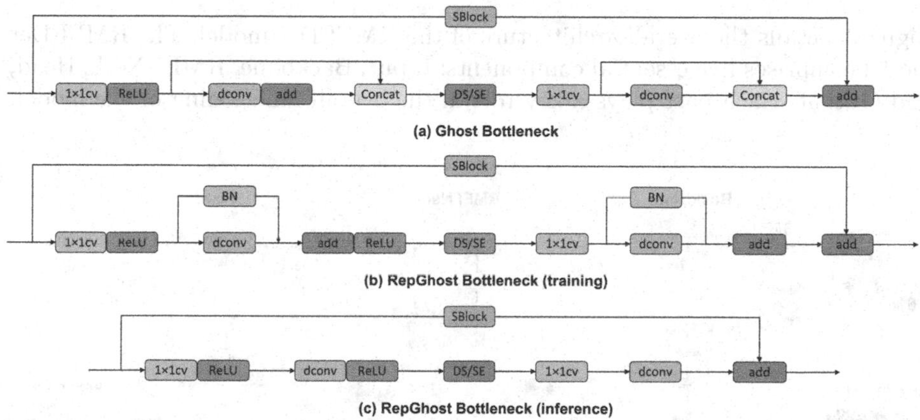

Fig. 2. Ghost Bottleneck and RepGhost Bottleneck. SBlock: shortcut block, DS: downsample layer, SE: SE block, 1x1cv: 1x1 conv.

3.3 WIoU

Compared to the CIoU [25] loss function in YOLOv8, this paper applies the Wise-IoU (WIoU) [20] in the bounding box regression loss. The WIoU incorporates a dynamic non-monotonic focusing mechanism to cope with the low quality of the training data labelling. It avoids over-penalizing the model by geometric factors such as distance and aspect ratio. When the prediction frame is highly matched with the target frame, the WIoU improves the model's generalization ability by mitigating the penalty effect of geometric factors so that the model suffers less interference during training. The RMFFDet model improves performance under low-quality labelled data by introducing the Wise-IoU v3 loss function to better adapt to real-world logistics scenarios. $L_W IoU$ is as follows:

$$L_{WIoU} = \left(1 - \frac{W_i H_i}{S_u}\right) \exp\left(\frac{(x_p - x_{gt})^2 + (y_p - y_{gt})^2}{(W_g^2 + H_g^2)^*}\right) \gamma \tag{1}$$

$$\gamma = \beta / \delta \alpha^{\beta - \delta} \tag{2}$$

where ß denotes the anomaly degree of the prediction frame, a more minor anomaly degree means a higher quality of the anchor frame. α and δ are hyperparameters. H and W denote the width and height of the two frames, respectively. x_p and y_p represent the coordinate values of the prediction frame, and x_{gt} and y_{gt} indicate the coordinate values of the ground truth value.

3.4 RMFFDet

Figure 3 details the overall architecture of the RMFFDet model. The RMFFDet model comprises five essential components: Input, Backbone, RMFFNeck, Head, and Output. Each part plays a key role in the overall functioning of the model.

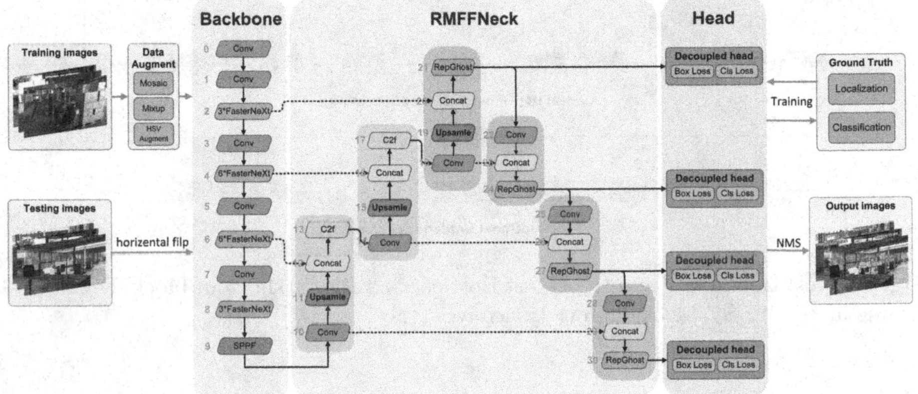

Fig. 3. The architecture of the RMFFDet model. a) The Backbone is optimized by the FasterNeXt module. b) The prediction heads use the feature maps from the RepGhost module.

Input. This paper combines Mosaic, MixUp, and traditional methods for online data augmentation to make the model more robust for images in complex logistics scenarios. These enhanced images are averaged to improve the prediction results of the model.

Backbone. The Backbone in RMFFDet is mainly responsible for extracting features from the input image. It's also accountable for gradually improving the semantic representation through the hierarchical structure. In this paper, the Backbone is optimized by the designed FasterNeXt module, which makes the Backbone reduce a large amount of computational costs.

RMFFNeck. RMFFNeck is proposed to make better use of the features extracted from the Backbone. The feature maps extracted by the Backbone are reprocessed and used rationally at different stages. Typical Necks include FPN [10], PANet [11], and BiFPN [18]. The similarity of these Necks is through the iterative use of various upsampling and stitching to design. Through structural re-parameterization techniques, RMFFNeck can learn complex convolutional operations during training and compute with equivalent simple convolutional operations during inference, resulting in a lightweight and high-performance model. Then, RMFFNeck enhances the feature extraction of small targets by a Four-scale fusion structure. This design allows RMFFNeck to have good potential for application in resource-constrained environments.

Head. In the Head, the feature information of each prediction head is taken from the RepGhost module. And the object classification prediction and bounding box regression prediction are decoupled and performed separately.

Output. During the training process of RMFFDet, Non-Maximum Suppression (NMS) is used to select the best detection results among the overlapping bounding boxes. The primary purpose of NMS is to eliminate redundant detection results to retain the most accurate and representative bounding boxes and obtain the output image.

4 Experiments

4.1 Datasets

Forklift-3k Dataset. We have collected and produced the Forklift-3k dataset of complex logistics scenarios for forklift safety driving. Figure 4 shows the various types of forklifts. The dataset contains two main categories of targets, Forklift (4,572 numbers) and Person (4,190 numbers), with a total of 3,342 images.

KITTI Dataset. KITTI [6] is one of the enormous datasets available for driverless applications. It encompasses various challenging scenarios, including small distant targets and abundant occlusions, closely resembling the complexities encountered in complex logistics scenarios. This paper uses KITTI to verify the generalization of the RMFFDet model. To fit the engineering scenario, this paper merges Car, Van, Truck and Tram in the source data into Car, Pedestrian and Person sitting into Person, and removes Misc and DontCare. A total of

Fig. 4. The image samples in the Forklift-3k dataset.

7,481 images and three categories are obtained, including Car (33,261 numbers), Person (4,709 numbers) and Cyclist (1,627 numbers).

4.2 Experimental Results

The experiments are conducted on a computational environment comprising a Windows 11 operating system, an i9-12900k CPU, and an NVIDIA GEFORCE RTX 4090 processor with 24GB of video memory. Several experimental hyperparameters are as follows: training epochs are set to 200 and 300 for the Forklift-3k and KITTI datasets, respectively; batchsize is set to 32; optimizer uses SGD, momentum is set to 0.937. The datasets are divided into a 6:2:2 ratio for training, validation, and testing purposes. All data augmentation techniques are turned off during the last ten training epochs. Figure 5 illustrates the training progress of the YOLOv8s and the proposed RMFFDet on the Forklift-3k dataset. Notably, RMFFDet exhibits a remarkable improvement over YOLOv8s. This is attributed to the effect of the WIoU and the RMFFNeck.

To gain insight into the contributions of FasterNeXt and RepGhost, this paper performs a comprehensive analysis using representative samples from the Forklift-3k dataset. By visualizing the feature maps during the detection process, we aim to elucidate the specific roles played by FasterNeXt and RepGhost in the model's feature extraction capabilities. As shown in Fig. 6, FasterNeXt primarily focuses on extracting low-level feature information, capturing fine-grained details and spatial features from the input data. Conversely, RepGhost is designed to specialize in extracting high-level feature information, allowing the model to capture more abstract and semantic representations of the target objects.

Figure 7 illustrates the commendable ability of RMFFDet to capture significant feature information of objects within complex logistics scenes. The utilization of GradCAM further facilitates a comprehensive understanding of the model's attention and decision-making process about various targets.

4.3 Comparison of Detection Performance

Table 1 presents a comprehensive analysis of the performance of various lightweight models on the Forklift-3K and KITTI datasets. As a lightweight

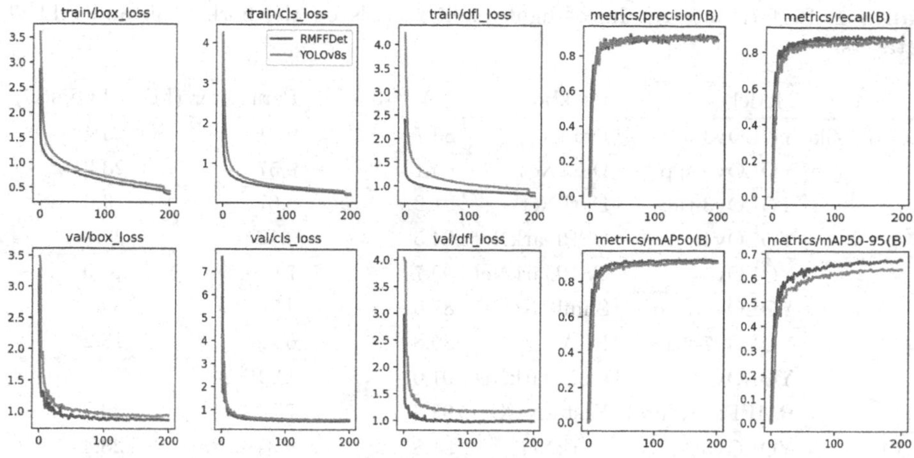

Fig. 5. Training process for YOLOv8 (baseline) and RMFFDet (our).

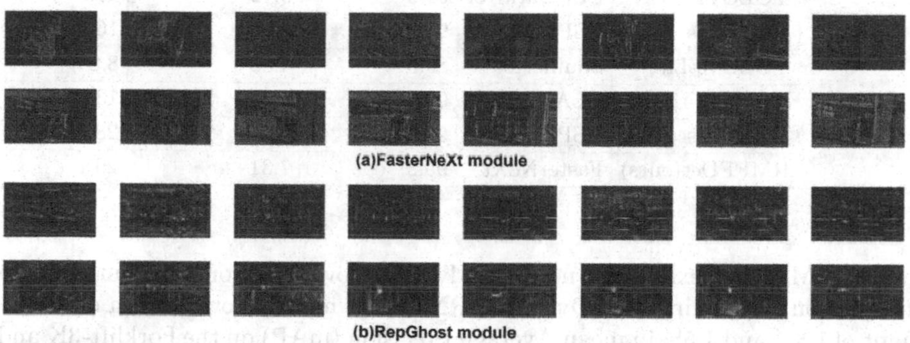

Fig. 6. Visualization of the feature map of the FasterNeXt and RepGhost modules.

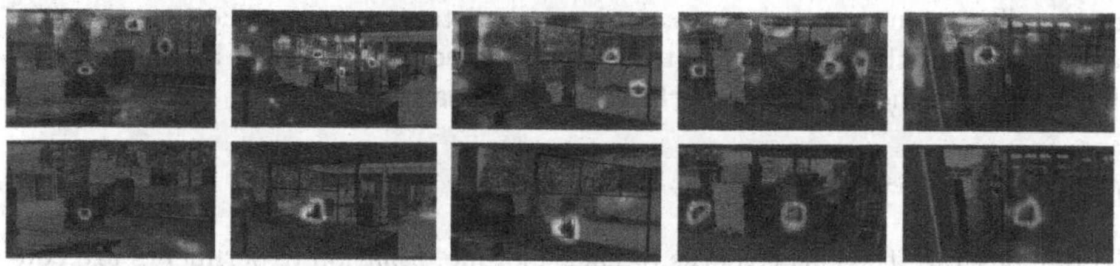

Fig. 7. The Grad-CAM graph of the proposed RMFFDet model.

Table 1. Performance results of lightweight models on the Forklift-3k and KITTI datasets.

Dataset	Model	Backbone	mAP@0.5(%)	Parameters(M)	FLOPs(G)
Forklift-3k	YOLOv3	DarkNet	83.7	9.31	23.4
	YOLOv3-spp	DarkNet	85.1	9.57	23.6
	YOLOv3-tiny	DarkNet	81.3	8.67	13.0
	YOLOv4	CSPDarkNet	84.5	9.12	20.8
	YOLOv5s	CSPDarkNet	90.7	7.02	16.0
	YOLOv5-Lite	ShuffleNet	87.0	4.38	8.8
	YOLOv7-tiny	ELAN	89.8	6.02	13.2
	YOLOv8s	CSPDarkNet	91.0	11.16	28.8
	RMFFDet(ours)	FasterNeXt	92.8	7.31	29.7
KITTI	YOLOv3	DarkNet	91.8	9.31	23.4
	YOLOv3-spp	DarkNet	92.3	9.57	23.6
	YOLOv3-tiny	DarkNet	84.3	8.67	13.0
	YOLOv4	CSPDarkNet	92.6	9.12	20.8
	YOLOv5s	CSPDarkNet	93.1	7.02	16.0
	YOLOv5-Lite	ShuffleNet	89.9	4.38	8.8
	YOLOv7-tiny	ELAN	90.0	6.02	13.2
	YOLOv8s	CSPDarkNet	93.7	11.16	28.8
	RMFFDet(ours)	FasterNeXt	95.2	7.31	29.7

model, RMFFDet exhibits a notable advantage over the compared models. In comparison to baseline YOLOv8s, the RMFFDet model showcases an enhancement of 1.8% and 1.5% in mean Average Precision (mAP) on the Forklift-3K and KITTI datasets, respectively, while reducing the model parameters by 34.5%.

Figure 8 presents the visualized detection results of different models in logistics scenes, shedding light on the challenges encountered, including uneven lighting conditions, many occlusions, and small distant targets. The experimental results confirm the robust detection performance and practical applicability of the RMFFDet model in real-world logistics scenarios.

4.4 Ablation Experiment

To demonstrate the effectiveness of each module, ablation experiments are conducted on the KITTI dataset in a consistent experimental setting. As shown in Table 2, the proposed improvement strategies have proven to effectively enhance the detection performance of RMFFDet. Specifically, FasterNeXt plays a crucial role in significantly reducing the computational costs in the Backbone. RMFF-Neck provides approximately 2% improvement in mAP through structural re-parameterization techniques and efficient multiscale fusion. Moreover, the WIoU enhances detection performance without increasing computational costs.

Fig. 8. Comparison of detection results: (a) YOLOv7-tiny; (b) YOLOv8s; (c) RMFFDet.

Table 2. Ablation experiments on the KITTI dataset.

Baseline	FasterNeXt	RMFFNeck	WIoU	mAP@0.5(%)	Parameters(M)	FLOPs(G)
√				93.7	11.16	28.4
√	√			93.3	9.48	24.7
√		√		95.6	9.24	35.4
√			√	94.3	11.16	28.4
√	√	√		94.7	7.31	29.7
√	√		√	93.9	9.48	24.7
√	√	√	√	95.2	7.31	29.7

4.5 Edge Platform Deployments

In this paper, YOLOv8s and RMFFDet are selected for inference acceleration experiments using TensorRT on the Jetson Nano edge platform. Figure 9 shows the experimental results for four different input sizes on 300 images. The application of TensorRT acceleration greatly accelerates the inference speed of the model. The RMFFDet model achieves an inference time of only 100.2ms per image when the image size is 640×640.

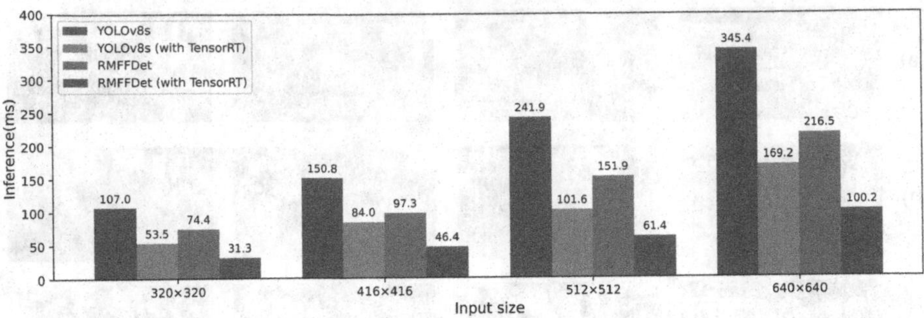

Fig. 9. Comparison of inference speed of YOLOv8s and RMFFDet before and after TensorRT acceleration at different input sizes.

5 Conclusion

This paper proposes the RMFFDet model, a novel detector model specifically designed for forklift driving. The FasterNeXt module is designed to improve the computational efficiency in the Backbone, while the proposed RMFFNeck module enhances the feature representation. RMFFDet's inference speed and detection accuracy are better than YOLOv8 to meet the industrial requirements of logistics scenarios. On the Jetson Nano edge platform, RMFFDet uses TensorRT acceleration to reduce inference time significantly. Furthermore, future research will explore advanced model compression techniques for RMFFDet to facilitate its deployment in practical applications.

References

1. Bochkovskiy, A., Wang, C.Y., Liao, H.Y.M.: Yolov4: optimal speed and accuracy of object detection. arXiv preprint arXiv:2004.10934 (2020)
2. Chen, C., Guo, Z., Zeng, H., Xiong, P., Dong, J.: Repghost: a hardware-efficient ghost module via re-parameterization. arXiv preprint arXiv:2211.06088 (2022)
3. Chen, J., Kao, S.h., He, H., Zhuo, W., Wen, S., Lee, C.H., Chan, S.H.G.: Run, don't walk: Chasing higher flops for faster neural networks. In: Proceedings of the IEEE/CVF Conference on Computer Vision and Pattern Recognition, pp. 12021–12031 (2023)
4. Ding, X., Zhang, X., Ma, N., Han, J., Ding, G., Sun, J.: Repvgg: Making vgg-style convnets great again. In: Proceedings of the IEEE/CVF Conference on Computer Vision and Pattern Recognition, pp. 13733–13742 (2021)
5. Ge, Z., Liu, S., Wang, F., Li, Z., Sun, J.: Yolox: exceeding yolo series in 2021. arXiv preprint arXiv:2107.08430 (2021)
6. Geiger, A., Lenz, P., Urtasun, R.: Are we ready for autonomous driving? the kitti vision benchmark suite. In: 2012 IEEE Conference on Computer Vision and Pattern Recognition, pp. 3354–3361. IEEE (2012)
7. Guan, L., Jia, L., Xie, Z., Yin, C.: A lightweight framework for obstacle detection in the railway image based on fast region proposal and improved yolo-tiny network. IEEE Trans. Instrum. Meas. **71**, 1–16 (2022)

8. Han, K., Wang, Y., Tian, Q., Guo, J., Xu, C., Xu, C.: Ghostnet: more features from cheap operations. In: Proceedings of the IEEE/CVF Conference on Computer Vision and Pattern Recognition, pp. 1580–1589 (2020)
9. Hurtik, P., Molek, V., Hula, J., Vajgl, M., Vlasanek, P., Nejezchleba, T.: Poly-yolo: higher speed, more precise detection and instance segmentation for yolov3. Neural Comput. Appl. **34**(10), 8275–8290 (2022)
10. Lin, T.Y., Dollár, P., Girshick, R., He, K., Hariharan, B., Belongie, S.: Feature pyramid networks for object detection. In: Proceedings of the IEEE Conference on Computer Vision and Pattern Recognition, pp. 2117–2125 (2017)
11. Liu, S., Qi, L., Qin, H., Shi, J., Jia, J.: Path aggregation network for instance segmentation. In: Proceedings of the IEEE conference on computer vision and pattern recognition. pp. 8759–8768 (2018)
12. Ma, N., Zhang, X., Zheng, H.T., Sun, J.: Shufflenet v2: Practical guidelines for efficient cnn architecture design. In: Proceedings of the European conference on computer vision (ECCV). pp. 116–131 (2018)
13. Mathew, M.P., Mahesh, T.Y.: Leaf-based disease detection in bell pepper plant using yolo v5, pp. 1–7. Signal, Image and Video Processing pp (2022)
14. Redmon, J., Divvala, S., Girshick, R., Farhadi, A.: You only look once: Unified, real-time object detection. In: Proceedings of the IEEE conference on computer vision and pattern recognition. pp. 779–788 (2016)
15. Redmon, J., Farhadi, A.: Yolo9000: better, faster, stronger. In: Proceedings of the IEEE conference on computer vision and pattern recognition. pp. 7263–7271 (2017)
16. Redmon, J., Farhadi, A.: Yolov3: An incremental improvement. arXiv preprint arXiv:1804.02767 (2018)
17. Ren, S., He, K., Girshick, R., Sun, J.: Faster r-cnn: Towards real-time object detection with region proposal networks. Advances in neural information processing systems 28 (2015)
18. Tan, M., Pang, R., Le, Q.V.: Efficientdet: Scalable and efficient object detection. In: Proceedings of the IEEE/CVF conference on computer vision and pattern recognition. pp. 10781–10790 (2020)
19. Terven, J., Cordova-Esparza, D.: A comprehensive review of yolo: From yolov1 to yolov8 and beyond. arXiv preprint arXiv:2304.00501 (2023)
20. Tong, Z., Chen, Y., Xu, Z., Yu, R.: Wise-iou: Bounding box regression loss with dynamic focusing mechanism. arXiv preprint arXiv:2301.10051 (2023)
21. Wang, G., Ding, H., Yang, Z., Li, B., Wang, Y., Bao, L.: Trc-yolo: A real-time detection method for lightweight targets based on mobile devices. IET Comput. Vision **16**(2), 126–142 (2022)
22. Wu, W., Guo, L., Gao, H., You, Z., Liu, Y., Chen, Z.: Yolo-slam: A semantic slam system towards dynamic environment with geometric constraint. Neural Computing and Applications pp. 1–16 (2022)
23. Yan, W., Gu, M., Ren, J., Yue, G., Liu, Z., Xu, J., Lin, W.: Collaborative structure and feature learning for multi-view clustering. Information Fusion **98**, 101832 (2023)
24. Zhao, C., Shu, X., Yan, X., Zuo, X., Zhu, F.: Rdd-yolo: A modified yolo for detection of steel surface defects. Measurement **214**, 112776 (2023)
25. Zheng, Z., Wang, P., Liu, W., Li, J., Ye, R., Ren, D.: Distance-iou loss: Faster and better learning for bounding box regression. In: Proceedings of the AAAI conference on artificial intelligence. vol. 34, pp. 12993–13000 (2020)
26. Zhou, X., Koltun, V., Krähenbühl, P.: Simple multi-dataset detection. In: Proceedings of the IEEE/CVF Conference on Computer Vision and Pattern Recognition. pp. 7571–7580 (2022)

RTMDet-R2: An Improved Real-Time Rotated Object Detector

Haifeng Xiang, Naifeng Jing, Jianfei Jiang, Hongbo Guo, Weiguang Sheng, Zhigang Mao, and Qin Wang(✉)

Shanghai Jiao Tong University, Shanghai, China
{xianghf666,hongboguo,qinqinwang}@sjtu.edu.cn

Abstract. Object detection in remote sensing images is challenging due to the absence of visible features and variations in object orientation. Efficient detection of objects in such images can be achieved using rotated object detectors that utilize oriented bounding boxes. However, existing rotated object detectors often struggle to maintain high accuracy while processing high-resolution remote sensing images in real time. In this paper, we present RTMDet-R2, an improved real-time rotated object detector. RTMDet-R2 incorporates an enhanced path PAFPN to effectively fuse multi-level features and employs a task interaction decouple head to alleviate the imbalance between regression and classification tasks. To further enhance performance, we propose the ProbIoU-aware dynamic label assignment strategy, which enables efficient and accurate label assignment during the training. As a result, RTMDet-R2-m and RTMDet-R2-l achieve 79.10% and 79.46% mAP, respectively, on the DOTA 1.0 dataset using single-scale training and testing, outperforming the majority of other rotated object detectors. Moreover, RTMDet-R2-s and RTMDet-R2-t achieve 78.43% and 77.27% mAP, respectively, while achieving inference frame rates of 175 and 181 FPS at a resolution of 1024 × 1024 on an RTX 3090 GPU with TensorRT and FP16-precision. Furthermore, RTMDet-R2-t achieves 90.63/97.44% mAP on the HRSC2016 dataset. The code and models are available at https://github.com/Zcba-Xie/RTMDet-R2.

Keywords: Remote Sensing Images Object Detection · Rotated Object Detection · Feature Fusion · Label Assignment

1 Introduction

Target detection tasks commonly use horizontal bounding boxes (HBB) to locate and classify objects [7]. However, existing HBB algorithms like the R-CNN series [24, 25] and YOLO series [13, 26, 27] are not directly applicable to object detection in remote sensing images. Challenges arise due to the similarity in shapes and limited visible features of objects in remote sensing images, as well as complex backgrounds and variations, particularly for small and densely-packed objects. The remote sensing perspective further complicates the representation of angle diversity [5].

© The Author(s), under exclusive license to Springer Nature Singapore Pte Ltd. 2024
Q. Liu et al. (Eds.): PRCV 2023, LNCS 14436, pp. 352–364, 2024.
https://doi.org/10.1007/978-981-99-8555-5_28

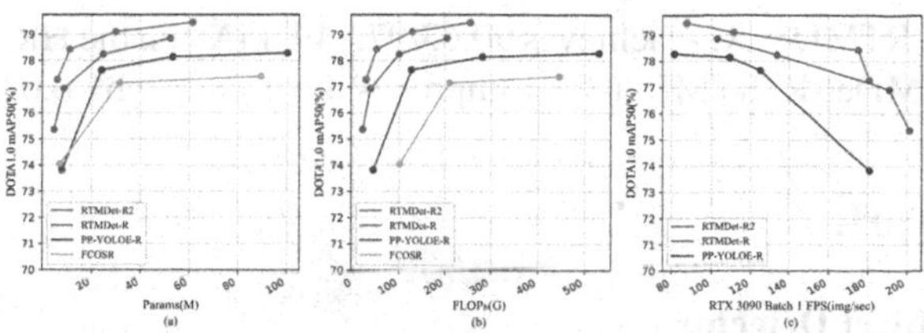

Fig. 1. Comparison of RTMDet-R2 and other state-of-the-art real-time rotated object detectors with single-scale training and testing on the DOTA 1.0 dataset. (a) Comparison of parameters and accuracy. (b) Comparison of FLOPs and accuracy. (c) Comparison of FPS and accuracy.

To achieve real-time object detection with high detection accuracy in remote sensing images, HBB is transformed into OBB (oriented bounding boxes) by incorporating rotation angles. Several models, including FCOSR [7], PP-YOLOE-R [8], and RTMDet-R [9], have been proposed for this purpose. These models enhance the balance between speed and accuracy by improving the model architecture, training strategies, and parameter scaling.

However, these models still have the following several issues. To begin with, both PP-YOLOE-R and RTMDet-R utilize the path aggregation feature pyramid network (PAFPN) [11] for multi-level features fusion, but PAFPN faces challenges in effectively integrating precise localization information from lower levels and semantic information from higher levels in the top-down path. Furthermore, there still exist issues of task misalignment and imbalance. Moreover, the sample label assignment strategies used in these models during training are primarily adapted from HBB algorithms, lacking optimization for OBB. To address these issues, we propose RTMDet-R2: an improved real-time rotated object detector. Our main contributions can be summarized as follows:

1. We introduce the enhanced path PAFPN (EP-PAFPN), which efficiently fuses positional and semantic information from multiple-level features by introducing the enhanced path (EP).
2. We propose the task interaction decuple head (TID-Head), which incorporates a task interaction module (TIM) to enable interaction between the classification and regression tasks, alleviating the imbalance of tasks.
3. We introduce the ProbIoU-aware dynamic label assignment strategy (PA-DLA), which replaces RIoU with ProbIoU and utilizes an IoU truncation strategy to achieve efficient and accurate label assignment.

As a result, RTMDet-R2 has achieved state-of-the-art performance in terms of the speed and accuracy trade-off on two datasets, DOTA 1.0 and HRSC2016. Specifically, as shown in Fig. 1., during single-scale training and testing on DOTA 1.0, RTMDet-R2-l and RTMDet-R2-m achieve 79.46% and 79.10% mAP[1], respectively. RTMDet-R2-s and RTMDet-R2-t achieve 78.43% and 77.27% mAP while achieving frame rates of 175 FPS and 181 FPS, respectively, at a resolution of 1024×1024. Compared to RTMDet-R, RTMDet-R2-l/m/s/t improves the accuracy by 0.61/0.86/1.50/1.91% mAP.

[1] We follow the latest metrics from the DOTA evaluation server, original voc format mAP is now mAP50.

Furthermore, RTMDet-R2-t achieves 90.63/97.44% mAP on the HRSC2016 dataset. Compared to RTMDet-R, RTMDet-R2-t improves the accuracy by 0.03/0.34% mAP.

2 Related Work

2.1 Rotated Object Detector

Rotated Object Detectors (ROD) can be divided into two types: Anchor-based and Anchor-free.

Anchor-based RODs can be further categorized as one-stage and two-stage methods. As a two-stage ROD, Oriented R-CNN [1] designs a lightweight and simpler oriented RPN on top of RROI to generate high-quality oriented proposals. ReDet [2] introduces rotation-invariant convolutions in the model and utilizes rotation-invariant RoI alignment to extract features. R3Det [3] is a one-stage ROD that utilizes a feature refinement module for feature reconstruction and alignment. S2ANet [4] generates high-quality OBB anchors using a feature alignment module, encodes the orientation information through the oriented detection module, and produces orientation-sensitive and orientation-invariant features to alleviate the inconsistency between classification scores and localization accuracy.

Anchor-free RODs are predominantly based on one-stage methods, often using prior points or key points. FCOSR [7] improves the label assignment strategy based on FCOS to enhance network performance. PP-YOLOE-R [8] introduces rotated task alignment learning (RTAL) and decoupled angle prediction head on top of PP-YOLOE. RTMDet-R [9] employs large-kernel CSP convolutional blocks to capture global context effectively. It also proposes a dynamic soft label assignment strategy to reduce the bias towards high-quality samples in the cost matrix.

2.2 Multi-level Features

SSD [28] and MS-CNN [29] utilize multi-level feature maps for prediction but do not aggregate them across different feature levels [12]. To improve the utilization efficiency of high-level low-resolution feature maps with strong semantic information and low-level high-resolution feature maps with weak semantic information, FPN [10] introduces a multi-level features fusion structure with lateral connections and top-down pathways, significantly enhancing model accuracy. PAFPN [11] builds upon the FPN architecture by introducing bottom-up pathways to shorten the information path between lower-level and top-level features. FPG [12], on the other hand, presents Feature Pyramid Grids, which is a deep multi-pathway feature pyramid. It represents the feature scale space as a regular grid of parallel bottom-up pathways and fuses them through multi-directional lateral connections.

2.3 Label Assignment

The purpose of label assignment is to differentiate between positive and negative samples [8]. There are two forms of label assignment: based on priors and based on predictions. Based on priors, label assignment distinguishes positive and negative samples using IoU, L2 distance, or predefined rules, such as ATSS [15]. ATSS adaptively calculates the threshold for positive and negative samples based on information like L2 distance and IoU. Based on predictions, label assignment differentiates positive and negative samples based on the model's predicted outputs, as seen in TAL [14] and SimOTA [13], among others. SimOTA dynamically assigns labels during the training process by computing the cost matrix between the model's outputs and the ground truth boxes. TAL calculates an alignment metric based on the network's output scores and the IoU between predicted and ground truth boxes, selecting the top M predictions with the highest alignment metric as positive samples and the rest as negative samples.

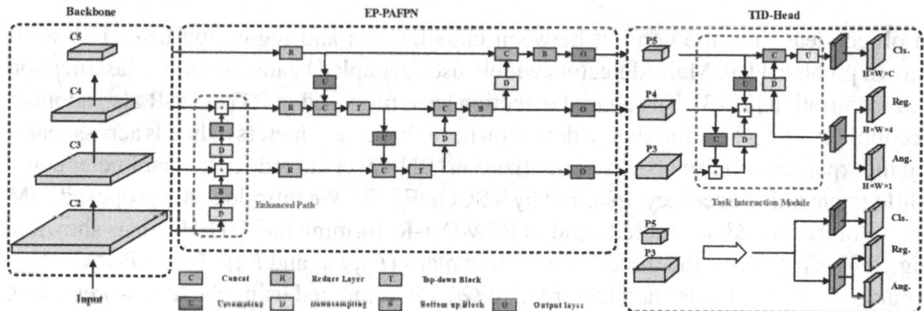

Fig. 2. The overall architecture of RTMDet-R2 consists of three main components: the Backbone, EP-PAFPN, and TID-Head. The Backbone is CSPNeXt. EP-PAFPN is composed of the CSPNeXtPAFPN and the enhanced path. TID-Head is composed of the R-SepBNHead from RTMDet-R and the task interaction module. The Reduce Layer is a 1 × 1 convolutional layer, the Output Layer and Downsampling are 3 × 3 convolutional layers, the Top-down Block and Bottom-up Block represent the CSP-blocks in RTMDet-R, Concat denotes channel-wise concatenation, and Upsampling refers to upsampling the feature map using the nearest neighbor interpolation with a scale factor of 2. In the TID-Head, the orange and green blocks represent stacked convolutional layers for the classification and regression branches, respectively.

3 Methodology

The overall structure of our proposed RTMDet-R2 is illustrated in Fig. 2. In this section, we will provide a detailed description of the introduced modules and strategy.

3.1 Enhanced Path PAFPN

FPN and PAFPN were proposed to efficiently utilize multi-scale features. However, PAFPN struggles to effectively integrate precise localization information from lower levels and semantic information from higher levels in the top-down path. To address this issue, we propose the enhanced path to strengthen the feature fusion in the top-down path, as shown in Fig. 2. The structure of the enhanced path is straightforward. It takes the feature map from a lower level of the backbone and processes it through a 3×3 downsampling convolution and a CSP block. The resulting feature map is then added to the feature map from the current level of the backbone, serving as the actual input to the top-down pathway of PAFPN. By incorporating the lower-level features with an acceptable computational cost, the enhanced path enables PAFPN to generate more comprehensive fused multi-level feature maps.

3.2 Task Interaction Decuple Head

In object detection, the conflict between classification and regression tasks is a well-known problem [16]. Many detectors widely use decoupled heads to handle classification and localization [7, 13]. The shared detection head proposed in RTMDet-R also adopts a decoupled form, where the shared detection head shares parameters of heads across scales but incorporates different batch normalization (BN) layers to reduce the parameter count while maintaining accuracy. Inspired by TSCODE [17], We introduce the proposed TIM on top of the shared decoupled head in RTMDet-R, forming the TID-Head, as shown in Fig. 2. The TIM takes the three-layer feature maps (P_3, P_4, and P_5) of the EP-PAFPN as inputs. The regression branch feature map G_{loc4} is computed using the detail-preserving encoding module [17], described by the following formula:

$$G_{loc4} = P_4 + u(P_5) + DConv(u(P_4) + P_3) \tag{1}$$

where $u(.)$ represents a 2x upsampling and $DConv(.)$ represents a downsampling convolution.

Unlike TSCODE, G_{cls4} is obtained by downsampling G_{loc4} through a 3×3 convolutional layer, concatenating it with P_5, and then performing upsampling. The computation can be written as:

$$G_{cls4} = u(Concat(DConv(G_{loc4}), P_5)) \tag{2}$$

Finally, G_{loc4} and G_{cls4} are individually fed into stacked convolutional layers for regression and classification. To fully utilize the multi-level feature maps, P_3 and P_5 continue to participate in classification and regression tasks as in RTMDet-R. Through TIM, interactions between multi-level features and between classification and regression tasks are achieved, mitigating the imbalance between tasks and improving network performance with relatively low computational cost.

3.3 ProbIoU-aware Dynamic Label Assignment

Label Assignment is one of the most crucial components in the training strategy of object detection networks, and it has a significant impact on the final prediction accuracy of the network. The Dynamic Label Assignment (RTM-DLA) strategy of RTMDet-R is an excellent approach; however, it lacks optimization for OBB tasks. Therefore, we propose two improvements to address this issue. Firstly, we innovatively introduce ProbIoU [18] to replace RIoU. Secondly, we introduce an IoU truncation strategy.

Traditionally, most object detection methods represent the shape and position of objects using bounding boxes. Recent research has explored the use of Gaussian distributions to represent objects in a more flexible and fuzzy manner. This approach has been adopted by various methods such as GWD [19] and KFIoU [20]. Similarly, ProbIoU introduces Gaussian bounding boxes to represent rotated bounding boxes and utilizes the Bhattacharyya coefficient between two Gaussian distributions to measure the similarity between them. In PA-DLA, ProbIoU is incorporated into the calculation of cost function and threshold truncation.

The cost function of PA-DLA can be represented as follows:

$$C = \begin{cases} \lambda_1 C_{cls} + \lambda_2 C_{reg} + \lambda_3 C_{center}, & ProbIoU > T \\ Inf, & Others \end{cases} \tag{3}$$

C_{cls}, C_{center}, and C_{reg} denote the classification cost, region prior cost, and regression cost, respectively. The default weights assigned to these costs are $\lambda_1 = 1$, $\lambda_2 = 3$, and $\lambda_3 = 1$. T represents the threshold for ProbIoU truncation, which is set to a default value of 0.075. The calculation for the three costs is described below.

C_{cls} and C_{center} are consistent with RTM-DLA and are defined as follows:

$$C_{cls} = CE(P, Y_{soft}) \times (Y_{soft} - P)^2 \tag{4}$$

$$C_{center} = \alpha^{|x_{pred} - x_{gt}| - \beta} \tag{5}$$

where α and β control the soft center region, with default values set to $\alpha = 10$ and $\beta = 3$. Y_{soft} represents the product of the one-hot label and ProbIoU. The soft classification cost in assignment not only reweights the matching costs with different regression qualities but also avoids the noisy and unstable matching caused by binary labels [9].

The regression cost C_{reg} is calculated by

$$C_{reg} = -\log(ProbIoU) \tag{6}$$

4 Experiments

4.1 Datasets

We evaluated our method on the DOTA 1.0 and HRSC 2016 datasets.

DOTA1.0 [21] is a large-scale remote sensing dataset for oriented object detection, consisting of 2806 aerial images with diverse scales, orientations, and shapes, encompassing 15 categories including plane, ship, and bridge. For single-scale training and testing, all images are cropped into 1024 × 1024 patches with a gap of 200.

HRSC2016 [22] is a challenging ship detection dataset with OBB annotations, which contains 1061 aerial images with size ranges from 300 × 300 to 1500 × 900.

Table 1. The settings for training the RTMDet-R2 series models on the DOTA dataset and HRSC2016 dataset. O, L, W, B, E, S, and A denote Optimizer, Base Learning Rate, Weight Decay, Batch Size, Training Epochs, Input Size, and Augmentation, respectively.

Model	Dataset	O	L	W	B	E	S	A
large	DOTA1.0	AdamW	2.5e−4	0.05 × 2.375	8	36	1024 × 1024	random flip and rotate
medium	DOTA1.0	AdamW	2.5e−4	0.05 × 2.200	8	36	1024 × 1024	random flip and rotate
small	DOTA1.0	AdamW	2.5e−4 × 1.25	0.05 × 3.000	8 × 1.25	36	1024 × 1024	random flip and rotate
tiny	DOTA1.0	AdamW	2.5e−4 × 1.75	0.05 × 2.325	8 × 1.75	36	1024 × 1024	random flip and rotate
tiny	HRSC2016	AdamW	2.0e−3 × 0.99	0.05	8	108	800 × 800	random flip and rotate

4.2 Implement Details

The RTMDet-R2 models are implemented based on MMRotate [23]. The hyperparameters of the models are shown in Table 1. Except for the medium and large models, which are trained on two RTX 3090 GPUs, all other models are trained on a single RTX 3090 GPU. The latency of all models is tested using the half-precision floating-point format (FP16) on an RTX 3090 GPU with TensorRT 8.4.3.1 and cuDNN 8.3.2. The inference batch size is set to 1.

4.3 Ablation Studies

We conducted a series of experiments on the DOTA 1.0 dataset using the RTMDet-R-t model to evaluate the effectiveness of the proposed method, which were trained and tested in a single-scale manner.

Table 2. Design of the detector neck. *indicates that EP will be connected to the highest-level feature map C5 from the backbone. The best results are in bold.

Neck	mAP50(%)↑	mAP75(%)↑	mmAP(%)↑	Params(M)↓	FLOPs(G)↓	Latency(ms)↓
PAFPN	75.36	50.64	47.37	**4.88**	**20.45**	**4.97**
EP-PAFPN*	75.66	50.74	47.30	8.57	29.15	5.61
EP-PAFPN	**75.91**	**51.34**	**47.97**	5.77	26.29	5.20

As shown in Table 2, EP-PAFPN effectively improves the network performance through stronger feature fusion capability and acceptable computational cost. Compared to the PAFPN in RTMDet-R, EP-PAFPN shows an improvement of 0.55% in mAP50.

When the EP is connected to the highest-level feature map C5 from the backbone, there is no significant improvement in network accuracy, but there is a noticeable increase in latency.

Table 3. Design of the detector head. †indicates the use of semantic context encoding (SCE) in the head. The best results are in bold.

Head	mAP50(%)↑	mAP75(%)↑	mmAP(%)↑
Baseline	75.36	50.64	47.37
TID-Head†	75.24	50.62	47.59
TID-Head	**75.98**	**50.90**	**47.73**

As shown in Table 3, TID-Head improves the network performance effectively by enhancing the interaction between the classification and regression tasks. Initially, we introduced the semantic context encoding (SCE) from TSCODE [17] into the Head, but experimental results showed that its impact was not significant. However, our proposed TID-Head improves mAP50 by 0.62%, while reducing the latency to 4.92 ms.

Table 4. Comparison of label assign strategies and ProbIoU loss. The best results are in bold.

Method	mAP50(%)↑	mAP75(%)↑	mmAP(%)↑
RTM-DLA	75.36	**50.64**	**47.37**
ProbIoU loss	76.60	49.33	46.44
PA-DLA	**76.75**	50.06	47.07

As shown in Table 4, PA-DLA achieves efficient and accurate label assignment by introducing a more suitable object similarity measurement method. Compared to the original DLA, PA-DLA improves mAP50 by 1.39%. We also compared PA-DLA with ProbIoU loss and found that although ProbIoU loss can improve mAP50 to some extent, it results in significant decreases in mAP75 and mmAP, possibly due to the fuzzy Gaussian representation [30]. While PA-DLA also leads to slight decreases in mAP75 and mmAP, the magnitude of the decrease is much smaller than that caused by ProbIoU loss. As shown in Table 5, we further investigated the impact of the threshold T in PA-DLA on network performance and found that $T = 0.075$ achieves a good overall performance.

Table 5. Selection of ProbIoU threshold T in PA-DLA. The best results are in bold.

T	mAP50(%)↑	mAP75(%)↑	mmAP(%)↑
0.001	**76.86**	49.10	46.72
0.010	76.58	49.85	**47.14**
0.025	75.48	**50.43**	46.71
0.075	76.75	50.06	47.07
0.100	76.81	49.80	47.06

Our step-by-step improvements are shown in Table 6. The joint optimization of network architecture and training strategy in RTMDet-R2 results in a significant improvement of 1.91% in mAP50 compared to the baseline.

Table 6. Step-by-step improvements from RTMDet-R-t to RTMDet-R2-t on the DOTA1.0 dataset.

Model	mAP50(%)↑	mAP75(%)↑	mmAP(%)↑	Params(M)↓	FLOPs(G)↓	Latency(ms)↓
Baseline	75.36	50.64	47.37	4.88	20.45	4.97
+ EP-PAFPN	75.91 (+0.55)	51.34 (+0.70)	47.97 (+0.60)	5.77 (+0.89)	26.29 (+5.84)	5.20 (+0.23)
+ TID-Head	76.15 (+0.24)	51.03 (−0.31)	47.68 (−0.29)	6.19 (+0.42)	27.74 (+1.45)	5.53 (+0.33)
+ PA-DLA	77.27 (+1.12)	51.00 (−0.03)	47.71 (+0.03)	6.19	27.74	5.53

4.4 Comparison with State-of-the-Arts

We compared RTMDet-R2 with previous state-of-the-art methods on the DOTA 1.0 dataset, as shown in Table 7. Under single-scale training and testing, RTMDet-R2-m and RTMDet-R2-l achieve 79.10% and 79.46% mAP50, respectively, outperforming almost all previous anchor-free and anchor-based methods. Moreover, RTMDet-R2-m and RTMDet-R2-l achieve 55.03% and 56.18% mAP75, surpassing RTMDet-R-m and RTMDet-R-l by 0.56% and 0.97% respectively, indicating that the predicted boxes generated by RTMDet-R2 exhibit higher quality. RTMDet-R2-s and RTMDet-R2-t also achieve a relative improvement of 1.50% and 1.91% in mAP50 compared to RTMDet-R-s and RTMDet-R-t, reaching 78.43% and 77.27% respectively. We visualize a portion of the DOTA 1.0 test set results in Fig. 3 (a).

Table 7. Comparison of RTMDet-R2 with previous rotated object detection methods on DOTA 1.0. R50 and X50 denote ResNet-50 and ResNeXt-50 (likewise for R101 and X101). Re50 denotes ReResNet-50, MV2 denotes MobileNet v2 and CRN denotes CSPRepResNet. The bold fonts indicate the best performance.

Method	Backbone	mAP50(%)↑	mAP75(%)↑	mmAP(%)↑
Anchor-based				
ReDet [2]	ReR50	76.25	–	–
Mask OBB [31]	R50	74.86	–	–
Oriented R-CNN [1]	R101	76.28	–	–
S2ANet [4]	R50	74.12	–	–
R3Det [3]	R101	73.80	–	–
Anchor-free				
CFA [33]	R101	75.05	–	–
ProbIoU [18]	R50	70.04	–	–
Oriented RepPoints [32]	Swin-tiny	77.63	–	–
FCOSR-s [7]	MV2	74.05	–	–
FCOSR-m [7]	X50	77.15	–	–
FCOSR-l [7]	X101	77.39	–	–
PPYOLOE-R-s [8]	CRN-s	73.82	–	–
PPYOLOE-R-m [8]	CRN-m	77.64	–	–
PPYOLOE-R-l [8]	CRN-l	78.14	–	–
PPYOLOE-R-x [8]	CRN-x	78.28	–	–
RTMDet-R-t [9]	CSPNeXt-t	75.36	50.64	47.37
RTMDet-R-s [9]	CSPNeXt-s	76.93	50.59	48.16
RTMDet-R-m [9]	CSPNeXt-m	78.24	54.47	50.56
RTMDet-R-l [9]	CSPNeXt-l	78.85	55.21	51.01
RTMDet-R2-t	CSPNeXt-t	77.27	51.00	47.71
RTMDet-R2-s	CSPNeXt-s	78.43	51.66	48.81
RTMDet-R2-m	CSPNeXt-m	79.10	55.03	51.08
RTMDet-R2-l	CSPNeXt-l	**79.46**	**56.18**	**51.13**

As shown in Fig. 1., we compare RTMDet-R2 with previous state-of-the-art real-time RODs on the DOTA 1.0 dataset in terms of parameters, FLOPs, and FPS. RTMDet-R2 t/s/m achieves comprehensive improvements in mAP50 compared to RTMDet-R s/m/l. Additionally, the parameters and FLOPs of RTMDet-R2 t/s/m are reduced by 30.14/54.72/42.91% and 26.26/49.38/37.05% respectively compared to RTMDet-R s/m/l. The mAP50 of RTMDet-R2-s even surpasses that of PP-YOLOE-R-x, which has

10x the FLOPs. RTMDet-R2 avoids the use of special operations like Deformable Convolution or Rotated RoI Align, making it deployable on various hardware platforms. After FP16 TensorRT deployment on a 3090 GPU, RTMDet-R2 t/s/m/l achieves inference frame rates of 181/175/111/88 FPS at a resolution of 1024 × 1024.

Table 8. Comparison with state-of-the-art methods on the HRSC2016 dataset. mAP07 and mAP12 indicate that the results were evaluated under VOC2007 and VOC2012 metrics (%), respectively. We report both results for a fair comparison. The best results are in bold.

Method	Input shape	mAP07	mAP12
R3Det [3]	800 × 800	89.26	96.01
GWD [19]	800 × 800	89.85	97.37
CSL [34]	800 × 800	89.62	96.10
S2ANet [4]	512–800	90.17	95.01
ReDet [2]	512–800	90.46	**97.63**
Oriented RCNN [1]	800–1333	90.50	97.60
RTMDet-R-tiny [9]	800 × 800	90.60	97.10
RTMDet-R2-tiny	800 × 800	**90.63**	97.44

As shown in Table 8, we compared RTMDet-R2 with previous SOTA methods on the HRSC2016 dataset. Among the models, RTMDet-R2-tiny achieves the highest mAP07, reaching 90.63%. Additionally, RTMDet-R2-tiny achieves 97.44% mAP12, which is 0.34% higher than RTMDet-R-tiny. The visualization of the detection results is shown in Fig. 3 (b).

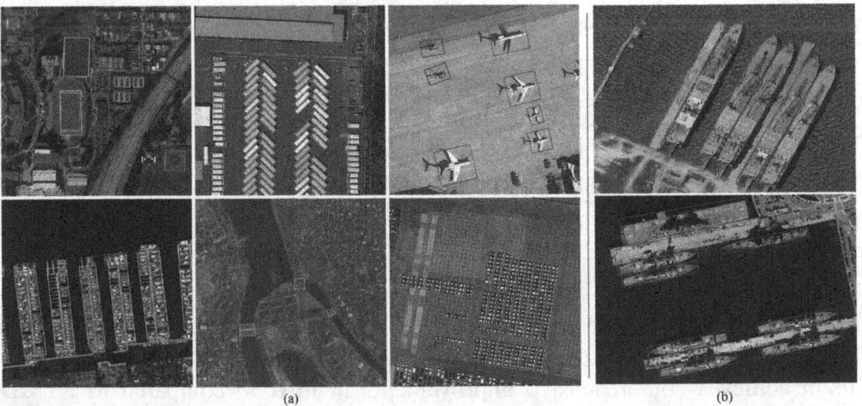

Fig. 3. The RTMDet-R2-m detection result on DOTA1.0 (a) and HRSC2016 (b). The confidence threshold is set to 0.3 when visualizing these results.

5 Conclusion

In this paper, we present RTMDet-R2, an improved real-time rotated object detector. RTMDet-R2 incorporates the enhanced path PAFPN for effective feature map fusion and utilizes the task interaction decouple head to address the imbalance between regression and classification tasks. Additionally, we introduce the ProbIoU-aware dynamic label assignment strategy, which improves training efficiency. The simplicity of the RTMDet-R2 structure enables easy and fast deployment. Extensive experiments on the DOTA and HRSC2016 datasets validate the effectiveness of our approach. Future work will focus on exploring stronger and lighter network architectures and optimizing training strategies to further enhance performance and efficiency.

References

1. Xie, X., Cheng, G.: Oriented R-CNN for object detection. In: ICCV, pp. 3520–3529 (2021)
2. Han, J., Ding, J., Xue, N., Xia, G.: ReDet: a rotation-equivariant detector for aerial object detection. In: CVPR, pp. 2786–2795 (2021)
3. Xue, Y., Qing, L., Junchi, Y.: R3det: Refined single-stage detector with feature refinement for rotating object. arXiv preprint arXiv:1908.05612 (2019)
4. Jiaming, H., Jian, D., Jie, L.: Align deep features for oriented object detection. IEEE Trans. Geosci. Aerial **60**, 1–11 (2021)
5. Youtian, L., Pengming, F.: IENet: Interacting Embranchment One Stage Anchor Free Detector for Orientation Aerial Object Detection. arXiv preprint arXiv: 1912.00969 (2019)
6. Steven, L., Fabrizio, V., Kristian, K.: Dafne: A one-stage anchor-free deep model for oriented object detection. arXiv preprint arXiv:2109.06148 (2021)
7. Zhonghua, L., Biao, H., Zitong, W.: FCOSR: A simple anchor-free rotated detector for aerial object detection. arXiv preprint arXiv:2111.10780 (2021)
8. Wang, X., Wang, G., Dang, Q.: PP-YOLOE-R: An Efficient Anchor-Free Rotated Object Detector. arXiv preprint arXiv:2211.02386 (2022)
9. Lyu, C., Zhang, W., Huang, H.: RTMDet: An Empirical Study of Designing Real-Time Object Detectors. arXiv preprint arXiv:2212.07784 (2022)
10. Lin, T., Dollár, P., Girshick, R., He, K.: Feature pyramid networks for object detection. In: CVPR, pp. 2117–2125 (2017)
11. Liu, S., Qi, L., Qin, H.: Path aggregation network for instance segmentation. In: CVPR, pp. 8759–8768 (2018)
12. Chen, K., Cao, Y.: Feature pyramid grids. arXiv preprint arXiv:2004.03580 (2020)
13. Zheng G., Songtao L., Jian S.: YOLOX: Exceeding YOLO series in 2021. arXiv preprint arXiv:2107.08430 (2021)
14. Chengjian, F., Yujie, Z., Yu G.: TOOD: Task-aligned one-stage object detection. In: ICCV, pp. 3490–3499 (2021)
15. Shifeng, Z., Cheng, C., Yongqiang, Y.: Bridging the gap between anchor-based and anchor-free detection via adaptive training sample selection. In: CVPR, pp. 9759–9768 (2020)
16. Yue, W., Yinpeng, C., Lu, Y.: Rethinking classification and localization for object detection. In: CVPR, pp. 10186–10195 (2020)
17. Zhuang, J., Qin, Z., Yu, H., Chen, X.: Task-Specific Context Decoupling for Object Detection. arXiv preprint arXiv:2303.01047 (2023)
18. Jeffri, M.L., Luis, F.Z., Lucas, N.K., Claudio, J.: Gaussian bounding boxes and probabilistic intersection-over-union for object detection. arXiv preprintarXiv:2106.06072 (2021)

19. Xue, Y., Junchi, Y., Qi, M.: Rethinking rotated object detection with Gaussian Wasserste in distance loss. In: ICML, pp. 11830–11841 (2021)
20. Xue, Y., Yue, Z., Gefan, Z.: The kfiou loss for rotated object detection. arXiv preprint arXiv: 2201.12558 (2022)
21. GuiSong, X., Xiang, B., Jian, D.: Dota: a large-scale dataset for object detection in aerial images. In: CVPR, pp. 3974–3983 (2018)
22. Zikun, L., Liu, Y., Lubin, W.: A high resolution optical satellite image dataset for ship recognition and some new baselines. In ICPRAM, pp. 324–331 (2017)
23. Zhou, Y., Yang, X., Zhang, G.: Mmrotate: a rotated object detection benchmark using pytorch. In: Proceedings of the 30th ACM International Conference on Multimedia, pp. 7331–7334 (2022)
24. Shaoqing R., Kaiming H., Ross G., Jian S.: Faster R-CNN: towards real-time object detection with region proposal networks. Adv. Neural Inf. Process. Syst. **28** (2015)
25. Kaiming, H., Georgia, G., Piotr, D.: Mask R-CNN. In: ICCV, pp. 2980–2988 (2017)
26. Joseph R., Ali F.: Yolov3: An incremental improvement. arXiv preprint arXiv:1804.02767 (2018)
27. ChienYao, W., Alexey, B.: Yolov7: trainable bag-of-freebies sets new state-of-the-art for real-time object detectors. In: CVPR, pp. 7464–7475 (2022)
28. Liu, W., Anguelov, D., Erhan, D.: SSD: single shot multibox detector. In: ECCV, pp. 21–37 (2016)
29. Cai, Z., Fan, Q., Feris, R.S., Vasconcelos, N.: A unified multi-scale deep convolutional neural network for fast object detection. In: ECCV, pp. 354–370 (2016)
30. Yu, Y., Da, F.: Phase-shifting coder: predicting accurate orientation in oriented object detection. In: CVPR, pp. 13354–13363 (2023)
31. Wang, J., Ding, J., Guo, H., Cheng, W., Pan, T.: Mask OBB: a semantic attention-based mask oriented bounding box representation for multi-category object detection in aerial images. Remote Sens. **11**(24), 2930 (2019)
32. Li, W., Chen, Y., Hu, K., Zhu, J.: Oriented reppoints for aerial object detection. In: CVPR, pp. 1829–1838 (2022)
33. Guo, Z., Liu, C., Zhang, X., Jiao, J.: Beyond bounding-box: convex-hull feature adaptation for oriented and densely packed object detection. In: CVPR, pp. 8792–8801 (2021)
34. Yang, X., Yan, J.: Arbitrary-oriented object detection with circular smooth label. In: ECCV, pp. 677–694 (2020)

Boosting Object Detection in Foggy Scenes via Dark Channel Map and Union Training Strategy

Zhanqiang Huo, Sensen Meng, Yingxu Qiao(✉), and Fen Luo

School of Software, Henan Polytechnic University, Jiaozuo 454000, China
qiaoyingxu@hpu.edu.cn

Abstract. Most existing object detection methods in real-world hazy scenarios fail to handle the heterogeneous haze and treat clear images and hazy images as adversarial while ignoring the latent information beneficial in clear images for detection, resulting in sub-optimal performance. To alleviate the above problems, we propose a new dark channel map-guided detection paradigm (DG-Net) in an end-to-end manner and provide an interpretable idea for object detection in hazy scenes from an entirely new perspective. Specifically, we design a unique dark channel map-guided feature fusion (DGFF) module to handle the adverse impact of the heterogeneous haze, which enables the model to focus on potential regions that may contain detection objects adaptively, assign higher weights to these regions, and thus improve the network's ability to learn and represent the features of hazy images. To more effectively utilize the latent features of clear images, we propose a new simple but effective union training strategy (UTS) that considers the clear images as a complement to the hazy images, which enables the DGFF module to work better. In addition, we introduce Focal loss and Self-calibrated convolutions to enhance the performance of the DG-Net. Extensive experiments show that DG-Net outperforms the state-of-the-art detection methods quantitatively and qualitatively, especially in real-world hazy datasets.

Keywords: Object detection · Foggy scenarios · Dark channel map · Feature fusion · Training strategy

1 Introduction

Deep learning models based on data-driven have achieved promising performance in various computer vision tasks. However, these models can often perform well only in no-degraded conditions. The captured images in adverse weather will be impaired at different levels [2], causing detection networks designed for ideal conditions to generalize poorly in real-world adverse scenarios, severely limiting

This Work Is Supported by the National Science Foundation of China (No. 62273292).
Supplementary Material Is Available at https://github.com/ssmemg/DG-Net-SI.

© The Author(s), under exclusive license to Springer Nature Singapore Pte Ltd. 2024
Q. Liu et al. (Eds.): PRCV 2023, LNCS 14436, pp. 365–377, 2024.
https://doi.org/10.1007/978-981-99-8555-5_29

the further application and development of detection models. Therefore, how to improve the performance of detectors in real-world adverse scenes has attracted more and more attention from academia to industry. In this paper, we take a typical hazy scene as an example and attempt to improve the detection accuracy of the detectors under foggy conditions.

Existing methods of object detection in hazy conditions can be roughly divided into two branches: the former is a one-stage detection network [8,9, 15,17,25,26] in an end-to-end manner. The latter is the two-stage detection network [7,10,12,19], which first performs image dehazing as a prerequisite step on the hazy images and then feeds the dehazed images into a pre-trained detection model based on clear images to perform the detection task. Although the two-stage methods achieve exciting visual perception, images that satisfy the human visual experience are not definitely beneficial to the detection networks. The defogging models introduce noise and defects that are difficult to visible by the human visual system, further degrading the detection performance of detectors. Recently, some works [8,9,26] have focused on one-stage detection networks. Huang et al. [9] optimizes the model by introducing the joint loss of repair and detection. Although this strategy is logical to the human mind, the dehazing task and the object detection task have different purposes, and there is a potential conflict between them, often leading to unsatisfactory detection results. Some studies [8,26] use generative adversarial networks to generate domain invariant features by domain adaptation strategy, often causing inefficient feature utilization and sub-optimal detection performance. Therefore, how to design a simple and friendly detection model under hazy conditions with high performance and fast inference speed is a challenging task.

To this end, we present a one-stage detection framework in haze scenes DG-Net based on dark channel map guidance and union training strategy. After the guidance (see Fig. 1), our method focuses more on potential regions (red boxes) that may contain detection objects, achieving better detection performances. We elaborate a dark channel map-guided feature fusion (DGFF) module to improve the feature extraction capability of the network from hazy images. In addition, we propose a simple but effective union training strategy (UTS), which can leverage those features from clear and hazy images. In a nutshell, the main highlights of this work can be summarized as follows:

- We design a novel feature fusion guidance module based on a dark channel map called DGFF for detection in hazy conditions, which enables the model to focus more on potential regions that may contain detection objects. To the best of our knowledge, we are the first to directly use the haze density information for detection tasks in hazy conditions.
- We introduce a simple but effective union training strategy that achieves friendly detection in hazy scenarios, improving object detection performance.
- Extensive experiments on detection benchmarks demonstrate the outstanding performance of our method against state-of-the-art methods.

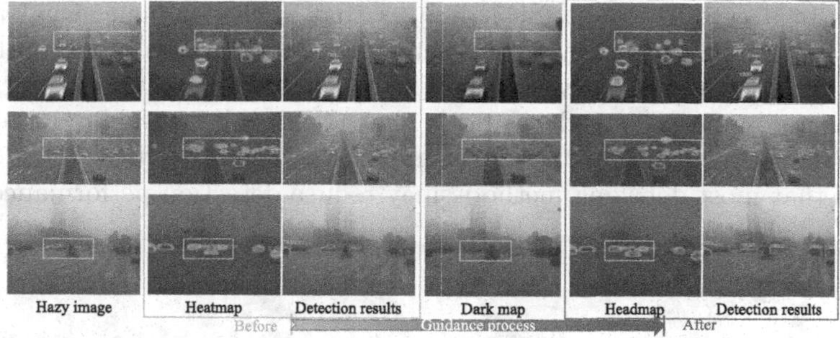

Fig. 1. Example guided results of the DG-Net on foggy scenarios. (Zoom in to see the details. (Color figure online))

2 Related Work

2.1 Object Detection

Object detection models based on data-driven algorithms can be classified into anchor-based and anchor-free. Anchor-based object detection models [20,21] need to set anchors covering as many detection objects as possible. The addition of anchors has improved the detectability of object detectors while introducing redundant parameters. Therefore, some works [3,5,23] have proposed anchor-free methods to solve this obstacle. YOLOX [5] series intelligently introduces the size information from the downsampling process of the previous backbone into the detection heads, thus enabling the localization of the object and significantly reducing the number of parameters in the detector.

2.2 Object Detection in Adverse Scenarios

Currently, object detection algorithms in adverse scenarios can be divided into one-stage and two-stage methods. In one-stage detection algorithms [8,9,15,17, 25,26], due to the lack of large-scale object detection datasets in hazy weather, early work [2] directly uses synthetic degraded datasets to train the detectors, which makes the models perform weakly on real-world degraded images, see Fig. 2(b). Several efforts [8,26] introduce generative adversarial networks to align features between domains, but such methods inevitably discard some beneficial features that improve detection accuracy, as shown in Fig. 2(c). Yang et al. [26] propose a domain adaptation framework that preserves depth information and background details during the feature alignment. A few recent works [9,17,25] combine image restoration and object detection to optimize the network and achieve remarkable performance simultaneously. Wang et al. [25] design a practical and joint detection framework that bridges dehazing and detection tasks via a unified learning architecture.

For two-stage detection algorithms [7,10,12,18,19], these methods achieve a significant visual enhancement, which improves the detection performance of detectors to a certain extent. Still, images that satisfy human visual perception are not definitely friendly to the detectors. Because the evaluation metrics of dehazing do not positively correlate with the final detection performance, the dehazing operation introduces noise that is invisible to the human eye but weakens the performance of the detectors, as shown in Fig. 2(a). Pei et al. [18] indicates that most dehazing methods may reduce detection performance.

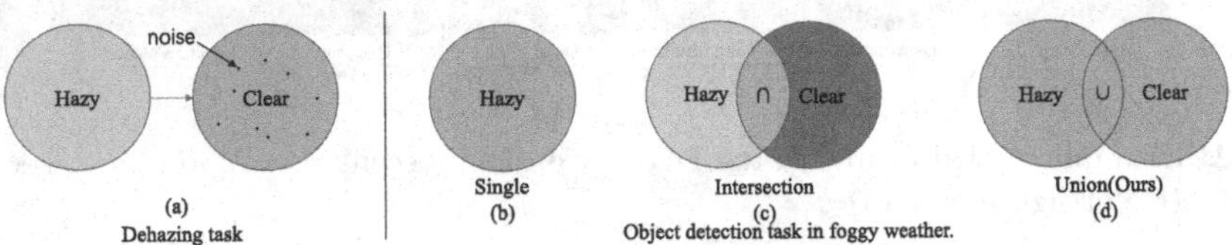

Fig. 2. The difference between image dehazing and foggy object detection tasks. (a) Dehazing task. (b), (c) and (d) Object detection task in foggy weather.

2.3 Attention Mechanism and Dark Channel Prior

Attention mechanisms [1] have been introduced to computer vision tasks inspired by the phenomenon that the human visual system can naturally and efficiently discover critical information in complex environments.

He et al. [7] found by counting the channel pixel values of haze-free images that in most non-sky areas, most local patches in haze-free outdoor images have very low-intensity values of at least a color channel in the RGB images. For hazy images, the intensity of these dark pixels in this channel is mainly caused by scattered light in the air. Therefore, the dark channel map can directly provide an accurate haze estimate of the hazy images.

3 Methodology

3.1 Overview of DG-Net

Figure 3 shows the overview structure of the DG-Net. DG-Net consists of four essential parts: a feature extraction module (Backbone), a dark channel map-guided feature fusion module (DGFF), a feature fusion network (PAFPN), and the detection heads module (Heads). Note that the DGFF module is well-designed for the detection task under hazy conditions. PAFPN and Heads are reasonable improvements on the advanced object detector YOLOXs [5].

Specifically, given a hazy image I_h, we first introduce the DGFF module to generate a weighted map F_{att}. Then, we multiply the feature maps F_b extracted by the Backbone and the weighted map F_{att} element by element to obtain the weighted feature maps F_w. Finally, we feed the weighted feature map into the PAFPN and Heads to produce the final prediction.

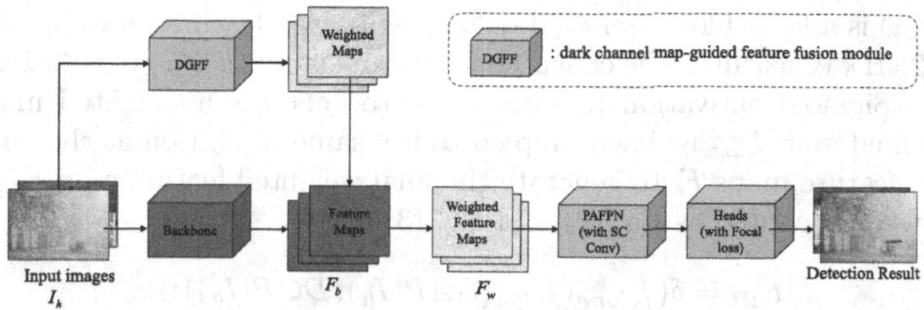

Fig. 3. Overview structure of the DG-Net. DGFF refers to Dark channel map-Guided Feature Fusion module. SC Conv refers to self-calibrated convolutions.

3.2 Dark Channel Map-Guided Feature Fusion Module

Under the hazy weather, the haze density information in images is essential for the detection task. Previous works have yet to consider it, and a few works [6,26] have only acted on haze density information as a constraint on the enhancement or reconstruction of the subtasks.

To this end, we propose a well-designed dark channel map-guided feature fusion module (DGFF) that guides the fusion of multi-scale feature maps, which embeds the prior information related to haze density into the network. Such an approach not only suggests differences in haze density at different spatial locations but also provides clues for the network to find those patches where potential detection objects may exist, significantly improving the feature modeling capability of the detector. Figure 4 shows the pipeline of our DGFF module.

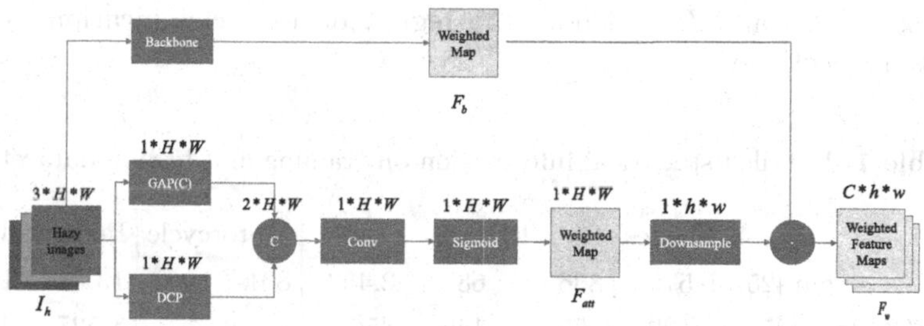

Fig. 4. The pipeline of our DGFF module.

In detail, the hazy image is subjected to global average pooling and dark channel prior pooling operation to obtain two pooled features. The dark channel prior pooling are denoted as Eq. (1).

$$DCP(I_h) = \min_{y \in \Omega(x)} ((\min_{c \in \{r,g,b\}} I_h{}^c(y)) \tag{1}$$

where $\Omega(x)$ is a local block centered on x; The pooled features are concatenated to get a 2×H×W feature. The concatenate feature enters a 7*7 convolution layer. Finally, a Sigmoid activation function layer to generate a weighted map F_{att}. The weighted map F_{att} is downsampled to the same resolution as the backbone extracted feature maps F_b to generate the final weighted feature map F_w. These operations are denoted as Eq. (2) and Eq. (3).

$$F_{att} = \delta(f_{Conv}(f_{Con}(GAP(I_h), DCP(I_h)))) \tag{2}$$

$$F_w = F_{att\downarrow} \otimes F_b \tag{3}$$

where F_{att} denotes the generated weighted map; DCP denotes the dark channel prior process; GAP denotes global average pooling; f_{Con} denotes the Concat function; f_{Conv} denotes 7*7 convolution layer; δ denotes the Sigmoid activation function layer; F_b denotes the backbone extracted original feature maps; F_{w} denotes the weighted feature maps; $\otimes$ denotes the element-wise multiplication; $\downarrow$ denotes the downsampled operation.

3.3 Focal Loss and Self-calibrated Convolutions

Focal Loss. Due to the specificity of the hazy scenes, the class imbalance is even more prominent for detection tasks in hazy conditions, as confirmed by the statistical information of the datasets in Table 1. Inspired by RetinaNet [13], the focal loss is introduced in our network to attenuate the impact of class imbalance. The DG-Net loss function can be defined as:

$$L_{total} = L_{IoU} + L_{Cls} + L_{FL} \tag{4}$$

Here, L_{IoU}, L_{Cls}, and L_{FL} denote the regression loss, classification loss, and focal loss, respectively.

Table 1. Detailed statistical information on training and testing datasets.

Dataset	Images	Bicycle	Bus	Car	Motorcycle	Person	Total
VOC-FOG-train [25]	9,578	836	684	2,453	801	13,519	18,293
VOC-FOG-test [25]	2,129	155	156	857	131	3,527	4,826
FDD [22]	101	17	17	425	9	269	737
RTTS [11]	4,332	534	1,838	18,413	862	7,950	29,597

Self-calibrated Convolutions. Most hazy images have a non-uniform distribution of haze, with significant differences in haze density between the foreground and background, and different areas of the same object are affected by the haze at different levels. We mitigate the negative impact of local haze overload on the objects by expanding the receptive fields of the convolution layers. For this purpose, we introduce an improved CNNs structure in the proposed

DG-Net network called the self-calibrated convolutions [16]. It can sense long-distance dependencies between spaces and channels around each spatial patch, thus vastly enlarging the receptive fields of the convolutional layer and improving the network's representation capabilities.

3.4 Union Training Strategy

There are differences between object detection in hazy conditions and dehazing tasks. The goal of image dehazing is to push the dehazing image as far away from the original hazy image as possible to pull as close as possible to the hazy-free image. This difference between the hazy and clear images in detecting under hazy conditions is not necessarily a confrontation but can even as a complement. The object features in the real-world hazy images are not guaranteed to be closer to the hazy or hazy-free features.

Existing approaches use the features from a single dataset or the intersection of clear and hazy datasets to train the detection networks. None of the above strategies can fully utilize clear and hazy features. Inspired by [24], we propose a union training strategy (UTS) for the detection task in hazy conditions to address these issues, as shown in Fig. 2(d). In detail, we use a mixed dataset X_h of hazy synthetic images with 10% clear images as the training dataset. It substantially increases the diversity and richness of features in the training dataset on the one hand and makes our model more robust on the other. More details about the union training strategy(UTS) can be found in Algorithm 1 and supplementary material.

Algorithm 1: Union Training Strategy

input : Hybrid datasets X_h with 10% clear images, total epoch N=200.
output: Best detection model in hazy conditions P^{best}.

Initialize model P^0, in addition, the backbone of the detection framework with pretrained weights on the COCO[14] dataset;
for $i \leftarrow 0$ **to** N **do**
  if $i < 100$ **then**
    Freeze pretrained backbone parameters;
    Optimize non-freezing parameters of the current model P^{now} by minimizing detection loss;
    Compare and update with the previous best model P^{best};
  else
    Unfreeze pretrained backbone parameters;
    Optimize all parameters of the current model P^{now} by minimizing detection loss;
    Compare and update with the previous best model P^{best}.

4 Experiments

4.1 Datasets

We choose RTTS [11] and Foggy Driving Dataset(FDD) [22] as real-world test datasets. In addition, to facilitate the training and validation of our model, we also introduce a synthetic hazy dataset VOC-FOG [25] based on the VOC [4] dataset. Table 1 shows the statistical details of these datasets.

4.2 Implementation Details

We used the Adam optimizer with an initial rate of 10^{-3}. The learning rate was adjusted using a cosine annealing and the minimum rate of 10^{-5}. The batch is set to 16. We empirically set the total number of epochs as 200, with the first 100 epochs frozen for training and the last 100 epochs unfrozen for fine-tuning. See the supplementary material for more experimental details and parameter settings.

4.3 Comparison with the SOTAs

For a fair comparison, we retrained these compared algorithms on the VOC-FOG [25] synthetic dataset for the one-stage methods according to the setting in the original paper. For the two-stage methods, we first defogged using the different dehazing methods. Then we fed the dehazed images into a baseline YOLOXs* [5] detection model, trained on haze-free images of the VOC-FOG [25] dataset.

The fifth column of Table 2 shows the mAP metrics of our DG-Net and the current advanced object detection algorithms in hazy on the VOC-FOG [25] test dataset. The comparison reveals that our DG-Net achieves better performance on the synthetic dataset, reaching the upper-middle level of current advanced object detection methods in hazy conditions. We analyze that this may be related to our experimental methods, where we slightly sacrifice the performance of the model on the synthetic dataset to improve the generalization ability in real-world hazy scenarios.

Columns 6 and 7 of Table 2 show the results for all compared methods on two real-world datasets. Our method achieves the highest mAP on FDD [22] and RTTS [11] datasets compared to the current SOTA methods, and our DG-Net obtains the best detection performance in real-world scenes, especially on the RTTS [11] dataset, where our method achieves a 4.25% accuracy improvement, significantly ahead of the SOTA methods.

For qualitative comparison, our method is visually compared to the existing three best-performing methods YOLOXs [5], DCP-YOLOXs* [7], TogetherNet [25]. Figure 5 shows the detection results of the different methods on the real-world hazy images. As observed, our method DG-Net can detect more object instances compared to the SOTA methods. More detection examples can be seen in the supplementary material.

4.4 Ablation Study

To evaluate the validity of our method, we conducted adequate ablation studies to analyze the impact of different module combinations. Table 3 shows the results of our ablation experiments, and as observed, each module of our approach contributes to improving detection performance. Note that we are surprised the DGFF can work better with a union training strategy.

Table 2. Quantitative comparison with state-of-the-art methods on test datasets. The best result in each column is in red, and the second is in blue. Paired denotes whether paired images are required for training network. * indicates that the detection model is trained with clean images of the VOC-FOG dataset.

Method	Publication	Paired	VOG-FOG-test	FDD	RTTS
YOLOXs [5]	arXiv'21	NO	80.07	31.54	51.23
YOLOXs* [5]	arXiv'21	NO	72.88	30.07	50.44
DCP-YOLOXs* [7]	TPAMI'11	NO	80.86	29.43	50.81
AOD-YOLOXs* [10]	ICCV'17	YES	77.21	31.18	47.51
Semi-YOLOXs* [12]	TIP'20	YES	78.56	31.26	50.01
FFA-YOLOXs* [19]	AAAI'20	YES	73.62	26.65	50.48
MS-DAYOLO [8]	ICIP'21	YES	83.42	33.7	52.41
DS-Net [9]	TPAMI'21	YES	65.89	29.74	32.71
IA-YOLO [17]	AAAI'22	NO	64.77	18.34	35.66
TogetherNet [25]	PG'22	YES	85.90	34.93	61.55
DG-Net(Ours)	/	NO	79.84	35.52	65.80

Fig. 5. Detection results by different methods on real-world foggy datasets. Zoom in for best view. (Color figure online)

Figure 6 shows that the noticeable improvements in the detection results and heatmaps after adding the DGFF module. We can find that the DG-Net can detect more objects with higher confidence scores after adding the DGFF module. More example results can be shown in Fig. 1 and the supplementary material.

Table 3. Ablation experiment results with different module combinations on RTTS dataset.

Baseline	UTS	Focal loss	SC-Conv	DGFF	mAP
✓					51.23
✓	✓				61.20
✓		✓	✓	✓	62.65
✓	✓	✓			63.64
✓	✓	✓	✓		64.01
✓	✓	✓	✓	✓	65.80

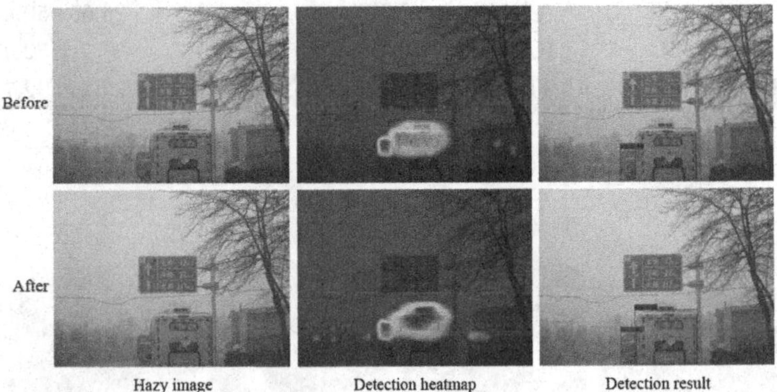

Fig. 6. Ablation studies on our DGFF module.

4.5 Efficiency Analysis

Table 4 shows the results of efficiency analysis for different models. We use a single NVIDIA A100 GPU to test the images with 640 × 640 × 3 resolution. As a result, while the real-time performance of our DG-Net is excellent, the detection performance of our method has also been improved by a wide margin.

Table 4. Efficiency comparison with different models on images of 640 × 640 pixels.

Model	Params(M)	FPS/Time	GFLOPS(G)	mAP
YOLOXs [5]	8.94	68.13/14ms	26.77	51.23
TogetherNet [25]	15.78	32.36/31ms	61.93	61.55
DG-Net(Ours)	10.02	58.68/17ms	29.64	65.80

5 Conclusion

In this paper, we present a novel dark channel map-guided detection network DG-Net for foggy scenarios. DG-Net exploits the latent information in the dark channel map to enhance the network's feature extraction and perception capabilities through a well-designed dark channel map-guided feature fusion (DGFF) module. Besides, we solve the detection problem of hazy images from a new perspective by a union training strategy, which effectively improves the richness of the object features and vastly improves the performance of detectors. Furthermore, we enhance DG-Net performance with Focal loss and Self-calibrated convolutions in the network. Note that the DG-Net does not need a corresponding haze-free image as ground truth so that it can be friendly applied to real-world hazy scenarios. In summary, the DG-Net can perform well by only adding a few parameters and has a faster detection speed. Experimental results prove that DG-Net achieves superior detection performance on both real and synthetic datasets.

In the future, we intend to design a more effective prior information-guided feature fusion module. Besides, it is also a valuable research direction to ascertain the most suitable ratio of hybrid datasets for object detection tasks in hazy conditions.

References

1. Brauwers, G., Frasincar, F.: A general survey on attention mechanisms in deep learning. IEEE Trans. Knowl. Data Eng. **35**(4), 3279–3298 (2023)
2. Cui, Z., Zhu, Y., Gu, L., Qi, G.J., Li, X., Zhang, R., Zhang, Z., Harada, T.: Exploring resolution and degradation clues as self-supervised signal for low quality object detection. In: Computer Vision - ECCV 2022, pp. 473–491. Springer, Cham (2022). https://doi.org/10.1007/978-3-031-20077-9_28
3. Duan, K., Bai, S., Xie, L., Qi, H., Huang, Q., Tian, Q.: Centernet: keypoint triplets for object detection. In: 2019 IEEE/CVF International Conference on Computer Vision (ICCV), pp. 6568–6577 (2019)
4. Everingham, M., Van Gool, L., Williams, C.K.I., Winn, J., Zisserman, A.: The pascal visual object classes (voc) challenge. Int. J. Comput. Vision **88**(2), 303–338 (2010)
5. Ge, Z., Liu, S., Wang, F., Li, Z., Sun, J.: Yolox: exceeding yolo series in 2021. arXiv preprint arXiv:2107.08430 (2021)

6. Guo, C., Yan, Q., Anwar, S., Cong, R., Ren, W., Li, C.: Image dehazing transformer with transmission-aware 3d position embedding. In: 2022 IEEE/CVF Conference on Computer Vision and Pattern Recognition (CVPR), pp. 5802–5810 (2022)
7. He, K., Sun, J., Tang, X.: Single image haze removal using dark channel prior. IEEE Trans. Pattern Anal. Mach. Intell. **33**(12), 2341–2353 (2011)
8. Hnewa, M., Radha, H.: Multiscale domain adaptive yolo for cross-domain object detection. In: 2021 IEEE International Conference on Image Processing (ICIP), pp. 3323–3327 (2021)
9. Huang, S.C., Le, T.H., Jaw, D.W.: Dsnet: joint semantic learning for object detection in inclement weather conditions. IEEE Trans. Pattern Anal. Mach. Intell. **43**(8), 2623–2633 (2021)
10. Li, B., Peng, X., Wang, Z., Xu, J., Feng, D.: Aod-net: all-in-one dehazing network. In: 2017 IEEE International Conference on Computer Vision (ICCV), pp. 4780–4788 (2017)
11. Li, B., et al.: Benchmarking single-image dehazing and beyond. IEEE Trans. Image Process. **28**(1), 492–505 (2019)
12. Li, L., Dong, Y., Ren, W., Pan, J., Gao, C., Sang, N., Yang, M.H.: Semi-supervised image dehazing. IEEE Trans. Image Process. **29**, 2766–2779 (2020)
13. Lin, T.Y., Goyal, P., Girshick, R., He, K., Dollár, P.: Focal loss for dense object detection. In: 2017 IEEE International Conference on Computer Vision (ICCV), pp. 2999–3007 (2017)
14. Lin, T.Y., Maire, M., Belongie, S., Hays, J., Perona, P., Ramanan, D., Dollár, P., Zitnick, C.L.: Microsoft coco: Common objects in context. In: Computer Vision - ECCV 2014, pp. 740–755. Springer, Cham (2014)
15. Liu, H., Jin, F., Zeng, H., Pu, H., Fan, B.: Image enhancement guided object detection in visually degraded scenes. IEEE Trans. Neural Networks Learn. Syst., 1–14 (2023)
16. Liu, J.J., Hou, Q., Cheng, M.M., Wang, C., Feng, J.: Improving convolutional networks with self-calibrated convolutions. In: 2020 IEEE/CVF Conference on Computer Vision and Pattern Recognition (CVPR), pp. 10093–10102 (2020)
17. Liu, W., Ren, G., Yu, R., Guo, S., Zhu, J., Zhang, L.: Image-adaptive yolo for object detection in adverse weather conditions. Proceedings of the AAAI Conference on Artificial Intelligence 36, pp. 1792–1800 (2022)
18. Pei, Y., Huang, Y., Zou, Q., Lu, Y., Wang, S.: Does haze removal help cnn-based image classification? In: ECCV 2018, pp. 697–712. Springer, Cham (2018)
19. Qin, X., Wang, Z., Bai, Y., Xie, X., Jia, H.: Ffa-net: feature fusion attention network for single image dehazing. In: Proceedings of the AAAI Conference on Artificial Intelligence, vol. 34, pp. 11908–11915 (2020)
20. Redmon, J., Farhadi, A.: Yolov3: An incremental improvement (2018)
21. Ren, S., He, K., Girshick, R., Sun, J.: Faster R-CNN: towards real-time object detection with region proposal networks. IEEE Trans. Pattern Anal. Mach. Intell. **39**(6), 1137–1149 (2017)
22. Sakaridis, C., Dai, D., Van Gool, L.: Semantic foggy scene understanding with synthetic data. Int. J. Comput. Vision **126**(9), 973–992 (2018)
23. Tian, Z., Shen, C., Chen, H., He, T.: Fcos: fully convolutional one-stage object detection. In: 2019 IEEE/CVF International Conference on Computer Vision (ICCV), pp. 9626–9635 (2019)
24. Wang, W., Li, B., Gou, Y., Hu, P., Peng, X.: Relationship quantification of image degradations. ArXiv abs/2212.04148 (2022)

25. Wang, Y., et al.: Togethernet: bridging image restoration and object detection together via dynamic enhancement learning. Comput. Graph. Forum **41**(7), 465–476
26. Yang, X., Mi, M.B., Yuan, Y., Wang, X., Tan, R.T.: Object detection in foggy scenes by embedding depth and reconstruction into domain adaptation. In: Computer Vision - ACCV 2022, pp. 303–318. Springer, Cham (2023)

Object Centric Body Part Attention Network for Human-Object Interaction Detection

Zhuang Liu and Xiaowei Zhang(✉)

School of Computer Science and Technology, Qingdao University, Qingdao, China
xiaowei19870119@sina.com

Abstract. The current transformer-based human object interaction (HOI) detection methods have achieved great progress, however, these methods adopt the same structre of decoder to detect human and object, which limits the accuracy of object feature extraction, thereby limiting the accuracy of HOI detection. And due to the distribution differences of multi-granularity features between human and object, the key of HOI detection is object centric interaction with the correlative human body parts. To address this issue, we propose an Object Centric Body Part Attention Network for Human-Object Interaction. First, we introduce a dual-branch decoder for human and object detection named Object Centric Decoder (OCD), where one focuses on quering objects and another pay attention to catch human who interacts with them. Secondly, in order to exploit more fine-grained human body information centered around object, we propose a Body Part Attention (BPA) module to obtain the interactive human body part features for HOI detection. We evaluated our proposed OBPA on the HICO-DET and V-COCO datasets, which significantly outperforms existing counterpart (1.7 mAP on V-COCO, and 0.9 mAP on HICO-DET compared to GEN-VLKT). Code will be available on https://github.com/zhuang1iu/OBPA-NET.

Keywords: Human-Object Interaction · Decoupling Human-Object Decoder · BodyPart Attention

1 Introduction

Human-Object Interaction (HOI) detection is a task of recognizing "a set of interactions" in an image. As a downstream task of object detection [1], HOI detection has received increasing attention in recent years due to its significant application potential. The HOI detection task involves locating the interacting subjects (i.e., humans) and interaction targets (i.e., objects), classifying the interaction labels, and outputting the triplets of humans, objects, and interactions. HOI detection requires a deeper understanding of the semantic information in images to accurately distinguish human activities.

Intuitively, human object interaction (HOI) detection first needs to determine the position of human and objects, as well as the class of objects. Then,

© The Author(s), under exclusive license to Springer Nature Singapore Pte Ltd. 2024
Q. Liu et al. (Eds.): PRCV 2023, LNCS 14436, pp. 378–391, 2024.
https://doi.org/10.1007/978-981-99-8555-5_30

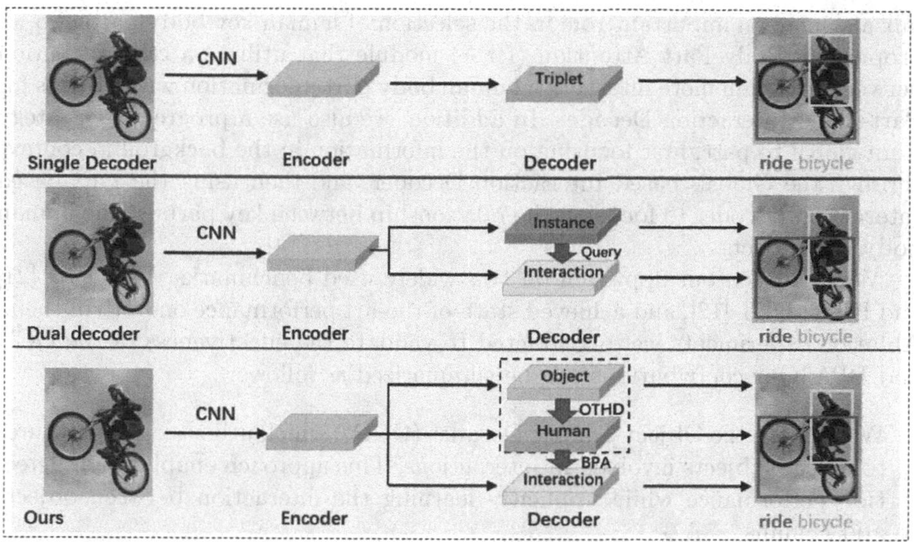

Fig. 1. Architecture comparison of different transformers-based HOI method. Single-decoder method adopts a single decoder to directly detect HOI triplets. Dual-decoder method utilizes separate decoders detect individual objects and interactions. Ours further decouples the instance detection branch and utilizes OCD to learn implicit connections between human and object pairs, then Body Part Attention (BPA) is used to select fine-grained features of human key body parts for interaction detection.

the interaction class is determined by the class information of objects, the overall features of human, and key body parts features. There are two traditional CNN based methods: the two-stage method first using a pre trained object detector to detect human and objects, and then inputting the generated person object pairs into the interactive classifier. One-stage methods propose predicting both the human-object offset vectors and action classes simultaneously by utilizing interaction points between humans and objects. With the recent success of Transformers in object detection, Transformer-based HOI detection methods have been actively developed and have become the main architectural foundation for this task. However, Existing transformer-based methods rely on a single decoder to handle human and object detection tasks, which limits their ability to adapt to different subtasks in multi task learning, resulting in poor object detection performance.

Based on this, we propose an object centric dual branch decoder (OCD), where one branch focuses on querying objects while others focus on human in contact with the objects. As shown in Fig. 1, our method divides human and object detection into two related branches, improving detection performance while also implicitly learning the interactive relationships between humans and objects. In addition, in the process of human object interaction (HOI) detection, the features of objects not only affect the determination of interaction categories, but also play an important role in the selection of human key body parts. So we propose an Body Part Attention (BPA) module that utilizes a cross attention network to obtain more interactive human body part information with objects for Part-based Interaction Decoder. In addition, we also use a progressive strategy from global to part, first focusing on the information in the background context through the Global- based Interaction Decoder, and then using the Part-based Interaction Decoder to focus on the relationship between key parts of the human body and objects.

We evaluated our approach on two widely used benchmarks, V-COCO [21] and HICO-DET [12], and achieved state-of-the-art performance on both of them. Ablation experiments were conducted to validate the effectiveness of the OCD and BPA. Our contributions can be summarized as follows:

- We propose the Object Centric Decoder (OCD), which utilizes object features to detect subjects involved in interactions. This approach enables high detection performance while implicitly learning the interaction between objects and humans.
- We introduce the Body Part Attention (BPA) module, which flexibly selects key body parts of humans. By leveraging the relationship between object features and human features, this module determines the interactive body parts of humans and models the relationship between fine-grained human body part features and object features.
- Our approach has achieved a 1.7mAP gain on V-COCO and a 0.9mAP promotion on HICO-Det compared with the previous state-of-the-art method GEN-VLKT [15].

2 Related Work

2.1 Human-Object Interaction.

Two-Stage Methods. Early methods for HOI detection stylishly employed two-stage approaches [9]. In the first stage, object detection methods were used to locate humans and objects. In the second stage, features of humans and objects were extracted and fed into classifiers to predict their interactions. Some early methods emphasized the second stage by introducing models that captured contextual information [26] or structural messages to model the relationship between humans and objects [14]. However, the main challenges of two-stage methods lie in effectively integrating human-object pairs and complex semantic information. Additionally, the efficiency of two-stage methods is constrained by the sequential architecture.

One-Stage Methods. A new trend in HOI detection is the adoption of one-stage methods, which leverage strong feature representations to perform human-object pair detection and interaction prediction in parallel. Liao et al. [17] proposed a proposal-free method, PPDM, which utilizes keypoints as the key. These keypoints represent the center points of minimum enclosing bounding boxes that encapsulate the human-object pairs involved in the predicted interaction. Currently, several Conv Transformer based HOI methods deploy the detection transformer [18–20, 25] architecture and directly predict HOIs as quintuplets (human, interaction class, object class, human box, object box). These methods share similarities to a large extent but differ in backbone networks or detection heads, such as HOTR [3], HOI Trans [7], and QPIC [10]. Zhang et al. proposed CDN [6], which utilizes an additional transformer decoder to predict interactions based on instance feature tokens.

2.2 Part Based Interaction Detection

Different parts of humans provide more detailed information for HOI detection. Fang et al. [27] utilized pairwise body part attention models to learn the attention on key parts and their correlations. Wan et al. [14] proposed a multi-level relation detection strategy that utilizes human pose cues to capture the global spatial configuration of relationships and serves as an attention mechanism to learn the attention on key part features. Tin++ [28] combines human body pose and body part features to extract deeper visual clues of interactions for learning interactions. These methods indicate that fine-grained features of human body parts are important for HOI detection.

3 Method

3.1 Overview

As shown in Fig. 2, our pipeline consists of three main modules: an *image feature extraction*, an *object centric decoder*, and an *interaction decoder*. Following the previous work [6,10,30], for a given input image $x \in \mathbb{R}^{3\times H_0 \times W_0}$, we initially used ResNet-50 to extract image features, and then input them into the transformer encoder along with position encoding. The feature map output by the transformer encoder is $f_{enc} \in \mathbb{R}^{C\times H\times W}$, where C is the number of channels and H, W are the size of the feature map.

Subsequently, the feature map f_{enc} is fed into the OCD, which comprises two branches: the first branch produces out^{obj}, responsible for predicting the bounding box and class of the object, while the second branch produces out^{hum}, responsible for predicting the bounding box of the humans. Then we use feed forward networks (FFNs) to obtain human bbox, obj bbox, and obj class. Next, we utilize the feature map f_{enc} alongside the outputs of the OCD as inputs for the Interaction Decoder. By traversing through the global-based interaction decoder (GBID) and the part-based interaction decoder(PBID), we obtain out^{int}. This output is further processed by a feed-forward network to predict the specific interaction class.

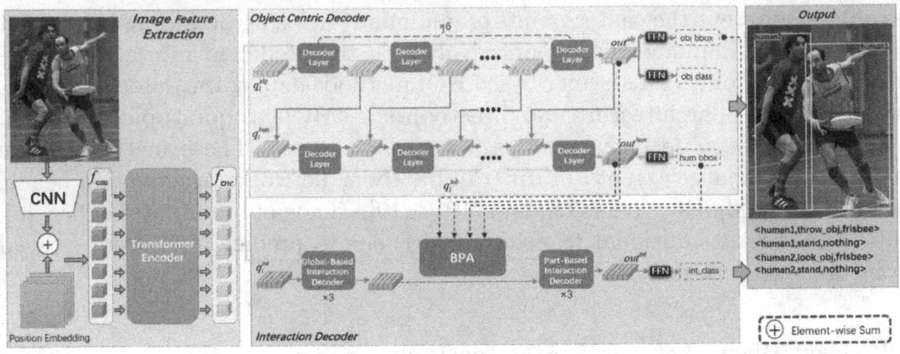

Fig. 2. The overall framework of our proposed method. BPA refer to the Body Part Attention module

3.2 Object Centric Decoder

Previous methods have employed a single Transformer Decoder to simultaneously detect humans and objects, implicitly capturing the interaction between them. However, the object detection performance of this approach is inferior to that of a standalone object detection branch. The category and spatial information of objects play a vital role in determining the interaction categories. Based on this idea, we designed three different branch architectures for human and object detection, as shown in Fig. 3.

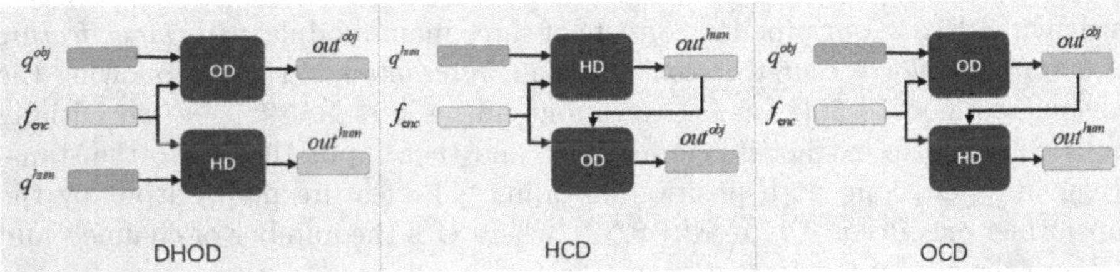

Fig. 3. DHOD, HCD and OCD denotes the decoupling human and object decoder module, human centric decoder module and object centric decoder module. DHOD is to directly decouple the human decoder and object decoder modules. HCD is the human centric decoder, where the output of the human decoder is used as the query embedding input for the object decoder. Finally, OCD utilize the object decoder output as query embedding for human decoder, implicit learning of the interactive relationship between the two.

After different experiments, we ultimately adopted the object centric decoder which is a dual-branch architecture. Each branch, denoted as B, consists of L layers and takes learnable embeddings $\mathbf{Q} = \left\{q_i^B\right\}_{i=1}^N$ and image features f_{enc} as input, where $B \in \{O, H\}$ represents objects and humans, the learnable embeddings is inherited from DETR and interacts with image features through cross attention. In each layer of the object detection branch, $\mathbf{Q}^O$ is refined through a transformer decoder layer. The output of object detection branch serves as the input for the query of the subsequent layer $\mathbf{Q}^O_{i+1}$ and also acts as the query embedding $\mathbf{Q}^H_i$ for the corresponding layer in the human detection branch. The result of two branches (out^{obj}, out^{hum}) can be represented as follows:

$$out^{obj}_{(l)} = \mathrm{Dec}^O_{(l)}(\mathbf{Q}^O_{(l-1)}, f_{enc}) \tag{1}$$

$$out^{hum}_{(l)} = \mathrm{Dec}^H_{(l)}(\mathbf{Q}^O_{(l)} + \mathbf{Q}^H_{(l-1)}, f_{enc}) \tag{2}$$

where $\mathrm{Dec}(q, kv)$ denotes a transformer decoder layer and l denotes the num of decoder layers. Finally, we used FFN networks to predict object classification, object bounding boxes, and human bounding boxes.

3.3 Interaction Decoder

Interaction Detector from Global to Part. The determination of many interaction categories requires not only the consideration of the global features of entire image but also the fine-grained features of key body parts. Therefore, we design an interaction decoder, $\mathrm{Dec}(q, kv)$, that operates from global to local. We set different learnable query embeddings $\mathbf{Q}^I_g, \mathbf{Q}^I_p$ for the two decoder processes to focus on both the global features of the image and the important features of the key body parts.

First, we use $\mathbf{Q}^I_g$, along with the output f_{enc} from the encoder, as the input for the global-based interaction decoder. After passing through L decoder layers for updating, the final output out^I_g serves as the query embedding input for the part-based interaction decoder. Subsequently, in the part-based interaction decoder, we utilize the attention mask matrix functionality of the transformer. Only image features containing key body parts are retained for attention computation, while features from other regions are masked out. The attention mask is pre-computed based on the results of body part bounding box detection.

Body Part Attention. In a human object interaction relationship, the interaction between various human body parts and objects is different, and it is more important to focus on the interaction between more fine-grained key parts of the human body and objects. Based on this, we utilized the Cross Attention structure to predict key parts with stronger interaction with objects, and designed a Body Part Attention (BPA) module as shown in Fig. 4.

We employ a cross-attention layer network to compute the regions of the body that exhibit higher interaction with objects and utilize a feed-forward network to output the importance scores, denoted as P_{score}, for the six body parts. Any body part with a $P^n_{score}(n < 6)$ greater than k is retained, resulting in the generation of the key body part mask M_{part} for all individuals in the image.

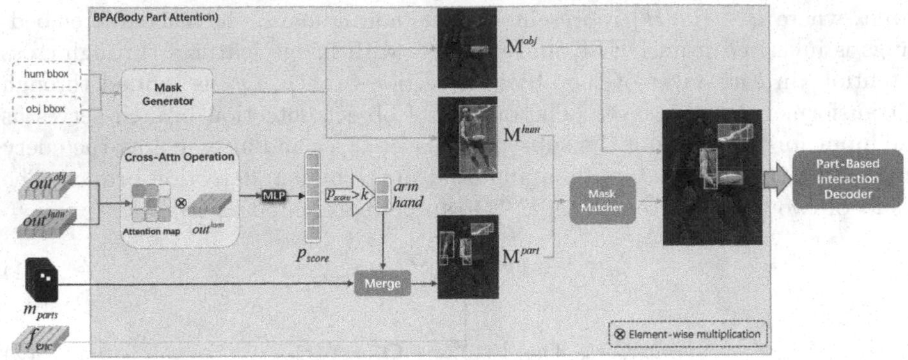

Fig. 4. The process of BPA. Where hum bbox, obj bbox,out^{obj} and $out^{hum}$are the outputs of the OCD, and m_{parts} is the masks of various parts of the human body. f_{enc} is the output of the encoder. P_{score} is the interaction score of various parts of the human body, and k is the threshold of key body part.

$$P_{score} = MLP(CrossAttn(out^{obj}, out^{hum})) \tag{3}$$

where $CrossAttn(q, kv)$ is a cross-attention operation, MLP is a multi-layer perceptron composed of multilayer Linear layers.

Part-Based Interaction Decoder. After obtaining key body parts using BPA, we propose a Part-Based Interaction Decoder that only focuses on key body parts and objects using the Transformer's mask mechanism. To obtain the key body part masks for each person, we first utilize pose estimation methods [22,23] to obtain the positions of the human body poses. Then, following the approach in HAKE [31], we divide the human body into six body parts: feet, legs, hip, hands, arms, and head. Set the n-th body part box of each person in the picture to $P^n_{box} = [w^n_1, h^n_1, w^n_2, h^n_2]$, then the overall body part mask matrix $m^n_{part}$is calculated as follows:

$$m^{n(xy)}_{part} = \begin{cases} 1 & h^n_1 \leq x * (H_0/H) \leq h^n_2, w^n_1 \leq y * (W_0/W) \leq w^n_2 \\ 0 & otherwise \end{cases} \tag{4}$$

where x, y is the index of the matrix. The scaling factor H_0/H and W_0/W are used here because the size of the feature map f_{enc} is scaled down from the original image.

As shown in Fig. 4, with the output of the object centric decoder, we generate human mask M_{hum} and object mask M_{obj} using the Mask Generator, which have the same size as the feature map f_{enc}. Subsequently, we match M_{hum} with M_{part}, preserving the overlapping regions between them. Additionally, in cases of pose estimation failure, we employ the Mask matcher to filter out instances where the shared region's area is zero. In such instances, we directly utilize M_{hum} as the subsequent Attention mask for further computations. The key body parts mask matrix M_{part} is calculated as:

$$M_{part} = \begin{cases} m_{part}^{n(xy)} \vee M_{hum} & min(m_{part}^{n(xy)} \vee M_{hum}) = 0, P_{score}^{n} > k \\ M_{hum} & min(m_{part}^{n(xy)} \vee M_{hum}) = 1, P_{score}^{n} > k \end{cases} \quad (5)$$

where $min()$ is a function for finding the minimum value of a matrix. Finally, we will perform logical AND operations on M_{part} and M_{obj} to obtain the mask for the Part based Interaction Decoder operation.

3.4 Training and Inference

To train our proposed method, we followed the transformer-based approaches [6,10] that were previously introduced. Firstly, we employed bipartite matching to assign predicted and ground truth HOI instances, followed by the calculation of matching pair loss. Our multitask loss consists of four components:

$$L = \lambda_1 L_{box} + \lambda_2 L_{iou} + \lambda_3 L_{obj} + \lambda_4 L_{int} \quad (6)$$

where L_{box} and L_{iou} are l_1 and GIoU loss applied to both human and object bounding boxes,L_{obj} is a cross entropy loss for object class prediction, and L_{int} is a focal loss for interaction class prediction.λ_1,λ_2,λ_3, and λ_4 are hyperparameters used to weigh each loss. Additionally, we incorporated intermediate supervision to enhance the representation learning. Specifically, the same feed-forward network (FFN) was attached to each decoder layer to compute intermediate losses. The computation of these auxiliary losses follows the same approach as L.

4 Experiments

4.1 Experimental Settings

Datasets. We adopt two datasets HICO-DET [12] and V-COCO [21]. V-COCO [21] provides 10,346 images (2,533 for training, 2,867 for validating, and 4,946 for testing) and 16,199 person instances. Each person has labels for 29 action categories (five of them have no paired object). HICODET [12] is a much larger dataset than V-COCO [21]. It includes 47,776 images (38,118 in train set and 9658 in test set), 600 HOI categories on 80 object categories (same with [16]) and 117 verbs, and provides more than 150k annotated human-object pairs.

Evaluation Metrics. We follow the standard settings in [8], reporting mean Average Precision (mAP) for evaluation. Prediction of a HOI triplet is considered as a true positive when both predicted human and object bounding boxes have IoUs larger than 0.5 compared to the ground truth boxes, and HOI category prediction is accurate. For V-COCO, we report mAP for two scenarios. For HICO-DET, we report mAP over two evaluation settings (Default and Known Object), with three HOI category subsets: all 600 HOI triplets (Full), 138 HOI triplets with fewer than 10 training samples (Rare), and 462 HOI triplets with 10 or more training samples (Non-Rare).

4.2 Implementation Details

In our deployment, we utilize ResNet-50 as the initial component of our visual feature extractor, followed by a six-layer transformer encoder. Both the object and human detection decoder layers consist of six layers, while the global-based interaction decoder and part-based interaction decoder consist of three layers each. The network parameters are initialized using the weights from DETR [18], which was pre-trained on the COCO dataset. During the training process, we set the number of queries to 100 for V-COCO and 64 for HICO-DET, following the methodology in [6]. We employ AdamW optimizer with a weight decay of 2e−5. The loss weight coefficients (λ_1, λ_2, λ_3, and λ_4) are set as 1, 2.5, 1, and 1 respectively. The model is trained for 90 epochs with a learning rate of 2e−5, which is decreased by a factor of 10 at the 60th epoch.

5 Results

5.1 Comparison to State-of-the-Art

We conducted experiments on the V-COCO and HICO-Det benchmarks to validate the effectiveness of our proposed method.

Table 1. Performance comparison on the V-COCO test set. $AP^{role}(S1)$, $AP^{role}(S2)$ denotes the performance under Scenario1 and Scenario2 in V-COCO, respectively.

Method	Backbone	$AP^{role}(S1)$	$AP^{role}(S2)$
CNN-based Methods			
PMFNet [14]	ResNet50	52.0	-
PD-Net [4]	ResNet50	53.3	-
AS-Net [2]	ResNet50	53.9	-
GG-Net [5]	ResNet50	54.7	-
Single-decoder Methods			
HOTR [3]	ResNet50	55.2	64.4
QPIC [10]	ResNet50	58.8	61.0
Dual-decoder Methods			
FGAHOI [13]	Swin-T	60.5	61.2
CDN [6]	ResNet50	61.7	63.8
GEN-VLKT [15]	ResNet50	62.4	64.5
MSTR [24]	ResNet50	62.0	65.2
BPI [29]	ResNet50	63.0	65.1
GEN-VLKT [15]	ResNet101	63.5	65.9
CDN [6]	ResNet101	63.9	65.8
OBPA-Net(ours)	ResNet50	**64.1**	**65.9**

As shown in Table 1, OBPA-Net outperforms all state-of-the-art methods on V-COCO. The table indicates that OBPA-Net not only surpasses CNN-based methods but also outperforms Transformer-based methods with both single-decoder and dual-decoder architectures. This demonstrates the importance of decoupling object detection and human detection and explicitly defining subtask objectives. Additionally, under the same backbone model, OBPA-Net achieves a 1.1% higher mAP compared to one of the latest transformer-based methods, Body Part Interactiveness [29]. It uses a fixed number of key body parts to predict the interactivity of human object pairs. This highlights the significance of adaptively selecting body part features and accurately extracting fine-grained human body part features in a more flexible manner (Table 2).

Table 2. Performance comparison on the HICO-DET. "Default" means that the average precision (AP) is calculated across all testing images for each HOI class. "Known Object" means that the AP is calculated for an HOI class over the images that specifically contain the object involved in that HOI class.

			Default			Known Object	
Method	Backbone	Full	Rare	Non-Rare	Full	Rare	Non-Rare
PD-Net [4]	Res152	22.37	17.61	23.79	26.86	21.70	28.44
AS-Net [2]	Res50	28.87	24.25	30.25	31.74	27.07	33.14
GG-Net [5]	Res50	29.17	22.13	30.84	33.50	26.67	34.89
HOTR [3]	Res50	25.10	17.34	27.42	-	-	-
QPIC [10]	Res101	29.90	23.92	31.69	32.38	26.06	34.27
MSTR [24]	Res50	31.17	25.31	32.92	34.02	28.83	35.57
CDN [6]	Res101	32.07	27.19	33.53	34.79	29.48	36.38
RLIP-ParSe [31]	Res50	32.84	26.85	34.63	-	-	-
GEN-VLKT [15]	Res50	33.75	29.25	35.10	36.78	32.75	37.99
BPI [29]	Res50	35.15	33.71	35.58	37.56	35.87	38.06
OBPA-Net(ours)	Res50	34.63	32.83	35.16	36.78	35.38	38.04

On HICO-DET dataset, compared with state-of-the-art one-stage methods, our method is 2.28% mAP higher than DEFR [11] and 1.79% mAP higher than RLIP-ParSe [31] under default full settings. Both single decoder and dual decoder methods use a single decoder to detect human and objects, and our method outperforms these methods to demonstrate the effectiveness of the decoupling strategy.

5.2 Ablation Study

To assess the effectiveness of OBPA-Net, we perform a series of ablation studies using the V-COCO dataset.

Impact of Object Centric Decoder. We first validate the effectiveness of the object centric decoder. The detection tasks of humans and objects are completed by the same decoder, which limits their ability to adapt to different subtasks of multi-task learning simultaneously. We propose three approaches to decouple the human decoder and object decoder. The first approach is to directly decouple the human decoder and object decoder modules. This results in many mismatched human-object pairs and a decrease in performance. The second approach is the human centric decoder, where the output of the human decoder is used as the query embedding input for the object decoder. In Table 3, it can be seen that the detection results of this method on $AP^{role}(S1)$ are very poor. The performance degradation is mainly due to the inclusion of some objects with a label of *nothing*, we speculate that this is because using human features to detect objects conflicts with detecting *nothing*, which means not detecting objects. Finally, we utilize the object centric decoder to achieve the decoupling of the human decoder and object decoder, and Table 3 demonstrates its effectiveness.

Table 3. Impact of object centric decoder. DSO Decoder, HC Decoder and OC Decoder denotes the decoupling human and object decoder, human centric decoder and object centric decoder.

Single Decoder	DHO Decoder	HC Decoder	OC Decoder	$AP^{role}(S1)$	$AP^{role}(S2)$
✓	-	-	-	63.2	64.5
-	✓	-	-	63.5	64.7
-	-	✓	-	62.7	64.6
-	-	-	✓	**64.1**	**65.9**

Impact of BPA. In Table 4, we validated the effectiveness of the BPA and different body part selection strategies. The results indicate that using BPA can help predict interaction categories, and flexibly selecting different numbers of human key parts is also very important compared to selecting a fixed number of human key parts.

Impact of Interaction Decoder from Global to Part. In Table 5, we validated the effectiveness of the impact of interactive decoder layers from global to local. Through experiments with different levels of GBID and PBID, we ultimately validated the method with the highest performance.

Table 4. Impact of BPA and interaction decoder from global to part. BPA refer to the Body Part Attention module. Adaptive represents the flexible selection of different numbers of key human body parts

BPA	Num of parts	$AP^{role}(S1)$	$AP^{role}(S2)$
-	-	63.4	64.9
✓	3	63.9	65.6
✓	6	63.9	65.5
✓	Adaptive	**64.1**	**65.9**

Table 5. Impact of interaction decoder from global to part. GBID and PBID refer to the global based interaction decoder and part based interaction decoder.

GBID layers	PBID layers	$AP^{role}(S1)$	$AP^{role}(S2)$
1	5	63.3	64.9
3	3	**64.1**	**65.9**
5	1	63.4	64.7

6 Conclusion

In this article, we explored the importance of object detection in HOI and proposed a new framework called OBPA-Net. From the perspective of improving object detection performance, OBPA-Net utilized the Object Centric Decoder (OCD) to implicitly learn the interactivity between human object pairs, and then utilized the Body Part Attention (BPA) module to extract fine-grained features of key body parts to improve HOI detection. We conducted extensive evaluations on two benchmark datasets: V-COCO and HICO-DET, indicating that our model outperformed current state-of-the-art methods in terms of performance. In the future, we will explore more flexible feature selection modules for human body parts to utilize more accurate fine-grained features for HOI detection.

References

1. Girshick, R.: Fast R-CNN. Computer Science (2015)
2. Chen, M., et al.: Reformulating HOI Detection as Adaptive Set Prediction (2021)
3. Kim, B., et al.: HOTR: End-to-End Human-Object Interaction Detection with Transformers (2021)
4. Zhong, X., et al.: Polysemy deciphering network for robust human-object interaction detection. Int. J. Comput. Vis. **129**(6), 1910–1929 (2021)
5. Zhong, X., et al.: Glance and gaze: inferring action-aware points for one-stage human-object interaction detection (2021)
6. Zhang, A., et al.: Mining the Benefits of Two-stage and One-stage HOI Detection (2021)

7. Zou, C., et al.: End-to-End Human Object Interaction Detection with HOI Transformer (2021)
8. Gao, C., Zou, Y., Huang, J.B.: iCAN: Instance-Centric Attention Network for Human-Object Interaction Detection (2018). https://doi.org/10.48550/arXiv.1808.10437
9. Gkioxari, G., et al.: Detecting and recognizing human-object interactions. In: 2018 IEEE/CVF Conference on Computer Vision and Pattern Recognition (CVPR). IEEE (2018)
10. Tamura, M., Ohashi, H., Yoshinaga, T.: QPIC: query-based pairwise human-object interaction detection with image-wide contextual information. https://doi.org/10.48550/arXiv.2103.05399
11. Jin, Y., et al.: The Overlooked Classifier in Human-Object Interaction Recognition (2021)
12. Chao, Y.W., et al.: Learning to detect human-object interactions. In: 2018 IEEE Winter Conference on Applications of Computer Vision (WACV) IEEE (2018)
13. Ma, S., et al.: FGAHOI: fine-grained anchors for human-object interaction detection, January 2023
14. Wan, B., et al.: Pose-aware multi-level feature network for human object interaction detection (2019)
15. Liao, Y., et al.: GEN-VLKT: simplify association and enhance interaction understanding for HOI Detection (2022)
16. Kim, B., et al.: UnionDet: union-level detector towards real-time human-object interaction detection (2020)
17. Liao, Y., et al.: PPDM: parallel point detection and matching for real-time human-object interaction detection (2019)
18. Carion, N., et al.: End-to-end object detection with transformers (2020)
19. Zhu, X., et al.: Deformable DETR: deformable transformers for end-to-end object detection (2020)
20. Chen, J., Yanai, K.: QAHOI: query-based anchors for human-object interaction detection (2021)
21. Gupta, S., Malik, J.: Visual Semantic Role Labeling (2015). https://doi.org/10.48550/arXiv.1505.04474
22. Li, J., et al.: CrowdPose: efficient crowded scenes pose estimation and a new benchmark (2018)
23. Zhang, Y., et al.: Exploring structure-aware transformer over interaction proposals for human-object interaction detection (2022)
24. Kim, B., et al.: MSTR: Multi-Scale Transformer for End-to-End Human-Object Interaction Detection (2022)
25. Zhou, D., et al.: Human-Object Interaction Detection via Disentangled Transformer (2022)
26. Wang, T., et al.: Deep contextual attention for human-object interaction detection. In: IEEE International Conference on Computer Vision 2019. IEEE (2019)
27. Fang, H.-S., Cao, J., Tai, Y.-W., Lu, C.: Pairwise body-part attention for recognizing human-object interactions. In: Ferrari, V., Hebert, M., Sminchisescu, C., Weiss, Y. (eds.) ECCV 2018. LNCS, vol. 11214, pp. 52–68. Springer, Cham (2018). https://doi.org/10.1007/978-3-030-01249-6_4
28. Li, Y.L., et al.: Transferable interactiveness knowledge for human-object interaction detection. IEEE Trans. Pattern Anal. Mach. Intell., **99**, 1 (2021)
29. Wu, Xiaoqian, et al. "Mining Cross-Person Cues for Body-Part Interactiveness Learning in HOI Detection." European Conference on Computer Vision arXiv, 2022

30. Li, Y.L., et al.: HAKE: human activity knowledge engine (2019)
31. Yuan, H., et al.: RLIP: Relational Language-Image Pre-Training for Human-Object Interaction Detection, September 2022

Salient Feature Enhanced Multi-object Tracking with Soft-Sparse Attention in Transformer

Caihua Liu[1,2], Xu Qu[1,2](✉), Xiaoyi Ma[1,2], Runze Li[1], Xu Li[1], and Sichu Chen[1]

[1] The College of Computer Science and Technology, Civil Aviation University of China, Tianjin 300300, China
2021052054@cauc.edu.cn

[2] Key Laboratory of Intelligent Airport Theory and System, CAAC, Tianjin, China

Abstract. Most existing transformer-based Multi-object tracking (MOT) methods use Convolutional Neural Network (CNN) to extract features and then use a transformer to detect and track objects. However, feature extract networks in existing MOT methods cannot pay more attention to the salient regional features and capture their consecutive contextual information, resulting in the neglect of potential object areas during detection. And self-attention in the transformer generates extensive redundant attention areas, resulting in a weak correlation between detected and tracking objects during the tracking. In this paper, we propose a salient regional feature enhancement module (SFEM) to focus more on salient regional features and enhance the continuity of contextual features, it effectively avoids the neglect of some potential object areas due to occlusion and background interference. We further propose soft-sparse attention (SSA) in the transformer to strengthen the correlation between detected and tracking objects, it establishes an exact association between objects to reduce the object's ID switch. Experimental results on the datasets of MOT17 and MOT20 show that our model significantly outperforms the state-of-the-art metrics of MOTA, IDF1, and IDSw.

Keywords: Multi-Object Tracking · Salient Regional Feature Enhancement · Soft-Sparse Attention · Vision Transformer

1 Introduction

Multiple-object tracking (MOT) aims to distinguish each object from the others by assigning an ID to each object and recording their trajectories in continuous image sequences [1,6]. It is widely used in various aspects of vision, such as visual surveillance [4], autonomous driving [2], and virtual reality [3].

This work was supported by the Scientific Research Project of Tianjin Educational Committee under Grant 2021KJ037.

© The Author(s), under exclusive license to Springer Nature Singapore Pte Ltd. 2024
Q. Liu et al. (Eds.): PRCV 2023, LNCS 14436, pp. 392–404, 2024.
https://doi.org/10.1007/978-981-99-8555-5_31

Compared to traditional separate tracking-by-detection (TBD) paradigm [6], existing methods are inclined to integrate the detector and embedding model into a unified Joint-Detection-Embedding (JDE) paradigms [5]. Benefiting from the long-range dependency modeling and interpretability of the transformer [10], it is widely used in MOT tasks, following the JDE paradigm. Most of the existing transformer-based MOT methods [7,9] implemented by a CNN backbone to extract features and an encoder-decoder transformer to detect and track objects.

However, feature extract networks in existing MOT methods cannot pay more attention to the salient regional features and capture their consecutive contextual information, resulting in the neglect of some detailed information in the feature map during detection. Besides, self-attention in the transformer calculates each pixel attention with all pixel values of the input features, which generates extensive redundant attention areas, resulting in a weak correlation between detected and tracking queries during tracking.

To solve the problems aforementioned, we propose a salient regional feature enhancement module (SFEM), it utilizes spatial attention [11] and an adaptive scaling dilated convolution (ASC) on the feature maps extracted from the backbone, the spatial attention used to focus more on salient regional features, and the ASC is guided by spatial attention feature weights to enhance the continuity of contextual information. Besides, we propose the soft-sparse attention (SSA) in the transformer, it effectively avoids the problems of lacking the ability to focus on the most relevant information between queries in a self-attention and exactly builds an association between queries.

To summarize, our contributions are listed as follows:

- We propose a salient regional feature enhancement module (SFEM) to pay more attention to salient regional features and capture their consecutive contextual information, which can accurately detect the objects occluded and interfered by background as well as improve the accuracy in MOT.
- We first propose the soft-sparse attention in the transformer, it could accurately catch the correlation between detected and tracking objects to reduce ID switch during tracking.
- Extensive experimental results on MOT17 and MOT20 indicate that our model outperforms the state-of-the-art methods on several metrics.

2 Related Works

2.1 Feature Enhancement in Tracking

Feature enhancement is designed to further refine the important contextual information in the feature maps, it enables to accurately detect the objects in a complex scenes. Several research studies have focused on feature enhancement in tracking, Zhao et al. [15] proposed an algorithm which incorporates the spatial and temporal attention to take full advantage of the hierarchical convolution features for tracking. Huang et al. [16] proposed a self-attention-based feature fusion and a classification enhancement structure, which highlight the target information and assist the anchor-free strategy respectively. Hu et al. [17] introduce a new multi-frequency feature representation method to improves feature expression ability of highly dynamic targets. The above mentioned works mainly consider fusing different scales and patterns feature or enhancing multi-frequency features, but most of them are scale-uniform enhancements for all regions and cannot adaptively select enhancement strategies.

2.2 Transformers in Tracking

Transformer [10] was first applied in natural language processing, which is a deep neural network based on self-attention mechanisms. Thanks to its powerful representation capabilities, researchers have applied it to a variety of computer vision tasks, including target detection, target tracking, image segmentation, etc. There are also many works in the MOT field with transformer, TrackFormer [9] and MOTR [7] realize the object detection and tracking simultaneously by concatenating the object and autoregressive track queries as inputs to the Transformer decoder in the next frame. Besides, TransCenter [12] and TransTrack [8] only use Transformers as feature extractors and iteratively pass track features to learn aggregated embedding of each object. TransMOT [14] still uses CNNs as object detectors and then learns an affinity matrix with Transformers for tracking. The above works explore the existing dominant MOT methods with Transformer. However, they lack the ability to focus on salient regional features and establish an exact correlation between detecting and tracking queries.

2.3 Attentions Applied in Transformer

The core of the attention mechanism is to selectively choose the important information from a large amount of information and capture the important information that is useful for the task at hand. Transformer [10] uses multi-head self-attention to capture richer features information by focusing on information from different representation subspaces at different locations. SparseTT [13] utilizes traditional sparse attention [20] in the transformer to highlight potential targets in the search area. However, self-attention in mostly transformer-based MOT methods lacks the ability to focus on the most relevant information between pixel values. In the original sparse attention [20], each pixel value of attention features is only determined by k pixel values that are most similar to it, which results in neglecting some potential important correlations.

3 Method

3.1 Overall Architecture

The overall architecture of our model is shown in Fig. 1. Multi-frame images are fed into the convolutional neural network (CNN) (e.g. ResNet-50 [18]) to extract features, then features are fed into salient regional feature enhancement

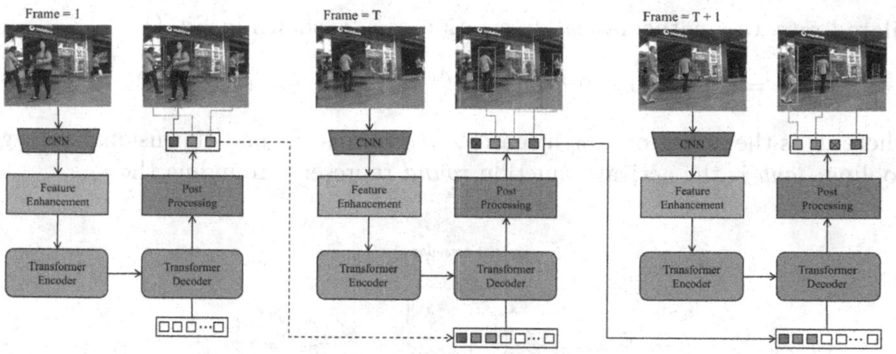

Fig. 1. The overall architecture of our model. At frame t = 1, the decoder transforms N learnable **object queries** (white) to output embeddings either initializing new autoregressive **track queries**. On subsequent frames, the decoder processes the joint set of N_{object} + N_{track} queries to follow or remove (cyan) existing tracks as well as initialize new tracks (purple).

module to obtain enhanced salient features, and the features are embedded with position information and flattened before being fed into the encoder. For the first frame, there is no track query, so we feed the fixed-length learnable detect queries into the decoder to get object prediction, and go through the post-processing to activate high confidence detection as track queries for the next frame. For successive frames, we feed the concatenation of track queries from the previous frame and the learnable detect queries into the decoder, which also feeds the track queries generated by the current frame for the next frame.

3.2 Salient Feature Enhancement Module

In this section, the salient regional feature enhancement module (SFEM) is described in detail, as shown in Fig. 2, which mainly includes an adaptive scaling dilated convolution (ASC) and spatial attention.
Spatial Attention. Feature maps F extracted from the backbone are the input of our feature enhancement network. We first make the max and average pooling on the F, and then a 3 × 3 convolution is used after concatenation of pooling to obtain the attention weights W. W has two functions, on the one hand it multiplies with the F to obtain the output of spatial attention F_1, on the other hand it is used to guide the ASC.
Adaptive Scaling Dilated Convolution. Adaptive scaling dilated convolution (ASC) take the W from spatial attention to guide dilated convolution. Specifically, we divide the W and F into N small patches, each w_i and f_i have a one-to-one correspondence since they have the same scale. Next, we perform an average pooling and a gating activation function on w_i to obtain the dilated rate r_i. Since the salient regions have a greater attention weights they get a greater

dilated rate, the mathematical description of r_i is shown in Eq. (1),

$$r_i = round(w_i \cdot tanh(\ln(1 + e^{Avg(w_i)}))), \quad i = 1, 2, 3, ..., N, \tag{1}$$

where w_i is the attention weight patch, Avg means the one-dimensional average pooling, $tanh$ is the activate function, $round$ represents rounding the r_i.

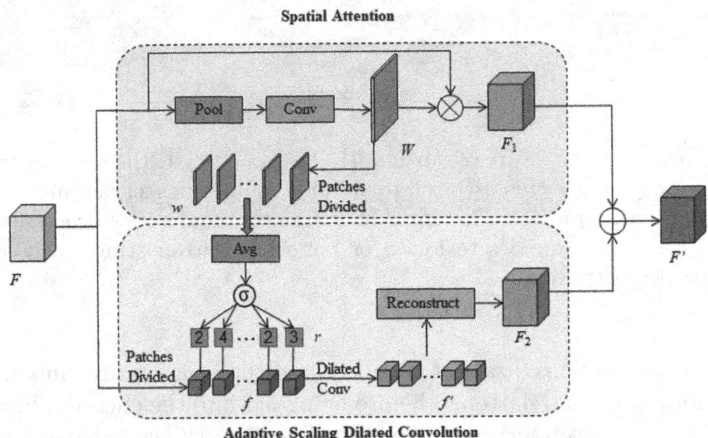

Fig. 2. The illustration of SFEM. F is the input feature map tensor, W means the attention weights in spatial attention [11], w and f mean the patches divided by weights and input feature tensor, r indicates dilated rates of convolution. σ means the gating activation function, $\otimes$ stands for matrix multiplication, $\oplus$ represents the matrix summation.

Finally, we do the dilated rate equal to r convolution for each f and then reconstruct all the f as the output of ASC F_2. The main function of this branch is to enhance the continuity context features in salient regions . After that, we merge the F_1 and F_2 as the output $F^{'}$ of our feature enhancement module, the mathematical description of $F^{'}$ is shown in Eq. (2),

$$F^{'} = RC(DC_{3\times3}(w_i, r_i)) + F_1), \quad i = 1, 2, 3, ..., N, \tag{2}$$

where $F1$ denotes the output of spatial attention, DC means the dilated convolution operation, and the RC means reconstruct patches operation.

3.3 Encoder

Encoder is an important but not essential component in the proposed method, which is responsible for encoding the frame features. AS shown in Fig. 3, it is composed of N encoder layers where each encoder layer takes the outputs of its previous encoder layer as input. Note that, in order to enable the network to have the perception of spatial position information, we add a spatial position embedding to each feature point. Thus, the first encoder layer takes the flattened features extracted from SFEM with spatial position embedding as input. In short, it can be formally denoted as Eq. (3),

$$encoder(\boldsymbol{Z}) = \begin{cases} f_{enc}^{i}(\boldsymbol{Z} + \boldsymbol{P}_{enc}), \ i = 1 \\ f_{enc}^{i}(\boldsymbol{Y}_{enc}^{i-1}), \quad 2 \leq i \leq N, \end{cases} \tag{3}$$

where $\boldsymbol{Z} \in \mathbb{R}^{HW \times C}$ represents the flattened frame feature. $\boldsymbol{P}_{enc} \in \mathbb{R}^{HW \times C}$ represents the spatial position encoding, f_{enc}^{i} represents the i -th encoder layer.

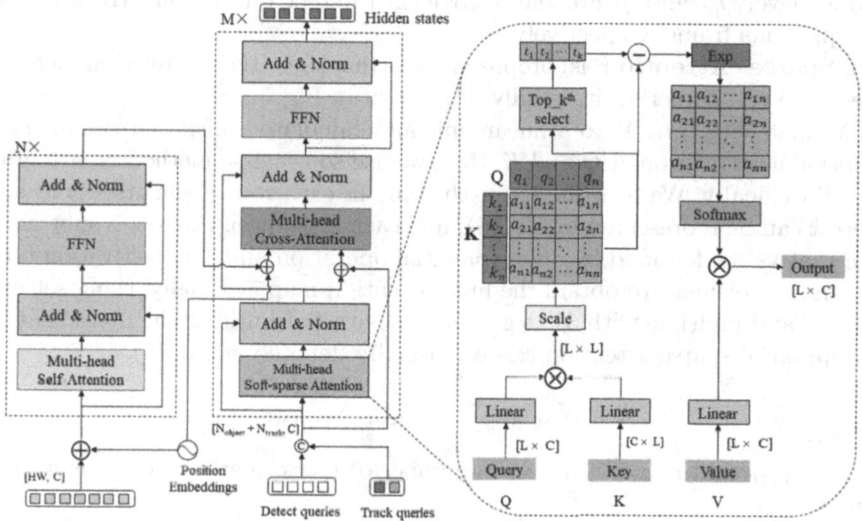

Fig. 3. left): The architecture of Transformer. We indicate the tensor dimensions in squared brackets, ⊕ represents the matrix summation, and C represents the concatenation of queries. **right)**: The illustration of soft-sparse attention. L and C represent the length and dimension of tokens respectively, ⊗ stands for matrix multiplication, ⊖ represents the matrix subtraction.

3.4 Decoder with Soft-Sparse Attention

Decoder is an essential component in our proposed model, it is responsible for decoding the features of queries to generate the detection and tracking. Similar to the encoder, the decoder is also composed of M layers as shown in Fig. 3. However, different from the encoder layer, the input of decoder contain the encoded features with spatial position embedding and concatenation of detect queries and track queries. Specifically speaking, we first use soft-sparse attention (SSA) on the concatenation of detect queries and track queries to catch the correlation between queries. Track queries are fed autoregressively from the previous frame output embedding of the last decoding layer (before the post-process). Next, we make a cross-attention between the output of SSA and encoded features to build spatial associations between feature maps and queries. In a word, it can be formally denoted as Eq. (4),

$$decoder(\boldsymbol{Q}, \mathbf{Y}_{enc}^{N}) = \begin{cases} f_{dec}^{i}(\boldsymbol{Q}, \mathbf{Y}_{enc}^{N} + \boldsymbol{P}_{enc}), & i = 1 \\ f_{dec}^{i}(\mathbf{Y}_{dec}^{i-1}, \mathbf{Y}_{enc}^{N} + \boldsymbol{P}_{enc}), & 2 \leq i \leq M, \end{cases} \tag{4}$$

where $\boldsymbol{Q} \in \mathbb{R}^{len(q_d + q_t) \times C}$ represents the concatenation of track queries and detect queries, $\mathbf{Y}_{enc}^{N} \in \mathbb{R}^{HW \times C}$ represents the output of the encoder, $\boldsymbol{P}_{enc} \in \mathbb{R}^{HW \times C}$ represents the spatial position encoding, f_{dec}^{i} represents the i -th encoder layer, q_d and q_t are the fixed length detect queries and track queries from previous frame, respectively.

Soft-Sparse Attention is proposed to strengthen the correlation between detect and track queries, it detailly illustrated in Fig. 3 right.

We first fed Q, K, V to a linear unit and calculate the similarity matrix as attention map a through Q and K, then we use soft-sparse method on attention map. Specifically, We pick the top k-th value in each row of the attention map, where k can be represented by Eq. (5), and each element of the row is subtracted from that value, followed by an exponential operation and *tanh* activation function on each element to obtain the final attention map $\overline{a}$. Finally, using softmax on the $\overline{a}$ and multiple with V to get the output of attention. The mathematical description of sparse-attention can be formally denoted as Eq. (6),

$$k = N_{track} + 10 \times \log_{10} N_{queries}, \tag{5}$$

where N_{track} and $n_{queries}$ mean the number of track queries and number of all queries.

$$output = Softmax(tanh(e^{a - Topk(a,k)})), \tag{6}$$

where a means the attention map, k is got from the Eq. (5), and *topk* represents the top k-th value in each row of the attention map a.

3.5 Loss Function

We follow [9] set final MOT prediction loss computed over all $N = N_{object} + N_{track}$ output predictions as Eq. (7),

$$\mathcal{L}_{\text{MOT}}(y, \hat{y}, \pi) = \sum_{i=1}^{N} \mathcal{L}_{\text{query}}(y, \hat{y}_i, \pi), \tag{7}$$

where the y and $\hat{y}$ represent the ground truth and predictions respectively, π is the mapping from Hungarian algorithm. $\mathcal{L}_{query}$ is defined as Eq. (8),

$$\mathcal{L}_{\text{query}} = \begin{cases} -\log \hat{p}_i(c_{\pi=i}) + \mathcal{L}_{\text{box}}\left(b_{\pi=i}, \hat{b}_i\right), & \text{if } i \in \pi \\ -\log \hat{p}_i(0), & \text{if } i \notin \pi, \end{cases} \tag{8}$$

where the output embeddings which were not matched via track ID or assignment are not part of the mapping π and will be assigned to the background class c_i = 0. $\mathcal{L}_{box}$ uses to compute the bounding box loss, which composed by a linear combination of a ℓ_1 distance and a generalized intersection over union (IoU) loss, it can be denoted as Eq. (9),

$$\mathcal{L}_{\text{box}} = \lambda_{\ell_1} \left\| b_i - \hat{b}_{\sigma(i)} \right\|_1 + \lambda_{\text{iou}} \mathcal{L}_{\text{iou}} \left(b_i, \hat{b} \right). \quad (9)$$

4 Experiments

4.1 MOT Benchmarks and Metrics

Datasets. The tracking result of our model is presented on two MOTChallenge benchmarks, MOT17 [21] and MOT20 [22] respectively. MOT17 contains 7 training sequences and 7 test sequences, each set has 5316 frames and 5919 frames with pedestrians annotated with bounding boxes respectively. To evaluate the tracking robustness independently, three sets of public detections are provided, namely, DPM, Faster R-CNN, and SDP. MOT20 [22] is set for highly crowded challenging scenes, with 8931 frames for training and 4479 frames for testing.
Metrics. We follow the standard evaluation protocols to evaluate our method. The common essential metrics include Multi-Object Tracking Accuracy (MOTA), Identity F1 Scores (IDF1), Mostly Tracked (MT), Mostly Lost (ML), False Positive (FP), False Negative (FN), and Identity Switches (IDSw).

4.2 Implementation Details

We follow up ResNet50 [18] CNN feature extraction and Transformer encoder-decoder architecture presented in Deformable DETR [19]. For the data processing, we follow the Trackformer [9], adopting several data augmentation methods that are random flip and crop. Besides, we restricted the maximum size of the input to 1380 and 800 for the shorter and longer side respectively. We follow Trackformer [9] use the joint of the MOT dataset and CrowdHuman [23] person detection dataset for training, we generate the adjacent training frames t-1 and t by applying random spatial augmentation to a signal image.

For the training procedure, we follow the trackformer [9], the backbone and encoder-decoder are trained with individual learning rates of 0.00001 and 0.0001, respectively. All the experiments are conducted on PyTorch with a Tesla V100 GPU. For MOT17 public detections model training, we train it upon [19] pre-trained 50 epochs on COCO [26] for a total of 40 epochs with a learning rate drop by a factor of 10 after the first 10 epochs, which takes about 60 h. For the private detections model training, we first train the model for 80 epochs on the CrowdHuman dataset, and then fine-tuned it on MOT17 and MOT20 with reduced learning rates for additional 30 epochs, and the whole training process takes about 6 d on each datasets.

4.3 Benchmark Results

MOT17. The MOT17 [21] benchmark is evaluated on a private and public detection setting. The latter allows for a comparison of tracking methods independent of the underlying object detection performance. MOT17 provides three sets of public detections with varying quality. We report the results of the evaluation on public and private detections in Table 1.

MOT20. The MOT20 [22] benchmark is evaluated in a private detection setting. MOT20 includes more crowded scenes, severe object occlusion and small objects than MOT17, which brings more challenges for object detecting and tracking. Thus, all methods show lower performance on MOT20 than MOT17. We report the results of the evaluation on private detections in Table 2.

Table 1. Comparison of existing multi-object tracking methods evaluated on the MOT17 test set. We report private as well as public detections results and separate between online and offline approaches. We compare the private result with Transformer-Based architecture methods. (Best results are shown in **bold**)

	methods	MOTA↑	IDF1 ↑	MT ↑	ML↓	FN ↓	FP ↓	IDSw ↓
private								
Online	Centertrack [25]	67.8	64.7	816	579	160332	**18498**	3039
	TransTrack [8]	74.5	63.9	s	/	112137	28323	3663
	TransCenter [12]	73.2	62.2	/	/	123738	23112	4614
	Trackformer [9]	74.1	68.0	**1113**	**246**	108777	34602	2829
	Ours	**74.8**	**68.7**	1047	265	**101247**	33564	**2541**
public								
Offline	JCC [27]	51.2	54.5	493	872	247822	25937	1082
	FWT [28]	51.3	47.6	505	830	247921	24101	2648
	TT [29]	54.9	63.1	575	897	233295	20235	**1088**
	MPNTrack [30]	58.8	61.7	**679**	**788**	213594	17413	1185
	Lif_T [24]	**60.5**	**65.6**	637	791	**206617**	**14966**	1189
Online	FAMNet [31]	52.0	48.7	450	787	253616	14138	3072
	Tracktor++ [1]	56.3	55.1	498	831	235449	**8866**	1987
	GSM [32]	56.4	57.8	523	813	230174	14379	**1485**
	CenterTrack [25]	60.5	55.7	580	777	208577	11599	2540
	TrackFormer [9]	62.5	60.7	**702**	**632**	174921	32828	3917
	Ours	**63.2**	**60.9**	698	640	**169511**	21254	3001

4.4 Ablation Study

Model Components. Table 3 shows the impact of integrating different components, we implement an ablation study on the MOT17 public detections. We did separate experiments on our model using SFEM and SSA individually, as well as

Table 2. Comparison of multi-object tracking methods evaluated on the MOT20.

methods	MOTA↑	IDF1 ↑	MT ↑	ML↓	FN ↓	FP ↓	IDSw ↓
GSDT [33]	67.1	**67.5**	/	/	135395	31507	3230
TransCenter [12]	58.3	46.8	/	/	174893	35959	4947
TransTrack [8]	64.5	59.2	/	/	151377	28566	3565
TrackFormer [9]	68.6	65.7	**666**	**181**	140375	**20384**	1532
Ours	**69.3**	66.8	651	189	**130375**	20684	**1386**

combining the two components to verify the effectiveness of the modules. Experimental results show that paying more attention to features in salient regions and associating contextual information with continuity can greatly improve the performance, and also show that enhancing the correlation between tracking and detected objects can handle the challenge of ID switch effectively in MOT.

Table 3. Ablation study on our proposed components of model

Baseline	SFEM	SSA	MOTA↑	IDF1 ↑	IDSw ↓
✓			62.3	58.6	4018
✓	✓		62.8	59.4	3308
✓		✓	62.9	59.8	3128
✓	✓	✓	**63.2**	**60.9**	**3001**

Table 4. Ablation study on our proposed Salient regional Feature Enhancement Module (SFEM) and Soft-Sparse Attention (SSA) in the decoder of transformer.

(a) Comparing Merging Multiple dilated rates Convolution(MMC) and Adaptive Scaling dilated Convolution(ASC) in SFEM

method	MOTA↑	IDF1 ↑	IDSw ↓
×	62.3	58.6	4018
MMC	62.8	59.4	3308
ASC	**62.9**	**59.8**	**3128**

(b) Comparing different attentions in the decoder of transformer,including self-attention in the baseline, traditional sparse attention and soft-sparse attention.

method	MOTA↑	IDF1 ↑	IDSw ↓
Self	62.3	58.6	4018
Sparse	62.9	59.7	3262
Soft-sparse	**63.0**	**60.3**	**3102**

Different Combinations of Dilated Convolution. Table 4a reports the ablation study on the SFEM without soft-sparse attention, we compare using attention weights guided adaptive scaling dilated convolution(ASC) and merging multiple dilated rates convolution to the feature enhancement. We implement it on the MOT17 public detections, and the result shows that broadening the receptive field of salient regions to enhance contextual feature continuity is better than fusing multiple dilated convolution feature maps.

Different Attentions in Transformer. Table 4b reports the ablation study on the attention in transformer without SFEM, we compare using self-attention, traditional sparse-attention, and soft-sparse attention in the decoder. We implement it on the MOT17 public detections, and the result shows that our method is able to reduce ID switches by strengthening the correlation between queries and focusing on potentially important areas.

4.5 Case Study

To illustrate the algorithm performance in real-world scenarios visually, a case study of a complex scene in the dataset is shown in Fig. 4. The left side of the figure shows the undetected target and ID switches due to occlusion in the baseline model. The left side of the figure shows that our method accurately detects the occluded object and assigns identities after occlusion.

Fig. 4. Case study of baseline and our model. The same box color represents the same identity. (baseline) expresses fail to detect object and ID switch (light blue and yellow turn to blue and green) due to occlusion (b) express success to identify object occluded and ID keeping (Color figure online)

5 Conclusion

In this paper, we propose a salient feature enhanced transformer-based MOT method, it utilizes a feature enhancement module during feature extraction and soft-sparse attention in the transformer during detection and tracking, which makes the metrics of MOTA, IDF1, and IDSw on datasets MOT17 and MOT20 significantly improvement than our baseline model. Extensive experiments show that our methods can cope well with objects occluded, interfered by background and some complex scenes. However there are some shortcomings in our model, the query passing in our model is performed frame-by-frame, limiting the efficiency of model learning during training, which is what we are aiming to improve.

References

1. Bergmann, P., Meinhardt, T., Leal-Taixé, L.: Tracking without bells and whistles. Int. Conf. Comput. Vis. (2019)
2. Zhou, X., Cui, J., Qu, M.: An improved multi-object tracking algorithm for autonomous driving based on DeepSORT. In: ICITE (2022)
3. Wang, X.: Intelligent multi-camera video surveillance: a review. Pattern Recognit. Lett. **34**(1), 3–19 (2013)
4. Uchiyama, H., Marchand, E.: Object detection and pose tracking for augmented reality: recent approaches. In: Proc (2012)
5. Wang, Z., Zheng, L., Liu, Y., Li, Y., Wang, S.: Towards real-time multi-object tracking. In: Vedaldi, A., Bischof, H., Brox, T., Frahm, J.-M. (eds.) ECCV 2020. LNCS, vol. 12356, pp. 107–122. Springer, Cham (2020). https://doi.org/10.1007/978-3-030-58621-8_7
6. Zhang, Y., Wang, C., Wang, X., Zeng, W., Liu, W.: Fairmot: on the fairness of detection and re-identification in multiple object tracking. In: IJCV, pp. 1–19 (2021)
7. Zeng, F., Dong, B., Zhang, Y., Wang, T., Zhang, X., Wei, Y.: MOTR: end-to-end multiple-object ttracking with transformer. In: ECCV (2022)
8. Sun, P., et al.: Transtrack: multiple-object tracking with transformer. arXiv preprint arXiv:2012.15460 (2020)
9. Meinhardt, T., Kirillov, A., Leal-Taixé, L., Feichtenhofer, C.: Trackformer: multi-object tracking with transformers. In: CVPR (2022)
10. Vaswani, A., et al.: Attention is all you need. In: NeurlPS (2017)
11. Jaderberg, M., Simonyan, K., Zisserman, A., Kavukcuoglu, K.: Spatial transformer networks. In: NIPS (2015)
12. Xu, Y., Ban, Y., Delorme, G., Gan, C., Rus, D., Alameda-Pineda, X.: TransCenter: transformers with dense representations for multiple-object tracking. In: IEEE Transactions on Pattern Analysis and Machine Intelligence (2023)
13. Fu, Z., Fu, Z., Liu, Q., Cai, W., Wang, Y.: SparseTT: visual tracking with sparse transformers. In: IJCA (2022)
14. Chu, P., Wang, J., You, Q., Ling, H., Liu, Z.: TransMOT: spatial-temporal graph transformer for multiple object tracking. In: WACA (2023)
15. Zhao, D., Zeng, Y.: Dynamic fusion of convolutional features based on spatial and temporal attention for visual tracking. In: IJCNN (2019)
16. Huang, D., Yang, M., Duan, J., Yu, S., Liu, Z.: Siamese network tracking based on feature enhancement. IEEE Access (2023)
17. Bi, F., Sun, J., Lei, M., Wang, Y., Sun X.: Remote sensing target tracking for UAV aerial videos based on multi-frequency feature enhancement. In: IGARSS (2020)
18. He, K., Zhang, X., Ren, S., Sun, J.: Deep residual learning for image recognition. In: CVPR (2016)

19. Zhu, X., Su, W., Lu, L., Li, B., Wang, X., Dai, J.: Deformable detr: deformable transformers for end-to-end object detection. In: ICLR (2020)
20. Zhao, G., Lin, J., Zhang, Z., Ren, X., Su Q., Sun X.: Explicit sparse transformer: concentrated attention through explicit selection. arXiv preprint arXiv:1912.11637 (2019)
21. Milan, A., Leal-Taixé, L., Reid, I., Roth, S., Schindler, K.: Mot16: a benchmark for multi-object tracking. arXiv preprint arXiv:1603.00831 (2016)
22. Dendorfer, P., et al.: Mot20: A benchmark for multi object tracking in crowded scenes. arXiv preprint arXiv:2003.09003 (2020)
23. Shao, S., et al.: Crowdhuman: a benchmark for detecting human in a crowd. arXiv:1805.00123 (2018)
24. Hornakova, A., Henschel, R., Rosenhahn, B., Swoboda, P.: Lifted disjoint paths with application in multiple object tracking. In: International Conference on Machine Learning (2020)
25. Zhou, X., Koltun, V., Krahenbuhl, P.: Tracking objects as points. In: ECCV, pp. 474–490 (2020)
26. Lin, T., et al.: Microsoft coco: common objects in context. In: ECCV (2014)
27. Keuper, M., Tang, S., Andres, B., Brox, T., Schiele, B.: Motion segmentation and multiple object tracking by correlation co-clustering. IEEE Trans. Pattern Anal. Mach. Intell. (2018)
28. Henschel, R., Leal-Taixé, L., Cremers, D., Rosenhahn, B.: Improvements to Frank-Wolfe optimization for multi-detector multi-object tracking. In: CVPR (2017)
29. Zhang, Y., et al.: Long-term tracking with deep tracklet association. IEEE Trans. Image Process. (2020)
30. Braso, G., Leal-Taixé, L.: Learning a neural solver for multiple object tracking. IEEE Conf. Comput. Vis. Pattern Recogn. (2020)
31. Chu, P., Ling, H.: Famnet: joint learning of feature, affinity and multi-dimensional assignment for online multiple object tracking. In: CVPR (2019)
32. Liu, Q., Chu, Q., Liu, B., Yu, N.: GSM: Graph similarity model for multi-object tracking. Int. Joint Conf. Art. Int. (2020)
33. Wang, Y., Weng, X., Kitani, K.: Joint detection and multi-object tracking with graph neural networks. In: ICRA (2021)

A BiGRU Based Adaptive Gain Estimation for Radar Multi-target Tracking

Long Liu[1], Qing Xu[1], Mengxuan Zhang[2](✉), Hongbing Ji[1], and Qiubo Zhao[1]

[1] School of Electronic Engineering, Xidian University, Xi'an 710071, China
[2] School of Artificial Intelligence, Xidian University, Xi'an 710071, China
mxzhang@xidian.edu.cn

Abstract. The currently available multi-target tracking algorithms were developed based on ideal tracking settings, which are unsuitable for actual combat situations. To handle the challenge of unknown measurement noise and low detection probability, this paper presents a novel adaptive gain estimation (AGE) approach with a survival likelihood estimation method. The former utilizes the bidirectional gate recurrent unit (BiGRU) to assign weights to both measurement and prediction in the absence of priori information. The latter leverages a binary classification network to determine the likelihood of target survival. Furthermore, the AGE can be seamlessly integrated with prediction and data association modules to form an end-to-end model known as MTT-AGE (Multi-target Tracking with Adaptive Gain Estimation). The results of MIT trajectory dataset, simulated scenarios and real-world data confirm the efficiency and stability of the MTT-AGE. Furthermore, the ablation experiments are conducted to verify the effectiveness of AGE based on MIT trajectory dataset and simulated scenarios where there is only a single objective.

Keywords: multi-target tracking · BiGRU · unknown target · survival likelihood estimation · AGE

1 Introduction

In recent years, it has achieved better estimation effect and has developed a large number of novel methods in radar multi-target tracking (MTT) applications. The traditional tracking algorithms are commonly based on data association, such as the global nearest neighbor data association (GNN), the joint probabilistic data association (JPDA) [1], and multiple hypothesis tracking (MHT) [2]. GNN based on Hungarian algorithm can select a measurement using distance. However, it might not work when the cluster is the closest measurement. The JPDA computes the association probability between targets and measurements. The MHT has been established to evaluate the likelihood for the radar tracking systems but it is heavily reliant on the priori information. Meanwhile, these above methods are time consuming, especially when there are quite a bit of clutter and a bunch of measurements. Mahler [3] presented some recast works in the Bayesian filtering paradigm using random finite set (RFS), such as probability hypothesis density (PHD), cardinalized PHD (CPHD) [4], multi-target multi-Bernoulli (MeMBer). Vo

© The Author(s), under exclusive license to Springer Nature Singapore Pte Ltd. 2024
Q. Liu et al. (Eds.): PRCV 2023, LNCS 14436, pp. 405–417, 2024.
https://doi.org/10.1007/978-981-99-8555-5_32

demonstrated Gaussian mixture PHD (GM-PHD) [5], which is based on Gaussian distributions. Then sequential Monte Carlo PHD (SMC-PHD) was proposed to process the non-Gaussian distributions while it has high computational complexity. The labelled multi-Bernoulli filter (LMB) and generalized LMB (GLMB) [6] can perform track to track association without high signal to noise ratio (SNR). However, those introduced methods are difficult to implement without a priori motion model. Since the measurement is usually cluttered and time-varying, the complex movement will make them inaccurate.

To handle the above problems, some algorithms based on deep learning were proposed. Li [7] presented a multi-step track prediction based on long and short-term memory (LSTM), which uses the encoder and decoder architecture to implement a motion model for trendless tracks. It only has the predicting process and lacks measurement updating process. Mehryart used convolution and LSTM (CONVLSTM) [8] to establish motion models, which defines probability density difference (PDD) maps. But it has high computational complexity and does not implement an end-to-end network structure. Steffen Jung proposed memory Kalman filter (MKF) [9], which is mainly free from the Markov model and overcomes the linearity limitation to predict state. It only addresses the problems of motion models based on Kalman filter (KF). An end-to-end network structure was presented in [10] that uses recurrent neural network and LSTM (RNN-LSTM) to solve the problem of data association. However, it will fail when targets are undetected. After that, KalmanNet [11] utilizes the power of deep learning for state estimation which focus only on gain estimation for single targets. All the above algorithms mainly focus on implementing motion models and use KF in the update step. Furthermore, they pay less attention to unknown targets or poor detection probability.

To model an end-to-end framework, this paper uses the LSTM [12] to obtain prediction and deep Hungarian algorithm (DHN) [13] to realize data association process. The MTT-AGE is introduced to deal with the problems of unknown measurement noise and low detection probability. AGE analyzes the relationship between prediction and measurement. Through it, the Kalman gain can be calculated without covariance and measurement noise for unknown targets. The survival likelihood estimation uses the characteristic of the whole track to determine whether a track exists or not. This method can handle the problem of low detection probability and miss-detection. Tracks will be remained when its measurements disappear temporarily.

This paper is organized as follows. Section 2 gives the motivation and the structure of the MTT-AGE, respectively. The experimental results are presented in Sect. 3. Finally, we draw some conclusions in Sect. 4.

2 The MTT-AGE Framework

The MTT-AGE consists of four parts and its structure is shown in Fig. 1. It is undoubtedly that the LSTM network predicts the state $\hat{x}_k^i$ of the ith target at time step k, which is based on the previous state $\boldsymbol{x}_{0:k-1}^i$. At the following step, the state $\hat{\boldsymbol{x}}_k^i$ will be amalgamated with measurements z_k. The data association module can assign the measurements z_k^i to the $\hat{\boldsymbol{x}}_k^i$ at time step k, and the likelihood module of survival estimation then comes next. Survival likelihoods are attached to both the associated and unassociated tracks. If the likelihood is 1, the unassociated tracks will set prediction $\hat{x}_k^i$ as estimation and the associated tracks may access the estimation $\boldsymbol{x}_k^i$ from the AGE module. If this probability is 0, it demonstrates that the tracks vanish. The unassociated measurements are considered as new tracks. Details of the MTT-AGE are clarified as follow.

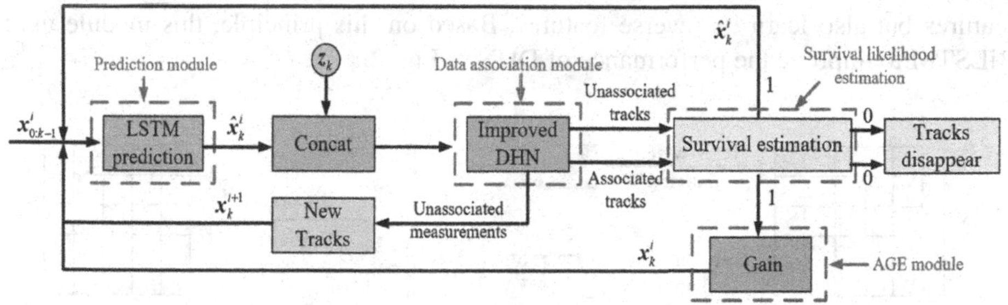

Fig. 1. The MTT-AGE framework.

2.1 Prediction Process

This module takes its main inspiration from by non-Markovian Chapman-Kolmogorov [9] which employs the previous state $\boldsymbol{x}_{0:k-1}^i$ from time step 0 to time step $k-1$. To leverage the information in the long-term tracking, it uses the state $\boldsymbol{x}_{0:k-1}^i$ from time step 0 to time step $k-1$ and obtains the current state $\hat{\boldsymbol{x}}_k^i$.

$$p_{k|k-1}^i\left(\hat{\boldsymbol{x}}_k^i\tilde{\boldsymbol{z}}\right) = \int f_{k|k-1}\left(\hat{\boldsymbol{x}}_k^i\boldsymbol{x}_{k-1}^i, \tilde{\boldsymbol{x}}\right)p_{k-1|k-1}^i\left(\hat{\boldsymbol{x}}_k^i\tilde{\boldsymbol{z}}\right)\mathrm{d}\boldsymbol{x}_k^i \tag{1}$$

$$\tilde{\boldsymbol{z}} = \boldsymbol{z}_{0:k-1}^i, \tilde{\boldsymbol{x}} = \boldsymbol{x}_{0:k-1}^i \tag{2}$$

Here, $\tilde{\boldsymbol{z}}$ denotes previous measurements from time step 0 to time step $k-1$, and $p_{k|k-1}^i$ denotes the posterior probability density. $f_{k|k-1}$ is the non-linear transition function at time $k-1$ which depends on the encoding and decoding of the LSTM.

The module is established via mean square error (MSE) loss function,

$$MSE^i = \sum\nolimits_{k=1}^{n}(\boldsymbol{x}_k^i - \hat{\boldsymbol{x}}_k^i)^2/n \tag{3}$$

2.2 Data Association

The task of this module is to categorize the corresponding measurement for each target. The data association denotes optimal match and its classical algorithms are shown in Sect. 1. In recent years, deep Hungarian algorithm (DHN) [13] is proposed to adds appearance feature to matrix $\boldsymbol{D}_k$ and employs the evaluation indicators (MOTA and MOTP). However, since the radar measurements only provide location information, the origin DHN network cannot be utilized directly. Thus, its distance matrix is replaced with Mahalanobis distance.

$$\boldsymbol{D}_k = d(\boldsymbol{x}, \boldsymbol{z}) = \sqrt{(\boldsymbol{x} - \boldsymbol{z})^T \Sigma^{-1} (\boldsymbol{x} - \boldsymbol{z})} \tag{4}$$

Here, $\boldsymbol{D}_k$ denotes distance matrix. DHN uses BiRNN that can learn not only its forward features but also learn its reverse features. Based on this principle, this module uses BiLSTM to improve the performance of DHN in Fig. 2.

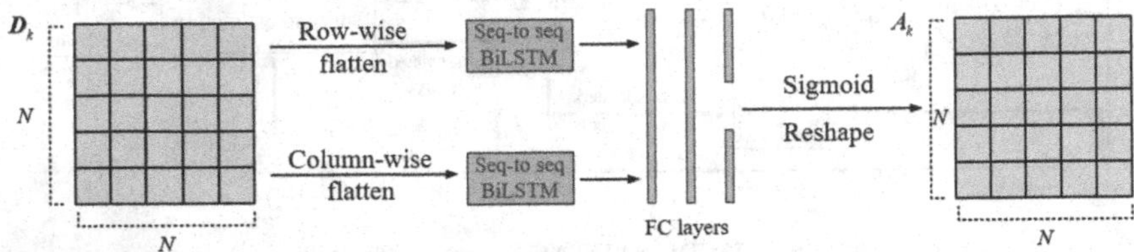

Fig. 2. Improved DHN algorithm. The flatten vectors of $\boldsymbol{D}_k$ in the row-wise and the column-wise are inputted to two BiLSTM networks. FC layers reshape the outputs which can obtain assignment matrix $\boldsymbol{A}_k$.

2.3 Survival Likelihood Estimation

The purpose of this module is to ascertain whether targets disappear or not. Our method draws inspiration from RNN-LSTM [10] which employs the RNN to estimate the initiation and termination likelihood of targets. The RNN-LSTM needs a survival likelihood threshold because the RNN makes poor judgements when the targets are absent. Hence, this module is presented to enhance the performance of survival probability estimation to deal with the problem of miss-detection issue depicted in Fig. 3. To train a binary classification network, unassociated tracks and existence time are taken into account. Additionally taken into consideration are the track's features. The network employs the binary cross-entropy (BCE) loss function shown below.

$$loss = -\sum_{i=1}^{n} \varepsilon_i \log \hat{\varepsilon}_i + \left(1 - \hat{\varepsilon}_i\right) \log(1 - \varepsilon_i) \tag{5}$$

Here, ε_i and $\hat{\varepsilon}_i$ denote the survival probability and true value, respectively.

As shown in Fig. 3, the weights, $\omega_s, \omega_d, \omega_h$, are utilized to calculate the corresponding probabilities. When the network output is 1, it means that the track still exists. When the network output is 0, it means that the track disappears.

This module can reduce the impact of miss-detection because this module employs the three factors to make estimation accurate. Meanwhile, it can also deal with the situations that the target will be obscured or disrupted.

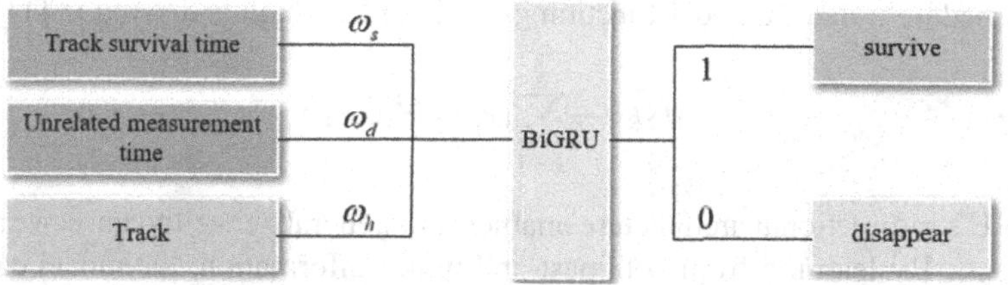

Fig. 3. Survival likelihood estimation network.

2.4 AGE Estimation

This module is to correct the state distribution [14]. The related measurement can be updated according to (6) and it can be represented below.

$$\boldsymbol{x}_{k|k} = \hat{\boldsymbol{x}}_{k|k-1} + \boldsymbol{K}_k\left(z_k - \boldsymbol{H}_k\hat{\boldsymbol{x}}_{k|k-1}\right) \tag{6}$$

$$\boldsymbol{x}_{k|k} = (\boldsymbol{I} - \boldsymbol{K}_k\boldsymbol{H}_k)\hat{\boldsymbol{x}}_{k|k-1} + \boldsymbol{K}_k z_k \tag{7}$$

$$\boldsymbol{x}_k^i = \boldsymbol{I} - \frac{\boldsymbol{H}_k^i\boldsymbol{P}_{k|k-1}^i(\boldsymbol{H}_k^i)^T}{\boldsymbol{H}_k^i\boldsymbol{P}_{k|k-1}^i(\boldsymbol{H}_k^i)^T + \boldsymbol{R}_k^i})\hat{\boldsymbol{x}}_k^i + \frac{\boldsymbol{H}_k^i\boldsymbol{P}_{k|k-1}^i(\boldsymbol{H}_k^i)^T}{\boldsymbol{H}_k^i\boldsymbol{P}_{k|k-1}^i(\boldsymbol{H}_k^i)^T + \boldsymbol{R}_k^i} z_k^i \tag{8}$$

$$\boldsymbol{x}_k^i = \omega_x^i\hat{\boldsymbol{x}}_k^i + \omega_z^i z_k^i \tag{9}$$

Here, $\boldsymbol{R}_k$ and $\boldsymbol{K}_k$ denote noise covariance and Kalman gain at time k, respectively. $\boldsymbol{P}_{k|k-1}$ is the current covariance and $\boldsymbol{H}_k$ is the observation matrix.

The likelihood function can enhance the interpretability of the network which implements the calculation without priori information. This network combines $\hat{\boldsymbol{x}}_k^i$ with corresponding measurement to form $\overline{\boldsymbol{x}}_k^i = [\hat{\boldsymbol{x}}_k^i, z_k^i]^T$. Then the application of the BiGRU can obtain the value ω_x^i and ω_z^i. As shown in Fig. 4, $\overline{\boldsymbol{x}}_k^i$ is send to the BiGRU with the hidden state h_{k-1} at time step $k-1$.

$$\boldsymbol{x}_k^i = \sigma\left(W' \cdot h_k\right) \tag{10}$$

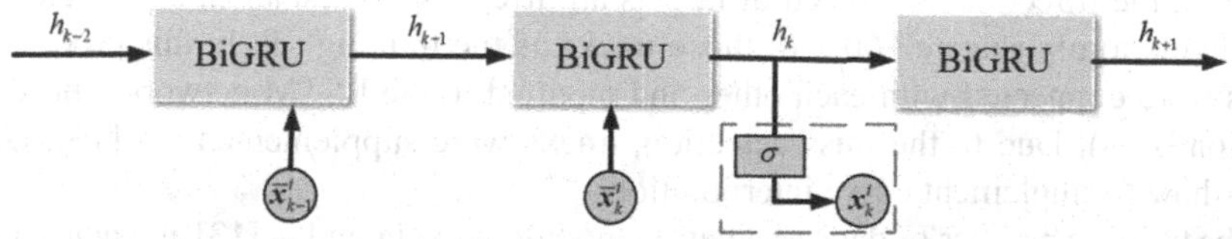

Fig. 4. AGE network.

Here, $\sigma(\cdot)$ denotes the sigmoid function, W' is a matrix of parameters of the full connected layer. Then, the current estimation $\boldsymbol{x}_k^i$ can be obtained by (10).

This module utilizes the loss function given in (11), which is derived in [11].

$$MSE^i = \sum_{k=1}^{n} (x_k^i - \tilde{x}_k^i)^2 / n \tag{11}$$

BiGRU's bidirectional architecture enables it to generalize well to unseen or out-of-domain data. By learning from both past and future information, the model can grasp the underlying relationships that transcend specific datasets, allowing it to make reliable predictions even on unfamiliar tracking.

3 Experiment

In this section, we conduct experiments on MIT Trajectory data[1], simulated dataset and Unmanned Aerial Vehicle (UAV) data. MIT Trajectory dataset consists of 40,453 tracks of vehicles and pedestrians from a parking lot scene. According to radar tracking algorithm effect detection method, we also design a multi-target tracking simulation experiment to verity the effectiveness of our algorithm. Additionally, we gathered UAV data while utilizing our lab's current configuration. For ablation studies, we utilize the MIT Trajectory data and simulated experiments to exemplify the effectiveness of AGE.

We use the metrics optimal subpattern assignment (OSPA) [15], containing tracking performance metric, localication error cmponent and cardinality error component, where $c = 100$ and $p = 2$. To prove the reliability of our innovative research, we compare with KF-DHN, LSTM-KF-DHN, GLMB, GMPHD, RNN-LSTM and KalmanNET for multi-target tracking. Meanwhile, we design different noise in our dataset and different orignial frame in order to proof its robustness. Moreover, the ablation studies that we utilize the mean squared error (MSE) in order to better validate the gap between the estimated state and turth. The LSTM, GRU, BiLSTM and KF are employed to update a single-target tracking to reveal the potency of our algorithm.

We implement our algorithm in python using TensorFlow toolbox, and test it on computer with 12th Gen Intel(R) Core(TM) i5-12490F CPU @ 3.00 GHz and a single NVIDIA GTX 2080Ti with 16G RAM. The implementation details are shown below.

3.1 Training

Since it is in a parking lot, the tracks has overlapping parts and certain route regulations. However, the direction and speed of targets are irregular. To establish a predictor, this module randomly selected 60% of those tracks as the training set. In this process, the tracks were connected with each other and inputted to the LSTM network whose loss function is (3). Due to the miss-detection, tracks were supplemented via [16], which shows how to implement curve interpolation.

The training process of data association module is explained in [13] and hence omitted here. The survival probability estimation module is a binary classification network which inputs the whole tracks and the duration of measurement disappearcance. This training process is implemented by (3).

---

[1] Http://www.ee.cuhk.edu.hk/~xgwang/mittrajsinglemulti.html

Figure 8(a) shows the training set for the update module, where the blue points are corresponding measurements z_k^i with different noise standard deviations and the red points are $\hat{x}_k^i$ obtained by the prediction module. The green points are the truth state $\hat{x}_k^i$ of the tracks. Furthermore, the z_k^i and $\hat{x}_k^i$ can obtain the current state x_k^i by (11).

3.2 Tracking on MIT Trajectory Data

The tracking benefit of the MTT-AGE is demonstrated in this section. It displays a frame of the tracking findings in Fig. 5(b). The blue dots correspond to measurements.

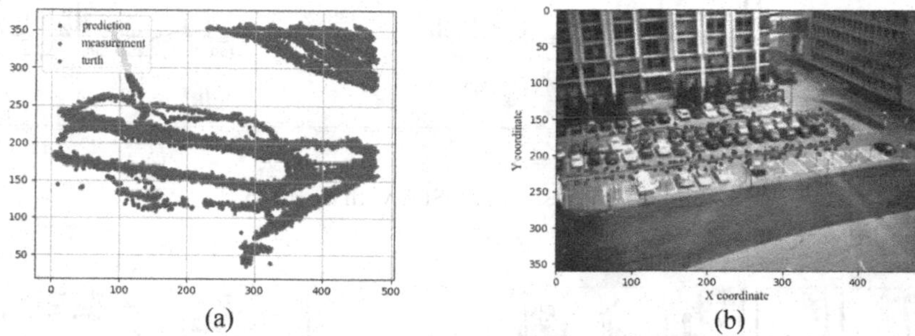

(a) (b)

Fig. 5. (a) training process, (b) frame 11732 tracking outcomes.

Additionally, we contrasted the experiments with standard deviations of measurement distortion of 0.5 and 1, respectively. In the comparison algorithm, the association algorithm implemented by the LSTM-KF and classic Kalman filters is the same as ours. They must also establish a priori information based on the statistical traits in the tracking situation, and the RFS based methods must do similarly. Thus, their measurement noise is set in view of the aforementioned scenario, and the motion-related parameter of KF is

$$x_k = F_k x_{k-1} + w_k \tag{12}$$

$$z_k = H_k x_k + v_k \tag{13}$$

$$F_k = \begin{bmatrix} 1\,1\,0\,0 \\ 0\,1\,0\,0 \\ 0\,0\,1\,1 \\ 0\,0\,0\,1 \end{bmatrix} \tag{14}$$

$$w_k \sim \mathcal{N}\left(0, \begin{bmatrix} \sigma_w^2 & 0 \\ 0 & \sigma_w^2 \end{bmatrix}\right) \tag{15}$$

$$H_k = \begin{bmatrix} 1\,0\,0\,0 \\ 0\,0\,1\,0 \end{bmatrix} \tag{16}$$

$$v_k \sim \mathcal{N}\left((\delta_v)^2, (\delta_v)^2\right) \tag{17}$$

Here, the σ_w and δ_v, which $\sigma_w = 0.5$ and $\delta_v = 0.5$, are standard deviation of process noise w_k and observation noise v_k, correspondingly. For GMPHD and GLMB, target detection and likelihood of survival are both 90%, while the state extraction weight threshold is set at 0.5. On the basis of this guideline, the outcomes of OSPA_dist (OSPA distance), OSPA_loc (OSPA localization), and OSPA_card (OSPA cardinality) are available from frame 0 to 2000 and shown in Figs. 6 and 7.

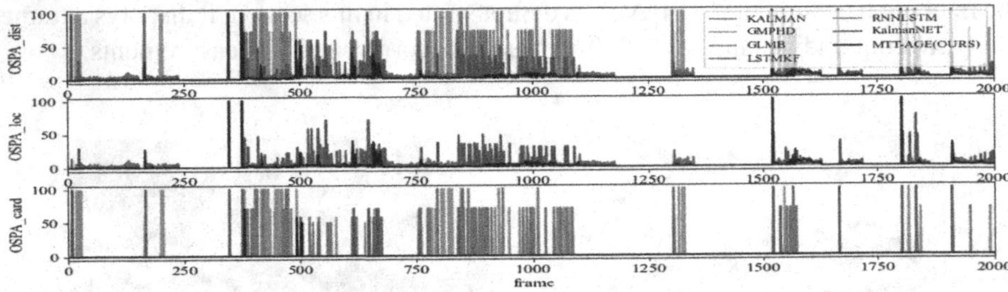

Fig. 6. The results of OSPA with $\delta_v = 0.5$.

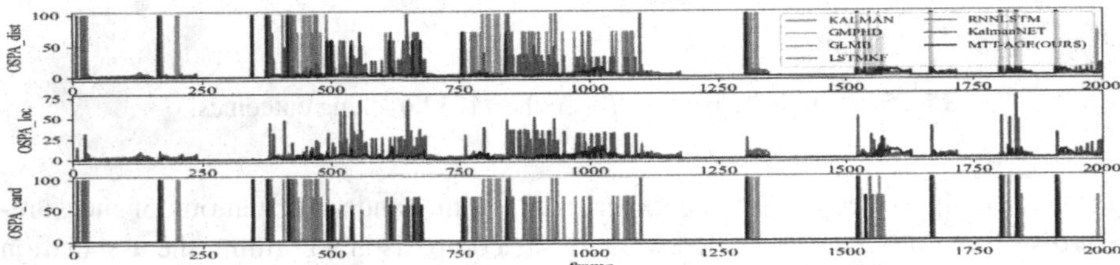

Fig. 7. The results of OSPA with $\delta_v = 1$.

The above illustration demonstrates how, for different levels of measurement noise, our performance is on par with that of competing methods. Since OSPA_dist is created by merging OSPA_loc and OSPA_card, the data alterations of these two are explored separately hereunder. For OSPA_loc, our method almost resembles the RFS based methods and LSTM-KF due to a very modest estimation error from frame 0 to 250 where there is a single target. The fact that it doesn't require initialization means that the initial fluctuations are very tiny. However, our cumulative error rapidly rises as the number of targets rises and the target motion becomes progressively more complex, resulting in fluctuations after frame 250. The positive aspect is that our tracking results are still superior to RNN-LSTM's because of the added features of our tracking system. We restrict the estimation between measurements and predictions in order to lessen the impact of network accuracy. Additionally, our errors are within allowable bounds. For OSPA_card, our algorithm and RNN-LSTM are much smoother than the RFS based methods, because their assessments of survival likelihood alleviate impact of lost information due to missing frames. It also shows that the accuracy of the cardinality estimation is higher than that of other methods, which is a clear indication that the survival estimation model is good.

To illustrate the robustness of our algorithm, we randomly select one moment as the starting moments with $\delta_v = 0.5$. The results are shown in Fig. 8.

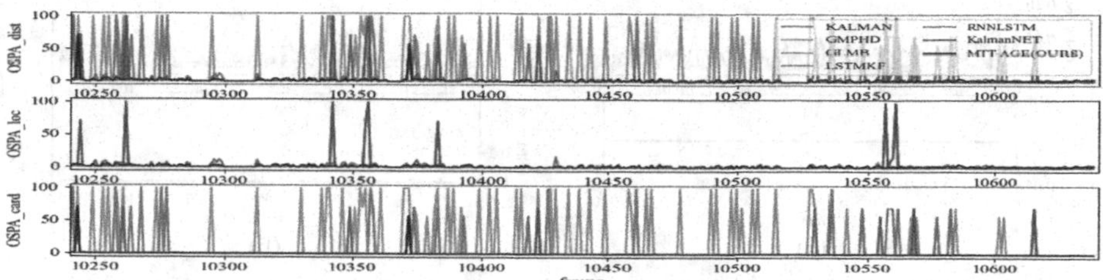

Fig. 8. The result of OSPA from frame 10240 to 10640.

Overall, MTT-AGE can achieve robust multi-target tracking in the case of unknown targets. Although our method fluctuates significantly at times, the LSTM-KF fluctuates as well, and preliminary analysis of these data leads to the conclusion that the general fluctuation is caused by the prediction error. The future work will concentrate on refining the prediction component.

3.3 Tracking on Simulated Scenarios

In order to ensure the reliability of MTT-AGE, this section examines that by simulated single-target and multi-target tracking experiments. For comparison, Monte Carlo was performed 100 times for each group of experiments. For the single-target experiments, the initial state of this target is $[50, 50, 0, 0]^T$, shown in Fig. 9(a). For the multi-target experiments, the initial states of three targets are $[50, 50, 0, 0]^T [50, 50, 0, 0]^T$, $[10, -10, 5, -5]^T$ and $[-20, -50, 0, 0]^T$, shown in Fig. 10(b). Their model is all shown in (12) and their parameters are shown in (14) and (15). Observation of a sensor is (13) and its parameters are shown from (16) to (17). The probability of targets detection is 98%.

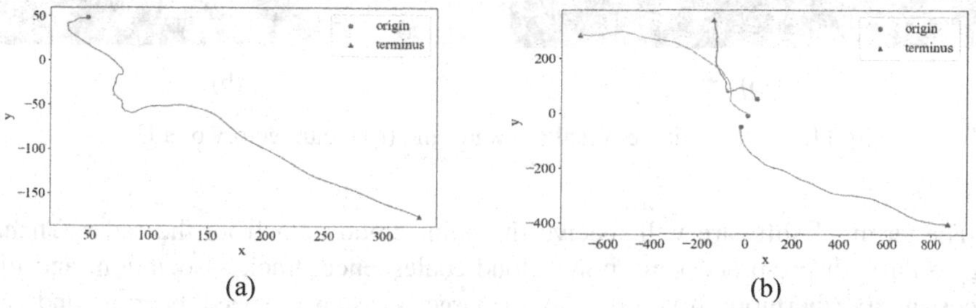

Fig. 9. (a) Simulated single-target trajectory, and (b) multi-target trajectories with $\delta_v = 0.5$.

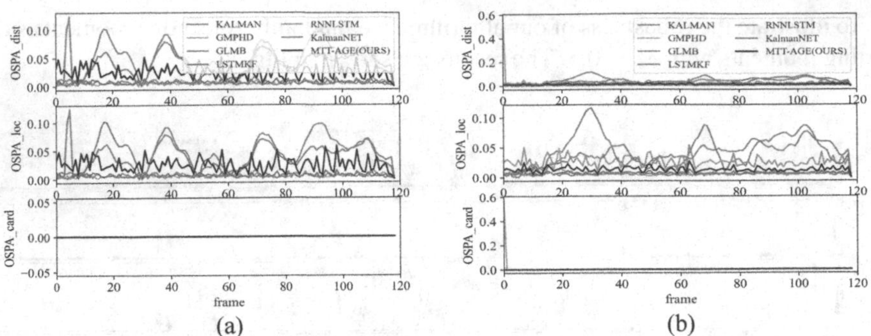

Fig. 10. (a) the results of OSPA of single-target dataset, and (b) multi-target dataset.

In Fig. 13, it reveals that the our method is reliable to multi-target tracking and can obtain accurate state estimates in the presence of known measurement noise.

3.4 Tracking on Real Dataset

With the already-developed radar terminal software in our laboratory, radar measurement data is gathered by following the detection of a UAV. The operator selects the options for controlling the radar in the terminal software, and sets the radar's working mode, horizontal pitch angle, rotation rate, data processing, and other relevant parameters to output target positioning. The terminal software interface shows real-time status feedback data for the radar's servo, signal processing, received channels, electronic compass, and GPS during the radar detection process, shown in Fig. 11.

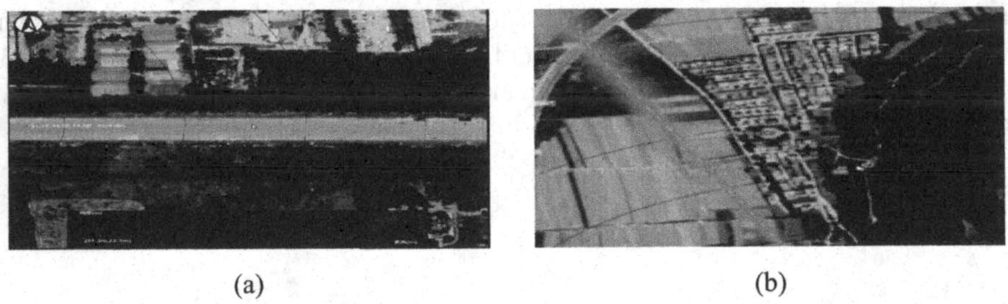

(a) (b)

Fig. 11. (a) the radar terminal software, and (b) the trajectory of a UAV.

The terminal software will receive the point cloud data from the radar. Then, it will go through pre-processing, point cloud coalescence, track association, and filter processing. Furthermore, it will display the target's distance, speed, bearing, and other information in real time on the software interface. Our intention is to more efficiently handle track association and filtering. The following provides an illustration of how well our algorithm performs in real-world settings while tracking UAVs.

In Fig. 11, it displays the UAV's flight route over specific area in Xi'an along with the outcomes of our radar detection in the polar coordinates. Based on the radar's rated error, it is determined that the ±0.5 m basis of the path's radius and of the servo azimuth and elevation angles are comparable on an ±0.3° basis.

The flight altitude of the UAV is about 100 m. The results are shown below.

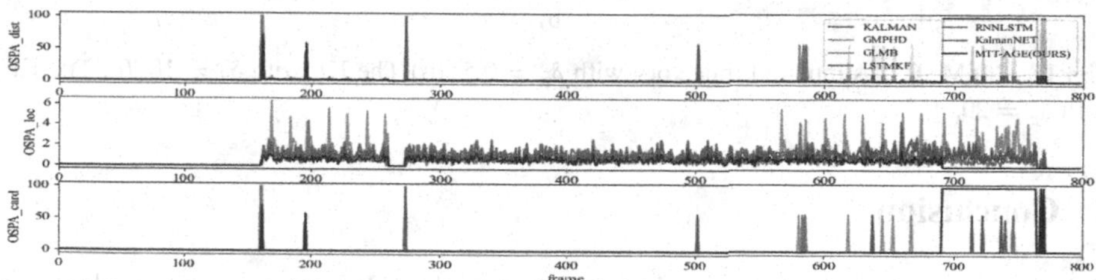

Fig. 12. The result of OSPA of UAVs.

In Fig. 12, it demonstrates the applicability of our technique to UAV tracking, particularly with uncertain radar detection noise.

In conclusion, it can be concluded that that our method is better suited to the case of unknown targets than other algorithms. Especially, KF and the RFS based methods need the priori information to establish the motion model and need covariance to obtain the Kalman gain which the LSTM-KF and KalmanNET also require.

3.5 The Verity of AGE Module

To ensure the reliability of gain estimation, this section examines AGE module separately with MIT Trajectory data from frame 0 to 100 and simulated experiments whose parameters are the same as the trajectory in Fig. 9(a) except for track time. In this section, we only consider the single-target tracking with the KF prediction module.

From the MSE results in Fig. 13(a), it can be seen that the AGE and KF have similar errors. Additionly, the LSTM, GRU and BiLSTM are pronounced variations due to an incorrect decision they made on the allocation of gains at the beginning. To lessen this impact, we intentionally raise the weight ω_z^i of measurement when training AGE. However, this causes an enormous error in AGE at certain moments, such as frame 10.

To proof the importance of known noise measurment, we design the following experiments. For simulated experiments, we adjust the priori measurement noise's standard deviation of KF to figure out its significance.

As seen in Fig. 13(b) and (c), KF will have a significant inaccuracy at beginning and require convergence time. Additionally, the convergence rate slows down as the δ_v gets greater. Therefore, KF must firstly estimate the measurement noise with $\alpha - \beta$ filtering [17] while tracking an unknown target. Although AGE also fluctuates with KF, it is mainly caused by the KF prediction, which indicates that AGE is a novelty method to solve this type of problem.

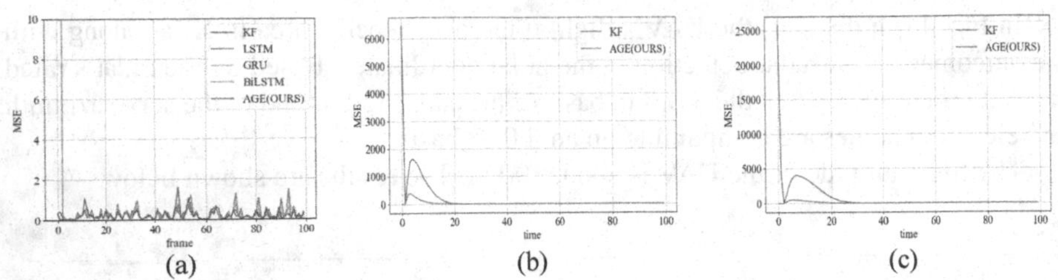

Fig. 13. (a) MSE of simulated trajectory with $\delta_v = 0.5$, (b) The KF with $\delta_v = 10$, (c) The KF with $\delta_v = 20$.

4 Conclusion

This paper presents an adaptive gain estimation method. It is an attempt to replace the computation of Kalman gain with a deep network structure and will find out the new relationship between prediction and measurement. Furthermore, this paper designs the survival likelihood estimation for the problem of low detection probability. The AGE also employs LSTM to obtain the trendless motion states by previous state information and it establishes an end-to-end tracking method for radar multi-target tracking.

The work in this paper centers on the application of AGE, but the accuracy of target tracking is also affected by prediction and association. The future work will focus on using spatio-temporal features [18] and attention mechanisms [19] to improve the accuracy of prediction. And we will enhance the association module by constructing a more sophisticated structure in order to adapt it to the high-clutter environment. In addition, the accuracy of estimating the state and survival likelihood of AGE can also be improved by adding attention mechanisms.

References

1. Wang D., Lian B., Liu Y., Gao B.: A cooperative UAV swarm localization algorithm based on probabilistic data association for visual measurement. IEEE Sens. J. (2022)
2. Wu L., Wang F., Xu Y., Jiang Y.and Wang J.: A parallel implementation of hypothesis-oriented multiple hypothesis tracking. In: 2020 IEEE 23rd International Conference on Information Fusion (FUSION), pp. 1–8 (2020)
3. Mahler R.: Advances in Statistical Multisource-Multitarget Information Fusion. Artech House, MA (2014)
4. Gao, L., Battistelli, G., Chisci, L., Farina, A.: Fusion-based multidetection multitarget tracking with random finite sets. IEEE Trans. Aero. Elec. Syst. **57**(4), 2438–2458 (2021)
5. Shi, K., Shi, Z., Yang, C., He, S., Chen, J., Chen, A.: Road-map aided gm-phd filter for multivehicle tracking with automotive radar. IEEE Trans. Ind. Inform. **18**(1), 97–108 (2022)
6. Park, W.J., Park, C.G.: Multi-target tracking based on gaussian mixture labeled multi-bernoulli filter with adaptive gating. In: 2019 First International Symposium on Instrumentation, Control, Artificial Intelligence, and Robotics (ICA-SYMP), pp. 226–229 (2021)
7. Li, Q.Y., He, B., Zhang, X.Y.: LSTM-based Encoder-Decoder multi-step track prediction technique. Air Weapon **28**(2), 49–54 (2021)

8. Emambakhsh, E., Bay, A., Vazquez, E.: Convolutional recurrent predictor: implicit representation for multi-target filtering and tracking. IEEE Trans. Signal Process. **67**(17), 4545–4555 (2019)
9. Jung, S., Schlangen, I., Charlish, A.: A mnemonic kalman filter for non-linear systems with extensive temporal dependencies. IEEE Signal Process. Lett. **27**, 1005–1009 (2020)
10. Milan, A., Rezatofighi, S.H., Dick, A.: Online multi-target tracking using recurrent neural networks. In: AAAI (2017)
11. Choi, G., Park, J., Shlezinger, N.: Split-KalmanNet: a robust model-based deep learning approach for state estimation. IEEE Trans. Vehicular Technology (2023)
12. Coskun, H., Achilles, F., DiPietro, R., Navab, N., Tombari, F.: Long short-term memory kalman filters: recurrent neural estimators for pose regularization. In: 2017 IEEE International Conference on Computer Vision (ICCV), pp. 5525–5533 (2017)
13. Xu, Y., Ban, Y., Alameda-Pineda, X.: Deepmot: A differentiable framework for training multiple object trackers. arXiv preprint arXiv:1906.06618 (2019)
14. Xie, B., Dai, S.: A comparative study of extended kalman filtering and unscented kalman filtering on lie group for stewart platform state estimation. In: 2021 6th International Conference on Control and Robotics Engineering (ICCRE), pp. 145–150 (2021)
15. Schuhmacher, D., Vo, B.-T., Vo, B.-N.: A consistent metric for performance evaluation of multi-object filters. IEEE Trans. Signal Process. **56**(8), 3447–3457 (2008)
16. Rongli, G., Yan, C.: Summary of spline Curve Interpolation. In: 2020 5th International Conference on Mechanical, Control and Computer Engineering (ICMCCE), pp. 1418–1421 (2020)
17. Li, Q., Chen, Z., Shi, W.: A novel state estimation approach for suspension system with time-varying and unknown noise covariance. Actuators **12**(2), 70–99 (2023)
18. Huang, X.: Interpretable local flow attention for multi-step traffic flow prediction. Neural Netw. **161**, 25–38 (2023)
19. Du, W., Côté, D., Liu, Y.: Saits: self-attention-based imputation for time series. Expert Syst. Appl. **219**, 119619 (2023)

Prompt Based Lifelong Person Re-identification

Chengde Yang, Yan Zhang, and Pingyang Dai(✉)

Key Laboratory of Multimedia Trusted Perception and Efficient Computing, Ministry of Education of China, School of Informatics, Xiamen University, Xiamen 361005, People's Republic of China

cdyang@stu.xmu.edu.cn, pydai@xmu.edu.cn

Abstract. In the real world, training data for person re-identification (ReID) comes in streams and the domain distribution may be inconsistent, which requires the model to incrementally learn new knowledge without forgetting the old knowledge. The problem is known as lifelong person re-identification (LReID). Previous work has focused more on the acquisition of task-irrelevant knowledge and neglected the auxiliary role of task-relevant information in alleviating catastrophic forgetting. To alleviating forgetting and improving the generalization ability, we introduced the prompt to learn task-relevant information, which can guide the model to perform task conditionally. We also proposed a special distillation module for the specific vision transformer structure, which further mitigated catastrophic forgetting. Extensive experiments on twelve person re-identification datasets outperforms other state-of-the-art competitors by a margin of 4.7% average mAP in anti-forgetting evaluation and 7.1% average mAP in generalising evaluation.

Keywords: Person re-identification · Lifelong learning · Knowledge distillation · Prompt

1 Introduction

The purpose of Person Re-Identification (Re-ID) is to retrieve the same pedestrian across disjoint camera views. With the development of deep learning in recent years, it has made great progress on multiple large-scale datasets. However, most methods assume that the training data can be accessed all at once or distributed uniformly, which is inconsistent with the realistic scenario. Its training process is mostly limited by fixed datasets, so it has poor generalization performance in real-world streaming data. Lifelong Person Re-ID is proposed to solve the problem of continuous input data stream, which requires the model to learn new knowledge while avoiding catastrophic forgetting of old knowledge.

Catastrophic forgetting is the biggest challenge of lifelong learning, and lifelong Person Re-ID is no exception. Since the deep learning model is highly dependent on parameters, the update of the model on the new datasets will lead

© The Author(s), under exclusive license to Springer Nature Singapore Pte Ltd. 2024
Q. Liu et al. (Eds.): PRCV 2023, LNCS 14436, pp. 418–431, 2024.
https://doi.org/10.1007/978-981-99-8555-5_33

to performance degradation on the old ones. Due to storage limitations and privacy protection, maintaining all data for retraining is not feasible. Most current methods use knowledge distillation and sample playback to mitigate catastrophic forgetting, which can effectively promote the learning of task-irrelevant knowledge, but ignore the guiding role of task-relevant information.

Extracting robust features is a crucial component of ReID, which has been dominated by CNN-based methods for a long time. These methods suffer from some conspicuous disadvantages, most importantly, CNN-based methods mainly focus on small discriminative regions due to a Gaussian distribution of effective receptive fields [5]. Recently, Vision Transformer (ViT) [6] and Data-efficient image Transformers (DeiT) [7] have shown that pure transformers can be as effective as CNN-based methods on feature extraction for image recognition. With the introduction of multi-head attention modules and the removal of convolution and downsampling operators, transformer-based models are suitable to solve the aforementioned problems in CNN-based ReID for the following reasons. Firstly, compared with CNN models, The multi-head self-attention captures long range dependencies and drives the model to attend diverse human-body parts. Secondly, without downsampling operators, transformer can keep more detailed information.

To this end, we propose a new lifelong person ReID method called PLReID. The main contributions of our work are summarized as follows:

1. We propose PLReID, a novel continual learning framework based on prompts for lifelong person Re-ID, providing a new method to tackle lifelong person Re-ID challenges through learning a prompt pool memory space, which provide specific guidance for different tasks.
2. We find that vision transformer has a natural advantage over CNN in the continual learning setting. We designed a distillation module for its unique structure, and subsequent experiments proved that it can effectively alleviate catastrophic forgetting.
3. Extensive experiments validate the proposed framework significantly outperformed the state-of-the-art methods, and it has quite good anti-forgetting ability and generalization ability.

2 Related Work

2.1 Person ReID

Person ReID has made remarkable progress in a variety of Settings. The fully supervised approach aims to learn robust feature representations from labelled data [1,12]. This method is suitable for data with stable distribution and requires multiple observations of the entire dataset. Later approaches study unsupervised domain adaptive [35,36] and sought unlabeled images to guide the learning process. Domain generalization (DG) is an open-set problem. Lately, DG ReID task is explored by [16]. However, these approaches assume that the data will be acquired at once and thus fail to address the challenge of lifelong learning.

Lifelong Person ReID has been proposed by AKA [15] to tackle the challenge of long-term visual search. This paradigm continuously consolidates knowledge from distribution-changing data in an incremental manner, which matches the real-world application. AKA constructs learnable knowledge graphs which adaptively accumulate knowledge. GwFReID [14] considers multiple consistency losses to constrain the training process. PKD [33] uses adaptively-chosen patches to pilot the forgetting-resistant distillation. However, The current methods ignore the guiding role of domain information in learning different tasks. We introduce prompt learning to learn domain knowledge, so as to provide guidance for learning different tasks, which is helpful to further alleviate catastrophic forgetting and improve generalization ability.

2.2 Lifelong Learning

Lifelong learning aims to adapt models to new tasks using knowledge learned in the past while maintaining stable performance in the old tasks. Current methods can be divided into three categories: rehearsal-based methods [5,6], architecture-based methods [4] and regularization-based methods [2,3].

Rehearsal-based methods alleviate catastrophic forgetting by recalling on stored data of previous tasks [17,18]. However, this approach is costly and has a large demand on storage space. Architecture-based methods aim at having separate components for each task. They usually attend to task-specific subnetworks. However, most of these methods require task identity to condition the network at test-time, are not applicable to more realistic class-incremental and task-agnostic settings when task identity is unknown. Regularization-based methods add regularization terms to limit large changes of model parameters, which has been shown to be one of the most effective ways to mitigate catastrophic forgetting. In our method, we designed a special distillation module for the specific vision transformer structure to alleviate forgetting.

2.3 Knowledge Distillation

The goal of Knowledge distillation is to transfer knowledge from a large teacher model to a small student model. It has been widely used to compress large-scale pre-training deep learning models. Knowledge distillation can be summarized into three categories: logit distillation [21,22], feature distillation [13] and relation distillation [10,19] which matches the final predictions, intermediate representations and inter-sample relations, respectively.

Compared with traditional image-level distillation, Kim et al. [20] proposed patch-level distillation, which is beneficial to learning fine-grained information. Sun et al. [33] use a differentiable patch sampler to select patches, which achieve more efficient patch-based knowledge distillation. Aiming at the special structure of vision transformer, we designed a distillation module combining logit and feature, which is helpful for learning more robust feature representation.

2.4 Prompt Learning

Prompt learning was originally developed for NLP to add additional text to the input to make better use of the knowledge of the pre-trained model. The general idea of prompting is to learn a function to modify or assist the input texts or images, such that the language or image model obtains additional information about the task.

A number of prompt methods have been proposed recently [8,9], including image-end prompts, language-end prompts and joint language-image prompts. Wang et al. [7] proposed a paradigm to learn prompts independently across domains with pre-trained, which can be applied to typical continual learning scenarios. L2P [11] exploits dependent prompting methods for continual learning, which learns to dynamically prompt a pre-trained model to learn tasks sequentially under different task transitions. However, in the training process of these methods, only prompt is learned and the parameters of the pre-training model are frozen, so it is extremely dependent on the generalization ability of the pre-training model and can not be well adapted to various downstream tasks. In our method, prompt was trained to learn task-relevant information and to instruct the model to perform task conditionally. This can effectively improve the representation of the model and significantly alleviate forgetting.

3 Method

3.1 Problem Definition and Formulation

In terms of LReID, we need to learn S domains in an incremental fashion. Suppose we have a stream of datasets $D = \{D^1, \ldots, D^S\}$. Each dataset D^s consists of training images D^s_{tr} and test images D^s_{te}, where $D^s_{tr} = \{(x_i, y_i)\}_{i=1}^{N_s}$ and N^s represents the number of images in dataset D^s_{tr}. According to the setting of person ReID, the classes of training and testing data are disjoint, namely $Y^s_{tr} \cap Y^s_{te} = \emptyset$.

At the s-th training step, only D^s_{tr} is available, which means the data from previous domains can not be available any more. In the test phase, the model is evaluated for anti-forgetting ability on the test split of all seen datasets and generalization ability on unseen datasets.

3.2 Baseline Approach

We use the graph-based model proposed by AKA [15] as our baseline solution and its backbone has been replaced with vision transformer (ViT). The baseline model consists of three parts, namely, a ViT feature extractor $h(\cdot;\theta)$ with parameters θ, a classifier $f(\cdot;\phi)$ with parameters ϕ, and a graph-based constraint module $g(\cdot;\varphi)$ with parameters φ. Given an input x, it will be fed into feature extractors h and classifiers f to get confidence scores $f(h(\boldsymbol{x};\theta);\phi)$. And the parameters θ and ϕ is optimized by a cross-entropy loss,

$$L_c = -\sum_{(\boldsymbol{x},\boldsymbol{y})\in D} ylog\left(\sigma(f\left(h(\boldsymbol{x};\theta);\phi\right))\right) \tag{1}$$

The graph-based constraint module $g(\cdot;\varphi)$ uses a stability loss L_s and a plasticity loss L_p to constrain the feature extractor h to generate more robust features. A detailed description can be found in AKA [15]. The total loss function of baseline method is:

$$L_{base} = L_c + \lambda_s L_s + \lambda_p L_p \tag{2}$$

where λ_s and λ_p is a trade-off factor.

3.3 Distillation for Vision Transformer

We use ViT as a feature extractor. Considering its special structure, we design a module combining logit distillation and feature distillation to mitigate catastrophic forgetting. The overall architecture of distillation module can be seen in Fig. 1.

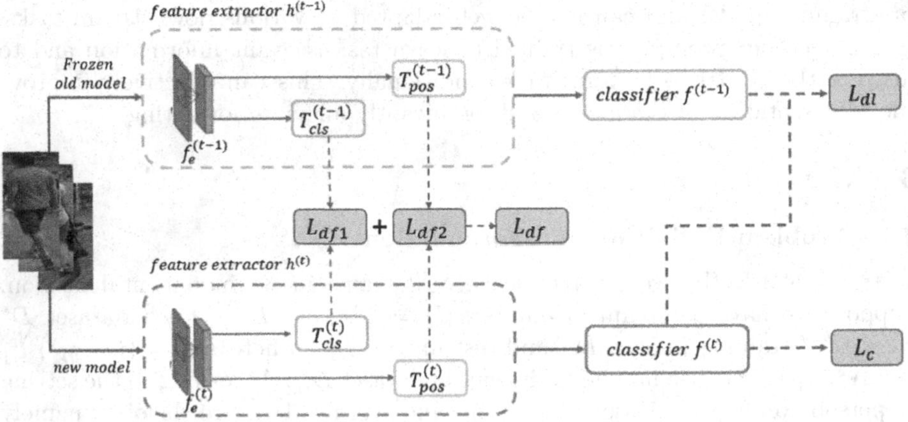

Fig. 1. Overall architecture of our proposed distillation module. The dotted line shows the calculation of the loss function. f_e donates embedding layers. $T_{cls}^{(t-1)}$ and $T_{pos}^{(t-1)}$ donate class token and position embedding generated by model trained in $(t-1)$-th step.

Logit Distillation. We use logit distillation to prevent large changes in the output logit of the current model and the previous model, which will mitigate catastrophic forgetting. The loss function can be described as:

$$L_{dl} = -\sum_{x \in D} \sum_{j=1}^{n} \sigma\left(f\left(h\left(x;\theta^{(t-1)}\right);\phi^{(t-1)}\right)\right)_j \log\left(\sigma\left(f\left(h\left(x;\theta^{(t)}\right);\phi^{(t)}\right)\right)_j\right) \tag{3}$$

where $\theta^{(t-1)}$ and $\phi^{(t-1)}$ are parameters of the previous step and are frozen in the current step, n is the number of old classes.

Feature Distillation. In vision transformer, class token aggregates the global features, and the position embedding encodes the location information of each token. We found that distilling the class tokens T_{cls} and position embedding T_{pos} generated by the old and new models further mitigated forgetting. We use mean square error loss to achieve optimization:

$$L_{df} = \frac{1}{D}\sum_{i}^{D}\left(\left(T_{cls}^{(t)} - T_{cls}^{(t-1)}\right)^2 + \left(T_{pos}^{(t)} - T_{pos}^{(t-1)}\right)^2\right) \tag{4}$$

where $T_{cls}^{(t-1)}$ and $T_{pos}^{(t-1)}$ are embedding generated by the previous model, D is the flattened dimension of class token and position embedding.

The total distillation loss function is:

$$L_d = L_{dl} + \lambda_d L_{df} \tag{5}$$

where λ_d is a trade-off factor. And the overall loss function is:

$$L_{tol} = L_{base} + \lambda L_d \tag{6}$$

3.4 Prompt Learning

Why Prompt. In setting of LReID, the data is continuously increasing from different domains, and the task identity is unknown at the test time. Previous work has focused more on the acquisition of task-irrelevant knowledge and neglected the auxiliary role of task-relevant information in alleviating catastrophic forgetting. L2P [11] demonstrated the effectiveness of prompt in continual learning tasks, and inspired by which, we introduce prompt into LReID.

During the training phase, we hope prompt learn task-relevant knowledge, and during the testing phase, the task-relevant prompt guides the model to perform the corresponding task. This will promote the enhancement of both anti-forgetting ability and generalization ability.

Structure Design. We represent the vision transformer (ViT) $f = f_r \circ f_e$, where f_e is the input embedding layers and f_r represents a stack of self-attention layers [11,38]. Given an input of 2D image $\boldsymbol{x} \in R^{H\times W\times C}$, we reshape it to a sequence of flattened 2D patches $x_p \in R^{L\times D}$, where L and D is the token length and embedding dimension. The embedding layer f_e projects the patched images to embedding features $\boldsymbol{x}_e = f_e(x) \in R^{L\times D}$. Prompt is essentially learnable parameters P_e, which we prepend to the embedding feature $\boldsymbol{x}_p = [P_e; \boldsymbol{x}_e]$. The extended sequences were fed to the self-attention layers for performing downstream tasks. The structure is shown in Fig. 1.

Prompt Pool and Selection Strategy. Given that task ids are unknowable during the testing phase, we maintain a prompt pool to share knowledge between similar tasks. The prompt pool is defined as $\boldsymbol{P} = \{P_1, \ldots, P_N\}$, where $P_i \in R^D$

is a single prompt with the same embedding size D as $\boldsymbol{x}_e$ and N is the total number of prompts. During the training phase, a fixed number (set T) of prompt groups are learned for each task and added to the prompt pool.

During the testing phase, we need to select the prompts that best matches the current task to provide task information. Details are shown in Fig. 2. We assume that class token will carry task-relevant information, so we determine which prompts will be selected based on their similarity to the class token. Denoting $\{s_i\}_i^T$ as a subset of T indices from $[1, N]$, the input embedding can be described as $\boldsymbol{x}_p = [P_{s_1}; \ldots; P_{s_T}; \boldsymbol{x}_e]$, where ; represents concatenation along the token length dimension.

An end-to-end approach is used to train the prompt, which is optimized with the same total loss function constraints as the embedding features.

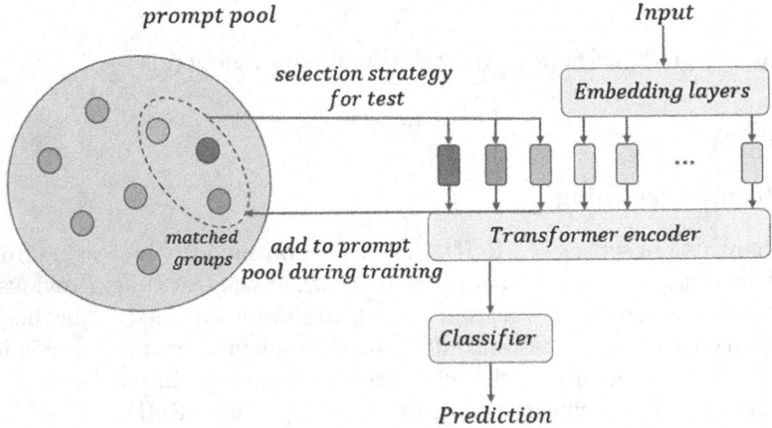

Fig. 2. Illustration of prompt module. We refer to the practice of L2P [10], selecting a group of prompts according to the selection strategy in Sect. 3.4 and prepend them to the input tokens. In the training phase, we use randomly initialized prompts for training, which are indirectly constrained by the total loss.

4 Experiments

4.1 Datasets and Evaluation Metrics

Dataset. We conduct extensive experiments on the LReID benchmark [15]. It consists of five seen datasets (Market-1501 [39], CUHK-SYSU [40], DukeMTMC-reID [41], MSMT17_V2 [42] and CUHK03 [37]) for anti-forgetting evaluation and seven unseen datasets (VIPeR [23], PRID [24], GRID [25], iLIDS [26], CUHK01 [27], CUHK02 [28] and SenseReID [29]) for generalization evaluation. In order to compare performance under equal conditions, we followed AKA [15] in dataset processing and experimental setting. To tackle the problem of unbalanced class

Table 1. Dataset statistics of the LReID benchmark [15]. '–' denotes that the dataset is only used for test.

Type	Dataset	Scale	#train IDs	#test IDs
Seen	Market-1501 [39]	large	500/751	750
	CUHK-SYSU [40]	mid	500/942	2900
	DukeMTMC-reID [41]	large	500/702	1110
	MSMT17_V2 [42]	large	500/1041	3060
	CUHK03 [37]	mid	500/700	700
Unseen	VIPeR [23]	small	–	316
	PRID [24]	small	–	649
	GRID [25]	small	–	126
	i-LIDS [26]	small	–	60
	CUHK01 [27]	small	–	486
	CUHK02 [28]	mid	–	239
	SenseReID [29]	mid	–	1718

number among datasets, 500 identities were randomly sample from each seen dataset for training. In the test phase, 3,594 different identities from seven unseen datasets were merged as a unified test set to evaluate the generalization ability on unseen domain. More detailed statistics for these datasets are provided in Table 1.

In order to simulate the real scenario and easily compare the experimental performance with previous work, we used the experimental setup of PKD [33]. There are two training orders, the training order-1 represents Market-1501 → CUHK-SYSU → DukeMTMC-reID → MSMT17_V2 → CUHK03, and the training order-2 represents DukeMTMC-reID → MSMT17_V2 → Market-1501 → CUHK-SYSU → CUHK03.

Evaluation Metrics. We use mean Average Precision (mAP) and Rank-1 accuracy (R1) to evaluate the performance on each ReID dataset. Moreover, we calculate the average of several experiments to evaluate the corresponding performance.

4.2 Implementation Details

We utilize ViT pretrained on ImageNet [34] as the feature extractor. In each training batch, we follow [15] to select 32 identities and sample 4 images for each identity randomly. All images are resized to 256 × 128. Adam optimizer with learning rate 3.5×10^{-4} is used. The model is trained for 50 epochs, and decrease the learning rate by ×0.1 at the 25th and 35th epoch. The retrieval of testing data is based on Euclidean distance of feature embeddings. For all experiments, we repeat three times and report mean performance. The whole architecture is implemented with PyTorch and trained on single NVIDIA A100 GPU.

Table 2. Comparison with the state-of-the-art methods on the LReID benchmark [15]. '*' represents the results reported in [33].

(a) Training order-1.

Method	Market		SYSU		Duke		MSMT17		CUHK03		Seen-Avg		Unseen-Avg	
	mAP	R1	mAP	R1	mAP	R1	mAP	R1	mAP	R1	mAP	R1	mAP	R1
Finetune	32.7	58.3	58.0	60.6	25.2	43.8	4.5	13.1	41.3	43.4	32.3	43.9	38.4	34.4
SPD [30]	35.6	61.2	61.7	64.0	27.5	47.1	5.2	15.5	**42.2**	**44.3**	34.4	46.4	40.4	36.6
LwF [31]	56.3	77.1	72.9	75.1	29.6	46.5	6.0	16.6	36.1	37.5	40.2	50.6	47.2	42.6
CRL [32]	58.0	78.2	72.5	75.1	28.3	45.2	6.0	15.8	37.4	39.8	40.5	50.8	47.8	43.5
AKA [15]	51.2	72.0	47.5	45.1	18.7	33.1	16.4	37.6	27.7	27.6	32.3	43.1	44.3	40.4
AKA [15]*	58.1	77.4	72.5	74.8	28.7	45.2	6.1	16.2	38.7	40.4	40.8	50.8	47.6	42.6
PKD [33]	68.5	85.7	75.6	78.6	33.8	50.4	6.5	17.0	34.1	36.8	43.7	53.7	49.1	45.4
Baseline	66.9	85.4	81.5	82.1	41.7	60.8	15.3	38.2	24.4	23.9	46.0	58.1	54.8	45.7
Ours	**69.1**	**87.1**	**82.3**	**83.9**	**44.0**	**63.6**	**17.5**	**40.6**	29.1	29.6	**48.4**	**61.0**	**56.2**	**48.4**
SingleTrain	76.5	90.4	82.3	84.3	69.2	83.5	37.8	62.6	50.5	52.9	63.3	74.7	–	–

(b) Training order-2.

Method	Duke		MSMT17		Market		SYSU		CUHK03		Seen-Avg		Unseen-Avg	
	mAP	R1	mAP	R1	mAP	R1	mAP	R1	mAP	R1	mAP	R1	mAP	R1
Finetune	26.1	45.7	3.3	10.3	29.1	54.1	57.2	60.0	40.3	40.9	31.2	42.2	36.1	32.0
SPD [30]	28.5	48.5	3.7	11.5	32.3	57.4	62.1	65.0	**43.0**	**45.2**	33.9	45.5	39.8	36.3
LwF [31]	42.7	61.7	5.1	14.3	34.4	58.6	69.9	73.0	34.1	34.1	37.2	48.4	44.0	40.1
CRL [32]	43.5	63.1	4.8	13.7	35.0	59.8	70.0	72.8	34.5	36.8	37.6	49.2	45.3	41.4
AKA [15]*	42.2	60.1	5.4	15.1	37.2	59.8	71.2	73.9	36.9	37.9	38.6	49.4	46.0	41.7
PKD [33]	58.3	74.1	6.4	17.4	43.2	67.4	74.5	76.9	33.7	34.8	43.2	54.1	48.6	44.1
Baseline	58.7	76.3	16.4	38.2	43.8	67.5	78.7	81.4	22.1	22.9	43.9	57.3	53.2	44.8
Ours	**61.4**	**78.2**	**17.1**	**39.5**	**45.7**	**69.7**	**80.7**	**82.9**	26.7	27.1	**46.3**	**59.5**	**56.1**	**48.6**
SingleTrain	69.2	83.5	37.8	63.6	76.5	90.4	82.3	84.3	50.5	52.9	63.3	74.7	–	–

4.3 Performance Evaluation

we compare our method to six lifelong learning methods that do not rely on exemplar memory: Finetune, SPD [30], LwF [31], CRL [32], AKA [15] and PKD [33]. Finetune donates fine-tuning model on new datasets without knowledge distillation. SPD is an advanced feature distillation method while LwF, CRL and AKA uses logit distillation. PKD proposed a novel patch-level distillation method. For convenience, we used off-the-shelf results of Finetune, SPD, LwF, and CRL, which were reported by PKD. The final result of each method on the LReID benchmark is shown in Table 2. We also report the upper-bound for each setting estimated by SingleTrain, where we trained each dataset individually with our method. Figure 4 is a visualization of prompt embedding during the training phase and selected prompts during the testing phase, indicating that the prompts did learn task-relevant information and instructed the model to perform task conditionally during the testing phase.

Anti-forgetting Performance on Seen Datasets. As shown in Table 2, we improve the average mAP of sort-of-the-art method PKD by 4.7% and 5.4% under training order-1 and order-2 on seen datasets, respectively. And the performance improvement on average R1 is also significant. We achieved optimal performance on all seen datasets except CUHK03. Unlike other related methods, we use SingleTrain methods to get upper-bound performance, which better measures the performance degradation caused by incremental learning. Table 2 shows that most datasets have small performance gaps, suggesting that our approach effectively mitigated catastrophic forgetting. Figure 3(a) illustrates the mAP and R1 curves on the first seen dataset after each training step. We can observe that our method is more stable and achieves the overall best performance.

Generalization Ability on Unseen Datasets. We improve the average mAP of sort-of-the-art method PKD by 7.1% and 7.5% under training order-1 and order-2 on unseen datasets, respectively. Figure 3(b) depicts the trend of average mAP and R1 on all unseen datasets during training process. It can be seen that performance is significantly better than that of alternative methods. This shows that our method has more outstanding generalization ability.

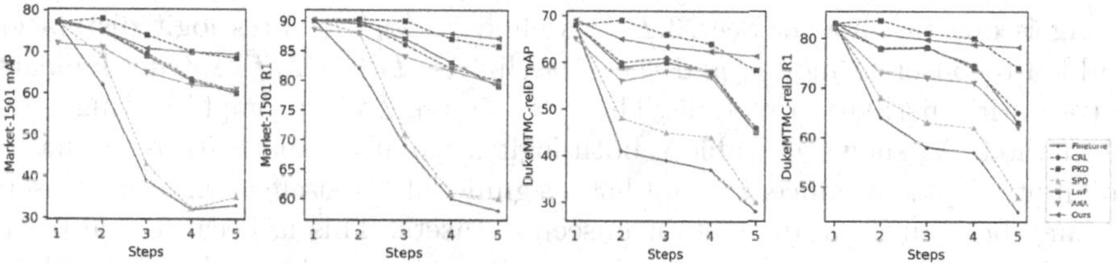

(a) Evolution of anti-forgetting performance on the first seen dataset during training process.

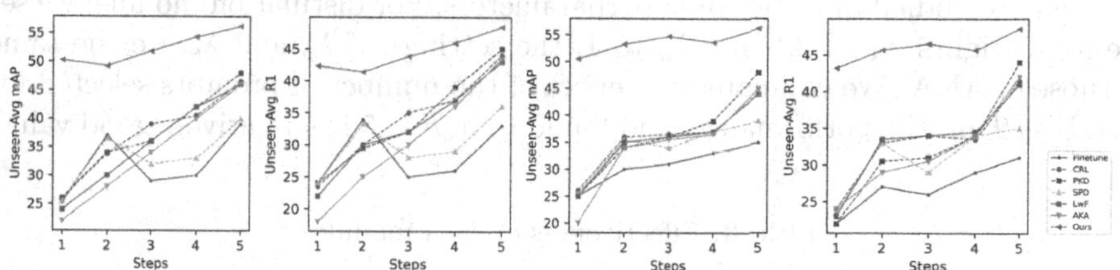

(b) Evolution of generalization ability on unseen datasets during training process.

Fig. 3. Evaluation of performance during training process. The training order of the two figures on the left is order-1, and that of the two figures on the right is order-2.

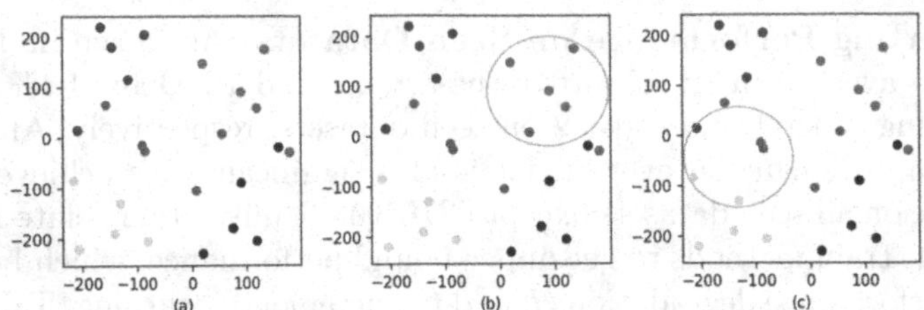

Fig. 4. Visualization for prompt. (a) shows t-SNE for prompt embedding, which comes from the final prompt pool after the training phase. The red, blue, yellow, green and black dots represent the trained prompts obtained from the corresponding datasets in training order-1, respectively. (b) shows the selected prompts for seen dataset Market during the testing phase. (c) shows the selected prompts for unseen dataset CUHK01 during the testing phase. (Color figure online)

4.4 Ablation Studies

In this section, we conducted experiments under training order-1 to verify the effectiveness distillation module and prompt strategy. In Table 3, "Baseline" setting is our baseline in Sect. 3.2. "Baseline + L_{dl}" denotes logit distillation is added to baseline model, and the "Baseline + *Ldl* + *Ldf*" setting indicates we use entire distillation module. The "Baseline + *Ld* + prompt" setting is our full method. As shown in Table 3, both logit distillation and feature distillation contribute to performance. Prompt has a significant boost effect on seen datasets, but may be counterproductive on unseen datasets. This is because due to the limitation of the number of tasks during the training phase, the accumulated prompts in prompt pool cannot cover the entire task space.

We also studied influence of hyperparameters. For distillation module, we set the loss weights λ_d to 2.5 and λ_d to 1, the settings of λ_p and λ_s are the same as those of AKA. We compared the effect of the number of prompts selected for each task T on the performance, and found that $T = 5$ is a relatively good value.

Table 3. Effectiveness of each module.

Setting	Seen-Avg		Unseen-Avg	
	mAP	R1	mAP	R1
Baseline	46.0	58.1	54.8	45.7
Baseline+*Ldl*	46.8	58.7	55.6	46.9
Baseline+*Ldl*+*Ldf*	47.2	59.3	**56.4**	48.1
Baseline+*Ld*+prompt(Full)	**48.4**	**61.0**	56.2	**48.4**

4.5 Discussion

Compared to recent works, our advantage is obvious, that is, we take into account the prompt role of task-relevant information, which has a significant effect on performance. However, the limitation is that distillation alone does not guarantee that the old knowledge can be updated effectively. Constructing pseudo feature maps by mapping is a good option, which will be considered in future work.

5 Conclusion

In this paper, we study lifelong person ReID problem, which is significant for real-world application but remains under-explored. Catastrophic forgetting is one of the most difficult challenges in LReID. We proposed prompt based method (PLReID) to solve this problem, which consists of a prompt module and a ViT based distillation module. Specifically, the prompt is trained to learn task-irrelevant information during the training phase, while during the testing phase, it was used to instruct the model to perform task conditionally. The distillation module is designed according to the special structure of the ViT, which has been shown to further alleviate forgetting. Extensive experiments on the LReID benchmark demonstrate that our method outperforms state-of-the-art methods in both anti-forgetting and generalization evaluations.

Acknowledgements. This work was supported by National Key R&D Program of China (No. 2022ZD0118202), the National Science Fund for Distinguished Young Scholars (No. 62025603), the National Natural Science Foundation of China (No. U21B2037, No. U22B2051, No. 62176222, No. 62176223, No. 62176226, No. 62072386, No. 62072387, No. 62072389, No. 62002305 and No. 62272401), and the Natural Science Foundation of Fujian Province of China (No. 2021J01002, No. 2022J06001).

References

1. Li, W., Zhu, X., Gong, S.: Harmonious attention network for person re-identification. In: CVPR, pp. 2285–2294 (2018)
2. Li, Z., Hoiem, D.: Learning without forgetting. IEEE Trans. Pattern Anal. Mach. Intell. **40**(12), 2935–2947 (2018)
3. Zhao, B., Xiao, X., Gan, G., Zhang, B., Xia, S.: Maintaining discrimination and fairness in class incremental learning. In: CVPR, pp. 13208–13217 (2020)
4. Serrá, J., Surís, D., Miron, M., Karatzoglou, A.: Overcoming catastrophic forgetting with hard attention to the task. In: ICML, pp. 4548–4557 (2018)
5. Rebuffi, S.-A., Kolesnikov, A., Sperl, G., Lampert, C.H.: Incremental classifier and representation learning. In: CVPR, pp. 5533–5542 (2017)
6. Hou, S., Pan, X., Loy, C.C., Wang, Z., Lin, D.: Learning a unified classifier incrementally via rebalancing. In: CVPR, 831–839 (2019)
7. Wang, Y., Huang, Z., et al. S:-prompts learning with pre-trained transformers: an Occam's Razor for domain incremental learning. In: NeurIPS, pp. 5682–5695 (2022)

8. Bahng, H., Jahanian, A., Sankaranarayanan, S.: Visual prompting: modifying pixel space to adapt pre-trained models. arXiv preprint arXiv:2203.17274 (2022)
9. Huang, T., Chu, J., Wei, F.: Unsupervised prompt learning for vision-language models. arXiv preprint arXiv:2204.03649 (2022)
10. Park, W., Kim, D., Lu, Y., Cho, M.: Relational knowledge distillation. In: CVPR, pp. 3967–3976 (2019)
11. Wang, Z., Zhang, Z., Lee, C.-Y., Zhang, H., et al.: Learning to prompt for continual learning. In: CVPR, pp. 139–149 (2022)
12. He, S., Luo, H., Wang, P., Wang, F., Li, H., Jiang, W.: Transreid: transformer-based object re-identification. In: ICCV, pp. 15013–15022 (2021)
13. Romero, A., Ballas, N., Kahou, S.E., Chassang, A., et al.: Fitnets: hints for thin deep nets. arXiv preprint arXiv:1412.6550 (2014)
14. Pu, N., Chen, W., Liu, Y., Bakker, E.M., Lew, M.S.: Generalising without forgetting for lifelong person re-identification. In: AAAI, pp. 2889–2897 (2021)
15. Pu, N., Chen, W., Liu, Y., Bakker, E.M., Lew, M.S.: Lifelong person re-identification via adaptive knowledge accumulation. In: CVPR, pp. 7901–7910 (2021)
16. Song, J., Yang, Y., Song, Y.-Z., Xiang, T.: Generalizable person re-identification by domain-invariant mapping network. In: CVPR, pp. 719–728 (2019)
17. Castro, F.M., Marín-Jiménez, M.J., Guil, N., Schmid, C., Alahari, K.: End-to-end incremental learning. In: ECCV, pp. 233–248 (2018)
18. Lopez-Paz, D., Ranzato, M.A.: Gradient episodic memory for continual learning. In: NeurIPS, pp. 6470–6479 (2017)
19. Liu, Y., Cao, J., Li, B., Yuan, C., et al.: Knowledge distillation via instance relationship graph. In: CVPR, pp. 7096–7104 (2019)
20. Kim, Y., Park, J., Jang, Y., Ali, M., et al.: Distilling global and local logits with densely connected relations. In: ICCV, pp. 6290–6300 (2021)
21. Hinton, G., Vinyals, O., Dean, J., et al.: Distilling the knowledge in a neural network. arXiv preprint arXiv:1503.02531 (2015)
22. Ba, J., Caruana, R.: Do deep nets really need to be deep? In: NeurIPS, pp. 2654–2662 (2014)
23. Gray, D., Tao, H.: Viewpoint invariant pedestrian recognition with an ensemble of localized features. In: Forsyth, D., Torr, P., Zisserman, A. (eds.) ECCV 2008. LNCS, vol. 5302, pp. 262–275. Springer, Heidelberg (2008). https://doi.org/10.1007/978-3-540-88682-2_21
24. Hirzer, M., Beleznai, C., Roth, P.M., Bischof, H.: Person re-identification by descriptive and discriminative classification. In: Heyden, A., Kahl, F. (eds.) Image Analysis, pp. 91–102. Springer, Heidelberg (2011). https://doi.org/10.1007/978-3-642-21227-7_9
25. Loy, C.C., Xiang, T., Gong, S.: Time-delayed correlation analysis for multi-camera activity understanding. In: IJCV, pp. 106–129 (2010)
26. Zheng, W.-S., et al.: Associating groups of people. In: BMVC, pp. 1–11 (2009)
27. Li, W., Zhao, R., Wang, X.: Human re-identification with transferred metric learning. In: ACCV, pp. 31–44 (2012)
28. Li, W., Wang, X.: Locally aligned feature transforms across views. In: CVPR, pp. 3594–3601 (2013)
29. Zhao, H., Tian, M., et al.: Spindle net: person re-identification with human body region guided feature decomposition and fusion. In: CVPR, pp. 1077–1085 (2017)
30. Tung, F., Mori, G.: Similarity-preserving knowledge distillation. In: ICCV, pp. 1365–1374 (2019)

31. Li, Z., et al.: Learning without forgetting. In: TPAMI, pp. 2935–2947 (2017)
32. Zhao, B., Tang, S., Chen, D., Bilen, H., Zhao, R.: Continual representation learning for biometric identification. In: WACV, pp. 1198–1208 (2021)
33. Sun, Z., Mu, Y.: Patch-based Knowledge distillation for lifelong person re-identification. In: ACM MM, pp. 696–707 (2022)
34. Russakovsky, O., Deng, J., Su, H., Krause, J., et al.: Imagenet large scale visual recognition challenge. In: IJCV, pp. 211–252 (2015)
35. Dai, Y., Liu, J., Sun, Y., Tong, Z., et al.: IDM: an intermediate domain module for domain adaptive person re-id. In: ICCV, pp. 11864–11874 (2021)
36. Ge, Y., et al.: Mutual mean-teaching: pseudo label refinery for unsupervised domain adaptation on person re-identification. arXiv preprint arXiv:2001.01526 (2020)
37. Li, W., Zhao, R., Xiao, T., Wang, X.: Deepreid: deep filter pairing neural network for person re-identification. In: CVPR, pp. 152–159 (2014)
38. Dosovitskiy, A., Beyer, L., Kolesnikov, A., Weissenborn, D., et al.: An image is worth 16x16 words: transformers for image recognition at scale. In: ICLR (2021)
39. Zheng, L., Shen, L., Tian, L., Wang, S., Wang, J., Tian, Q.: Scalable person re-identification: a benchmark. In: ICCV, pp. 1116–1124 (2015)
40. Xiao, T., Li, S., Wang, B., Lin, L., Wang, X.: End-to-end deep learning for person search. arXiv preprint arXiv:1604.01850 (2016)
41. Ristani, E., Solera, F., Zou, R., Cucchiara, R.: Performance measures and a data set for multi-target, multi-camera tracking. In: ECCV, pp. 17–35 (2016)
42. Wei, L., Zhang, S., Gao, W., Tian, Q.: Person transfer gan to bridge domain gap for person re-identification. In: CVPR, pp. 79–88 (2018)

Hierarchical Focused Feature Pyramid Network for Small Object Detection

Siwei Wang, Zhiwei Chen, Haoyang Ding, and Liujuan Cao(✉)

Key Laboratory of Multimedia Trusted Perception and Efficient Computing, Ministry of Education of China, Xiamen University, 361005 Xiamen, People's Republic of China

caoliujuan@xmu.edu.cn

Abstract. Small object detection has been a persistently practical and challenging task in the field of computer vision. Advanced detectors often utilize a feature pyramid network (FPN) to fuse the features generated from various receptive fields, which improve the detection ability of multi-scale objects, especially for small objects. However, existing FPNs typically employ a naive addition-based fusion strategy, which neglects crucial details that may exist only at specific levels. These details are vital for accurately detecting small objects. In this paper, we propose a novel Hierarchical Focused Feature Pyramid Network (HFFPN) to enhance these details while ensuring the detection performance for objects of other scales. HFFPN consists of two key components: Hierarchical Feature Subtraction Module (HFSM) and Feature Fusion Guidance Attention (FFGA). HFSM is first designed to selectively amplify the information important to small object detection. FFGA is devised to focus on effective features by utilizing global information and mining small objects' information from high-level features. Combining these two modules contributes greatly to the original FPN. In particular, the proposed HFFPN can be incorporated into most mainstream detectors, such as Faster RCNN, Retinanet, FCOS, *etc.* Extensive experiments on small object datasets demonstrate that HFFPN achieves consistent and significant improvements over the baseline algorithm while surpassing the state-of-the-art methods.

Keywords: Small object detection · Feature pyramid network · Self-attention

1 Introduction

Object detection is a widely studied task that aims to locate and classify the objects of interest. In recent years, object detection has achieved remarkable progress due to the powerful ability of Convolutional Neural Networks (CNNs) and the availability of an enormous amount of data [4]. However, as an important branch of object detection, small object detection has always been a bottleneck for detector performance. Small objects, typically refer to objects with a pixel

© The Author(s), under exclusive license to Springer Nature Singapore Pte Ltd. 2024
Q. Liu et al. (Eds.): PRCV 2023, LNCS 14436, pp. 432–444, 2024.
https://doi.org/10.1007/978-981-99-8555-5_34

size of less than 1024 (32×32) [18], have very important research significance in practical scenarios such as remote sensing detection [1,14], disaster rescue [22, 38], and intelligent transportation system [20,31]. Unfortunately, the features of small objects are extremely limited, making them susceptible to background and noise interference. Moreover, these weak features are likely to be lost during the feature extraction and downsampling process, leading to a noticeable drop in detection performance when dealing with small objects. For example, Faster R-CNN [24] achieves an mAP of 41.0% and 48.1% for medium and large objects on the COCO dataset [18], respectively, but the result for small objects drops significantly to only 21.2%. Therefore, as a task with both theoretical significance and practical demand, how to effectively enhance the detection performance on small objects is an urgent and important problem to be solved.

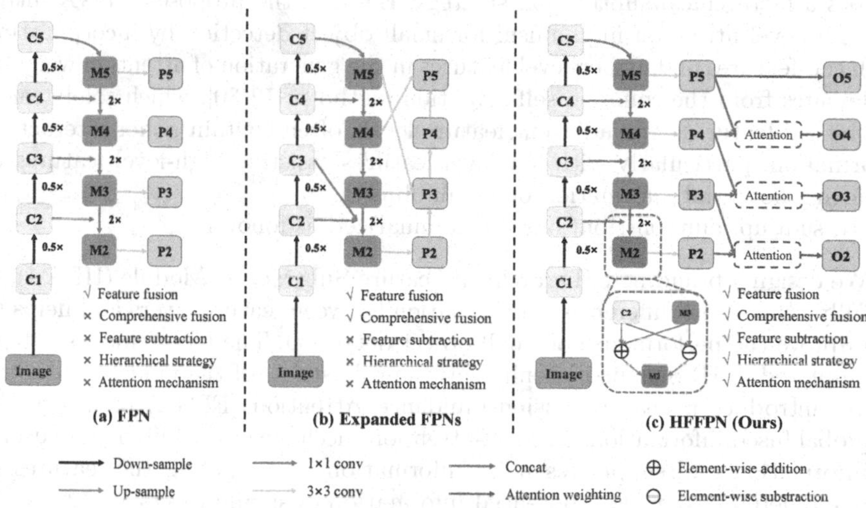

Fig. 1. Pictorial demonstrations of existing feature pyramid networks.

In order to detect objects of various sizes, advanced detectors often adopt a divide-and-conquer approach that utilizes larger receptive fields to detect large objects and smaller ones to detect small objects. This principle is usually reflected in the Feature Pyramid Network (FPN) [16]. As shown in Fig. 1, many studies have noticed the importance of FPN and attempted to fuse low-level and high-level features in a more effective manner to obtain better detection results. Consequently, many FPN variants have been devised to achieve more comprehensive feature fusion [7,19,27]. We collectively refer to them as the Expanded FPNs. However, the fusion strategies of expanded FPNs are generally accomplished by the element-wise addition operation, and the only difference between them is the level of the fused features. In contrast, to extract detailed features that are conducive to small object detection, the element-wise subtraction between the corresponding levels may better obtain edge information [28]. It should be noted that in the high-level feature layers, the information of small objects is almost submerged in the frequent downsampling process, and subtracting such features cannot extract small object features. Instead, it may lead to the loss of main body features. Therefore, a hierarchical feature fusion strategy is necessary. On the other hand, we notice that the fused features have information of different scales. Using global information can help to guide the refinement of each level of features, thus improving detection performance [15].

Based on the above observations, we propose a novel approach called Hierarchical Focused Feature Pyramid Network (HFFPN). HFFPN mainly consists of two parts: Hierarchical Feature Subtraction Module (HFSM) and Feature Fusion Guidance Attention (FFGA). HFSM leverages the feature subtraction operation to obtain the edge information of objects. To avoid erasure effects on main body information caused by subtraction operations at higher semantic levels, HFSM adopts a hierarchical subtraction strategy. Besides, the proposed FFGA introduces a novel attention mechanism for small object detection by incorporating both self-features and higher-level features in the generation of attention weights. It deviates from the common self-attention methods [12,30], which solely relies on the self-features. The adjacent feature levels often contain richer interaction information, particularly with low-level features assisting high-level features in exploring potential information on small objects.

To sum up, our contributions are summarized as follows:

- We design a brand-new Hierarchical Feature Subtraction Module (HFSM). It fully utilizes the difference of information between feature layers and helps to improve the performance of small object detection. The hierarchical strategy employed in HFSM further enhances the robustness of the model.
- We introduce a Feature Fusion Guidance Attention (FFGA) to utilize the global fused information. The self-attention mechanism used highlights useful information and suppresses noise information by weighting the features of itself, helping to explore potential information of small objects.
- Extensive experiments on the DOTA and COCO datasets demonstrate that the proposed HFFPN significantly improves the performance of the baseline algorithm and surpasses the current state-of-the-art detectors.

2 Related Work

2.1 Small Object Detection

With the development of deep learning, extensive research has been carried out on small object detection. There have been numerous attempts to enhance the performance of small object detection from different perspectives, all with the common goal of increasing the exploitable features of small objects. SCRDet [37] achieves a more refined feature fusion network by introducing flexible downsampling strides, allowing for the detection of a broader spectrum of smaller objects with greater precision. R3Det [36] designs a feature refinement module to enhance the detection performance of small objects. Oriented RepPoints [13] captures features from adjacent objects and background noise for adaptive point learning, which utilizes contextual information to discover small objects.

2.2 Feature Pyramid Network

It is a consensus that the shallow layers are usually rich in detailed information but lack abstract semantic information, while the deeper layers are on the contrary due to the downsampling. Smaller objects predominantly rely on shallow features and can be more effectively detected by detectors with smaller receptive fields. Feature Pyramid Network [16] combines the deep layer and shallow layer features by building a top-down pathway to form a feature pyramid. PAFPN [19] enriches the feature hierarchy by adding a bottom-up path, enhancing deeper features without losing information from the shallow layers. HRFPN [27] utilizes multiple cross-branch convolution to enhance feature expression. NAS-FPN [7] searches for the optimal combination method for feature fusion in each layer.

2.3 Self-Attention

The attention mechanism exhibits an impressive capability to quickly concentrate on and distinguish objects within a scene, while effectively ignoring irrelevant aspects. And self-attention is also a powerful technique in deep learning that allows a model to selectively focus on different parts of input, effectively capturing dependencies and relationships within it. Spatial self-attention and channel self-attention are two common kinds of self-attention. SENET [12] is the first proposed channel attention. It uses a SE block to gather global information through channel-wise relationships and enhance the representation capacity. CBAM [30] can sequentially generate attention feature maps in both channel and spatial dimensions for adaptive feature refinement, resulting in the final feature map. Self-attention mechanism has shown outstanding performance in handling small objects to some extent. SCRDet [37] utilizes pixel attention and channel attention to highlight small object regions while mitigating the impact of noise interference. CrossNet [14] develops a cross-layer attention module to enhance the detection of small objects by generating more pronounced responses.

3 Methodology

3.1 Overview

In order to fully utilize the information of small objects, we propose a novel feature pyramid network, named HFFPN, as shown in Fig. 2. The detector receives the input image I and sends it to the backbone network for feature extraction. The image feature C_i gradually becomes richer in semantic information during the subsampling process while losing detailed information. C_i is then passed through the proposed Hierarchical Feature Subtraction Module (HFSM) to obtain intermediate feature M_i in a top-down manner. Next, M_i is further fused through convolution with a kernel size of 3 to obtain fused feature P_i. Finally, P_i is sent to the proposed Feature Fusion Guidance Attention (FFGA) to obtain focused feature O_i, which are particularly focused on effective information, especially small objects. The focused feature O_i will be used by the model to predict the category and location of objects.

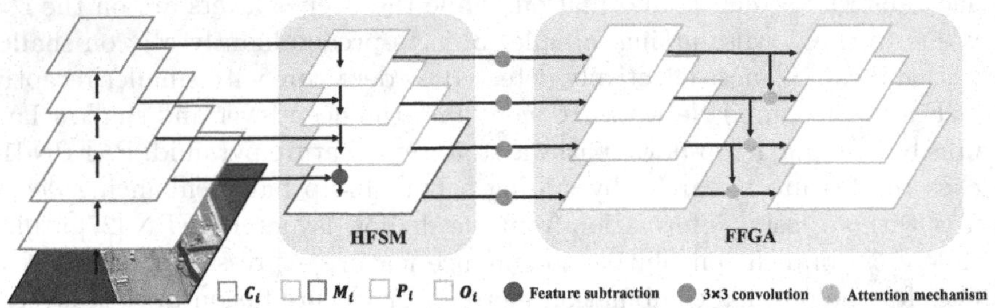

Fig. 2. Overview of the proposed HFFPN, which consists of HFSM and FFGA.

3.2 Hierarchical Feature Subtraction Module

The Hierarchical Feature Subtraction Module (HFSM) is designed to enhance the specific details of low-level features in the feature pyramid. Generally, features at the bottom of pyramid have higher resolution and smaller receptive fields, and contain local information such as edges, textures, and colors, which are crucial for detecting small objects. However, the widely used fusion strategy, *i.e.*, element-wise addition, fails to enhance the local information due to its uniqueness at each level. To cope with it, we propose HFSM that adopts the subtraction operation with hierarchy to highlight the local information, thereby alleviating the above-mentioned problem. The specific process of HFSM is as follows.

Firstly, the input image I passes through the backbone network to obtain the image feature C_i:

$$C_i = \begin{cases} I, & i = 0, \\ \mathcal{F}(C_{i-1}), & i = 1, \dots, t \end{cases}, \tag{1}$$

where $\mathcal{F}(\cdot)$ denotes the convolution block in backbone and t is the number of feature layers.

Secondly, C_i is then processed by HFSM to obtain the intermediate feature M_i. The proposed HFSM aims to better extract detailed information from different feature levels. The subtraction operation can capture the differential information between two feature levels, which often includes fine-grained or edge information, crucial for detecting small objects. Afterwards, the intermediate features are further fused through a 3×3 convolutional layer. These processes can be represented by the following equations:

$$M_i = \begin{cases} \sigma(C_i), & i = t, \\ \sigma(C_i) \oplus \texttt{UP}(M_{i+1}), & i = l+1, \ldots, t-1 \\ \frac{1}{2}(\sigma(C_i) \oplus \texttt{UP}(M_{i+1})) \oplus |\sigma(C_i) \ominus \texttt{UP}(M_{i+1})|, & i = 2, \ldots, l, \end{cases}, \quad (2)$$

$$P_i = \texttt{conv}_{3\times3}(M_i). \quad (3)$$

where $\sigma(\cdot)$ denotes a 1×1 convolution, and $\texttt{UP}(\cdot)$ represents upsampling with ratio of 2. $\oplus$ and $\ominus$ denote element-wise addition and element-wise subtraction, respectively. $|\cdot|$ indicates the operation of taking absolute values. l is a hyperparameter for hierarchical strategy.

3.3 Feature Fusion Guidance Attention

Feature Fusion Guidance Attention (FFGA) is a generalized self-attention mechanism that can effectively focus on useful information, especially small object information. In the feature pyramid, the fused features contain multi-scale information from different levels, and adjacent levels have stronger complementary abilities in feature distribution due to their similar receptive fields. Based on the features between adjacent levels, self-attention is designed to guide the current level of features to focus on useful parts, which can effectively improve the quality of each feature layer and thus improve detection performance. Specifically, the process of FFGA guiding feature focusing can be expressed as Fig. 3.

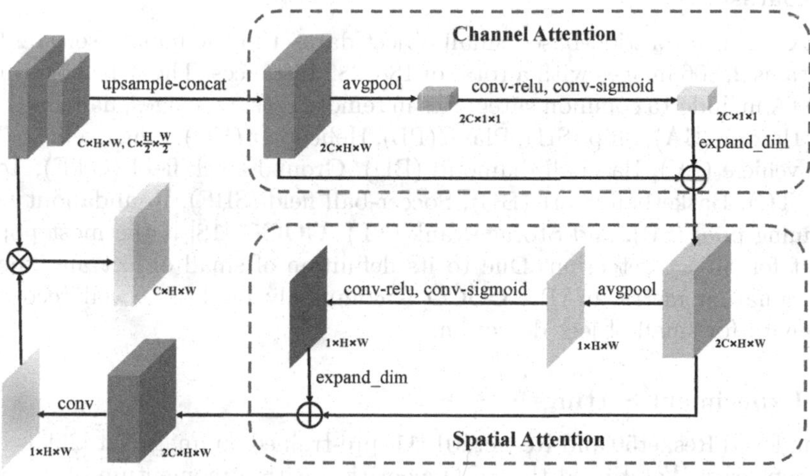

Fig. 3. Diagram of the FFGA.

Firstly, the input of FFGA are the current layer feature $P_i \in \mathbb{R}^{C\times H\times W}$ and the up one $P_{i+1} \in \mathbb{R}^{C\times\frac{H}{2}\times\frac{W}{2}}$. These two features are concatenated along the channel dimension to obtain the guided feature $F_g \in \mathbb{R}^{2C \times H \times W}$. F_g is then sequentially fed into the channel attention (CA) and spatial attention (SA) modules, and we obtain the attention feature $F_a \in \mathbb{R}^{2C \times H \times W}$. Afterwards, F_a is passed through a 1×1 convolution to generate the attention map $W_a \in \mathbb{R}^{1 \times H \times W}$. This map is multiplied as attention weight with the current layer feature P_i to obtain the focused feature $O_i \in \mathbb{R}^{C \times H \times W}$ after attention guidance. This process can be represented by the following formulas:

$$F_g = \texttt{concat}(P_i, \texttt{UP}(P_{i+1})), \tag{4}$$

$$F_a = SA(CA(F_g)), \tag{5}$$

$$W_a = \texttt{conv}_{1\times1}(F_a), \tag{6}$$

$$O_i = \begin{cases} P_i \otimes W_a, & i = 2, \ldots, t-1, \\ P_i, & i = t, \end{cases}, \tag{7}$$

where the composition of channel attention and spatial attention has been detailed in Fig. 3. They have a similar structure that mainly consists of an average pooling layer, a 1×1 convolution layer followed by a ReLU activation, and a 1×1 convolution layer followed by a sigmoid activation. The input feature generates attention focusing on channel and spatial dimensions in the two modules respectively. After dimension expansion, they are element-wise added to the original feature, allowing the original feature to obtain a different degree of attentional gain in the channel and spatial dimensions.

4 Experiments

4.1 Datasets

DOTA [32] is a rotation-based small object dataset in the remote sensing field. It contains 2,806 images with a total of 188,282 instances. The detection targets in DOTA include 15 common categories in remote sensing images, namely Bridge (BR), Harbor (HA), Ship (SH), Plane (PL), Helicopter (HC), Small vehicle (SV), Large vehicle (LV), Baseball diamond (BD), Ground track field (GTF), Tennis court (TC), Basketball court (BC), Soccer-ball field (SBF), Roundabout (RA), Swimming pool (SP), and Storage tank (ST). **COCO** [18] is the most popular dataset for object detection. Due to its definition of small object and specialized evaluation metric mAP_s, COCO is commonly used as a well recognized benchmark for small object detection.

4.2 Experiment Settings

We employed Resnet50 and Resnet101 [11] pre-trained on ImageNet [25] as backbone networks. We utilized the SGD algorithm with a momentum of 0.9 and a weight decay of 0.0001 for network optimization. The initial learning rate warms up at a rate of 0.001 per iteration for the first 500 iterations. The training schedule for all experiments was consistent. We trained 12 epochs on the two datasets, and the learning rate decays at the epoch 8 and 11 with ratio of 0.1. The code for all experiments was built on the MMdetection [2] platform.

4.3 Comparison Results

Results on DOTA. We selected the RoI Transformer [5], a general method for aerial object detection, as the baseline algorithm. Table 1 reports the comparison result on DOTA test set. With Resnet50 as the backbone, our method obtains 76.64% mAP_{50}, improving the performance of baseline by approximately 1%, thereby surpassing the performance of the state-of-the-art algorithms. With Resnet101 as the backbone, HFFPN also increases the baseline's performance by 0.87% mAP_{50}, achieving the best result on the DOTA dataset. These results fully demonstrate HFFPN's advantages on small object detection and reflect its potential applications. Figure 4 provides a more intuitive visual comparison.

Table 1. Comparison with state-of-the-art methods on DOTA test set. The reported results come from AerialDetection [6] and OBBDetection [33]. ‡ indicates that it is the result of our re-implement. Note that we only list some classes for better display.

Methods	Backbone	GTF	SV	SH	SBF	HA	SP	mAP_{50}
Single-stage Methods								
RSDet [23]	R152-FPN	68.50	70.20	73.60	64.30	66.10	69.30	74.10
R^3Det [36]	R152-FPN	66.10	70.92	78.21	61.81	68.16	69.83	73.74
S^2A-Net [9]	R50-FPN	71.11	78.11	87.25	60.36	65.26	69.13	74.12
R^3Det-DCL [35]	R152-FPN	69.70	76.84	87.30	63.50	68.96	68.79	75.54
Two-stage Methods								
SCRDet [37]	R101-FPN	68.36	68.36	72.41	65.02	66.25	68.24	72.61
Gliding Vertex [34]	R101-FPN	77.34	73.01	86.82	59.55	72.94	70.86	75.02
ReDet [10]	ReR50-ReFPN	74.00	78.13	88.04	61.76	72.10	68.07	76.25
Oriented R-CNN [33]	R101-FPN	76.92	74.27	87.52	65.51	74.36	70.15	76.28
Anchor-free Methods								
PIoU‡ [3]	R50-FPN	68.90	77.58	81.57	60.47	57.68	65.12	69.68
O^2-DNet [29]	H-104	61.21	71.32	78.62	60.93	58.21	66.98	71.04
DRN [21]	H-104	64.10	76.22	85.84	57.65	69.30	69.63	73.23
CFA [8]	R101-FPN	67.17	79.99	84.46	54.86	73.04	70.24	75.05
Oriented RepPoints [13]	R101-FPN	71.76	**79.95**	87.33	59.15	75.23	73.75	76.52
RoI-Trans.‡ [5]	R50-FPN	76.65	78.40	87.55	60.12	74.89	69.70	75.70
RoI-Trans.‡	R101-FPN	75.75	78.11	87.46	63.80	**76.05**	71.35	76.02
RoI-Trans. (Ours)	R50-HFFPN	**78.11**	78.33	**87.71**	**65.24**	75.39	**72.99**	**76.64**
RoI-Trans. (Ours)	R101-HFFPN	79.84	77.97	87.68	65.00	76.36	71.67	76.89

Results on COCO. On the COCO dataset, we applied HFFPN to two-stage [24], one-stage [17], and anchor-free [26] detectors, respectively. Table 2 shows the performance gain brought by HFFPN. Although the overall mAP improvement is not significant due to the small proportion of small objects in the COCO dataset, the consistent and significant increase in the mAP_s metric indicates that HFFPN makes detectors more capable of detecting small objects while maintaining their detection capabilities of other scales of objects.

Table 2. Comparison experiment on COCO. The baseline results come from [2].

Methods	Backbone	mAP	mAP_{50}	mAP_{75}	mAP_s	mAP_m	mAP_l
Faster RCNN [24]	R50-FPN	37.4	58.1	40.4	21.2	41.0	48.1
Faster RCNN	R50-HFFPN	37.6	58.4	40.7	21.9 (+0.7)	40.9	48.3
RetinaNet [17]	R50-FPN	36.5	55.4	39.1	20.4	40.3	48.1
RetinaNet	R50-HFFPN	36.6	55.7	39.1	21.2 (+0.8)	40.3	48.0
FCOS [26]	R50-FPN	36.6	56.0	38.8	21.0	40.6	47.0
FCOS	R50-HFFPN	36.6	55.9	38.7	21.7 (+0.7)	40.2	47.2

4.4 Ablation Study

To further verify the advantages and effectiveness of the proposed method, we conduct a series of experiments on the DOTA dataset. The baseline algorithm is RoI Transformer with Resnet50.

Evaluation for Component Effectiveness. To evaluate the effects of HFSM and FFGA, we carry out several ablation experiments, and the experimental results are shown in the Table 3. Without any improvement schemes, the mAP_{50} detected by the baseline is 75.70%. The introduction of HFSM and FFGA gradually improves the detection accuracy to 76.24% and 76.64%. The results indicate that each combination in HFFPN brings improvement to the detector.

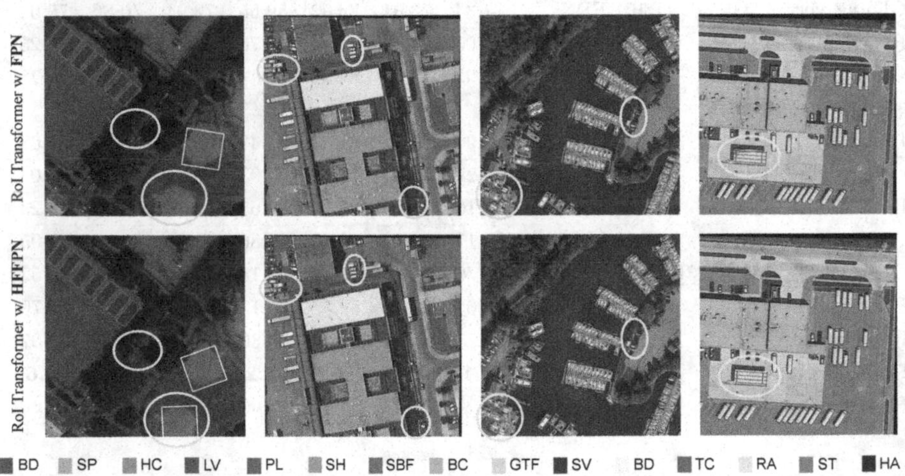

Fig. 4. Visualization on DOTA test set. The yellow circles highlight the difference of detection result. We can easily find that HFFPN (second row) can help detect more small objects and achieve higher accuracy in classification and regression. (Color figure online)

Evaluation on Different Settings of l in HFSM. Hierarchical level l, as a hyperparameter in HFSM, determines in which feature layers the operation of feature subtraction is performed. Specifically, the feature subtraction module will be introduced when the level lower than l. Table 4 shows the results under different values of l. When l is 2, the performance of baseline with HFFPN reaches the highest. Assuming we do not employ the hierarchical strategy by setting l equals to 5, where feature subtraction is performed between each level of features, we would observe a significant drop in results. The hierarchical strategy ensures that the subtraction is performed only on detailed features, making it applicable to a wide range of input images and thus enhancing the model's robustness.

Comparison with Other FPNs. Table 5 presents performance of the baseline algorithm with different FPNs. It can be observed that some expanded FPNs do enhance the detector's performance to some extent, but the improvements are not as significant as those of the proposed HFFPN.

Evaluation on Different Detectors. To verify that the proposed HFFPN is a common method for most detectors, experiments were conducted on several different detectors. Table 6 shows the comparison results of these detectors with or without using HFFPN. The experimental results show that the use of HFFPN has led to performance improvements for all detectors, strongly indicating the universality and effectiveness of the proposed method.

Table 3. Evaluation on the effectiveness of each component. FS, HS, CA and SA denote feature subtraction, hierarchical strategy, channel attention, and spatial attention, respectively.

Baseline	FS	HS	CA	SA	mAP_{50}
✓					75.70
✓	✓				76.13
✓	✓	✓			76.24
✓	✓	✓	✓		76.38
✓	✓	✓		✓	76.41
✓	✓	✓	✓	✓	**76.64**

Table 4. Results of differnent l.

l	2	3	4	5
mAP_{50}	**76.64**	76.33	76.07	75.59

Table 5. Comparison with other FPNs.

Backbone	mAP_{50}	mAP
R50-FPN [16]	75.70	46.27
R50-PAFPN [19]	76.26	46.56
R50-HRFPN [27]	76.33	46.85
R50-NASFPN [7]	73.71	45.03
R50-HFFPN	**76.64**	**47.07**

Table 6. Improvements on DOTA by applying HFFPN to different detectors.

Methods	Backbone	mAP_{50}	mAP
PIoU [3]	R-50-FPN	69.68	40.05
PIoU	R-50-HFFPN	70.26 (+0.58)	40.55 (+0.50)
Gliding Vertex [34]	R-50-FPN	72.65	40.93
Gliding Vertex	R-50-HFFPN	73.24 (+0.59)	41.42 (+0.49)
Oriented RCNN [33]	R-50-FPN	75.72	46.78
Oriented RCNN	R-50-HFFPN	76.14 (+0.42)	46.85 (+0.07)
RoI-Trans. [5]	R-50-FPN	75.70	46.27
RoI-Trans.	R-50-HFFPN	76.64 (+0.94)	47.07 (+0.80)

5 Conclusion

To better utilize the detailed information for small object detection, this paper proposes a hierarchical focused feature pyramid network. It mainly contains a hierarchical feature subtraction module and feature fusion guidance attention. This design overcomes the problem of neglecting edge information that exists in common FPN methods, thus improving the detection ability of small objects without affecting the detection performance of objects at other scales. Comparison and ablation experiments on multiple datasets demonstrate the excellent performance of the proposed method, fully verifying the effectiveness of HFFPN.

Acknowledgements. This work was supported by National Key R&D Program of China (No. 2022ZD0118201), the National Science Fund for Distinguished Young Scholars (No.62025603), the National Natural Science Foundation of China (No. U21B2037, No. U22B2051, No. 62176222, No. 62176223, No. 62176226, No. 62072386, No. 62072387, No. 62072389, No. 62002305 and No. 62272401), and the Natural Science Foundation of Fujian Province of China (No. 2021J01002, No. 2022J06001).

References

1. Bashir, S.M.A., Wang, Y.: Small object detection in remote sensing images with residual feature aggregation-based super-resolution and object detector network. Remote Sens. **13**(9), 1854 (2021)
2. Chen, K., et al.: Mmdetection: open MMLAB detection toolbox and benchmark. arXiv preprint arXiv:1906.07155 (2019)
3. Chen, Z., et al.: PIoU loss: towards accurate oriented object detection in complex environments. In: Vedaldi, A., Bischof, H., Brox, T., Frahm, J.-M. (eds.) ECCV 2020. LNCS, vol. 12350, pp. 195–211. Springer, Cham (2020). https://doi.org/10.1007/978-3-030-58558-7_12
4. Cheng, G., Yuan, X., Yao, X., Yan, K., Zeng, Q., Han, J.: Towards large-scale small object detection: survey and benchmarks. arXiv preprint arXiv:2207.14096 (2022)

5. Ding, J., Xue, N., Long, Y., Xia, G.S., Lu, Q.: Learning ROI transformer for oriented object detection in aerial images. In: CVPR, pp. 2849–2858 (2019)
6. Ding, J., et al.: Object detection in aerial images: a large-scale benchmark and challenges. TPAMI **44**(11), 7778–7796 (2021)
7. Ghiasi, G., Lin, T.Y., Le, Q.V.: NAS-FPN: learning scalable feature pyramid architecture for object detection. In: CVPR, pp. 7036–7045 (2019)
8. Guo, Z., Liu, C., Zhang, X., Jiao, J., Ji, X., Ye, Q.: Beyond bounding-box: convex-hull feature adaptation for oriented and densely packed object detection. In: CVPR, pp. 8792–8801 (2021)
9. Han, J., Ding, J., Li, J., Xia, G.S.: Align deep features for oriented object detection. IEEE Trans. Geosci. Remote Sens. **60**, 1–11 (2021)
10. Han, J., Ding, J., Xue, N., Xia, G.S.: Redet: a rotation-equivariant detector for aerial object detection. In: CVPR, pp. 2786–2795 (2021)
11. He, K., Zhang, X., Ren, S., Sun, J.: Deep residual learning for image recognition. In: CVPR, pp. 770–778 (2016)
12. Hu, J., Shen, L., Sun, G.: Squeeze-and-excitation networks. In: CVPR, pp. 7132–7141 (2018)
13. Li, W., Chen, Y., Hu, K., Zhu, J.: Oriented reppoints for aerial object detection. In: CVPR, pp. 1829–1838 (2022)
14. Li, Y., Huang, Q., Pei, X., Chen, Y., Jiao, L., Shang, R.: Cross-layer attention network for small object detection in remote sensing imagery. IEEE J. Select. Top. Appl. Earth Observ. Remote Sens. **14**, 2148–2161 (2020)
15. Liao, M., Zou, Z., Wan, Z., Yao, C., Bai, X.: Real-time scene text detection with differentiable binarization and adaptive scale fusion. TPAMI **45**(1), 919–931 (2022)
16. Lin, T.Y., Dollár, P., Girshick, R., He, K., Hariharan, B., Belongie, S.: Feature pyramid networks for object detection. In: CVPR, pp. 2117–2125 (2017)
17. Lin, T.Y., Goyal, P., Girshick, R., He, K., Dollár, P.: Focal loss for dense object detection. In: ICCV, pp. 2980–2988 (2017)
18. Lin, T.Y., et al.: Microsoft COCO: common objects in context. In: Fleet, D., Pajdla, T., Schiele, B., Tuytelaars, T. (eds.) ECCV 2014. LNCS, vol. 8693, pp. 740–755. Springer, Cham (2014). https://doi.org/10.1007/978-3-319-10602-1_48
19. Liu, S., Qi, L., Qin, H., Shi, J., Jia, J.: Path aggregation network for instance segmentation. In: CVPR, pp. 8759–8768 (2018)
20. Liu, Y., Sun, P., Wergeles, N., Shang, Y.: A survey and performance evaluation of deep learning methods for small object detection. Expert Syst. Appl. **172**, 114602 (2021)
21. Pan, X., et al.: Dynamic refinement network for oriented and densely packed object detection. In: CVPR, pp. 11207–11216 (2020)
22. Pi, Y., Nath, N.D., Behzadan, A.H.: Convolutional neural networks for object detection in aerial imagery for disaster response and recovery. Adv. Eng. Inform. **43**, 101009 (2020)
23. Qian, W., Yang, X., Peng, S., Yan, J., Guo, Y.: Learning modulated loss for rotated object detection. Proc. AAAI Conf. Artif. Intell. **35**(3), 2458–2466 (2021)
24. Ren, S., He, K., Girshick, R., Sun, J.: Faster r-cnn: towards real-time object detection with region proposal networks. NeurIPS **28** (2015)
25. Russakovsky, O., et al.: ImageNet large scale visual recognition challenge. Int. J. Comput. Vis. **115**(3), 211–252 (2015)
26. Tian, Z., Shen, C., Chen, H., He, T.: FCOS: fully convolutional one-stage object detection. In: ICCV, pp. 9627–9636 (2019)
27. Wang, J., et al.: Deep high-resolution representation learning for visual recognition. TPAMI **43**(10), 3349–3364 (2020)

28. Wang, L., Tong, Z., Ji, B., Wu, G.: TDN: temporal difference networks for efficient action recognition. In: CVPR, pp. 1895–1904 (2021)
29. Wei, H., Zhang, Y., Chang, Z., Li, H., Wang, H., Sun, X.: Oriented objects as pairs of middle lines. ISPRS J. Photogramm. Remote. Sens. **169**, 268–279 (2020)
30. Woo, S., Park, J., Lee, J.-Y., Kweon, I.S.: CBAM: convolutional block attention module. In: Ferrari, V., Hebert, M., Sminchisescu, C., Weiss, Y. (eds.) ECCV 2018. LNCS, vol. 11211, pp. 3–19. Springer, Cham (2018). https://doi.org/10.1007/978-3-030-01234-2_1
31. Wu, J., Zhou, C., Zhang, Q., Yang, M., Yuan, J.: Self-mimic learning for small-scale pedestrian detection. In: ACMMM, pp. 2012–2020 (2020)
32. Xia, G.S., et al.: Dota: a large-scale dataset for object detection in aerial images. In: CVPR, pp. 3974–3983 (2018)
33. Xie, X., Cheng, G., Wang, J., Yao, X., Han, J.: Oriented r-cnn for object detection. In: ICCV, pp. 3520–3529 (2021)
34. Xu, Y., et al.: Gliding vertex on the horizontal bounding box for multi-oriented object detection. TPAMI **43**(4), 1452–1459 (2020)
35. Yang, X., Hou, L., Zhou, Y., Wang, W., Yan, J.: Dense label encoding for boundary discontinuity free rotation detection. In: CVPR, pp. 15819–15829 (2021)
36. Yang, X., Yan, J., Feng, Z., He, T.: R3det: refined single-stage detector with feature refinement for rotating object. In: AAAI, vol. 35, pp. 3163–3171 (2021)
37. Yang, X., et al.: Scrdet: towards more robust detection for small, cluttered and rotated objects. In: ICCV, pp. 8232–8241 (2019)
38. Zhang, M., Yue, K., Zhang, J., Li, Y., Gao, X.: Exploring feature compensation and cross-level correlation for infrared small target detection. In: ACMMM, pp. 1857–1865 (2022)

JLInst: Boundary-Mask Joint Learning for Instance Segmentation

Xiaodong Zhao[1,2,3], Junliang Chen[2,3], Zepeng Huang[2,3], and Linlin Shen[2,3](✉)

[1] Department of Information Management, Guangdong Justice Police Vocational College, Guangzhou 510520, China
zhaoxiaodong2020@email.szu.edu.cn

[2] Computer Vision Institute, School of Computer Science and Software Engineering, Shenzhen University, Shenzhen, China
{chenjunliang2016,huangzepeng2021}@email.szu.edu.cn

[3] Guangdong Key Laboratory of Intelligent Information Processing, Shenzhen University, Shenzhen 518060, China
llshen@szu.edu.cn

Abstract. Lots of methods have been proposed to improve instance segmentation performance. However, the mask produced by state-of-the-art segmentation networks is still coarse and does not completely align with the whole object instance. Moreover, we find that better object boundary information can help instance segmentation network produce more distinct and clear object masks. Therefore, we present a simple yet effective instance segmentation framework, termed **JLInst** (Boundary-Mask Joint Learning for Instance Segmentation). Our methods can jointly exploit object boundary and mask semantic information in the instance segmentation network, and generate more precise mask prediction. Besides, we propose the Adaptive Gaussian Weighted Binary Cross-Entropy Loss (**GW loss**), to focus more on uncertain examples in pixel-level classification. Experiments show that JLInst achieves improved performance (**+3.0%** AP) than Mask R-CNN on COCO test-dev2017 dataset, and outperforms most recent methods in the fair comparison.

Keywords: Instance Segmentation · Boundary Information Enhancement

1 Introduction

Instance segmentation is a fundamental but challenging task in computer vision, which aims to generate a per-pixel mask with a category label for each instance in an image. With the rapid development of deep convolutional networks, many excellent approaches [7,9] have been proposed for instance segmentation. Previous methods are mainly based on object detection network, which provides box-level localization information.

© The Author(s), under exclusive license to Springer Nature Singapore Pte Ltd. 2024
Q. Liu et al. (Eds.): PRCV 2023, LNCS 14436, pp. 445–457, 2024.
https://doi.org/10.1007/978-981-99-8555-5_35

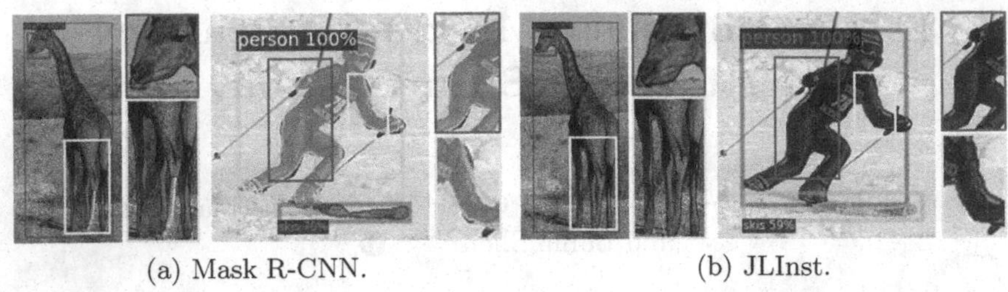

(a) Mask R-CNN. (b) JLInst.

Fig. 1. Comparison between Mask R-CNN and JLInst.

Mask R-CNN [7] is a successful two-stage instance segmentation framework, which first employs a Faster R-CNN [17] detector to detect objects in an image and further predicts binary mask within each detected bounding box. Other methods built upon Mask R-CNN also achieve good performance. Recently, inspired by the rapid development of one-stage detectors, a number of one-stage instance segmentation frameworks [1,18] have been proposed.

However, the quality of the predicted instance mask is still not satisfactory. One of the most important problems is the imprecise segmentation around instance boundaries. Mask R-CNN generates instance masks by performing pixel-level classification via fully convolutional networks (FCN). FCN treats all pixels in the proposal equally and ignores the object shape and boundary information. So the predicted instance masks of Mask R-CNN are coarse and not well-aligned with the real object boundaries. As shown in Fig. 1, Mask R-CNN easily predicts coarse and indistinct masks, while the boundary predicted by our JLInst matches better with ground truths.

To address this issue, we attempt to exploit instance boundary information to enhance the mask prediction in the instance segmentation network. We notice that boundary semantic information can guide the network to generate more distinct masks that are well aligned with their ground truths. Based on this motivation, we propose JLInst instance segmentation network. Based on Mask R-CNN, we replace the original mask branch with the proposed boundary-mask joint learning branch, which contains two sub-networks for jointly of learning object mask and boundary semantic information. We design a feature enhancement module to strengthen the connection between mask branch and boundary branch. At last, the final mask prediction combining the outputs from mask and boundary branch contains more abundant shape and localization information. Moreover, we design the Adaptive Gaussian Weighted Binary Cross-Entropy Loss (GW loss), to focus more on uncertain examples in the pixel-level classification and deeply exploit the semantic information while training the network.

We summarize our main contributions as follow.

- We propose **JLInst** instance segmentation framework, to jointly learn object boundaries and masks in the instance segmentation network, and generate more precise mask prediction by combining mask and boundary information.

- In order to focus more on uncertain examples, we propose the Adaptive Gaussian Weighted Binary Cross-Entropy Loss to reweight the pixel-wise classification loss.
- JLInst achieves improved performance (**+3.0%** AP) than Mask R-CNN on COCO test-dev 2017. Furthermore, our JLInst achieves state-of-the-art performance of 39.1% AP, which surpasses most of instance segmentation frameworks.

2 Related Work

Instance Segmentation. Recent studies on instance segmentation can be divided into two categories: two-stage and one-stage methods.

Two-stage methods usually adopt the classical detect-then-segment strategy. The dominant instance segmentation framework is still Mask R-CNN [7], which uses two-stage detector Faster R-CNN [17] to detect objects in an image and then generates binary segmentation mask within each detected bounding box. PANet [13] strengthens feature representation through bottom-up path augmentation based on Mask R-CNN. Mask Scoring R-CNN [9] adds an additional mask-IoU head to deal with the misalignment between mask quality and mask score. One-stage methods have been proposed due to the rapid development of one-stage detectors [11,19]. Some instance segmentation networks [1,18] still follow the detect-then-segment pipeline by replacing the detection networks with the one-stage detectors. YOLACT [1] produces instance masks by linearly combining a set of prototype masks and per-instance mask coefficients. BlendMask [2] extends this idea by assembling with attention maps. CondInst [18] also achieves remarkable performance with great efficiency.

Boundary Refinement in Segmentation. Some studies focus more on the post-processing scheme to refine the boundaries in the predicted masks from the segmentation network. Chen et al. [3] propose fully connected contditional random field (CRF) to capture spatial details and refine boundaries of predicted masks. Some recent works attempt to deeply exploit boundary information while training the network to enhance the predicted mask. BMask R-CNN [4] introduce an extra branch to strengthen the boundary awareness of mask features. It can generate more clear object masks than previous methods.

3 Methods

3.1 Overview

As shown in Fig. 2, JLInst still adopts the same two-stage procedure as Mask R-CNN [7]. We develop a boundary-mask joint learning branch, which contains the mask branch and boundary branch to jointly learn the mask and boundary information. Besides, we design a feature enhancement module to strengthen the semantic information between these two branches. The final mask output is generated by combining the predictions from two branches. In addition, we design GW loss to help optimize the pixel-wise classification while training the network.

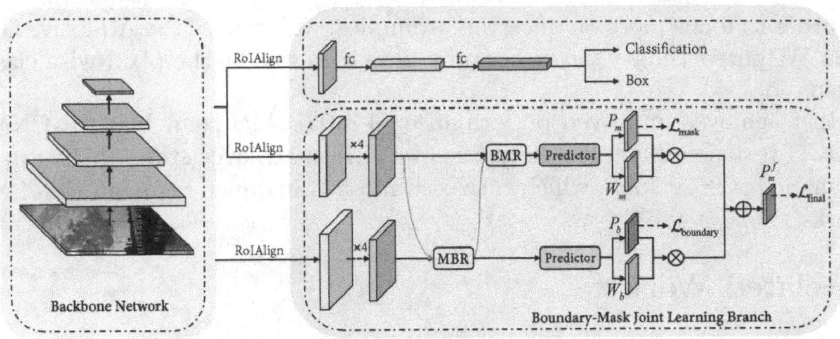

Fig. 2. The Overall architecture of **Boundary-Mask Joint Learning for Instance Segmentation Network (JLInst)**. In the boundary-mask joint learning branch, the dotted arrow denotes 3×3 convolution and the solid arrow denotes identity connection unless specified. "×n" denotes a stack of n consecutive convolutional kernels.

3.2 Boundary-Mask Joint Learning Branch

RoI Feature Extraction. We define R_m and R_b as the Region of Interest (RoI) features for mask branch and boundary branch respectively. The ouputs F_m is generated by passing R_m into four consecutive convolutions. However, to preserve better spatial information for boundary prediction, we upsample all the features from P_3-P_5 to the size of P_2 and add them as the input when performing RoIAlign. Then, the features R_b after RoIAlign is downsampled by the final strided 3×3 convolution and the output feature is denoted as F_b. F_b has the same resolution as F_m and is used for feature fusion.

Mask-Guided Boundary Refinement Module. To further exploit the semantic information and strengthen the connection between the mask branch and boundary branch, we propose the Mask-guided Boundary Refinement Module (MBR). As shown in Fig. 3, MBR consists of Fusion Block, Attention Block and Residual Block. We aim to use the mask information from the mask branch to help boundary branch produce more precise boundary prediction.
Fusion Block. The main feature F_b is from the boundary branch and the guide feature F_m is from the mask branch. Then we concat these two feature maps and put them into the 1×1 and 3×3 convolutional layers. The ouput feature G_b is created by adding the main feature F_b and the feature maps from convolutional layers, according to Eq. 1.

$$G_b = f_{3\times3}(f_{1\times1}(f_{concat}(F_b, F_m))) + F_b \tag{1}$$

where $f_{n\times n}$ denotes a $n \times n$ convolutional layer, f_{concat} means the concat operation.
Attention Block. We apply the attention mechanism, i.e. spatial attention and channel attention, to further capture the spatial and channel information in the fused feature G_b.

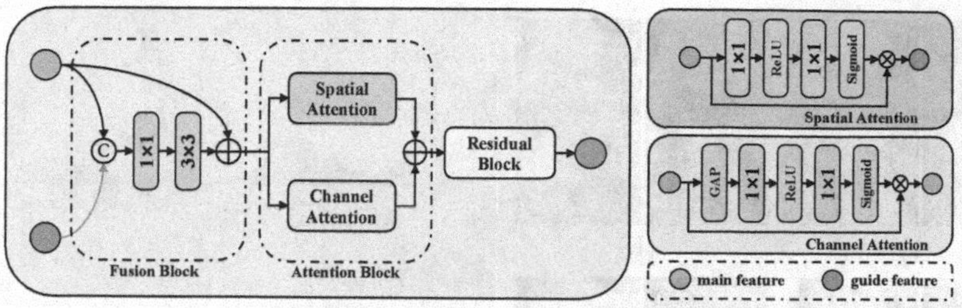

Fig. 3. The architecture of Mask-guided Boundary Refinement Module (MBR)

In the spatial attention block, we first adjust the weight w_s by following Eq. 2.

$$w_s = \sigma(f_{1\times1}(\delta(f_{1\times1}(G_b)))) \tag{2}$$

where δ is the ReLU function and σ is the sigmoid function.

Then we rescale G_b according to Eq. 3.

$$G_b^s = w_s \otimes G_b \tag{3}$$

where $\otimes$ denotes element-wise multiplication.

As for the channel attention block, we first obtain the global information w_c' by using a global average pooling (GAP) layer.

$$w_c' = GAP(G_b) \tag{4}$$

Then the new weight w_c of each channel is adjusted by using two 1×1 convolutional layers followed by the sigmoid function, according to Eq. 5.

$$w_c = \sigma(f_{1\times1}(\delta(f_{1\times1}(w_c')))) \tag{5}$$

After that, we rescale the features of each channel by channel-wise multiplication of the weights w_c and G_b. The output after rescaling is denoted as G_b^c and calculated by Eq. 6.

$$G_b^c = w_c \otimes G_b \tag{6}$$

At last, we generate the rescaled output G_b^{attn} by element-wise addition of the channel and spatial excitation.

$$G_b^{attn} = G_b^s \oplus G_b^c \tag{7}$$

where $\oplus$ denotes element-wise summation.

Residual Block. We finally generate the output features H_b by passing G_b^{attn} to a residual block (ResBlock) as follow.

$$H_b = ResBlock(G_b^{attn}) \tag{8}$$

The block consists of a 3×3 convolutional layer, a 1×1 convolutional layer and 3×3 convolutional layer in order. Each convolutional layer is followed by group normalization and ReLU activation.

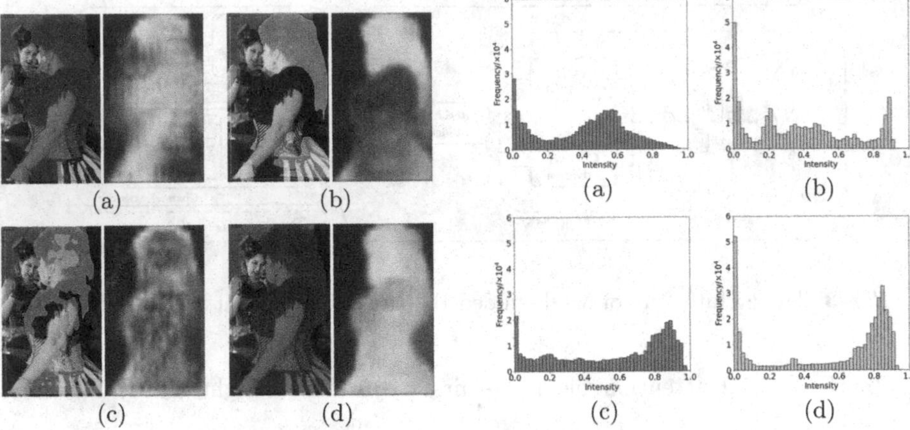

Fig. 4. Visualization results for analyzing the impact of BCE loss and GW loss in JLInst. (a) The coarse activation map from JLInst with BCE loss at the beginning of training. (b) with BCE loss after training. (c) with GW loss at the beginning of training. (d) with GW loss after training.

Fig. 5. Histogram of activation maps. (a) the histogram calculated from Fig. 4(a). (b) the histogram calculated from Fig. 4(b). (c) the histogram calculated from Fig. 4(c). (d) the histogram calculated from Fig. 4(d).

Boundary-Guided Mask Refinement Module. We integrate the boundary features with mask features so that boundary information can be used to enrich mask features and guide more precise mask prediction. The structure of Boundary-guided Mask Refinement Module (BMR) is the same as MBR. The only difference is that the main feature of BMR is F_m and the guide feature becomes H_b after being processed by MBR of boundary branch.

Final Output. To further exploit the semantic information from the mask branch and boundary branch, we generate the final mask prediction of the instance segmentation network by combining the outputs from these two branches. We define P_m and W_m as the mask feature and mask weight from mask branch respectively, while define P_b and W_b as the boundary feature and boundary weight from boundary branch respectively. The final mask prediction P_m^f is calculated by Eq. 9.

$$P_m^f = (W_m \otimes P_m) \oplus (W_b \otimes P_b) \tag{9}$$

3.3 Learning and Optimization

Adaptive Gaussian Weighted Binary Cross-Entropy Loss. The segmentation task in Mask R-CNN can be regarded as a binary classification task. Generally, BCE loss is used to train a pixel-level classification, which treats each pixel equally. However, for the segmentation task, the feature map we finally generate needs to be processed by image thresholding, that is, pixels with the value larger than the threshold are classified into the foreground, and those smaller than the threshold are classified into the background (We set the threshold as 0.5 by default). Hence, we believe that pixels near the threshold have great uncertainty. As shown in Fig. 4(a) and Fig. 4(b), we can notice that JLInst with the BCE loss can easily focus on discriminative regions during training. So the output mask is incomplete and there still exists lots of uncertain pixels in the output activation map. Therefore, we design Adaptive Gaussian Weighted Binary Cross-Entropy Loss (GW loss) to increase the network's attention to pixels around the threshold and reduce the penalty for pixels approaching 0 or 1. With the supervision of the GW loss, the JLInst tends to excite the uncertain pixels and generate exact and well-separated activation maps (Fig. 4(d)). Figure 5 shows the histograms of activation map learned with the BCE loss or GW loss. We can observe that the number of uncertain pixels (0.5) has been substantially reduced by the GW loss.

The BCE loss is defined as follows.

$$\mathcal{L}_{\text{bce}}(p, y) = -\frac{1}{N} \sum_{i=1}^{N} \left[y_i \log\left(p_i\right) + \left(1 - y_i\right) \log\left(1 - p_i\right)\right] \tag{10}$$

where $\mathcal{L}_{\text{bce}}$ indicates the pixel-wise binary cross-entropy loss between predicted foreground-likelihood p_i in [0,1] and ground-truth label y_i for each pixel i. N is the number of pixels in the predicted feature map.

Then, we use the gaussian distribution function to calculate the corresponding loss weight according to the pixel value. We aim to place the maximum weight for pixels with 0.5, and reduce the weight for pixels around 0 and 1.

$$W_g(p) = \frac{1}{\sqrt{2\pi\sigma^2}} e^{-\frac{(p-\mu)^2}{2\sigma^2}} \tag{11}$$

where μ denotes the mean value and σ denotes standard deviation. We set $\mu = 0.5$ and $\sigma = 0.5$ by default.

Then we design our GW loss by reweighting the BCE loss, which can supervise JLInst to produce more distinct mask outputs and better optimise the network during training.

$$\mathcal{L}_{\text{gw}}(p, y) = W_g(p) \cdot \mathcal{L}_{\text{bce}}(p, y) \tag{12}$$

Mask Loss. We consider the optimization of the mask branch as a binary classification problem and use our GW loss as the mask loss to optimize the mask learning. Y_m means the corresponding groundtruth mask.

$$\mathcal{L}_{\text{mask}} = \mathcal{L}_{\text{gw}}(P_m, Y_m) \tag{13}$$

Boundary Loss. According to [6], we use dice loss [15] and GW loss to help the boundary branch produce crisp boundary effectively. Dice loss is insensitive to the number of pixels and can well measure the overlap between boundary predictions and ground truths, which is calculated by Eq. 14.

$$\mathcal{L}_{\text{dice}}(p,y) = 1 - \frac{2\sum_{i}^{HW} p_i y_i + \epsilon}{\sum_{i}^{HW}(p_i)^2 + \sum_{i}^{HW}(y_i)^2 + \epsilon} \tag{14}$$

where p denotes the prediction and y denotes corresponding ground truth. H and W are height and width of the predicted feature map respectively. ϵ is a smooth term to avoid zero division (we set $\epsilon = 1$ by default).

Our boundary loss $\mathcal{L}_{\text{boundary}}$ is formulated as follows. Boundary ground truth Y_b is generated by using Laplacian operator from mask ground truth Y_m.

$$\mathcal{L}_{\text{boundary}}(P_b, Y_b) = \mathcal{L}_{\text{gw}}(P_b, Y_b) + \mathcal{L}_{\text{dice}}(P_b, Y_b) \tag{15}$$

Final Mask Loss. We generate the final mask prediction of JLInst by combining the outputs from the mask and boundary branches. Similar to Sect. 3.3, we use our GW loss as the final mask loss to optimize the final mask learning.

$$\mathcal{L}_{\text{final}} = \mathcal{L}_{\text{gw}}(P_m^f, Y_m) \tag{16}$$

Multi-task Learning. Multi-task learning has been proved effective in many works. In our JLInst, we define the multi-task loss as follows.

$$\mathcal{L} = \mathcal{L}_{\text{cls}} + \mathcal{L}_{\text{box}} + \mathcal{L}_{\text{mask}} + \mathcal{L}_{\text{boundary}} + \mathcal{L}_{\text{final}} \tag{17}$$

where the classification loss $\mathcal{L}_{\text{cls}}$, regression loss $\mathcal{L}_{\text{box}}$ are inherited from Mask R-CNN, and other loss functions are mentioned in previous sections.

4 Experiments

4.1 Dataset and Evaluation Metrics

Our experiments are conducted on COCO [12] dataset. We train our models with the data in train-2017 split containing around 115k images, and evaluate the performance of ablation studies on val-2017 split with about 5k images. Main results are reported on the test-dev split (20k images without available public annotations). We report all the results in the standard COCO-style average precision (AP).

Table 1. Comparison with state-of-the-art methods for instance segmentation on COCO *test-dev2017*

Method	Backbone	AP	AP_{50}	AP_{75}	AP_S	AP_M	AP_L
BMask R-CNN [4]	ResNet-101-FPN	37.7	59.3	40.6	16.8	39.9	54.6
MMask R-CNN [14]	ResNet-101-FPN	38.0	59.3	41.3	21.2	40.2	49.3
BoundaryFormer [10]	ResNet-101-FPN	37.7	58.8	40.5	20.4	40.2	49.0
CondInst [18]	ResNet-50-FPN	35.4	56.4	37.6	18.4	37.9	46.9
BlendMask [2]	ResNet-50-FPN	34.3	55.4	36.6	14.9	36.4	48.9
BMask R-CNN [4]	ResNet-50-FPN	35.9	57.0	38.6	15.8	37.6	52.2
MMask R-CNN [14]	ResNet-50-FPN	36.3	57.3	38.3	19.3	38.3	47.3
BoundaryFormer [10]	ResNet-50-FPN	36.4	57.2	39.0	19.6	38.6	47.9
Mask R-CNN [7]	ResNet-50-FPN	34.6	56.5	36.6	15.4	36.3	**49.7**
JLInst(Ours)	ResNet-50-FPN	**37.6**	**58.0**	**40.7**	**20.2**	**39.8**	48.9
Mask R-CNN [7]	ResNet-101-FPN	36.2	58.6	38.4	16.4	38.4	**52.1**
JLInst(Ours)	ResNet-101-FPN	**39.1**	**60.0**	**42.5**	**20.7**	**41.6**	51.5

4.2 Experimental Settings

For fair comparisons, we conduct our experiments on Detectron2 [20] platform in PyTorch [16] framework. If not specified, all the settings are the default settings in Detectron2. Unless specified, ResNet-50 [8] is used as our default backbone network. The backbone networks are initialized with the weight of the models pretrained on ImageNet [5]. Specifically, our network is trained with stochastic gradient descent (SGD) optimizer for 90K iterations with an initial learning rate of 0.02 and a mini-batch of 16 images. The learning rate is reduced by a factor of 10 at iteration 60K and 80K, respectively. The input images are resized such that the shorter side is 800 pixels and the longer is less than 1333 pixels. As for ablation experiments, we adopt 500 pixels for shorter side (the longer is less than 700 pixels). Due to memory limitation, the batch size (16 by default) is adjusted with a linearly scaled learning rate. Momentum and weight decay are set to 0.9 and $1e^{-4}$, respectively.

4.3 Comparisons with State-of-the-Art Method

We compare JLInst with some state-of-the-art instance segmentation methods. All models are trained with 1× schedule on COCO train2017 and evaluated on COCO test-dev2017. As shown in Table 1, JLInst with ResNet-50-FPN can surpass these methods with the same backbone. The combination of our JLInst and ResNet-101-FPN backbone achieves as high as 39.1% AP, which surpasses the AP of all other competitors.

Table 2. Experimental results on COCO *val2017* for the impacts of MBR and BMR.

MBR	BMR	AP	AP_{50}	AP_{75}
✗	✗	34.0	53.8	35.9
✓	7	34.3	53.8	36.5
✗	✓	34.5	54.1	36.8
✓	✓	34.7	53.9	37.0

Table 3. Experimental results on COCO *val2017* for the impacts of each component in MBR/BMR.

FB	AB	RB	AP	AP_{50}	AP_{75}
✗	✗	✗	34.0	53.8	35.9
✓	✗	✗	34.2	53.8	36.4
✓	✓	✗	34.4	53.8	36.8
✓	✓	✓	34.7	53.9	37.4

Table 4. Experimental results on COCO *val2017* for the impacts of different losses.

	AP	AP_{50}	AP_{75}
BCE loss	34.4	53.3	36.9
GW loss	**34.7**	53.9	37.0

Table 5. Experiment results on COCO *val2017* for the impacts of standard deviation σ in GW loss.

σ	AP	AP_{50}	AP_{75}
0.01	33.6	53.2	36.0
0.1	34.3	53.4	36.8
0.3	34.4	53.7	36.6
0.5	**34.7**	53.9	37.0
0.7	34.6	54.0	36.9
0.9	34.5	53.8	36.9

4.4 Ablation Studies

The Effectiveness of MBR and BMR. We now analyze whether each feature fusion module of our model is effective for improvement of segmentation performance. The results are listed in Table 2. We make different combinations of MBR and BMR on ResNet-50 FPN JLInst baseline. As shown in Table 2, MBR improves the AP from 34.0% to 34.3%. When combined with BMR, the AP is improved from 34.0% to 34.5%. When all the modules are used, the AP is further improved to 34.7%.

The Impacts of Each Component in MBR/BMR. In this section, we analyze whether each component of MBR/BMR is effective for performance improvement. As shown in Table 3, we gradually add Fusion Block (FB), Attention Block (AB) and Residual Block (RB) on ResNet-50 FPN JLInst baseline. FB improves the AP from 34.0% to 34.2%. When combined FB with AB, the AP is further increased from 34.0% to 34.4%. When FB, AB and RB are all used, the AP is further improved to 34.7%. In general, our ablation studies show that FB, AB and RB can effectively improve segmentation performance.

The Effectiveness of GW Loss. In JLInst, we propose GW loss to supervise the prediction. Compared with BCE loss, our GW loss can help JLInst focus more on those uncertain examples. So in this session, we design this experiment to compare the impact of BCE loss and our GW loss for final mask prediction. As shown in Table 4, using BCE loss brings the 34.4% AP. When we replaced BCE loss with GW loss, the performance of JLInst increases from 34.4% to 34.7%. Therefore, our GW loss is more suitable for JLInst and can bring better performance for instance segmentation.

The Impacts of Parameters in GW Loss. In this section, we analyse the impacts of different standard deviations σ in GW loss. As shown in Table 5, the performance gradually increases with the growth of σ. But when σ exceeds 0.5, the AP begins to decline. Therefore, we set $\sigma = 0.5$ based on the experiments.

Table 6. Experiment results on COCO *val2017* for the impacts of proposed losses in JLInst.

$\mathcal{L}_{\text{final}}$	$\mathcal{L}_{\text{mask}}$	$\mathcal{L}_{\text{boundary}}$	AP	AP_{50}	AP_{75}	AP_S	AP_M	AP_L
✓	✗	✗	34.0	53.9	36.1	12.8	35.7	53.9
✓	✓	✗	34.3	53.9	36.5	13.4	36.2	53.4
✓	✗	✓	34.4	53.8	36.6	13.3	36.2	54.3
✓	✓	✓	34.7	53.9	37.4	13.8	36.6	54.0

The Impacts of Proposed Losses in JLInst. In JLInst, we propose Mask Loss $\mathcal{L}_{\text{mask}}$, Boundary Loss $\mathcal{L}_{\text{boundary}}$ and Final mask Loss $\mathcal{L}_{\text{final}}$, to jointly learn the mask and boundary information and ensure that network produce more distinct mask prediction. So we design this experiment to analyze whether each loss function is effective to improve the performance of JLInst. The results are shown in Table 6. When only $\mathcal{L}_{\text{final}}$ is used, the AP of JLInst is 34.0%. When supervised by $\mathcal{L}_{\text{final}}$ and $\mathcal{L}_{\text{mask}}$, the AP is increased from 34.0% to 34.3%. When supervised by $\mathcal{L}_{\text{final}}$ and $\mathcal{L}_{\text{boundary}}$, the AP is improved from 34.0% to 34.4%. When all the losses are used, the performance of JLInst is further improved to 34.7%.

5 Conclusion

In this paper, we propose a simple but effective instance segmentation framework, named JLInst, to jointly learn object boundary and mask semantic information and generate more precise mask prediction by combining mask and boundary information. Moreover, we design GW loss to help JLInst focus more on examples with high uncertainty in the pixel-level classification. Experiments show that our JLInst achieves remarkable improvements on COCO dataset, and outperforms most recent methods in the fair comparison.

Acknowledgments. This work was supported by the National Natural Science Foundation of China under Grant 82261138629; Guangdong Basic and Applied Basic Research Foundation under Grant 2023A1515010688 and Shenzhen Municipal Science and Technology Innovation Council under Grant JCYJ20220531101412030.

References

1. Bolya, D., Zhou, C., Xiao, F., Lee, Y.J.: Yolact: real-time instance segmentation. In: Proceedings of the IEEE/CVF International Conference on Computer Vision, pp. 9157–9166 (2019)
2. Chen, H., Sun, K., Tian, Z., Shen, C., Huang, Y., Yan, Y.: Blendmask: top-down meets bottom-up for instance segmentation. In: Proceedings of the IEEE/CVF Conference on Computer Vision and Pattern Recognition, pp. 8573–8581 (2020)
3. Chen, L.C., Papandreou, G., Kokkinos, I., Murphy, K., Yuille, A.L.: Deeplab: semantic image segmentation with deep convolutional nets, atrous convolution, and fully connected CRFS. IEEE Trans. Pattern Anal. Mach. Intell. **40**(4), 834–848 (2017)
4. Cheng, T., Wang, X., Huang, L., Liu, W.: Boundary-preserving mask R-CNN. In: Vedaldi, A., Bischof, H., Brox, T., Frahm, J.-M. (eds.) ECCV 2020. LNCS, vol. 12359, pp. 660–676. Springer, Cham (2020). https://doi.org/10.1007/978-3-030-58568-6_39
5. Deng, J., Dong, W., Socher, R., Li, L.J., Li, K., Fei-Fei, L.: Imagenet: a large-scale hierarchical image database. In: 2009 IEEE Conference on Computer Vision and Pattern Recognition, pp. 248–255. IEEE (2009)
6. Deng, R., Shen, C., Liu, S., Wang, H., Liu, X.: Learning to predict crisp boundaries. In: Proceedings of the European Conference on Computer Vision (ECCV), pp. 562–578 (2018)
7. He, K., Gkioxari, G., Dollár, P., Girshick, R.: Mask r-cnn. In: Proceedings of the IEEE International Conference on Computer Vision, pp. 2961–2969 (2017)
8. He, K., Zhang, X., Ren, S., Sun, J.: Deep residual learning for image recognition. In: Proceedings of the IEEE Conference on Computer Vision and Pattern Recognition, pp. 770–778 (2016)
9. Huang, Z., Huang, L., Gong, Y., Huang, C., Wang, X.: Mask scoring r-cnn. In: Proceedings of the IEEE/CVF Conference on Computer Vision and Pattern Recognition, pp. 6409–6418 (2019)
10. Lazarow, J., Xu, W., Tu, Z.: Instance segmentation with mask-supervised polygonal boundary transformers. In: Proceedings of the IEEE/CVF Conference on Computer Vision and Pattern Recognition, pp. 4382–4391 (2022)
11. Lin, T.Y., Goyal, P., Girshick, R., He, K., Dollár, P.: Focal loss for dense object detection. In: Proceedings of the IEEE International Conference on Computer Vision, pp. 2980–2988 (2017)
12. Lin, T.-Y., et al.: Microsoft COCO: common objects in context. In: Fleet, D., Pajdla, T., Schiele, B., Tuytelaars, T. (eds.) ECCV 2014. LNCS, vol. 8693, pp. 740–755. Springer, Cham (2014). https://doi.org/10.1007/978-3-319-10602-1_48
13. Liu, S., Qi, L., Qin, H., Shi, J., Jia, J.: Path aggregation network for instance segmentation. In: Proceedings of the IEEE Conference on Computer Vision and Pattern Recognition, pp. 8759–8768 (2018)

14. Ma, L., Dong, B., Yan, J., Li, X.: Matting enhanced mask r-cnn. In: 2021 IEEE International Conference on Multimedia and Expo (ICME), pp. 1–6. IEEE (2021)
15. Milletari, F., Navab, N., Ahmadi, S.A.: V-net: fully convolutional neural networks for volumetric medical image segmentation. In: 2016 Fourth International Conference on 3D Vision (3DV), pp. 565–571. IEEE (2016)
16. Paszke, A., et al.: Pytorch: an imperative style, high-performance deep learning library. arXiv preprint arXiv:1912.01703 (2019)
17. Ren, S., He, K., Girshick, R., Sun, J.: Faster r-cnn: towards real-time object detection with region proposal networks. Adv. Neural. Inf. Process. Syst. **28**, 91–99 (2015)
18. Tian, Z., Shen, C., Chen, H.: Conditional convolutions for instance segmentation. In: Vedaldi, A., Bischof, H., Brox, T., Frahm, J.-M. (eds.) ECCV 2020. LNCS, vol. 8693, pp. 282–298. Springer, Cham (2020). https://doi.org/10.1007/978-3-030-58452-8_17
19. Tian, Z., Shen, C., Chen, H., He, T.: FCOS: fully convolutional one-stage object detection. In: Proceedings of the IEEE/CVF International Conference on Computer Vision, pp. 9627–9636 (2019)
20. Wu, Y., Kirillov, A., Massa, F., Lo, W.Y., Girshick, R.: Detectron2 (2019). https://github.com/facebookresearch/detectron2

Boosting One-Stage Multi Object Tracking with Attention Learning

Ying Cui[1], Hang Zheng[1], Yueqian Quan[1], Xiang Pan[1], Yike Wang[2], and Junxia Li[3(✉)]

[1] College of Computer Science and Technology, Zhejiang University of Technology, Hangzhou 310023, China
{cuiying,zhenghang,qyq,panx}@zjut.edu.cn

[2] University of California, Berkeley, USA
yike_wang@berkeley.edu

[3] School of Computer and Software, Nanjing University of Information Science and Technology, Nanjing 210044, China
junxiali99@163.com

Abstract. One-stage multi-object tracking methods have achieved promising results by showing their great balance between accuracy and speed. However, the internal differences and relationships between detection and re-identification (re-ID) lead to worse performance. In this work, we propose a one-stage multi-object tracking method with attention boosting, namely AeMOT, which can effectively improve the collaboration and performance in detection and re-ID. Specifically, a discriminability enhancement module is designed to enhance the discriminative feature representations for detection and tracking, and an identity preserving module is properly designed to preserve the semantic alignment of id-embedding and improve the adaptiveness of object matching with scale variation for re-ID association. Experimental results on challenging benchmarks including MOT17 and MOT20 demonstrate that our proposed method achieves leading performance and outperforms state-of-the-art trackers.

Keywords: AeMOT · multi-object tracking · one-stage · attention mechanism

1 Introduction

Multi-object tracking (MOT) is one of the most foundational and challenging tasks in computer vision, with wide applications in many practical fields such as autonomous driving, human-computer interaction and intelligent surveillance video, etc. MOT aims to track the multiple objects of interest in videos through

This work was supported in part by the National Natural Science Foundation of China (No. 62102364, No. 62272235), and in part by the Natural Science Foundation of Zhejiang Province (LY22F020016).

© The Author(s), under exclusive license to Springer Nature Singapore Pte Ltd. 2024
Q. Liu et al. (Eds.): PRCV 2023, LNCS 14436, pp. 458–469, 2024.
https://doi.org/10.1007/978-981-99-8555-5_36

estimating their trajectories from the temporal and spatial information. Different from single-object tracking (SOT) [1–3], multiple object tracking suffers from inter-target occlusions, interactions and ambiguities, which bring big challenge to the task. Accurate and real-time multi-object tracking is still the goal that researchers are constantly pursuing.

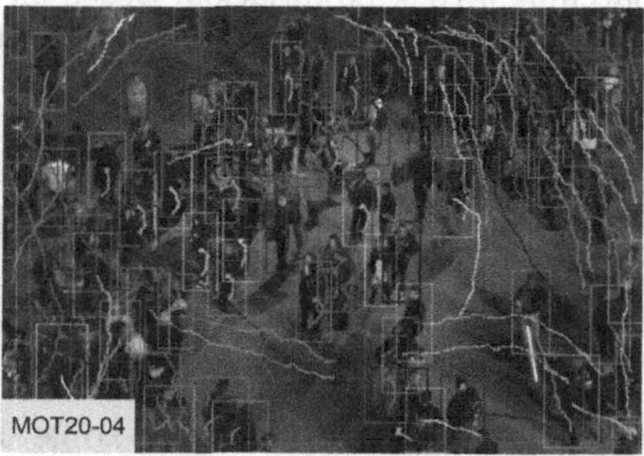

Fig. 1. Example tracking results of our AeMOT on MOT17 and MOT20 test sets. Different identity is shown in different colored bounding box and trajectory in the past 100 frames. Tracking under challenges such as crowded scenes can be robustly performed by our method.

Existing methods mainly adopt two major paradigms: the tracking-by-detection (TBD) paradigm [4–6], and the joint detection and tracking (JDT) paradigm [7–15]. In the past decades, the traditional methods mainly follow the tracking-by-detection strategy, which divides MOT into two models: a detector and an embedding (re-ID) model. They first utilize a detector to obtain bounding boxes of objects in each frame, and then associate the obtained bounding boxes with existing tracklets by matching the predicted location and extracted identification (ID) embedding of each bounding box across frames. Although they can benefit from the significant advances in recent object detection [16–18] and re-identification (reID) [19] development, these methods still suffer from massive computation costs. In recent years, the one-stage joint detection and tracking (JDT) methods that learn detection and association within a single network have demonstrated their ability to achieve good balance between the inference speed and tracking accuracy [7,20,21]. JDT has become a popular trend in multi-object tracking. Nevertheless, an inevitable problem of JDT methods is that sharing the same backbone feature between the detection task and re-ID task may lead to conflict about information requirement from the feature representations. Despite FairMOT [20] try to tackle the problem by balance the loss optimization of the two tasks, the negative transfer from learning two different tasks are still not well solved.

Encouraged by the advantage of attention mechanism in deep neural networks and Transformer frameworks [22,23], in this paper, we investigate to leverage attention mechanism to improve the feature representation for whole tracking and perceive the instance-specific information for re-identification. Specifically, a discriminative feature enhancement module is designed via spatial attention and channel attention that leverage the self-attention mechanism to capture the contextual information for better feature representations. Then, an identity preserving module is designed via channel re-calibration to preserve the semantic alignment of id-embedding and improve the adaptiveness of object matching with scale variation. Based on the two designed modules, we propose a new attention enhanced one-stage algorithm for multi-object tracking, termed as AeMOT that follows the joint-dectection-and-tracking approach. Although the introduced techniques are not mostly novel, we have show the proper adoption is important and valuable to MOT task. Extensive experimental results have demonstrated that our AeMOT achieve leading performance and outperforms existing state-of-the-art methods on challenging benchmarks such as MOT17 [24] and MOT20 [25], as shown by visual examples in Fig. 1.

The main contributions of our work are as follow:

- We propose AeMOT, a one-stage JDT method for multi-object tracking, which deals with the conflict between the detection and re-ID task.
- We introduce a discriminative feature enhancement module (DEM) to capture the contextual information for better feature representations and an identity preserving module (IPM) for improving the representation for re-ID task.
- Experiments on MOT-17 and MOT-20 demonstrate the superiority of our method compared with other one-stage MOT algorithms.

2 Related Work

Tracking-by-Detection. In the past decade, researchers usually leverage the advance of object detection and follow the tracking-by-detection (TBD) paradigm to solve the task [4–6]. Traditional methods mainly exploit two separate models that first utilize an object detection model to localize objects and obtain bounding-boxes, and then adopt an association model to link one detected object to a specific trajectory according to re-identification features of the objects. SORT [26] uses a Kalman filter to track bounding boxes and associates each bounding box with Hungarian Algorithm. STRN [5] proposes a similarity learning framework to encode various spatial-temporal relations between tracks and objects. DeepSORT [6] incorporates the usage of appearance features to SORT [26]. POI [27] leverages high score detection and deep learning-based appearance feature to tracking. ByteTrack [4] tracks objects by associating each detection box instead of only the high score detection box. Though these methods achieve good tracking results by using long-term trajectory to recover missed or occluded detection, they still follow the tracking-by-detection paradigm, which limits the tracking efficiency in practical application, and the two separate models make them suffer from a serious time-consuming problem that prevents them from real-time tracking.

Joint Detection and Tracking. In recent years, Joint detection and tracking (JDT) algorithms that learn detection and association within a single network have achieved good balance between the inference speed and tracking accuracy [7–15]. TransTrack [7] leverages the query-key mechanism of Transformer [22] to establish a new joint detection and tracking framework. It sequentially tracks the objects in the current frame through the association of feature queries from the previous frame. QDTrack [10] uses a dense similarity learning method for tracking. GSDT [12] proposes a joint tracking method using graph neural networks. MeMOT [21] further proposes a long-range spatio-temporal memory algorithm to enhance the ability of linking objects with a long time span. CenterTrack [13] proposes a point-based framework based on CenterNet [28] to perform joint detection and tracking. It treats the tracking in a local perspective that only associate objects in adjacent frames to simplify the association task and achieve To achieve high levels of both detection and tracking accuracy, FairMOT [20] proposes to equally treat the detection and association tasks. It constructs two parallel branches to respectively detecting objects and extracting re-ID features. However, since the JDT methods share the same backbone feature between the detection task and re-ID task, they may lead to conflict about information requirement from the feature representations. In this paper, to tackle the problem, we extend the attention mechanism in the framework of multi-object tracking, and carefully design two types of attention modules to capture rich contextual relationships for the two joint sub-tasks.

3 Methodology

In this section, we present the detailed description for the proposed AeMOT framework. Two newly designed modules, a discriminability enhancement module (DEM) and an identity preserving module (IPM) is incorporated with the backbone DLA-34 and Header network of the baseline FairMOT [20] to boost the tracking performance. An overview of the framework is illustrated in Fig. 2.

3.1 Discriminability Enhancement Module

Effective feature representation with sufficient spatial information is crucial for multi-object detection with only one single frame as input. Though recent Transformer based methods [14,15] that perform better feature extraction and spatio-temporal memory based methods [21] that store history states have made progress to solve the problem, they suffer from computation cost and memory cost cause by the strong architecture. In this work, we refer the essential philosophy of Transformer to solve the issue, leveraging the self-attention and Chanel attention to boost feature representations of normal backbone that trades off a balance between accuracy and efficiency. Encouraged by its success in scene segmentation, We introduce the self-attention mechanism based dual attention network [23] to perform the discriminability enhancement with spatial and contextual information.

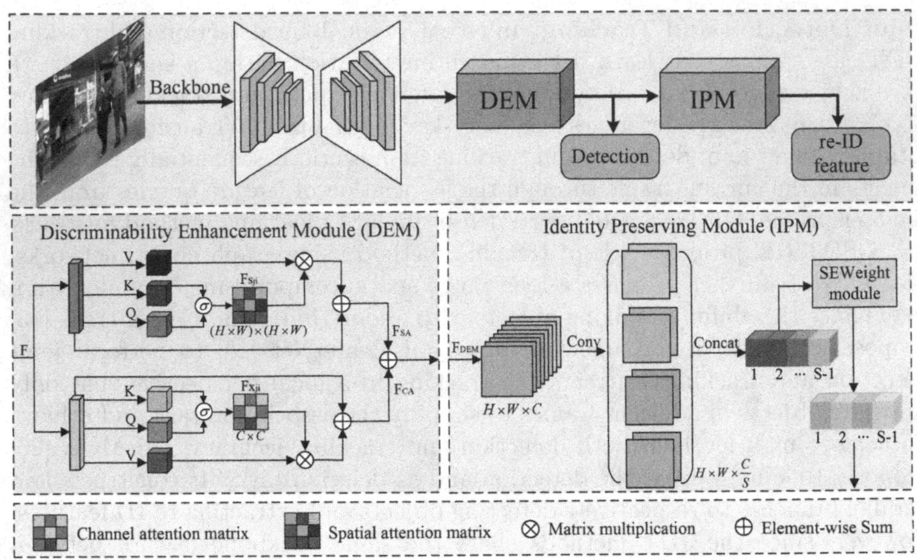

Fig. 2. Illustration of the proposed AeMOT, with two newly designed modules, the discriminability enhancement module (DEM) and identity preserving module (IPM).

Spatial Attention Enhancement. In order to enable the network to focus on foreground information and suppress background noise, spatial attention is implemented via self-attention to encode the long-range spatial and contextual information with local features. The common process of self-attention mechanism is written as:

$$SelfAttention(Q, K, V) = \mathrm{softmax}(K \otimes Q^T) \otimes V \quad (1)$$

Let $F \in \mathbb{R}^{C\times H\times W}$ denotes the feature obtained by the backbone, it then obtains $Q, K, V \in \mathbb{R}^{C\times H\times W}$ through three learnable weight matrices, a spatial attention map $F_{s_{ji}} \in \mathbb{R}^{N\times N}$ is calculated with the K and Q, where $N = H\times W$. It denotes the feature impact of the i-th position on the j-th position. The correlation of the two position correlation is positive to the value of $F_{s_{ji}}$.Then, a matrix multiplication is performed between the transpose of $F_{s_{ji}}$ and F to obtain a new feature map, which is then reshaped to $\mathbb{R}^{C\times H\times W}$, The final output $F_{SA} \in \mathbb{R}^{C\times H\times W}$ can be obtained by multiplying it with a scale parameter α and applying an element-wise sum operation with F:

$$F_{SA} = \alpha \sum_{i=1}^{N} (F_{s_{ji}} V) + F_j, \quad F_{s_{ji}} = K \otimes Q^T \quad (2)$$

In each position of the F_{SA}, the final feature is a weighted sum at all positions with the original features. Therefore, the final feature obtains a global contextual view and aggregation from the spatial attention map selectively, which improves the compactness within the object class.

Channel Attention Enhancement. Since each channel of the features can be regarded as an object-specific response map, channel attention can be leveraged to enhance the feature representation of object-specific information. As shown in Fig. 2, the feature $F \in \mathbb{R}^{C\times H\times W}$ obtained by backbone is first reshaped to $\mathbb{R}^{C\times N}$, where $N = H \times W$. Then the input matrices $Q, K, V \in \mathbb{R}^{C\times H\times W}$ are obtained by three learnable weight matrices, then a matrix multiplication between K and the transpose of Q is performed, followed by a softmax layer to obtain the channel attention map $F_{x_{ji}} \in \mathbb{R}^{C\times C}$, which measures the feature impact of the ith channel upon the jth channel. Next, a matrix multiplication between the transpose of $F_{x_{ji}}$ and F is performed and the result is reshaped to $\mathbb{R}^{C\times H\times W}$. It is then multiplied by a scale parameter β and summed with the feature map F in an element-wise manner to obtain the final output $F_{CA} \in \mathbb{R}^{C\times H\times W}$:

$$F_{CA} = \beta \sum_{i=1}^{C}(F_{x_{ji}} V) + F_j, \quad F_{x_{ji}} = K \otimes Q^T \tag{3}$$

For each channel of F_{CA}, it is a weighted sum of features at all channel and original features, which can boost the feature discriminability with long-range contextual information.

Finally, the features obtained from the two spatial and channel enhancements are add together to achieve the output F_{DEM} for detection and tracking.

3.2 Identity Preserving Module

The contextual spatial information boosted by the proposed DEM can mostly promote the detection. However, it is investigated that low-level features contain more object-specific information for re-ID learning. Thus, as a complementary, we propose the identity preserving module to re-calibrate channels to achieve the object-specific information and preserve the semantic alignment of id-embedding for re-ID.

We leverage the channel attention mechanism [29] to conduct the module. As illustrated by the right part of Fig. 2, it is mainly implemented with three steps. Firstly, the features $F_{DEM} \in \mathbb{R}^{C\times H\times W}$ output by DEM are split into multi-scale with different kernel sizes on each channel. With such operation, we can obtain richer information about the position of the input tensor and process it on multiple scales in a parallel manner. Next, we utilize the SEWeight module [30] further extract the channel attention of the input feature maps in multiple scales and generate the channel-wise attention weight vector, which can be represented as:

$$Z_i = SEweight(F_{DEM_i}), \quad i = 0, 1, 2 \cdots S - 1. \tag{4}$$

$Z_i \in \mathbb{R}^{C'\times 1\times 1}$ is the attention weight, and each feature map with different scales, F_{DEM_i} have the common channel dimension $C' = \frac{C}{S}$. Through the operation, the shallow features which are suitable for the re-ID tasks could be more concerned. By multiplying the weight with the feature in the corresponding scale F_{DEM_i}, we can obtain the enhanced feature with better information interaction between both local and global channels:

$$Y_i = F_{DEM_i} \odot Softmax(Z_i) \quad i = 1, 2, 3, \cdots S - 1. \tag{5}$$

Finally they are concatenated to generate the refined feature map for re-ID head:

$$F_{IPM} = Cat([Y_0, Y_1, \cdots, Y_{(S-1)}].) \tag{6}$$

The identity preserving module preserves the semantic alignment of id-embedding and improves the elasticity of objects of different scales during the association process, thus improves the adaptiveness of object matching with scale variation. Ablation study demonstrates that it complements to DEM for improving the re-ID performance.

4 Experiments

4.1 Implementation Details

Following the baseline FairMOT, DLA-34 is adopted as our backbone. The input image size is $1080 \times 608 \times 3$, and the output feature map resolution is 272×152. The Adam optimizer is utilized for training our model for 20 epochs on NVIDIA GeForce RTX 3090 GPU. The learning rate is set as 10^{-4}, using a factor 0.1 decay at 10-th epoch. The batch size is set to 12.

We use the following official metrics for MOT evaluation: the multi-object-tracking-accuracy (MOTA) which reflects the average tracking performance, the ratio of correctly identified detections over the average number of ground-truth and computed detections (IDF1) which emphasizes the association accuracy, the higher order tracking accuracy (HOTA) which is described as the geometric mean of detection accuracy (DetA) and association accuracy (AssA), the number of identity switches (IDs), the average precision (AP), and the true positive rate (TPR) at a false accept rate of 0.1 for evaluating re-ID features.

For fair comparison, the training datasets are the same with FairMOT, including ETH [31] and CityPerson [32], CalTech [33], CUHK-SYSU [34], PRW [35], MOT17 and CrowdHuman [36]. We evaluate our tracker on two tracking benchmarks: MOT17 [24] and MOT20 [25].

4.2 Ablation Study

Evaluation of Proposed Modules. To evaluate the proposed modules of our framework, we conduct ablation studies on the validation set of MOT17 [24], and the results are shown in Table 1. Benefiting from the spatial and channel attention aggregation via self-attention, the proposed DEM has a significant improvement in detection. As shown in the second row of Table 1, By incorporating DEM, the tracker outperforms the baseline FairMOT [20] by MOTA(+1.1), IDF1(+1.6) and AP(+0.9). However, the identity switch is rising. It is because

Table 1. Evaluation of the proposed modules on the validation set of MOT17. The best results are shown in **bold**.

No	Baseline	DEM	IPM	MOTA↑	IDF1 ↑	IDs↓	AP↑
1	✓			67.5	69.9	408	79.6
2	✓	✓		68.6	71.5	432	80.5
3	✓		✓	68.6	70.4	402	79.8
4	✓	✓	✓	**69.2**	**73.9**	**380**	**80.9**

Table 2. Evaluation of training on the validation set of MOT17. The best results are shown in **bold**.

Training Data	Method	MOTA↑	IDF1↑	IDs↓	AP↑	TPR↑
MOT17	FairMOT	67.5	69.9	408	79.6	93.4
	AeMOT (ours)	**69.2**	**73.9**	**380**	**80.9**	**94.9**
MOT17+MIX	FairMOT	69.1	72.8	299	81.2	94.4
	AeMOT (ours)	**70.0**	**74.4**	**282**	**82.8**	**96.0**

the enhanced features reduce intra-class differentiation, which can be solved by IPM. As can be seen in the third row of Table 1, compared with the baseline, IPM achieves performance gains of MOTA(+1.1) and IDF1(+0.5) and reduce the identity switch, which indicates that the representation of the object in the re-ID head can be enhanced by IPM.

By combining the two modules, the framework shows the highest performance gains and surpasses the baseline by MOTA(+1.7), IDF1(+4.0), IDs(−6.9%), and AP(+1.3), which demonstrates that the two modules can complement to each other, and achieve a good balancing in detection and re-ID feature learning. The gains of IDs(−6.9%) prove that DEM also contributes to re-ID task with adjustment of IPM.

Evaluation of Training Data. In comparison with the baseline, we follow the training data from multiple datasets. Table 2 shows the results of models with different training data, where"MIX" stands for the composed five datasets illustrated in the subsection of Implementation Details. Compared with the baseline, our method outperforms it on both training strategies. Also, our framework can benefit from larger training data.

4.3 Comparing with SOTA MOT Methods

We compare our method with the state-of-the-art trackers on the test set of MOT17 [24] and MOT20 [25], and the results are shown in Table 3. Our proposed AeMOT ranks first and outperforms the state-of-the-art trackers on the most important metrics. Compared with the baseline FairMOT, our method surpasses it by MOTA(+7.0), IDF1(+2.7), HOTA(+1.1), and IDs(−17.9%) on the MOT20 dataset, and surpasses it by MOTA(+0.2),

Table 3. Evaluation on the test set of MOT challenge. Comparisons with the state-of-the-art trackers are shown under the "private detector" method. The best results of each dataset are shown in **bold**.

Benchmark	Tracker	Device	MOTA↑	IDF1↑	HOTA↑	DetA↑	AssA↑	IDs↓	FPS↑
MOT17 [24]	GSDT [12]	–	66.2	68.7	55.2	60.0	51.1	3318	4.9
	CTrackerv1 [9]	–	66.6	57.4	49.0	53.6	45.2	5529	6.8
	CenterTrack [13]	Titanxp	67.8	64.7	52.2	53.8	51.0	3039	17.5
	QDTrack [10]	–	68.7	66.3	53.9	55.6	52.7	3378	20.3
	TraDeS [11]	2080Ti	69.1	63.9	52.7	55.2	50.8	3555	11.9
	MeMOT [21]	A100	72.5	69.0	56.9	–	55.2	2724	–
	MOTR [14]	V100	73.4	68.6	57.8	60.3	55.7	**2439**	–
	FairMOT [20]	2080Ti	73.7	72.3	59.3	60.9	58.0	3303	25.9
	AeMOT (Ours)	3090	**73.9**	**73.9**	**60.8**	**61.1**	**60.8**	2955	20.9
MOT20 [25]	FairMOT [20]	2080Ti	61.8	67.3	54.6	54.7	54.7	5243	13.2
	MeMOT [21]	A100	63.7	66.1	54.1	–	**55.0**	1938	–
	TransTrack [7]	V100	65.0	59.4	48.9	53.3	45.2	3608	7.2
	GSDT [12]	–	67.1	67.5	53.6	54.2	54.0	3131	0.9
	Trackformer [15]	–	68.6	65.7	54.7	56.7	53.0	**1532**	–
	AeMOT (Ours)	3090	**68.8**	**70.0**	**55.8**	**57.1**	54.7	4307	10.1

IDF1(+1.6), HOTA(+1.5), and IDs(−10.5%) on the MOT17 dataset. Compared with the newest Transformer-based and memory-based method MeMOT [21], our AeMOT achieve a substantial performance gains of MOTA(+1.4/+5.1), IDF1(+4.9/+3.9) and HOTA(+3.9/+1.7) on the MOT17/MOT20, respectively. It should be noticed that MOT20 is a much more challenging benchmark with serious occlusions and crowded scenarios. The larger performance gains on MOT20 further demonstrate that our AeMOT can achieve robust performance in intensive and more challenging scenes, which owing to the substantial discriminability enhancement and identity information preservation of our proposed framework. Compared with existing methods, our tracker also achieve a comparable real-time speed of 20.9 FPS.

To further demonstrate the performance, we show some visualization results on MOT17 testing set and MOT20 testing set in Fig. 3. It can be found that tracking under challenges such as crowded scenes can be robustly performed by our method, which can effectively handle large-scale change and maintain the right identity.

Fig. 3. Visual tracking results of our tracker on MOT17 and MOT20 test set. Different identity is shown in different colored bounding box and trajectory. The illustration shows the change every 100 frames.

5 Conclusion

In this paper, we present a new framework, namely AeMOT, for multiple object tracking. It is designed with a discriminability enhancement module to enhance discriminative feature representations for detection and tracking, and an identity preserving module to further preserve identification features for the re-ID association. Our design significantly boosts the capacity of detection and tracking, alleviating the problem of large target size variation and the overlap of the target in MOT. More importantly, it improves the elasticity of objects of different scales during the association process. The two modules complement each other. Experimental results demonstrate the effectiveness of our method.

References

1. Guo, D., Wang, J., Cui, Y., Wang, Z., Chen, S.: Siamcar: siamese fully convolutional classification and regression for visual tracking. In: Proceedings of the IEEE/CVF Conference on Computer Vision and Pattern Recognition, pp. 6269–6277 (2020)
2. Guo, D., Shao, Y., Cui, Y., Wang, Z., Zhang, L., Shen, C.: Graph attention tracking. In: Proceedings of the IEEE/CVF Conference on Computer Vision and Pattern Recognition, pp. 9543–9552 (2021)

3. Cui, Y., et al.: Joint classification and regression for visual tracking with fully convolutional siamese networks. Int. J. Comput. Vis. **130**(2), 550–566 (2022)
4. Zhang, Y., et al.: ByteTrack: multi-object tracking by associating every detection box. In: Avidan, S., Brostow, G., Cissé, M., Farinella, G.M., Hassner, T. (eds.) ECCV 2022. LNCS, vol. 13682, pp. 1–21. Springer, Cham (2022). https://doi.org/10.1007/978-3-031-20047-2_1
5. Xu, J., Cao, Y., Zhang, Z., Hu, H.: Spatial-temporal relation networks for multi-object tracking. In: Proceedings of the IEEE/CVF International Conference on Computer Vision, pp. 3988–3998 (2019)
6. Wojke, N., Bewley, A., Paulus, D.: Simple online and realtime tracking with a deep association metric. In: 2017 IEEE International Conference on Image Processing (ICIP), pp. 3645–3649. IEEE (2017)
7. Sun, P., et al.: Transtrack: multiple object tracking with transformer. arXiv preprint arXiv:2012.15460 (2020)
8. Xu, Y., et al.: TransCenter: transformers with dense representations for multiple-object tracking. IEEE Trans. Pattern Anal. Mach. Intell. **45**(6), 7820–7835 (2023)
9. Peng, J., et al.: Chained-tracker: chaining paired attentive regression results for end-to-end joint multiple-object detection and tracking. In: Vedaldi, A., Bischof, H., Brox, T., Frahm, J.-M. (eds.) ECCV 2020. LNCS, vol. 12349, pp. 145–161. Springer, Cham (2020). https://doi.org/10.1007/978-3-030-58548-8_9
10. Pang, J., et al.: Quasi-dense similarity learning for multiple object tracking. In: Proceedings of the IEEE/CVF Conference on Computer Vision and Pattern Recognition, pp. 164–173 (2021)
11. Wu, J., Cao, J., Song, L., Wang, Y., Yang, M., Yuan, J.: Track to detect and segment: an online multi-object tracker. In: Proceedings of the IEEE/CVF Conference on Computer Vision and Pattern Recognition, pp. 12352–12361 (2021)
12. Wang, Y., Kitani, K., Weng, X.: Joint object detection and multi-object tracking with graph neural networks. In: 2021 IEEE International Conference on Robotics and Automation (ICRA), pp. 13708–13715. IEEE (2021)
13. Zhou, X., Koltun, V., Krähenbühl, P.: Tracking objects as points. In: Vedaldi, A., Bischof, H., Brox, T., Frahm, J.-M. (eds.) ECCV 2020. LNCS, vol. 12349, pp. 474–490. Springer, Cham (2020). https://doi.org/10.1007/978-3-030-58548-8_28
14. Zeng, F., Dong, B., Zhang, Y., Wang, T., Zhang, X., Wei, Y.: MOTR: end-to-end Multiple-object tracking with tTransformer. In: Avidan, S., Brostow, G., Cissé, M., Farinella, G.M., Hassner, T. (eds.) ECCV 2022. LNCS, vol. 13687, pp. 659–675. Springer, Cham (2022). https://doi.org/10.1007/978-3-031-19812-0_38
15. Meinhardt, T., Kirillov, A., Leal-Taixe, L., Feichtenhofer, C.: Trackformer: multi-object tracking with transformers. In: Proceedings of the IEEE/CVF Conference on Computer Vision and Pattern Recognition, pp. 8844–8854 (2022)
16. Girshick, R.: Fast r-cnn. In: Proceedings of the IEEE International Conference on Computer Vision, pp. 1440–1448 (2015)
17. Redmon, J., Farhadi, A.: Yolov3: an incremental improvement. arXiv preprint arXiv:1804.02767 (2018)
18. Li, J., Pan, Z., Liu, Q., Cui, Y., Sun, Y.: Complementarity-aware attention network for salient object detection. IEEE Trans. Cybernet. **52**(2), 873–886 (2020)
19. Gu, X., Chang, H., Ma, B., Shan, S.: Motion feature aggregation for video-based person re-identification. IEEE Trans. Image Process. **31**, 3908–3919 (2022)
20. Zhang, Y., Wang, C., Wang, X., Zeng, W., Liu, W.: Fairmot: on the fairness of detection and re-identification in multiple object tracking. Int. J. Comput. Vision **129**, 3069–3087 (2021)

21. Cai, J., et al.: Memot: multi-object tracking with memory. In: Proceedings of the IEEE/CVF Conference on Computer Vision and Pattern Recognition, pp. 8090–8100 (2022)
22. Vaswani, A., et al.: Attention is all you need. Adv. Neural Inf. Process. Syst. **30** (2017)
23. Fu, J., et al.: Dual attention network for scene segmentation. In: Proceedings of the IEEE/CVF Conference on Computer Vision and Pattern Recognition, pp. 3146–3154 (2019)
24. Milan, A., Leal-Taixé, L., Reid, I., Roth, S., Schindler, K.: Mot16: a benchmark for multi-object tracking. arXiv preprint arXiv:1603.00831 (2016)
25. Dendorfer, P., et al.: Mot20: a benchmark for multi object tracking in crowded scenes. arXiv preprint arXiv:2003.09003 (2020)
26. Bewley, A., Ge, Z., Ott, L., Ramos, F., Upcroft, B.: Simple online and realtime tracking. In: 2016 IEEE International Conference on Image Processing (ICIP), pp. 3464–3468. IEEE (2016)
27. Yu, F., Li, W., Li, Q., Liu, Y., Shi, X., Yan, J.: POI: multiple object tracking with high performance detection and appearance feature. In: Hua, G., Jégou, H. (eds.) ECCV 2016. LNCS, vol. 9914, pp. 36–42. Springer, Cham (2016). https://doi.org/10.1007/978-3-319-48881-3_3
28. Zhou, X., Wang, D., Krähenbühl, P.: Objects as points. arXiv preprint arXiv:1904.07850 (2019)
29. Zhang, H., Zu, K., Lu, J., Zou, Y., Meng, D.: Epsanet: an efficient pyramid squeeze attention block on convolutional neural network. In: Proceedings of the Asian Conference on Computer Vision, pp. 1161–1177 (2022)
30. Hu, J., Shen, L., Sun, G.: Squeeze-and-excitation networks. In: Proceedings of the IEEE Conference on Computer Vision and Pattern Recognition, pp. 7132–7141 (2018)
31. Ess, A., Leibe, B., Schindler, K., Van Gool, L.: A mobile vision system for robust multi-person tracking. In: 2008 IEEE Conference on Computer Vision and Pattern Recognition, pp. 1–8. IEEE (2008)
32. Zhang, S., Benenson, R., Schiele, B.: Citypersons: a diverse dataset for pedestrian detection. In: Proceedings of the IEEE Conference on Computer Vision and Pattern Recognition, pp. 3213–3221 (2017)
33. Dollár, P., Wojek, C., Schiele, B., Perona, P.: Pedestrian detection: a benchmark. In: 2009 IEEE Conference on Computer Vision and Pattern Recognition, pp. 304–311. IEEE (2009)
34. Xiao, T., Li, S., Wang, B., Lin, L., Wang, X.: Joint detection and identification feature learning for person search. In: Proceedings of the IEEE Conference on Computer Vision and Pattern Recognition, pp. 3415–3424 (2017)
35. Zheng, L., Zhang, H., Sun, S., Chandraker, M., Yang, Y., Tian, Q.: Person re-identification in the wild. In: Proceedings of the IEEE Conference on Computer Vision and Pattern Recognition, pp. 1367–1376 (2017)
36. Shao, S., Zhao, Z., Li, B., Xiao, T., Yu, G., Zhang, X., Sun, J.: Crowdhuman: a benchmark for detecting human in a crowd. arXiv preprint arXiv:1805.00123 (2018)

TPNet: Enhancing Weakly Supervised Polyp Frame Detection with Temporal Encoder and Prototype-Based Memory Bank

Jianzhe Gao, Zhiming Luo(✉), Cheng Tian, and Shaozi Li

Department of Artificial Intelligence, Xiamen University, Xiamen, Fujian, China
zhiming.luo@xmu.edu.cn

Abstract. Polyp detection plays a crucial role in the early prevention of colorectal cancer. The availability of large-scale polyp video datasets and video-level annotations has spurred research efforts to formulate polyp detection as a weakly-supervised anomaly detection task, which leverages video-level labeled training data for detecting frame-level polyps. However, few studies have investigated the impact of specific properties within polyp videos, including temporal dynamics, ambiguity, and complex noise. In this work, we propose TPNet, a novel framework that addresses several challenges posed by colonoscopy videos, for weakly-supervised polyp frame detection. Specifically, we design a Temporal Encoder that effectively capturing the temporal dynamics and intricate patterns within polyp video segments to foster accuracy. Additionally, we introduce a Prototype-based Memory Bank that facilitates the storage and retrieval of significant discriminative information, which enhance the sensitivity and robustness in ambiguous and complicated conditions. Experiments conducted on one of the largest and most challenging colonoscopy datasets demonstrate that our proposed TPNet achieves state-of-the-art performance, surpassing the latest cutting-edge method with 6.19% in average precision (AP).

Keywords: Polyp Detection · Weakly Surpervised · Temporal Encoder · Prototype-based Memory Bank

1 Introduction

Colorectal cancer (CRC) is one of the most prevalent and lethal forms of cancer globally [11,15,16]. Gastrointestinal endoscopy, a widely employed procedure, plays a pivotal role in the early identification of gastric and colorectal cancers [2]. This procedure involves inserting a flexible tube with a miniature camera into the digestive tract, enabling the identification of precancerous lesions [7]. However, the miss rate of polyp detection remains alarmingly high [1,12,25]. Therefore, there is an urgent need for advanced techniques that can effectively address the

© The Author(s), under exclusive license to Springer Nature Singapore Pte Ltd. 2024
Q. Liu et al. (Eds.): PRCV 2023, LNCS 14436, pp. 470–481, 2024.
https://doi.org/10.1007/978-981-99-8555-5_37

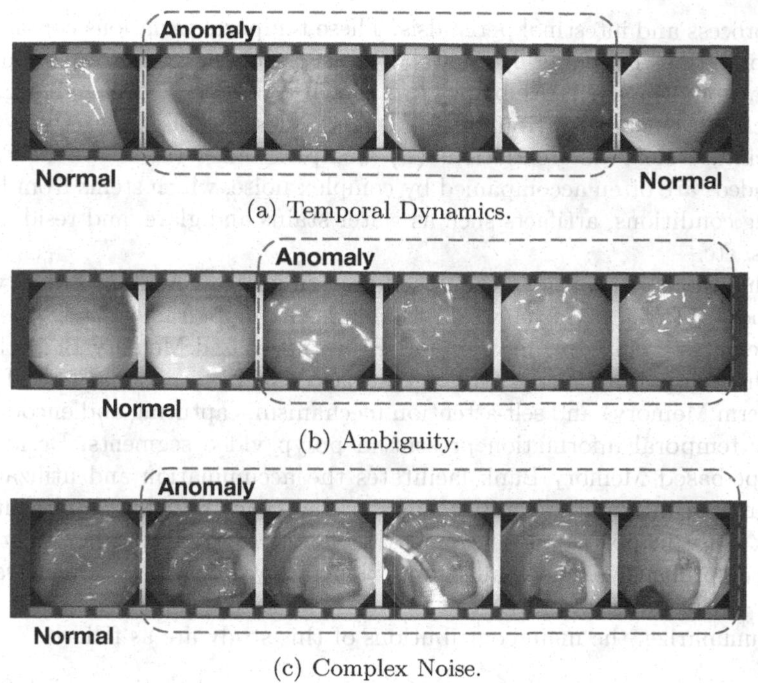

(a) Temporal Dynamics.

(b) Ambiguity.

(c) Complex Noise.

Fig. 1. The illustration of polyp frame detection task that aims to detect polpys within colonoscopy videos. Frames are selected from a continuous colonoscopy video segment, showcasing a series of typically ones in sequential order.

high miss rate and assist healthcare professionals in accurately detecting and diagnosing polyps within the gastrointestinal endoscopy process.

With the increasing availability of data sharing and the low cost of video-level annotations, Tian *et al.* [22] formulates polyp frame within colonoscopy video detection as a weakly-supervised video anomaly detection task. As shown in Fig. 1, given a series of colonoscopy videos, the objective is to learn a model that accurately localizes polyp frames within a video using only video-level annotations. However, the direct usage of weakly-supervised video anomaly detection methods [5,20,21,23,26,28] to this task has yielded sub-optimal results due to the significant differences between colonoscopy and real-world video. Therefore, there is a pressing need of developing more robust methods for improving the performance of weakly-supervised polyp frame detection. In the latest research, Tian *et al.* [22] combines a convolutional transformer-based multiple instance learning method with contrastive learning and results in 84.55% AP. Nevertheless, it does not fully consider the unique characteristics of colonoscopy videos.

Note that polyp videos focuses exclusively on the internal conditions of the colon during endoscopic examinations so that they exhibit unique characteristics that set them apart from real-world videos: **(1)** Temporal Dynamics (as shown in Fig. 1a): Polyps present morphological and positional changes during the examination process and intestinal peristalsis. These temporal variations contain valuable information, including dynamic patterns associated with both normal and abnormal conditions; **(2)** Ambiguity (as shown in Fig. 1b): In certain segments, polyps closely resemble surrounding normal tissues, lacking discernible features that facilitate clear differentiation; **(3)** Complex Noise (as shown in Fig. 1c): Polyp videos are often accompanied by complex noise, which stems from limited recording conditions, artifacts such as water stains and glare, and residual food particles, etc.

In this work, to consider the specific attributes inherent in polyp videos, we propose TPNet, a novel weakly-supervised polyp frame detection model that integrates **T**emporal Encoder and **P**rototype-based Memory Bank. Specifically, the Temporal Encoder is designed based on Bi-directional LSTM (Long Short-Term Memory) and self-attention mechanism, capturing and encoding the dynamic temporal information present in polyp video segments. Besides, the Prototype-based Memory Bank facilitates the accumulation and utilization of discriminative information, mitigating the challenges posed by ambiguity and complex noise in polyp video segments. Finally, we conduct experiments on a recent, comprehensive, and challenging colonoscopy datasets to demonstrate the superior performance of our TPNet.

To summarize, the main contributions of this study are as follows:

- We propose a novel weakly-supervised polyp frame detection model, TPNet, that combines Temporal Encoder and Protorype-based Memory Bank to effectively addressing the challenges associated with colonoscopy videos, including temporal dynamics, ambiguity, and complex noise.
- We introduce a Temporal Encoder based on Bi-directional LSTM and self-attention mechanism to capture the temporal variations and patterns within polyp video segments, enables a comprehensive understanding of the dynamic changes in polyp appearance and position over time.
- We design a Prototype-based Memory Bank that stores and updates information from polyp video segments, which facilitates better discrimination between normal and abnormal snippets and enhances sensitivity and robustness to intricate features.
- Extensive experiments demonstrate the remarkable capability of our model, outperforming state-of-the-art models with a significant improvement of 6.19% in Average Precision on one of the largest and most challenging colonoscopy video datasets.

2 Method

2.1 Overall Architecture

As illustrated in Fig. 2, given the input video segment $v \in \mathcal{R}^{T\times H\times W}$, where T is the temporal length that means the number of segments, and W and H

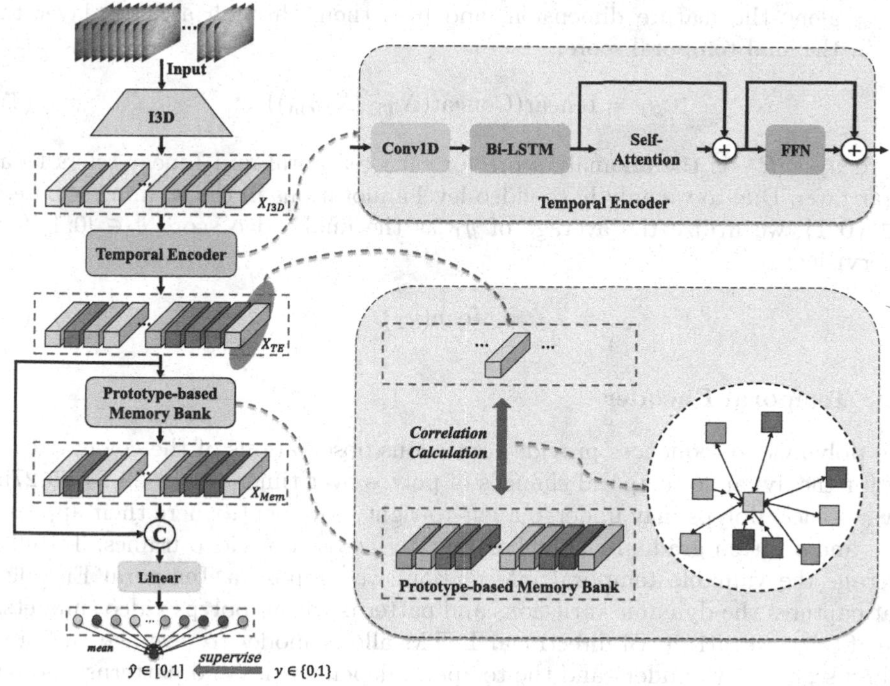

Fig. 2. The pipeline of our proposed TPNet.

denotes the spatial resolution of each segments. Similar to numerous weakly-supervised video anomaly detection methods [5,20–23,26,28], we extract video features using a pre-trained I3D model [4]:

$$X_{\text{I3D}} = \text{I3D}(v), \tag{1}$$

where $X_{\text{I3D}} \in \mathcal{R}^{T \times D_{I3D}}$, where D_{I3D} is the dimensional length. Then, the Temporal Encoder module is employed to model the dynamic relationships and interactions among X_{I3D}:

$$X_{\text{TE}} = \text{Temporal Encoder}(X_{\text{I3D}}), \tag{2}$$

where $X_{\text{TE}} \in \mathcal{R}^{T \times D}$ represents the temporal enhanced features that capture the temporal dynamics and dependencies within the video segments.

Next, X_{TE} is passed through the Prototype-based Memory Bank, which enables the temporal-enhanced features to interact with each memory block and obtain the memory-enhanced features:

$$X_{\text{Mem}} = \text{Memory}(X_{\text{TE}}), \tag{3}$$

where Memory denotes the Prototype-based Memory Bank and $X_{\text{Mem}} \in \mathcal{R}^{T \times D}$ represents the prototype-augmented features. Furthermore, we concat X_{TE} with

X_{Mem} along the feature dimension, and pass them through a linear layer to obtain the final temporal score:

$$y_{\text{T}} = \text{Linear}(\text{Concat}(X_{\text{TE}}, X_{\text{Mem}})), \tag{4}$$

where $y_{\text{T}} \in \mathcal{R}^{T}$ is the anomaly score for each temporal and Linear refers to a linear layer. Due to we only have video-level annotations in the training set, *i.e.*, $y \in \{0, 1\}$, we utilize the average of y_{T} as the final video score $\hat{y} \in [0, 1]$ for supervising:

$$\hat{y} = \text{Mean}(y_{\text{T}}). \tag{5}$$

2.2 Temporal Encoder

The polyp video sequences provide continuous observations of the colon, allowing for the dynamic temporal changes of polyps over time [8,9,18,22,24,25,27]. For instance, polyps may undergo a left-to-right movement, where their appearance and position gradually shift horizontally across the video frames. To fully leverage the valuable temporal information, we propose a Temporal Encoder that captures the dynamic variations and patterns within polyps video snippets. Specifically, we utilize Bi-directional LSTM allows model to leverage the preceding segments to understand the temporal dependencies and patterns leading up to the current segment, and also incorporate information from succeeding segments to anticipate future temporal dynamics. In addition, we use the self-attention mechanism to dynamically attend to the segments that carry more informative and discriminative temporal patterns.

Given $X_{I3D} \in \mathcal{R}^{T \times D_{I3D}}$ indicated features extracted from pre-trained I3D model, firstly, we employ a Bi-directional LSTM network to facilitate temporal dynamic modeling, the process is defined as:

$$X_{\text{LSTM}} = \text{Bi-LSTM}(\text{Conv1D}(X_{\text{I3D}})), \tag{6}$$

where Conv1D indicates a 1D convolutional layer to extract local features and capture spatial patterns within each video segments, and Bi-LSTM represents a Bi-directional LSTM layer. Furthermore, we apply linear transformations to X_{LSTM} to derive query (Q), key (K), and value (V) matrices, followed by performing global modeling using self-attention:

$$X_{\text{att}} = \text{Softmax}\left(\frac{QK^{T}}{\sqrt{d}}\right) V + X_{\text{LSTM}}, \tag{7}$$

where d represents the dimension of the query and key vectors, + denotes residual operation. The attention weights capture the importance of each element in the sequence with respect to other elements. Finally, we employ a feed-forward neural network (FFN) to further process the attended output X_{att}:

$$X_{\text{TE}} = \text{FFN}(X_{\text{att}}) + X_{\text{att}}, \tag{8}$$

where $X_{\text{TE}} \in \mathcal{R}^{T \times D}$ and FFN represents a linear layer followed by the GELU activation function.

By incorporating the Temporal Encoder into our model, we enable the consideration of information from both past and future segments in polyp video sequences. This allows for a more comprehensive understanding of the temporal dynamics and patterns present in polyp videos.

2.3 Prototype-Based Memory Bank

Memory banks [6,10,13,17,19] have gained extensive usage in the domain of unsupervised video anomaly detection, owing to their ability to store and retrieve crucial information, enabling the learning of latent patterns and regularities from unlabeled data. However, the complex data distribution in weakly-supervised scenarios poses challenges for memory banks to effectively learn and store key information, resulting in limited adoption in weakly-supervised video anomaly detection fields. Nevertheless, in the case of polyp videos with reduced data diversity compared to real-world data, we can efficiently leverage memory banks to store and retrieve these discriminatory features so that enhances the performance.

Prototype-based Memory Bank consists of multiple memory blocks, with each memory block storing a prototype that represents features of video segments. Formally, we define Prototype-based Memory Bank as a matrix $M \in \mathcal{R}^{N \times D}$, where N represents the number of memory blocks and D denotes the dimension of each prototype. We use a uniform distribution to randomly generate initial values for each prototype. Given $X_{\text{TE}} \in \mathcal{R}^{T \times D}$, we first calculate the correlation of X_{TE} to M:

$$W = \text{Sigmoid}(X_{\text{TE}} \cdot M^T), \tag{9}$$

where $\cdot$ denotes matrix multiplication and Sigmoid refers to the Sigmoid function. $W \in \mathcal{R}^{T \times N}$ is used to measure the correlation between input features and the memory bank. According to this relevance weight W, we retrieve the relevant prototypes from the memory bank to obtain the augmented features that incorporate previously stored information:

$$X_{\text{Mem}} = W \cdot M, \tag{10}$$

where $X_{\text{Mem}} \in \mathcal{R}^{T \times D}$ represents the memory-augmented features. This operation allows the model to selectively retrieve and integrate relevant information from the memory bank into the input features.

The Prototype-based Memory Bank enables the storage and retrieval of prototypes that capture various video segment features. By preserving these prototypes, the model not only learns additional fine-grained features but also captures long-term dependency patterns within polyp video segments, thereby improving sensitivity, accuracy, and robustness in polyp detection.

2.4 Loss Function

Given the final video score $\hat{y} \in [0, 1]$, we employ a binary cross-entropy loss function to supervise the training:

$$Loss = -\sum_{i=1}^{B}(y_i \log(\hat{y}_i) + (1 - y_i)\log(1 - \hat{y}_i)), \tag{11}$$

where $y \in \{0, 1\}$ represents label and B is the batch size.

3 Experiment

3.1 Experiment Settings

Training and Testing Settings: We use the dataset that combines Hyper-Kvasir [3] dataset and LDPolypVideo [14] dataset, which totally contains over one million frames and has diverse polyps with various sizes and shapes. Furthermore, similar with [22], the training set comprises 61 normal videos without polyps and 102 abnormal videos with polyps, while the testing set consists of 30 normal videos and 60 abnormal videos. Notably, the training set is annotated at the video-level, whereas the testing set is annotated at the frame-level.
Evaluation Metrics: Following [22], we adopt the frame-level area under the ROC curve (AUC) and average precision (AP) as the evaluation metrics, and higher AUC and AP values indicate better performance.
Implement Details: We adopted the methodology presented in [22], which divided each video into 32 video snippets and employed the pre-trained I3D model [4] to extract 2048D snippet features from the $mixed_{5c}$ layer. The implementation was conducted on the PyTorch platform and trained using a NVIDIA 3090 GPU. For the training process, we utilized a batch size of 16 and trained the model for 200 epochs. The optimization algorithm employed was Adam, with a learning rate of 1e-4. Furthermore, we performed end-to-end training without applying any data augmentation techniques. The number of memory blocks within Prototype-based Memory Bank is set to 80 and the dimension D in X_{TE} and X_{Mem} is set to 1024.

3.2 Comparison with Other Methods

Quantitive Evaluation: We selected the cutting-edge weakly-supervised video anomaly detection models [5,20,21,23,26,28] from recent years and the latest CTMIL [22] that specifically designed for polyp frame detection. The results are all reproduced from their respective open-source codes at the same setting and used the same feature extracted from pre-trained I3D model for a fair comparison.

As depicted in Table 1, the recent cutting-edge weakly-supervised video anomaly detection methods have showcased a wide range of outcomes when

Table 1. Comparison with other methods. Results recomputed from released source codes using the same features from I3D in the same setting. Results of CTMIL and ours are the average over 30 runs.

Method	Year	Feature	AUC(%)	AP(%)
DeepMIL [20]	2018	I3D(RGB)	89.41	68.53
GCN-Ano [28]	2019	I3D(RGB)	92.13	75.39
CLAWS [26]	2020	I3D(RGB)	95.62	80.42
AR-Net [23]	2020	I3D(RGB)	88.59	71.58
MIST [5]	2021	I3D(RGB)	94.53	72.85
RTFM [21]	2021	I3D(RGB)	96.30	77.96
CTMIL [22]	2022	I3D(RGB)	97.18	84.55
Ours	**2023**	**I3D(RGB)**	**98.97**	**90.74**

applied to polyp frame detection, owing to their tailored design for various real-world data distributions. Furthermore, beyond these methods, CTMIL stands out by achieving an impressive AP of 84.55%. This remarkable performance underscores the potential of integrating contrastive learning to enhance the discrimination between polyp and normal tissues. However, our proposed TPNet outperforms CTMIL with a significant advancement, achieving an outstanding AP of 90.74%. The substantial improvement serves as a testament to the heightened accuracy and robustness of our model. Moreover, this significant margin further reinforces the superiority of our methodology, which incorporates the guidance of temporal dynamics and leverages discriminative information for optimal polyp frame detection.

Qualitative Evaluation: To provide an intuitive demonstration of the effectiveness of our proposed TPNet, we conducted a series of qualitative experiments. In Fig. 3, we have selected several specific segments that exhibit characteristics commonly observed in polyp videos.

In Fig. 3a, we demonstrate the temporal changes of a polyp over several segments, highlighting the progressive movement that occur. The result showcases our proposed TPNet is capable of effectively tracking the temporal dynamics of polyps within videos and utilizing this contextual information about the temporal sequences to accurately identify polyps. In addition, we present snippets where the polyp appears visually similar to the surrounding normal tissues in Fig. 3b, which demonstrates the effectiveness of our approach in distinguishing subtle differences. Moreover, as illustrated in Fig. 3c and Fig. 3d, TPNet demostrates the capability to identify polyps in the presence of various sources of noise, which further confirms that our proposed method can effectively utilize the stored prototypes for a robust polyp detection.

Through these qualitative experiments, we highlighted the strengths of our proposed TPNet and demonstrated its efficacy in addressing various challenging scenarios encountered in polyp videos. The results showcase the model's ability to

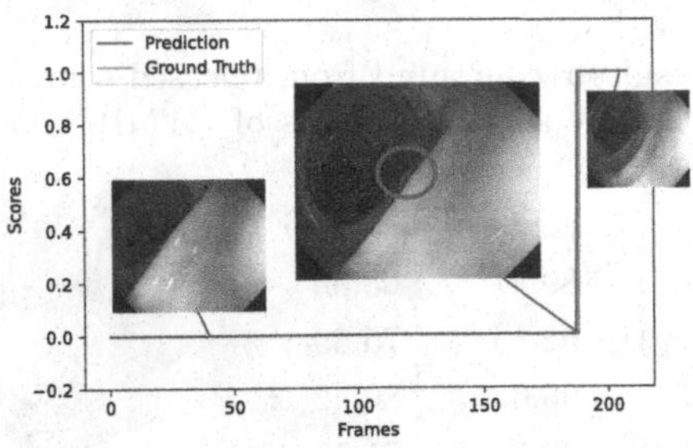

(a) The polyp evolution with carmera processing.

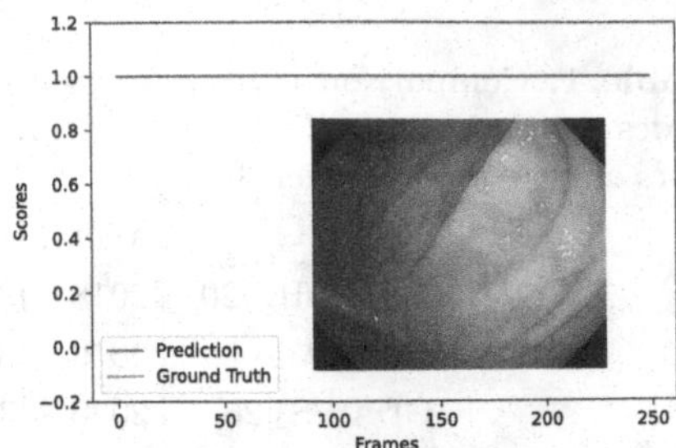

(b) Polyps that are hardly distinguishable from surrounding skin.

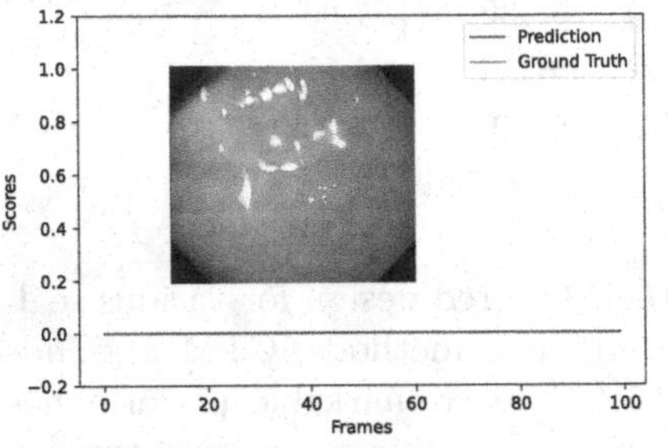

(c) Interference with water and light.

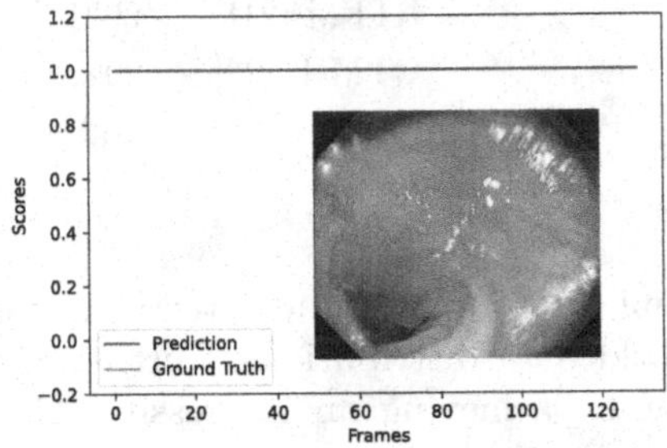

(d) Interference with water and light.

Fig. 3. Qualitative evaluation involves several complex scenarios.

accurately track the temporal pattern of polyps, effectively differentiate polyps from normal tissues, and handle noise interference. These promising findings could provide valuable insights for future research and development in this field.

3.3 Effectiveness of Each Component

To analyze the role of the Temporal Encoder (TE) and Prototype-based Memory Bank (PMB) within our proposed TPNet, we conduct a comprehensive ablation study. As indicated in Table 2, we can observe that: **(1)** The TE-only model

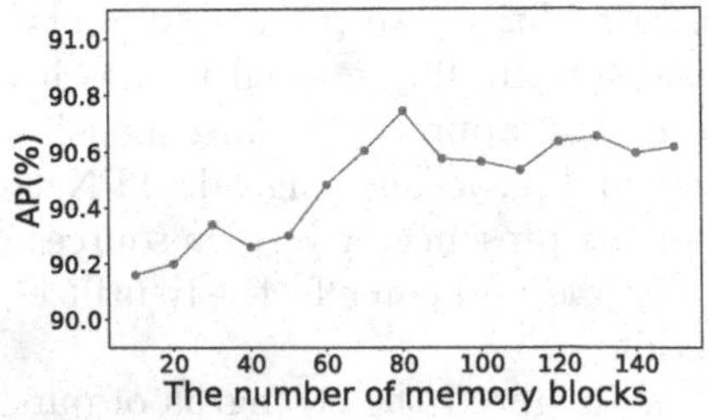

Fig. 4. Parametric experiments of the number of memory blocks.

Table 2. Ablation studies in our proposed TPNet. Results are the average over 30 runs.

TE	PMB	AUC(%)	AP(%)
–	–	94.41	75.53
✓	–	98.13	88.56
–	✓	95.42	77.74
✓	✓	98.97	90.74

(*Row 2*) achieves impressive performance, with an AUC of 98.13% and an AP of 88.56%, surpassing the state-of-the-art methods significantly. This result confirms the temporal dynamic guidance is essential for polyp frame detection; **(2)** When combined with the PMB, the performance of the vanilla model (*Row 1*) or the TE-only model (*Row 3*) both improves. This finding underscores the importance of the prototypes stored and retrieved by the PMB. In this manner, our model captures the distinctive characteristics of polyps, enhancing its ability to discriminate between polyp features and normal features, which further boosts the precision and overall performance; **(3)** The best results are achieved by the model (*Row 4*) that combines TE and PMB, which highlights the complementary nature and synergistic effects of these two components in TPNet. By leveraging both temporal dynamic guidance and prototype information, the model is able to better capture the temporal patterns and utilize typically features for more accurate and robust polyp frame detection.

3.4 Influence of the Number of Memory Blocks

We varied the number of memory blocks from 10 to 150 and measured the average precision (AP) as the performance metric. The results are summarized in Fig. 4, we can summarize these conclusions: **(1)** The more memory blocks there are, the greater the improvement in the model's performance, which highlights the effectiveness of utilizing a enough number of memory blocks in capturing and storing features; **(2)** The model achieves its peak performance, *i.e.*, 90.74% AP, when the number of memory blocks is set to 80, which provides robust guidance for our parameter settings; **(3)** As the number of memory blocks continues to increase, the performance starts to decline and eventually reaches a stable state. This observation suggests that setting an excessive number of memory blocks can introduce redundancy to the model, resulting in diminishing returns and potential computational inefficiency.

4 Conclusion

In this work, we present TPNet, a novel weakly-supervised polyp frame detection model that effectively addresses the challenges associated with colonoscopy videos. Specifically, we design a Temporal Encoder module that captures global relationships and interactions within polyp videos, which significantly captures the dynamic temporal changes and patterns within polyp video segments. In addition, we propose a Prototype-based Memory Bank, which enables the storage and retrieval of significant discriminative information, enhancing the model's sensitivity and robutsness to intricate polyps. Extensive experiments demonstrates that our proposed TPNet achieved the state-of-the-art performance in weakly-supervised polyp frame detection.

Acknowledgement. This work is supported by the National Natural Science Foundation of China (No. 62276221), the Natural Science Foundation of Fujian Province of China (No. 2022J01002), and the Science and Technology Plan Project of Xiamen (No. 3502Z20221025).

References

1. Ahn, S.B., Han, D.S., Bae, J.H., Byun, T.J., Kim, J.P., Eun, C.S.: The miss rate for colorectal adenoma determined by quality-adjusted, back-to-back colonoscopies. Gut Liver **6**(1), 64 (2012)
2. Ali, S., Dmitrieva, M., Ghatwary, N., Bano, S., Polat, G., Temizel, A., Krenzer, A., Hekalo, A., Guo, Y.B., Matuszewski, B., et al.: Deep learning for detection and segmentation of artefact and disease instances in gastrointestinal endoscopy. Med. Image Anal. **70**, 102002 (2021)
3. Borgli, H., et al.: Hyperkvasir, a comprehensive multi-class image and video dataset for gastrointestinal endoscopy. Sci. Data **7**(1), 283 (2020)
4. Carreira, J., Zisserman, A.: Quo vadis, action recognition? a new model and the kinetics dataset. In: CVPR, pp. 6299–6308 (2017)
5. Feng, J.C., Hong, F.T., Zheng, W.S.: Mist: multiple instance self-training framework for video anomaly detection. In: Proceedings of the IEEE/CVF Conference on Computer Vision and Pattern Recognition, pp. 14009–14018 (2021)
6. Gong, D., et al.: Memorizing normality to detect anomaly: memory-augmented deep autoencoder for unsupervised anomaly detection. In: Proceedings of the IEEE/CVF International Conference on Computer Vision, pp. 1705–1714 (2019)
7. Itoh, H., Misawa, M., Mori, Y., Kudo, S.E., Oda, M., Mori, K.: Positive-gradient-weighted object activation mapping: visual explanation of object detector towards precise colorectal-polyp localisation. Int. J. Comput. Assist. Radiol. Surg. **17**(11), 2051–2063 (2022)
8. Ji, G.-P., et al.: Progressively normalized self-attention network for video polyp segmentation. In: de Bruijne, M., et al. (eds.) MICCAI 2021. LNCS, vol. 12901, pp. 142–152. Springer, Cham (2021). https://doi.org/10.1007/978-3-030-87193-2_14
9. Ji, G.P., et al.: Video polyp segmentation: a deep learning perspective. Mach. Intell. Res. **19**, 1–19 (2022). https://doi.org/10.1007/s11633-022-1371-y
10. Kim, Y., Kim, M., Kim, G.: Memorization precedes generation: learning unsupervised GANs with memory networks. arXiv preprint arXiv:1803.01500 (2018)
11. Ladabaum, U., Dominitz, J.A., Kahi, C., Schoen, R.E.: Strategies for colorectal cancer screening. Gastroenterology **158**(2), 418–432 (2020)
12. Leufkens, A., Van Oijen, M., Vleggaar, F., Siersema, P.: Factors influencing the miss rate of polyps in a back-to-back colonoscopy study. Endoscopy **44**(05), 470–475 (2012)
13. Liu, Z., Nie, Y., Long, C., Zhang, Q., Li, G.: A hybrid video anomaly detection framework via memory-augmented flow reconstruction and flow-guided frame prediction. In: Proceedings of the IEEE/CVF International Conference on Computer Vision, pp. 13588–13597 (2021)
14. Ma, Y., Chen, X., Cheng, K., Li, Y., Sun, B.: LDPolypVideo benchmark: a large-scale colonoscopy video dataset of diverse polyps. In: de Bruijne, M., et al. (eds.) MICCAI 2021. LNCS, vol. 12905, pp. 387–396. Springer, Cham (2021). https://doi.org/10.1007/978-3-030-87240-3_37
15. Mathur, P., et al.: Cancer statistics, 2020: report from national cancer registry programme, India. JCO Glob. Oncol. **6**, 1063–1075 (2020)
16. Misawa, M., et al.: Development of a computer-aided detection system for colonoscopy and a publicly accessible large colonoscopy video database (with video). Gastrointest. Endosc. **93**(4), 960–967 (2021)
17. Park, H., Noh, J., Ham, B.: Learning memory-guided normality for anomaly detection. In: Proceedings of the IEEE/CVF Conference on Computer Vision and Pattern Recognition, pp. 14372–14381 (2020)

18. Podlasek, J., Heesch, M., Podlasek, R., Kilisiński, W., Filip, R.: Real-time deep learning-based colorectal polyp localization on clinical video footage achievable with a wide array of hardware configurations. Endosc. Int. Open **9**(05), E741–E748 (2021)
19. Santoro, A., Bartunov, S., Botvinick, M., Wierstra, D., Lillicrap, T.: Meta-learning with memory-augmented neural networks. In: International Conference on Machine Learning, pp. 1842–1850. PMLR (2016)
20. Sultani, W., Chen, C., Shah, M.: Real-world anomaly detection in surveillance videos. In: CVPR, pp. 6479–6488 (2018)
21. Tian, Y., Pang, G., Chen, Y., Singh, R., Verjans, J.W., Carneiro, G.: Weakly-supervised video anomaly detection with robust temporal feature magnitude learning. In: Proceedings of the IEEE/CVF International Conference on Computer Vision, pp. 4975–4986 (2021)
22. Tian, Y., et al.: Contrastive transformer-based multiple instance learning for weakly supervised polyp frame detection. In: Wang, L., Dou, Q., Fletcher, P.T., Speidel, S., Li, S. (eds.) Medical Image Computing and Computer Assisted Intervention - MICCAI 2022. MICCAI 2022. LNCS, vol. 13433, pp. 88–98. Springer, Cham (2022). https://doi.org/10.1007/978-3-031-16437-8_9
23. Wan, B., Fang, Y., Xia, X., Mei, J.: Weakly supervised video anomaly detection via center-guided discriminative learning. In: 2020 IEEE International Conference on Multimedia and Expo (ICME), pp. 1–6. IEEE (2020)
24. Wu, L., Hu, Z., Ji, Y., Luo, P., Zhang, S.: Multi-frame collaboration for effective endoscopic video polyp detection via spatial-temporal feature transformation. In: de Bruijne, M., et al. (eds.) MICCAI 2021. LNCS, vol. 12905, pp. 302–312. Springer, Cham (2021). https://doi.org/10.1007/978-3-030-87240-3_29
25. Xu, J., Zhao, R., Yu, Y., Zhang, Q., Bian, X., Wang, J., Ge, Z., Qian, D.: Real-time automatic polyp detection in colonoscopy using feature enhancement module and spatiotemporal similarity correlation unit. Biomed. Signal Process. Control **66**, 102503 (2021)
26. Zaheer, M.Z., Mahmood, A., Astrid, M., Lee, S.-I.: CLAWS: clustering assisted weakly supervised learning with normalcy suppression for anomalous event detection. In: Vedaldi, A., Bischof, H., Brox, T., Frahm, J.-M. (eds.) ECCV 2020, Part XXII. LNCS, vol. 12367, pp. 358–376. Springer, Cham (2020). https://doi.org/10.1007/978-3-030-58542-6_22
27. Zhao, X., et al.: Semi-supervised spatial temporal attention network for video polyp segmentation. In: Wang, L., Dou, Q., Fletcher, P.T., Speidel, S., Li, S. (eds.) Medical Image Computing and Computer Assisted Intervention - MICCAI 2022. MICCAI 2022. LNCS, vol. 13434, pp. 456–466 Springer, Cham (2022). https://doi.org/10.1007/978-3-031-16440-8_44
28. Zhong, J.X., Li, N., Kong, W., Liu, S., Li, T.H., Li, G.: Graph convolutional label noise cleaner: Train a plug-and-play action classifier for anomaly detection. In: Proceedings of the IEEE/CVF Conference on Computer Vision and Pattern Recognition, pp. 1237–1246 (2019)

Learning Frequency-Based Disentanglement and Filtering for Generalizable Person Re-identification

Pengpeng Song[1] and Jinjia Peng[1,2(✉)]

[1] School of Cyber Security and Computer, Hebei University, Baoding, Hebei, China
songpengpeng@stumail.hbu.edu.cn, pengjinjia@hbu.edu.cn
[2] Hebei Machine Vision Engineering Research Center, Baoding, China

Abstract. Domain Generalization (DG) in Person Re-identification (ReID) tackles the task of testing in unseen domains without using target domain data during training. Existing DG ReID methods achieve impressive performance with unified ensemble models or multi-expert hybrid networks. However, as the number of source domains increases, complex relationships between training samples result in domain-invariant characteristics with spurious correlations, impacting further generalization. To address this, we propose a **B**ilateral **F**requency-**A**ware **N**etwork(**BFAN**) that leverages spectral feature correlation learning for discriminative hybrid features. BFAN includes a Bilateral Frequency Component-guided Attention (BFCA) module to capture semantic information from diverse frequency features and fuse it with spatial features. Additionally, a Fourier Noise Masquerade Filtering (FNMF) module is introduced to suppress non-generalization-supporting components in the frequency domain. Extensive experiments on various datasets demonstrate our method's notably competitive performance.

Keywords: Domain Generalization · Person Re-identification · Frequency Domain Learning

1 Introduction

Person re-identification (ReID) [23,25] aims to match individuals across different camera views or frames, due to the maturation of clustering method [9,17,18], Unsupervised Person ReID [15] has achieved remarkable development but applying trained models to unseen domains often leads to performance degradation due to domain gaps [20]. With a substantial amount of unlabeled pedestrian data available, it is crucial to explore ReID models with robust generalization capabilities. Recent research has focused on domain adaptive [25] (DA) and domain

Supported by Central Government Guides Local Science and Technology Development Fund Projects (236Z0301G); Hebei Natural Science Foundation (F2022201009); Science and Technology Project of Hebei Education Department (QN2023186).

© The Author(s), under exclusive license to Springer Nature Singapore Pte Ltd. 2024
Q. Liu et al. (Eds.): PRCV 2023, LNCS 14436, pp. 482–494, 2024.
https://doi.org/10.1007/978-981-99-8555-5_38

generalization [23,26] (DG) ReID. While DA ReID adapts the model using target domain samples, DG ReID tackles the more challenging scenario of unseen target domains without fine-tuning. This paper specifically addresses the practical DG person ReID problem, which is important for real-world applications.

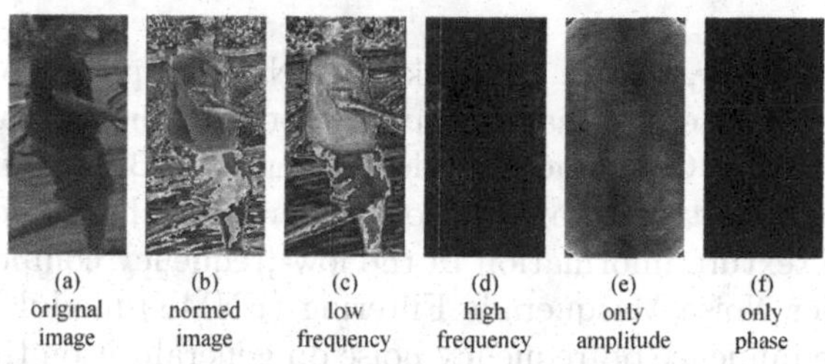

Fig. 1. Re-construction of frequency components of person samples: (a) original image; (b) normlized image; (c-d) reconstruction of high and low frequency components of the image after separation; (e-f) reconstructed image with amplitude and phase component information only by setting another component to a constant.

Existing DG ReID methods can be divided into two categories. One category [4,26] focuses on ensemble models that learn a shared feature space across different domains, while the other category [5,23] emphasizes correlations between domains and uses multi-expert networks to enhance generalization. Nonetheless, both of these approaches may encounter two potential issues: (1) The distribution of source domain data in the spatial domain follows the assumption of being independent and identically distributed (i.i.d.), but the spurious correlations caused by the similarity of instance-level features detrimentally impact model generalization. (2) When the training and test samples come from different distributions, the generalization behavior of CNNs is easily disturbed by frequency domain noise [2], which is also reflected in DG ReID models.

Recent theoretical study [24] has shown that the sample spectrum influences the model's generalization behavior. Models exhibit preferences for component information at different frequencies during training. Feature learning in the frequency domain can effectively address these issues. As shown in Fig. 1, low-frequency components capture texture structure and image-specific energy distributions, while high-frequency information describes pixel variations between object edges and the background, showing consistency across domains. The implicit high-level semantics of the image amplitude and phase further aid the model in avoiding local optima [2].

Therefore, inspired by previous studies, this paper proposes the **BFAN** network for multi-source DG ReID in the image frequency domain. It aims to enable the model to learn domain-invariant features without spatial domain bias and leverage complementary information from different frequency components, improving generalization to unseen domains. The BFAN applies fast

Fourier transform to obtain frequency representations, separates them into high and low-frequency components using Bilateral Frequency Component-guided Attention (BFCA), captures long-range dependencies with non-local attention, and employs Fourier Noise Masquerade Filtering (FNMF) to filter out non-generalization-supporting frequency information. Our contributions are summarized as follows:

- A Bilateral Frequency-Aware Network (BFAN) is proposed to learn domain-invariant features based on image frequency spectrum in the embedding space.
- Bilateral Frequency Component-guided Attention (BFCA) module is constructed to perceive the semantic information in the high-frequency components and the texture information in the low-frequency components.
- A novel Fourier Noise Masquerade Filtering (FNMF) module is designed to suppresses the influence of frequency noise on generalization through filtering operations in the spatial dimension.

2 Methods

2.1 Overview

In this study, the DG ReID problem is addressed by incorporating complementary information from the spatial and frequency domains. The structured training framework, depicted in Fig. 2, aims to improve the model's generalization in unseen domains. Frequency domain information is used as a supplement to spatial domain information for extracting robust domain-invariant representations. The backbone network receives input from multi-source domains with non-overlapping person identity tags. The BFCA module is introduced at the middle layer to acquire complementary frequency domain information, separating features into high-frequency and low-frequency components. To capture original image semantics and texture information, the FNMF module is positioned before the GAP at the end of the trunk. It effectively filters out frequency domain noise, enhancing the model's generalizability and overall performance.

In typical DG ReID methods [4,26], N_s source domains $D_s = D_s^1, D_s^2, ..., D_s^{N_s}$ are provided during training. Each D_s^k consists of N_k samples x_i^k with corresponding labels y_i^k in the k-th domain. The objective is to train models on D_s that perform well on unseen domain D_T without further updates using D_T data. In DG ReID, the source domains are assumed to follow independent identical distributions, with disjoint label spaces. The aim is to train a generalizable model using the source data. During testing, the model is directly evaluated on the unseen domain D_T.

The BFCA module applies Fourier transform to the embedding space features, separating them into different frequency components using a controllable-sized Euclidean distance mask. Non-local convolutional operations are then applied to the high/low frequency components in the frequency domain. This injects original semantic and texture information into the spatial domain features, aiding convergence to the local optimum. On the other hand, the FNMF

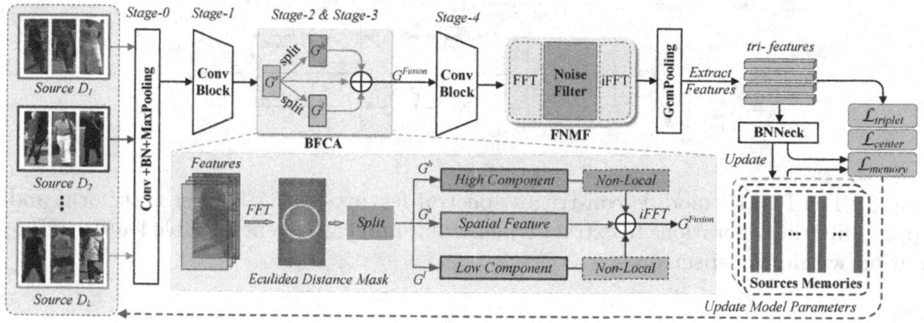

Fig. 2. The overall architecture of Bilateral Frequency-aware Network (BFAN) which unites multiple source domains for federated training and does not require the target domain to participate in the training process.

module performs frequency domain filtering deeper in the network using pooling operations and learns non-shared instance-level masks. It demonstrates the feasibility of frequency domain noise filtering without modifying the original data sources. Detailed descriptions of these modules will be provided in the following subsections.

2.2 Bilateral Frequency Component-Guided Attention

Most DG person ReID methods focus on capturing the semantic information embedded in spatial domain features while ignoring the potential complementary information in the frequency domain, which helps the auxiliary model to improve feature recognition. Specifically, since the ultimate goal of the domain generalized person ReID task is to train a model that performs well even in the face of unknown domains, the ability of the model to learn domain-invariant features becomes particularly important, as previous methods inevitably retain bias to the source domain after training is completed, and it becomes particularly important to learn the debiased complementary information in the frequency domain components to bridge the domain bias. Inspired by the semantic preservation and non-intuitive generalization of the Fourier frequency components [16], the Bilateral Frequency Component-guided Attention (BFCA) is designed containing the Euclidean distance mask.

As shown in Fig. 2, assuming the i-th sample $x_i^k \in \mathbb{R}^{C\times H\times W}$ from the k-th source domain mini batch, each channel of the latent layer feature x_i^k is transformed to the frequency domain space using the Fast Fourier transform, and the fast Fourier transform equation is

$$\mathcal{F}\left(x_i^k\right)(c,u,v)=\sum_{c=0}^{C-1}\sum_{h=0}^{H-1}\sum_{w=0}^{W-1} x_i^k\left(c,h,w\right)e^{-j2\pi\left(\frac{h}{H}u+\frac{w}{W}v\right)} \tag{1}$$

which C, H and W are the number of channels, width and height of the image, respectively, $\mathcal{F}^{-1}\left(x_i^k\right)$ defines the inverse Fourier transformation and $\otimes$ indi-

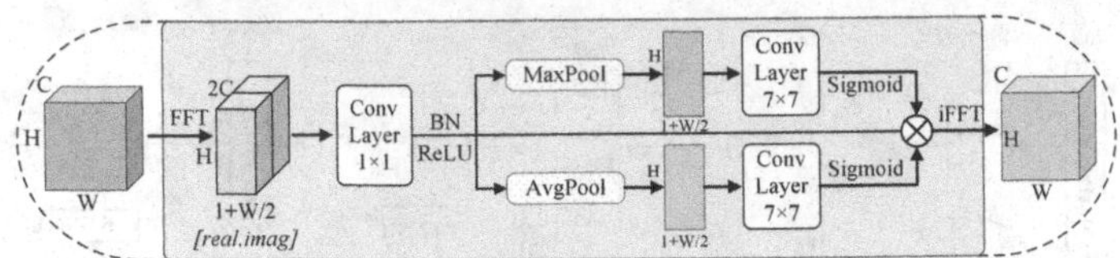

Fig. 3. The FNMF module constructs spectral features using Fourier transform and applies filtering operations to extract transferable and easily generalizable feature information within the spectrum.

cates the element-wise multiplication of two matrices. The transformed image frequency domain features are denoted as $G^{freq} = \mathcal{F}\left(x_i^k\right)$. The high frequency component and low frequency component can be obtained by decomposing the frequency domain features through the Euclidean distance mask with the following equation,

$$G^h, G^l = \mathcal{F}^{-1}\left(G^{freq} \otimes \hat{M}\left(r\right)\right) \tag{2}$$

Define $\hat{M}\left(\cdot\right)$ as the Euclidean distance mask construction threshold function, and construct the frequency component mask through the hyperparameter radius r to obtain the high frequency component $G^h \in \mathbb{R}^{C\times H\times W}$ and low frequency component $G^l \in \mathbb{R}^{C\times H\times W}$ of the image, the specific formula and construction process are as follows,

$$\hat{M}\left(i,j\right) = \begin{cases} 1, d\left(\left(i,j\right),\left(u_c,v_c\right)\right) \leq r \\ 0, otherwise \end{cases} \tag{3}$$

where $d\left(\cdot\right)$ refers to the Euclidean distance, use $\hat{\mathcal{M}}\left(i,j\right)$ to index the position of $\left(i,j\right)$ in the frequency component feature map, and use $\left(u_c,v_c\right)$ to denote the feature map center-of-mass location. Then, the weighted sum of all positions in the frequency domain is obtained from G^h and G^l using powerful non-local attention [19] to obtain a long-range representation G^{Fusion} covering the whole globe, expressed by the equation,

$$G^{Fusion} = \text{iFFT}\left(\varphi\left(G^h\right) + \varphi\left(G^l\right)\right) + G^s \tag{4}$$

where as φ denotes Non Local operation, iFFT denotes the Fourier Inversion (Fig. 3).

2.3 Fourier Noise Masquerade Filtering

Previous studies [16,24] have indicated that frequency characteristics affect a significant role in balancing robustness and accuracy. Additionally, ReID models exhibit varying preferences for specific frequency components during the acquisition of intermediate features. In this paper, an Fourier Noise Masquerade Filtering (FNMF) module was developed to improve the transferability of frequency components for DG ReID which through a straightforward filtering operation. Simultaneously, it suppresses the components that hinder cross-domain generalization. In contrast to the previous frequency-based approach implemented in the pixel feature space, the proposed filtering operation was applied in the latent feature space. Given the representation of potential frequency domain features as $X_{freq} \in \mathbb{R}^{H\times(\lfloor\frac{W}{2}\rfloor+1)\times 2C}$. The proposed frequency domain noise filtering mechanism is expressed as shown in the following equation,

$$X'_{freq} = X_{freq} \otimes M_s\left(X_{freq}\right) \tag{5}$$

where $\otimes$ denotes element-wise multiplication. $M_s(\cdot)$ refers to the attention module to filter frequency noise and learn the spatial mask with a resolution of $H \times \left(\lfloor\frac{W}{2}\rfloor + 1\right)$. The mask is applied to filter out frequency domain components along the channel dimension that do not contribute to generalization. The resulting filtered frequency domain features are denoted as X'_{freq}. For $X_{freq} \in \mathbb{R}^{2C\times H\times(\lfloor\frac{W}{2}+1\rfloor)}$ which represents the amplitude and phase parts after the FFT, first utilize a 1 × 1 convolutional layer, followed by Batch Normalization (BN) and ReLU activation, to project X_{freq} into an embedding space for subsequent filtration. After embedding, aggregate the information of X_{freq} over channels using both average-pooling and max-pooling operations along the channel axis, generating two frequency descriptors denoted by X_{freq}^{avg} and X_{freq}^{max} , respectively. These two descriptors can be viewed as two compact representations of X_{freq} in which the information of each frequency component is compressed separately by the pooling operations while the spatial discriminability is still preserved. Then concatenate X_{freq}^{avg} with X_{freq}^{avg} and use a large-kernel 7 × 7 convolution layer followed by a sigmoid function to learn the spatial mask. Mathematically, this instantiation can be formulated as,

$$\begin{aligned} X'_{freq} = X_{freq} \otimes \sigma\left(Conv_{7\times 7}\left(AvgPool\left(X_{freq}\right)\right)\right) + \\ X_{freq} \otimes \sigma\left(Conv_{7\times 7}\left(MaxPool\left(X_{freq}\right)\right)\right) \end{aligned} \tag{6}$$

where σ denotes the sigmoid function. $AvgPool(\cdot)$ and $MaxPool(\cdot)$ denote the average and max pooling operations, respectively. $Conv_{7\times 7}(\cdot)$ is a convolution layer with the kernel size of 7. Albeit using a large-size kernel, the feature $AvgPool(X_{freq})$, $MaxPool(X_{freq})$ has only two channels through the information squeeze by pooling operations such that this step is still very computationally efficient in practice.

2.4 Loss Function

To generalize to the unseen target domain, this paper uses memory-based and contrastive learning recognition loss, which is non-parametric and suitable for DG ReID. In addition, similar to [5], triplet loss [7] $\mathcal{L}_{tri}$ and center loss [21] $\mathcal{L}_{cent}$ are also used for parameter updates when training generalization models.

As shown in Fig. 2, a separate memory dictionary is maintained for each source domain, and the identity loss is calculated using the learned features and

Table 1. Summary of all the datasets.

Datasets	#IDs	#Images	#Cameras
Market1501 (MA) [27]	1,501	32,217	6
CUHK02 (C2) [11]	1,816	7,264	10
CUHK03 (C3) [12]	1,467	14,096	2
CUHK-SYSU (CS) [22]	11,934	34,574	1
MSMT17 (MS) [20]	4,101	126,441	15
VIPeR [6]	634	1264	2
iLIDs [28]	300	4515	2
GRID [14]	1025	1275	8
PRID [8]	749	949	2

Table 2. Evaluation protocols.

	Training Sets	Testing Sets
Protocol-1	Full-(MA+C2+C3+CS)	PRID,GRID, VIPeR,iLIDs
Protocol-2	MA+CS+MS	C3
	MA+CS+C3	MS
	MS+C3+CS	MA
Protocol-3	Full-(MA+CS+MS)	C3
	Full-(MA+C3+CS)	MS
	Full-(CS+C3+MS)	MA

memory centroids based on the frequency domain, so that the model gradient is updated in a way that is more in line with the potential hyper-distribution. Specifically, for the source domain D_i^S with n person identities, there are n_i slots in the memory dictionary M_c^i, and each slot is initialized to a feature centroids corresponding to the person identities. Then, given the feature $F_{req}(x_i)$ extracted by BFAN forward propagation, calculate the similarity between $F_{req}(x_i)$ and each centroid in the memory. The contrast recognition loss aims to classify f as the corresponding pedestrian identity. The calculation formula is as follows,

$$\mathcal{L}_{\mathsf{M}} = -\log \frac{\exp\left(\mathsf{M}_i^T \cdot F_{req}\left(x_i\right)/\tau\right)}{\sum_{c=1}^{n_i} \exp\left(\mathsf{M}_c^T \cdot F_{req}\left(x_i\right)/\tau\right)} \tag{7}$$

where τ is the temperature factor that controls the scale of distribution.In each iteration training, the centroid of the corresponding person in the memory is updated using the features extracted by BFAN in the current minibatch. A centroid in the memory is updated through as follows,

$$\mathsf{M}_c \leftarrow m \cdot \mathsf{M}_c + \frac{(1-m)}{|\mathcal{B}_c|} \cdot \sum\nolimits_{x_i \in \mathcal{B}_c} F_{req}\left(x_i\right) \tag{8}$$

where $m \in [0, 1)$ is the momentum coeffcient, which is set as 0.2, $\mathcal{B}_c$ denotes the person samples belonging to the k-th identity and $|\mathcal{B}_c|$ denotes the number of peroson samples for the k-th identity in current mini-batch.

$$\mathcal{L}_{total} = \mathcal{L}_{\mathsf{M}} + \mathcal{L}_{tri} + \mathcal{L}_{cent} \tag{9}$$

3 Experiments

3.1 Datasets and Evaluation Protocals

Datasets. As shown in Table 1, to evaluate the effectiveness of our proposed method, this paper conducts experiments on 9 standard person re-ID datasets including Market1501 [27], MSMT17 [20], CUHK02 [11], CUHK03 [12], CUHK-SYSU [22], PRID [8], GRID [14], VIPeR [6], iLIDs [28]. The evaluation metrics employed in our work include Cumulative Matching Characteristics (CMC) and mean Average Precision (mAP). For the sake of simplicity, Market1501 is referred to as MA, MSMT17 as MS, CUHK02 as C2, CUHK03 as C3, and CUHKSYSU as CS.

Evaluation Protocols. Due to the removal of DukeMTMC-reID, new protocols (Table 2) have been introduced for DG ReID. In protocol-1, all source domain images are used for training and evaluation follows [23]. Protocol-2 selects one domain for testing and the others for training. Protocol-3 uses all source domain images for training. Since the CS person search dataset comprises only 1 camera, it is excluded from the testing process.

3.2 Implementation Details

The person images are resized to 256 × 128. The backbone of BFAN is ResNet-50, pretrained on ImageNet. The batch size is set to 64, including 16 identities with 4 images per identity. Similar to [26], the data augmentation includes color jitter and random erasing. The model is trained for 60 epochs with a warmup strategy applied in the first 10 epochs. The learning rate is initialized and divided by 10 at the 30th and 50th epochs, respectively. The optimizer used is Adam with a momentum of 0.9 and weight decay of 0.0005. The initial learning rate is 3.5×10^{-4}, and it is decayed using a cosine annealing schedule. All experiments are conducted on hardware consisting of two Intel(R) Xeon(R) Silver 4214R CPUs @2.40 GHz and two NVIDIA A4000 GPUs.

Baseline. To establish the baseline, the original Non-local attention method [19] is utilized. It is important to note that the baseline does not take advantage of complementary feature information from frequency domain components and does not perform frequency domain filtering operations.

3.3 Comparison with State-of-the-art Methods

Genralizable Performances on Small-Scale Datasets. As presented in Table 3, the proposed method is extensively compared to state-of-the-art methods following protocol-1. Results of other methods that utilize DukeMTMC-reID in the source domains are reported for reference, but it is excluded from our training sets during the experiments. Despite using fewer source domains, our method achieves the highest performance.

Generalizable Performances Under Large-Scale Datasets. As shown in Table 4, our proposed method compare with other state-of-the-arts under protocol-2 and protocol-3,. 'Source' refers that only the training sets in the source domains are used for training and 'Full-' denotes that all images in the source domains (i.e. combining training and testing sets) are leveraged at training time. When generalizing to the unseen target domains, namely Market-1501, CUHK03, and MSMT17, our proposed BFAN outperforms the second-best approach by 5.3% in Market1501 and 6.7% in CUHK03 in terms of Rank-1 accuracy, and 9.7%, 4.5%, and 0.5% in terms of mAP (mean Average Precision), respectively. Moreover, when incorporating the testing sets of all visible source domains during model training, our BFAN surpasses the second-best approach,

Table 3. Performance (%) comparison with the state-of-the-art methods on the small-scale person ReID datasets under Protocol-1.

Method	Source	→VIPeR(V)		→PRID(P)		→GRID(G)		→iLIDS(I)		Average	
		mAP	R1	mAP	R1	mAP	R1	mAP	R1	mAP	R1
DDAN [3]	Full-(MA+C2 +C3+CS+D)	60.8	56.5	67.5	62.9	50.9	46.2	81.2	78.0	65.1	60.9
SNR [10]		58.0	49.2	60.4	47.3	49.0	39.4	84.0	77.3	62.9	53.3
CBN [29]		59.2	49.0	65.7	61.3	47.8	43.3	79.4	75.3	63.0	57.2
Person30K [1]		60.4	53.9	68.4	60.6	56.6	50.9	83.9	79.3	67.3	61.1
RaMoE [5]		64.6	56.6	67.3	57.7	54.2	46.8	90.2	85.0	69.1	61.5
MetaBIN [4]		66.0	56.2	79.8	72.5	58.1	49.7	85.5	79.7	72.4	64.5
$QAConv_{50}$ [13]	Full-(MA+C2 +C3+CS)	66.3	57.0	62.2	52.3	57.4	48.6	81.9	75.0	67.0	58.2
M^3L [26]		68.2	60.8	65.3	55.0	50.5	40.0	74.3	65.0	64.6	55.2
MetaBIN [4]		64.3	55.9	70.8	61.2	57.9	50.2	82.7	74.7	68.9	60.5
META [23]		68.4	61.5	71.1	61.9	60.1	52.4	83.5	79.2	70.9	63.8
Ours		**68.7**	**60.9**	**71.3**	**63.2**	**59.8**	**56.1**	**84.6**	**81.0**	**71.1**	**65.3**

Table 4. Performance (%) comparison with the state-of-the-art methods on the large-scale person ReID datasets under Protocol-2 and Protocol-3. The model denoted as M^3L^* indicates the usage of IBN-Net50 as the backbone architecture. In the absence of this superscript, the ResNet-50 is employed as the backbone.

Method	Source	Market-1501		Source	CUHK03		Source	MSMT17	
		mAP	R1		mAP	R1		mAP	R1
SNR [10]	MS+CS+C3	34.6	62.7	MS+CS+MA	8.9	8.9	CS+MA+C3	6.8	19.9
MetaBIN [4]		57.9	80.0		28.8	28.1		17.8	40.2
M^3L [26]		58.4	79.9		20.9	31.9		15.9	36.9
M^3L^* [26]		61.5	82.3		34.2	34.4		16.7	37.5
$QAConv_{50}$ [13]		63.1	83.7		25.4	24.8		16.4	45.3
Ours		**67.6**	**85.3**		**33.3**	**34.8**		**18.3**	**39.8**
SNR [10]	Full-(MS+CS+C3)	52.4	77.8	Full-(MS+CS+MA)	17.5	17.1	Full-(CS+MA+C3)	7.7	22.0
M^3L [26]		61.2	81.2		32.3	33.8		16.2	36.9
M^3L^* [26]		62.4	82.7		35.7	36.5		17.4	38.6
$QAConv_{50}$ [13]		66.5	85.0		32.9	33.3		17.6	46.6
MetaBIN [4]		67.2	84.5		43.0	43.1		18.8	41.2
Ours		**72.5**	**87.9**		**45.5**	**44.2**		**20.6**	**43.3**

by 3.4%, 1.1%, and 2.1% in terms of Rank-1 accuracy, and 5.3%, 2.5%, and 1.8% in terms of mAP. These results demonstrate that our proposed method significantly improves the generalization capability of learned features, even when learning complementary information and performing filtering operations in the frequency domain.

3.4 Ablation Studies

The effectiveness of BFCA and FNMF. A comprehensive analysis was conducted to evaluate the effectiveness of the proposed BFAN. Each module was individually compared to the Baseline model (ResNet+Non-local Attention). The performance of each module can be seen in Fig. 4. Ablation experiments in Protocol-2 specifically assessed the impact of BFCA and FNMF. "w/o

FNMF" excluded the filtering operation, using only the BFCA module, while "w/o BFCA" employed only the FNMF module. Results in Fig. 4 demonstrate that both the BFCA and FNMF modules outperformed the Baseline, indicating their superior generalization performance.

The effectiveness of High/Low Frequency Components. Additionally, ablation experiments on different frequency branches of BFCA were conducted to examine the impact on the DG ReID model's generalization performance, as shown in Fig. 5. "HFC" represents learning complementary information solely on the high-frequency component, while "LFC" signifies learning solely on the low-frequency component. Results in Fig. 5 indicate that both high-frequency and low-frequency components contribute to improving the DG ReID model's performance. However, the enhancement from low-frequency components is more significant compared to the relatively smaller enhancement from high-frequency components. These results confirm that learning texture information in the low-frequency range is easier than learning semantic information in the high-frequency range.

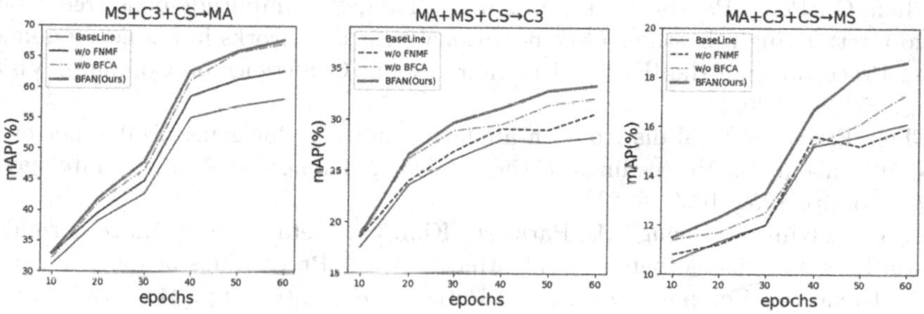

Fig. 4. Following Protocol-2, ablation studies were conducted on the individual components of BFAN to evaluate their impact and significance.

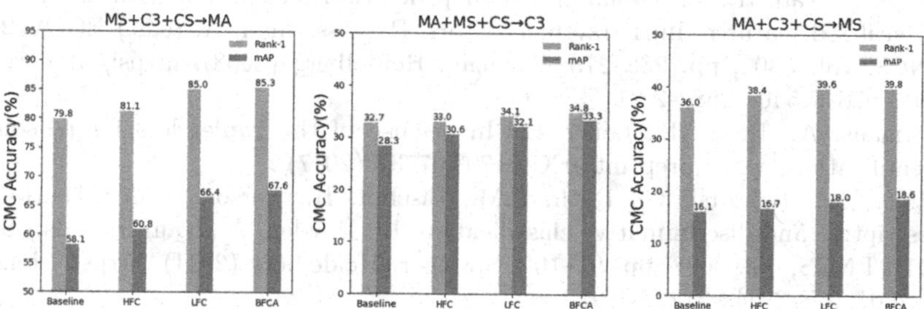

Fig. 5. Ablation studies on different frequency components of BFCA under the Protocol-2.

4 Conclusion

In this paper, a DG Person ReID method based on complementary learning of spectral features is proposed. This method enables real-time transmission of spectral feature information for domain invariant feature learning. The proposed learning framework captures complementary information between high-frequency and low-frequency components of images in the embedding space. It enhances the transferable frequency domain components while suppressing frequency domain noise that hinders generalization, resulting in more robust and consistent extracted features. Extensive experiments have verified the effectiveness of the various components within the proposed framework.

References

1. Bai, Y., et al.: Person30k: a dual-meta generalization network for person re-identification. In: Proceedings of the IEEE/CVF Conference on Computer Vision and Pattern Recognition, pp. 2123–2132 (2021)
2. Chen, G., Peng, P., Ma, L., Li, J., Du, L., Tian, Y.: Amplitude-phase recombination: rethinking robustness of convolutional neural networks in frequency domain. In: Proceedings of the IEEE/CVF International Conference on Computer Vision, pp. 458–467 (2021)
3. Chen, P., et al.: Dual distribution alignment network for generalizable person re-identification. In: Proceedings of the AAAI Conference on Artificial Intelligence, vol. 35, pp. 1054–1062 (2021)
4. Choi, S., Kim, T., Jeong, M., Park, H., Kim, C.: Meta batch-instance normalization for generalizable person re-identification. In: Proceedings of the IEEE/CVF Conference on Computer Vision and Pattern Recognition, pp. 3425–3435 (2021)
5. Dai, Y., Li, X., Liu, J., Tong, Z., Duan, L.Y.: Generalizable person re-identification with relevance-aware mixture of experts. In: Proceedings of the IEEE/CVF Conference on Computer Vision and Pattern Recognition, pp. 16145–16154 (2021)
6. Gray, D., Tao, H.: Viewpoint invariant pedestrian recognition with an ensemble of localized features. In: Forsyth, D., Torr, P., Zisserman, A. (eds.) ECCV 2008. LNCS, vol. 5302, pp. 262–275. Springer, Heidelberg (2008). https://doi.org/10.1007/978-3-540-88682-2_21
7. Hermans, A., Beyer, L., Leibe, B.: In defense of the triplet loss for person re-identification. arXiv preprint arXiv:1703.07737 (2017)
8. Hirzer, M., Beleznai, C., Roth, P.M., Bischof, H.: Person re-identification by descriptive and discriminative classification. In: Heyden, A., Kahl, F. (eds.) SCIA 2011. LNCS, vol. 6688, pp. 91–102. Springer, Heidelberg (2011). https://doi.org/10.1007/978-3-642-21227-7_9
9. Jiang, G., Wang, H., Peng, J., Chen, D., Fu, X.: Graph-based multi-view binary learning for image clustering. Neurocomputing **427**, 225–237 (2021)
10. Jin, X., Lan, C., Zeng, W., Chen, Z., Zhang, L.: Style normalization and restitution for generalizable person re-identification. In: proceedings of the IEEE/CVF Conference on Computer Vision and Pattern Recognition, pp. 3143–3152 (2020)
11. Li, W., Wang, X.: Locally aligned feature transforms across views. In: Proceedings of the IEEE Conference on Computer Vision and Pattern Recognition, pp. 3594–3601 (2013)

12. Li, W., Zhao, R., Xiao, T., Wang, X.: Deepreid: deep filter pairing neural network for person re-identification. In: Proceedings of the IEEE Conference on Computer Vision and Pattern Recognition, pp. 152–159 (2014)
13. Liao, S., Shao, L.: Interpretable and generalizable person re-identification with query-adaptive convolution and temporal lifting. In: Vedaldi, A., Bischof, H., Brox, T., Frahm, J.-M. (eds.) ECCV 2020, Part XI. LNCS, vol. 12356, pp. 456–474. Springer, Cham (2020). https://doi.org/10.1007/978-3-030-58621-8_27
14. Loy, C.C., Xiang, T., Gong, S.: Time-delayed correlation analysis for multi-camera activity understanding. Int. J. Comput. Vision **90**, 106–129 (2010)
15. Peng, J., Jiang, G., Wang, H.: Adaptive memorization with group labels for unsupervised person re-identification. IEEE Trans. Circ. Syst. Video Technol. **33**, 5802–5813 (2023)
16. Wang, H., Wu, X., Huang, Z., Xing, E.P.: High-frequency component helps explain the generalization of convolutional neural networks. In: Proceedings of the IEEE/CVF Conference on Computer Vision and Pattern Recognition, pp. 8684–8694 (2020)
17. Wang, H., Feng, L., Yu, L., Zhang, J.: Multi-view sparsity preserving projection for dimension reduction. Neurocomputing **216**, 286–295 (2016)
18. Wang, H., Yao, M., Jiang, G., Mi, Z., Fu, X.: Graph-collaborated auto-encoder hashing for multiview binary clustering. IEEE Trans. Neural Netw. Learn. Syst. (2023)
19. Wang, X., Girshick, R., Gupta, A., He, K.: Non-local neural networks. In: Proceedings of the IEEE Conference on Computer Vision and Pattern Recognition, pp. 7794–7803 (2018)
20. Wei, L., Zhang, S., Gao, W., Tian, Q.: Person transfer GAN to bridge domain gap for person re-identification. In: Proceedings of the IEEE Conference on Computer Vision and Pattern Recognition, pp. 79–88 (2018)
21. Wen, Y., Zhang, K., Li, Z., Qiao, Yu.: A discriminative feature learning approach for deep face recognition. In: Leibe, B., Matas, J., Sebe, N., Welling, M. (eds.) ECCV 2016, Part VII. LNCS, vol. 9911, pp. 499–515. Springer, Cham (2016). https://doi.org/10.1007/978-3-319-46478-7_31
22. Xiao, T., Li, S., Wang, B., Lin, L., Wang, X.: End-to-end deep learning for person search. arXiv preprint arXiv:1604.01850 2(2), 4 (2016)
23. Xu, B., Liang, J., He, L., Sun, Z.: Mimic Embedding via adaptive aggregation: learning generalizable person re-identification. In: Avidan, S., Brostow, G., Cissé, M., Farinella, G.M., Hassner, T. (eds.) Computer Vision - ECCV 2022. ECCV 2022, Part XIV, LNCS, vol. 13674, pp. 372–388. Springer, Cham (2022). https://doi.org/10.1007/978-3-031-19781-9_22
24. Xu, Z.-Q.J., Zhang, Y., Xiao, Y.: Training behavior of deep neural network in frequency domain. In: Gedeon, T., Wong, K.W., Lee, M. (eds.) ICONIP 2019, Part I. LNCS, vol. 11953, pp. 264–274. Springer, Cham (2019). https://doi.org/10.1007/978-3-030-36708-4_22
25. Zhai, Y., et al.: Ad-cluster: augmented discriminative clustering for domain adaptive person re-identification. In: Proceedings of the IEEE/CVF Conference on Computer Vision and Pattern Recognition, pp. 9021–9030 (2020)
26. Zhao, Y., et al.: Learning to generalize unseen domains via memory-based multi-source meta-learning for person re-identification. In: Proceedings of the IEEE/CVF Conference on Computer Vision and Pattern Recognition, pp. 6277–6286 (2021)
27. Zheng, L., Shen, L., Tian, L., Wang, S., Wang, J., Tian, Q.: Scalable person re-identification: a benchmark. In: Proceedings of the IEEE International Conference on Computer Vision, pp. 1116–1124 (2015)

28. Zheng, W.S., Gong, S., Xiang, T.: Associating groups of people. In: British Machine Vision Conference (2009)
29. Zhuang, Zijie, et al.: Rethinking the distribution gap of person re-identification with camera-based batch normalization. In: Vedaldi, Andrea, Bischof, Horst, Brox, Thomas, Frahm, Jan-Michael. (eds.) ECCV 2020, Part XII. LNCS, vol. 12357, pp. 140–157. Springer, Cham (2020). https://doi.org/10.1007/978-3-030-58610-2_9

Stereo3DMOT: Stereo Vision Based 3D Multi-object Tracking with Multimodal ReID

Chen Mao[1,2], Chong Tan[1], Hong Liu[1], Jingqi Hu[1,2], and Min Zheng[1(✉)]

[1] Shanghai Institute of Microsystem and Information Technology, Chinese Academy of Sciences, Shanghai 200050, China
{chen.mao,chong.tan,hong.liu,jingqihu,min.zheng}@mail.sim.ac.cn
[2] University of Chinese Academy of Sciences, Beijing 101408, China

Abstract. 3D Multi-Object Tracking (MOT) is a key component in numerous applications, such as autonomous driving and intelligent robotics, playing a crucial role in the perception and decision-making processes of intelligent systems. In this paper, we propose a 3D MOT system based on a cost-effective stereo camera pair, which includes a 3D multimodal re-identification (ReID) model capable of multi-task learning. The ReID model obtains the multimodal features of objects, including RGB and point cloud information. We design data association and trajectory management algorithms. The data association computes an affinity matrix for the object feature embeddings and motion information, while the trajectory management controls the lifecycle of the trajectories. In addition, we create a ReID dataset based on the KITTI Tracking dataset, used for training and validating ReID models. Results demonstrate that our method can achieve accurate object tracking solely with a stereo camera pair, maintaining high reliability even in cases of occlusion and missed detections. Experimental evaluation shows that our approach outperforms competitive results on the KITTI MOT leaderboard. Our code, dataset, and model are available at https://github.com/maomao279/Stereo3DMOT.

Keywords: 3D MOT · ReID · Stereo Vision · Point Cloud

1 Introduction

3D MOT is a crucial task in autonomous driving and robotics, playing an irreplaceable role in the continuous perception of the surrounding environment. It can accurately obtain information such as the volume, angle, and 3D location of the tracked objects. The goal is to identify the same object and continually record its status in consecutive video frames. The tracked object always maintains the same ID during movement, and the robustness of tracking methods needs to be maintained even in cases of occlusion and missed detections. Currently, most existing 3D MOT methods are based on the Tracking-by-Detection

© The Author(s), under exclusive license to Springer Nature Singapore Pte Ltd. 2024
Q. Liu et al. (Eds.): PRCV 2023, LNCS 14436, pp. 495–507, 2024.
https://doi.org/10.1007/978-981-99-8555-5_39

(TBD) paradigm, in which a 3D detector [1–3] first identifies the objects in the image, followed by tracking of these objects. Usually, the detection and tracking methods are used simultaneously to accomplish the 3D MOT task.

Some lightweight 3D MOT methods [4–7], which achieve tracking solely through the movement information of targets, demonstrate superior real-time performance. They save a significant amount of computational resources as they don't require inference through neural networks. However, these methods struggle to effectively address the challenges of object re-identification triggered by occlusion. To further enhance the accuracy of 3D MOT, [8–11] employ a combination of 2D and 3D detections, acquiring multimodal information from RGB images and LiDAR. However, the need to integrate both 2D and 3D detectors imposes a significant computational burden. Other methods [12,13] only utilize a 3D detector and a single camera, improving object re-identification rates by introducing a ReID model to leverage the neural network's capability of extracting appearance features. However, the information provided by a single RGB image is limited. In scenarios where lighting conditions rapidly change, the features extracted by these methods do not exhibit robustness.

Considering both the cost of equipment and computational complexity, we propose a 3D MOT method based on the TBD paradigm: Stereo3DMOT. Our method requires only a stereo camera pair as the sensing device and can integrate any 3D object detector. For feature extraction of objects during tracking, we design a 3D ReID neural network model: Stereo3D ReID. It integrates RGB appearance features and 3D point cloud features, compensating for the significant influence of lighting and color change on object appearance in RGB images. This multimodal ReID model is realized without the use of LiDAR, effectively ensuring the time-invariance of objects. We design multi-stage data association and trajectory management methods that fully utilize ReID feature embeddings and trajectory motion information. The results show that our method remains robust in scenarios of occlusion and missed detections. Our contributions can be summarized as follows:

- We propose a 3D multimodal ReID neural network model based on a stereo camera pair, which can simultaneously carry out multi-task learning for ReID and disparity matching. It integrates both RGB appearance and point cloud 3D features using only cameras.
- We design data association and trajectory management methods that can better reconstruct the motion process of trajectories and remain robust in situations of occlusion and missed detection.
- We create a 3D ReID dataset based on the KITTI Tracking dataset. We submit our results to the KITTI MOT leaderboard and achieve competitive results(see Fig. 1).

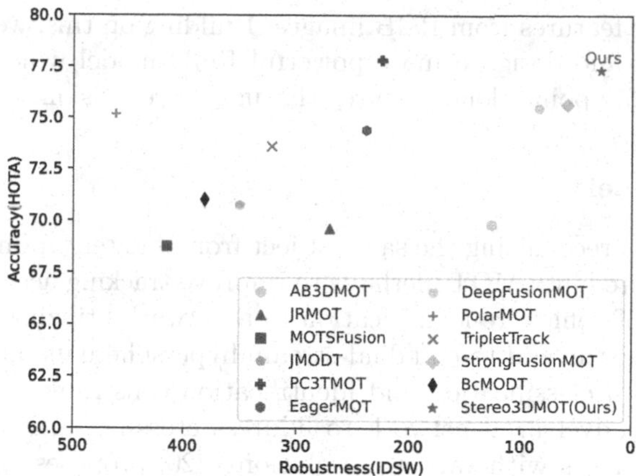

Fig. 1. The proposed method is compared with multiple tracking methods on the KITTI MOT leaderboard. Higher and further to the right is better. Our method achieves superior accuracy and robustness.

2 Related Work

2.1 3D MOT

3D MOT tasks aim to track the trajectory of each detected object in real-world 3D space. AB3DMOT [4] proposes a simple real-time 3D MOT system with noteworthy performance. Simpletrack [5] summarizes current 3D MOT methods into a unified framework by decomposing them into four constituent parts: pre-processing of detection, association, motion model, and life cycle management. The aforementioned methods provide a baseline, but they have a lower tracking accuracy. GNN3DMOT [14] proposes a novel feature interaction mechanism by introducing the graph neural network (GNN) instead of obtaining features for each object independently. EagerMOT [15] proposes a simple tracking formulation that integrates all available object observations from both sensor modalities to obtain a well-informed interpretation of the scene dynamics. DeepFusionMOT [16] proposes a camera-LiDAR fusion-based 3D real-time tracking framework and simultaneously proposes a novel deep association mechanism that makes full use of the characteristics of cameras and LiDARs. These methods that combine 2D and 3D detections show an improvement in accuracy, but they also introduce a substantial computational load. Fantrack [12] proposes a data-driven approach to online MOT that uses a convolutional neural network (CNN) for data association in a tracking-by-detection framework. TripletTrack [13] condenses triplet embeddings and motion descriptors through a Long Short-Term Memory (LSTM). Thesemethods solely utilize a 3D detector, focusing on extracting features from RGB images. Building on this, we propose a 3D MOT framework and design a more powerful ReID model which combines 2D appearance and 3D point cloud features through stereo vision.

2.2 ReID Model

ReID tasks aim at recognizing the same object from different frames. ReID models can be integrated into MOT methods to improve tracking accuracy. [17] tackles the problem of vehicle re-identification in a network utilizing triplet embeddings. [18] proposes an end-to-end dual-stream hypersphere manifold embedding network with both classification and identification constraints. [19] proposes a novel generative adversarial network to address cross-resolution person ReID, allowing query images with varying resolutions. [20] proposes an unsupervised ReID deep learning approach and an unsupervised tracklet association learning framework. BOT [21] collects and evaluates these effective training tricks in person ReID. AGW [22] designs a powerful baseline and introduces a new evaluation metric (mINP) for person ReID. To meet the increasing application demand for general instance re-identification, FastReID [23] presents a widely used software system in JD AI Research. When people appear in extreme illumination or change clothes, the RGB appearance-based ReID methods tend to fail. To overcome this problem, [24,25] exploit depth information to provide a more invariant body shape regardless of illumination and color change. Inspired by this, our ReID model merges the RGB and point cloud features of objects, which helps to maintain feature stability.

3 Method

Stereo3D MOT is a complete tracking system. In Sect. 3.1, we introduce our proposed Stereo3D ReID network structure and its functions, the output features provide data sources for object modeling. In Sect. 3.2, we detail the entire framework and expound upon our improvement methods module by module.

3.1 3D ReID Network Model

To conduct the training and testing of the ReID model on tasks related to autonomous driving, we create a ReID dataset based on the KITTI Tracking dataset [26], which includes object images from stereo RGB cameras and dense disparity maps. This ReID dataset only contains the object image region while ignoring the background, which is more conducive to directly training the ReID task.

The ReID network model is a crucial part of the tracking method introduced in this paper. The output of the ReID network integrates the RGB appearance features of the object in stereo images and the 3D features of the point cloud. The network architecture is divided into three sections: Backbone Network, Multi-task Head, and Losses. The network structure is illustrated in Fig. 2.

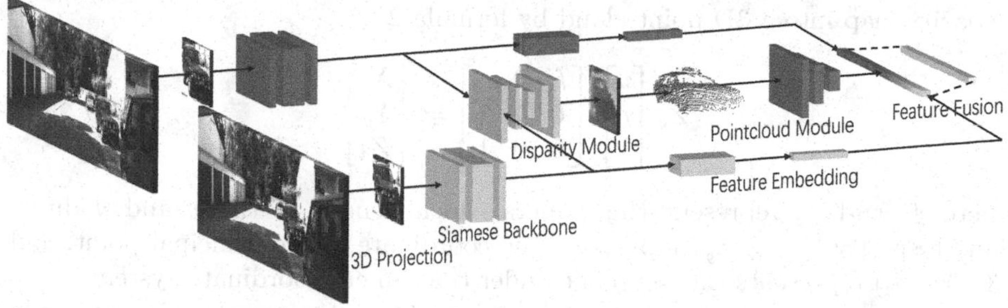

Fig. 2. Stereo3D ReID Architecture. First, the input 3D boxes are perspective-projected onto the 2D images, and the backbone network extracts RGB image features. Then, the disparity estimation module computes the disparity map, converts it to a point cloud, and the point cloud module extracts three-dimensional features. Ultimately, the RGB features and point cloud features are integrated and outputted.

Backbone Network. Upon receiving the 3D bounding boxes outputted by the 3D object detector, we project the 3D bounding boxes onto the left and right 2D images and feed them into the backbone network. We adopt a lightweight siamese MobileNetv3 [27] structure, where the left and right networks share weights. Embracing the concept of multi-task learning, the backbone network outputs features that are used in both object re-identification and stereo disparity matching tasks, thereby enhancing the overall inference efficiency of the network.

Multi-task Head. The features outputted from the backbone network are 2D adaptive average pooled into a tensor of dimension $(B, N_1, 1)$. At the same time, these features are fed into the disparity matching module to extract disparity information. The disparity module draws inspiration from Fast-ACVNet++ [28], and we restructure the feature extraction network to implement multi-task learning. As the foreground object region accounts for a small proportion, the disparity module only needs to calculate the object ROI regions in the left and right images, using dense disparity maps as supervision, which doesn't introduce excessive computation. Since the object region is cropped, the actual disparity of the object should be the sum of the output disparity map and the horizontal distance between the center points of the object in the left and right images. The computation formula is as follows:

$$d_f = d_{out} + (u_l - u_r) \tag{1}$$

where d_f is the actual disparity of the object region, d_{out} is the output of the disparity module, and u_l, u_r are the horizontal coordinates of the center points of the object in the left and right images, respectively. We transform the object's disparity map into a 3D point cloud by formula 2.

$$Z_c \begin{bmatrix} u \\ v \\ 1 \end{bmatrix} \begin{bmatrix} f_x & 0 & c_x \\ 0 & f_y & c_y \\ 0 & 0 & 1 \end{bmatrix} = \begin{bmatrix} X_c \\ Y_c \\ Z_c \end{bmatrix} \tag{2}$$

where f_x and f_y represent the camera's focal length in the x and y directions respectively, (c_x, c_y) represents the coordinate of the principal point, and (X_c, Y_c, Z_c) represents the 3D point under the camera coordinate system.

We incorporate the feature extraction portion of pointnet [29], randomly selecting N_2 points' spatial location points (x, y, z) from the point cloud to form a tensor of dimensions $[N_2, 3]$, which is then inputted into the point cloud feature extraction module. The global features of the point cloud are aggregated into a tensor of dimensions $[B, N_2, 1]$. Finally, the features are concatenated into a tensor of dimensions $[B, 2N_1 + N_2, 1]$, and passed through a 1×1 convolution layer and fully connected layer for feature mapping and fusion. During the inference process, this feature vector is directly outputted; during the training process, a classification layer needs to be connected to the last layer, supervised by the object IDs from the training set.

Losses. The Stereo3D ReID model uses three loss functions: cross entropy loss, triplet loss, and disparity matching loss. Cross entropy loss is a classification loss function that predicts the categories of images in the ReID task, helping the network distinguish between targets and non-targets better. For each object i, where the true ID label is y and the predicted category probability is p_i, the cross entropy loss is calculated as follows:

$$Loss_C = \sum_{i=1}^{N} -q_i log(p_i) \begin{cases} q_i = \varepsilon/N & y \neq i \\ q_i = 1 - \frac{N-1}{N} & y = i \end{cases} \tag{3}$$

where N is the number of categories, and ε is a very small constant used to encourage the model not to over-rely on the training set. The formula uses label smoothing [30] optimization for the value of q_i to prevent the model from overfitting during training.

The triplet loss function is a type of distance loss, which is used in ReID tasks to model the similarity between images, and measures their similarity by comparing the distances between different image feature vectors, as follows:

$$Loss_T = max((d(a, p) - d(a, n) + margin), 0) \tag{4}$$

where $d\,(a, p)$ represents the distance between the feature vectors of the object a and the positive instance p, and $d\,(a, n)$ represents the distance between a and the negative instance n, $margin$ is set to enlarge the distance between the positive and negative instance, ultimately clustering images of the same object in the distance space.

After the disparity d_{out} is output by the disparity matching module, it needs to be input into formula 2 to obtain the final estimated disparity. The difference is then calculated with the real disparity [28], as the following formula:

$$Loss_D = Smooth_{L_1}(d_{out} + (u_l - u_r) - d_{gt}) \tag{5}$$

where $Smooth_{L1}$ is the smooth L1 loss, d_{gt} is the ground-truth disparity map.

3.2 Stereo3DMOT Tracking Method

The Stereo3DMOT tracking method is divided into three parts: object modeling, data association, and trajectory management. The architecture diagram of the method is shown in Fig. 3.

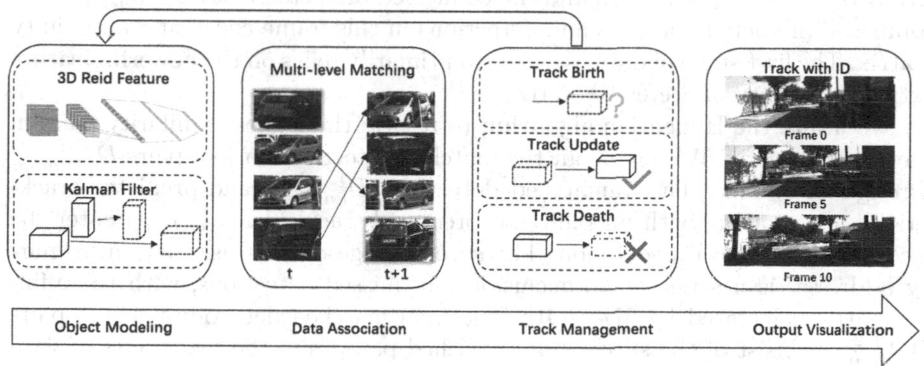

Fig. 3. Overall structure of the proposed Stereo3DMOT framework.

Object Modeling. In addition to the features obtained from Stereo3D ReID, we also employ motion information to model objects. We use 3D Kalman filtering [4] as a constant velocity model to approximate the frame-to-frame displacement of the object. The object's state in the next frame is predicted based on the coordinates of the object's center point (x, y, z), 3D size (l, w, h), and heading angle θ. For objects captured for the first time, we model them and create new trajectories. The states of the trajectories are set as unconfirmed, and the ID numbers are sequentially assigned from 0 to positive infinity.

Data Association. We propose a method for calculating the degree of correlation between 3D bounding boxes: 3D-EIOU. We find that in 3D scenes, The dimensions of the object's 3D bounding box are almost unaffected by changes in the object's pose. 3D-EIOU is more suitable than the conventional IOU method 6 for comparing the similarity of bounding boxes in a 3D scene. It consists of three parts: the volume intersection over the union between 3D bounding boxes($S_{3D\text{-}IOU}$), the L2 norm of center point coordinates, and the L1 norm of the 3D bounding box's length, width, height, and angle. The computation formula is as follows:

$$S_{3D\text{-}IOU} = \frac{3D\text{-}Box_d \cap 3D\text{-}Box_t}{3D\text{-}Box_d \cup 3D\text{-}Box_t} \tag{6}$$

$$S_{3D\text{-}EIOU} = S_{3D\text{-}IOU} - \alpha(\|Cd_{x,y,z} - Ct_{x,y,z}\|_2 + \|Sd_{l,h,w,r_y} - St_{l,h,w,r_y}\|_1) \tag{7}$$

where $Cd_{x,y,z}$ and $Ct_{x,y,z}$ are the center point coordinate vectors of the detection bounding box and the predicted tracklet bounding box, respectively. Sd_{l,h,w,r_y} and St_{l,h,w,r_y} are the vectors composed of the length, width, height, and angle of the detection and predicted tracklet bounding box, respectively.

For each detection in every frame, a two-stage data association with the predicted tracklets is performed. The first stage of data association calculates the cosine similarity of the ReID feature vectors and the weighted value of 3D-EIOU between each detection and the confirmed predicted tracklet $T_{confirm}$. All confirmed predicted tracklets and detections in this frame constitute an affinity matrix. The first stage of data association primarily relies on the powerful feature extraction ability of Stereo3D ReID.

We utilize the Hungarian algorithm to resolve the highest similarity pairs in the affinity matrix. We screen out the matched detection-tracklet pairs $D\text{-}T^1_{match}$ through thresholds. The unmatched detections $D^1_{unmatch}$ and predicted tracklets $T^1_{unmatch}$, along with unconfirmed predicted tracklets $T_{unconfirm}$, enter the second stage of data association. The second stage of data association primarily addresses issues related to occlusion and missed detections, with the affinity matrix calculated by 3D-EIOU. The final matched detection-tracklet pairs $D\text{-}T^{1,2}_{match}$ consist of the successfully matched pairs from the two stages of data association, and we output the final unmatched tracklets $T^{1,2}_{unmatch}$ and detections $D^{1,2}_{unmatch}$.

Track Management. For unmatched detections (detections in the first frame are all unmatched), we create new trajectories and model the objects, setting their state as unconfirmed. Once a predicted tracklet successfully matches with a detection, we update the ReID features of the trajectory as a weighted value of the current trajectory features and the detection features and continue this trajectory. If it's unconfirmed and successfully match three times, it switches to confirmed state. For unmatched tracklets, we propose a method to judge whether the lifespans of trajectories L_t reach the maximum life threshold L_{max}. The formula for calculating lifespan is as follows:

$$L_t = F_{miss} + \gamma L_{dis} \cdot \sqrt{F_{survival}} \tag{8}$$

where F_{miss} is the number of frames since the last successful match, L_{dis} is the distance between the bounding box's center point pixel and the image boundary, $F_{survival}$ is the existing frame number, and γ is a constant. We believe that the longer the disappearance and existing time, and the closer to the image boundary when disappeared, the more likely it is to have left the camera's range. We tend to give confirmed trajectories a larger life threshold.

Table 1. KITTI MOT leaderboard. Comparison with 3D MOT methods from the past three years by using the test set of the KITTI Car Tracking benchmark. The best results are shown in bold. For fair comparison, we mark the methods which use PointRCNN for detection with *.

Method	Year and Publication	Input Detection	HOTA(%) ↑	IDSW ↓	AssA(%) ↑	AssPr(%) ↑	MOTA(%) ↑	MOTP(%) ↑
AB3DMOT* [4]	2020(IROS)	3D	69.99	113	69.33	89.02	83.61	85.23
JRMOT [31]	2020(IROS)	2D+3D	69.61	271	66.89	88.95	85.10	85.28
MOTSFusion [32]	2020(RA-L)	2D+3D	68.74	415	66.16	85.49	84.24	85.03
GNN3DMOT* [14]	2020(CVPR)	2D+3D	—	142	—	—	82.40	84.05
JMODT [9]	2021(IROS)	2D+3D	70.73	350	68.76	88.02	85.35	85.37
PC3TMOT* [7]	2021(TITS)	3D	**77.80**	225	81.59	88.75	**88.81**	84.26
EagerMOT [15]	2021(ICRA)	2D+3D	74.39	239	74.16	**91.05**	87.82	85.69
DeepFusionMOT* [16]	2022(RA-L)	2D+3D	75.46	84	80.06	89.77	84.64	85.02
PolarMOT [33]	2022(ECCV)	3D	75.16	462	76.95	89.27	85.08	85.63
TripletTrack [13]	2022(CVPR)	3D	73.58	322	74.66	89.55	84.32	**86.06**
StrongFusionMOT* [10]	2022(IEEE Sensors)	2D+3D	75.65	58	79.84	89.81	85.53	85.07
BcMODT* [11]	2023(Remote Sensing)	2D+3D	71.00	381	69.14	88.70	85.48	85.31
Stereo3DMOT(Ours)*	—	3D	77.32	**28**	**81.86**	89.61	87.10	85.06

4 Experiments

4.1 Experimental Results

We conduct experiments using the KITTI Tracking dataset in this paper. KITTI provides testing standards of performance and a leaderboard for tracking methods. We adhere to the evaluation algorithm used by the KITTI MOT leaderboard. The primary evaluation method, HOTA, associates the metrics for detection, association, and localization into a unified index. We also adopt metrics such as ID Switches (IDSW), Multi-Object Tracking Accuracy (MOTA), and Multi-Object Tracking Precision (MOTP). The results on the KITTI Tracking test set are shown in Table 1. Compared with other methods, our method outperforms most current methods on the HOTA metric, leading among 3D MOT methods on the IDSW and ASSA metrics. It demonstrates that our method has higher robustness and accuracy, and can maintain stability in cases of occlusion or missed detection.

Figure 4 presents a visual example comparing our method with AB3DMOT. Our method demonstrates strong robustness when handling complex scenes, such as intersections with object occlusion and disappearance, without any ID switches.

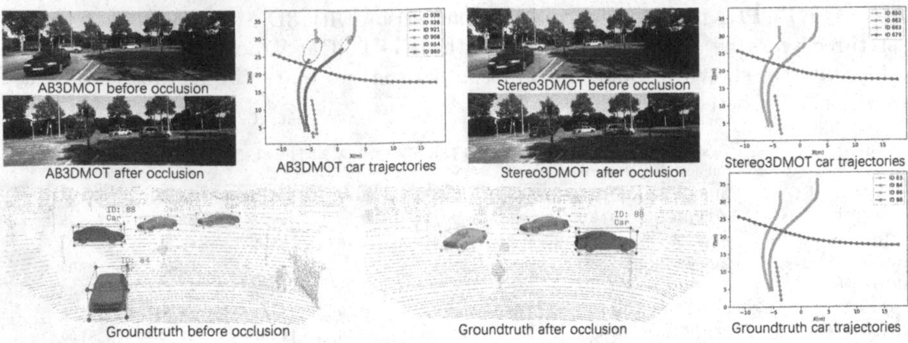

Fig. 4. A visual comparison example of bird's eye view trajectories between our method and AB3DMOT. The red circle in the figure indicates that there is an ID switch in the trajectory. (Color figure online)

4.2 Ablation Study

To explore the impact of different modules on the overall tracking performance, we conduct ablation experiments using 21 training set sequences from the KITTI Tracking dataset. We evaluate based on key indicators such as HOTA, MOTA, and IDSW. We use PointRCNN as the detector, only use Kalman filtering as the motion model and IOU data association (AB3DMOT [4]) as the baseline. We gradually incorporate our approaches from this starting point. As can be seen from Table 2, the methods we proposed significantly improve the accuracy and robustness of tracking.

Stereo3D ReID Model. Based on the baseline, we incorporate the Stereo3D ReID model to verify the impact of features extracted by the ReID neural network on tracking performance. We train both the car and pedestrian datasets for 300 epochs. To integrate the ReID model with the trajectory motion model, it is necessary to simultaneously incorporate the multi-stage data association algorithm, using the number of missing frames F_{miss} as the criterion for evaluating the life threshold.

Data Association. We divide the trajectories into confirmed and unconfirmed states, integrating them into the multi-stage data association. We replace IOU with 3D-EIOU as the association metric, where the thresholds are -0.2 and -0.1 for cars and pedestrians categories respectively.

Track Management. The lifecycle of trajectories is calculated by formula 8, and the trajectories are deleted once it reaches the threshold. We set different maximum life thresholds for unconfirmed and confirmed states of the trajectories, which are 3 and 22 respectively.0

Table 2. Ablation study. We conduct experiments on the impact of each method we proposed on the validation set of the KITTI Tracking dataset. ReID means the Stereo3D ReID that we proposed, DAAS means data association and TRMA means trajectory management. The best results are shown in bold.

Method	Car			Pedestrian		
	HOTA(%) ↑	MOTA(%) ↑	IDSW ↓	HOTA(%) ↑	MOTA(%) ↑	IDSW ↓
Motion(Baseline)	73.32	80.57	47	44.54	56.09	113
+ReID	76.05	83.25	35	46.10	**59.58**	98
+ReID+DAAS	76.11	83.53	32	47.71	59.52	78
+ReID+DAAS+TRMA	**77.03**	**84.22**	**15**	**49.22**	59.02	**49**

5 Conclusion

We propose Stereo3DMOT, a stereo vision based 3D MOT method. In this system, Stereo3D ReID is a multi-task learning model that extracts appearance information from stereo RGB images. It utilizes stereo disparity matching to obtain disparity maps and converts them into point clouds. Ultimately, it outputs multimodal features that combine 2D RGB appearance with 3D point clouds. We associate the ReID feature embeddings and motion information of the trajectories, and by designing trajectory management and data association algorithms, we further improve the tracking performance. Experiments show that our method maintains tracking stability even under conditions of occlusion and missed detections. Additionally, we create a dataset for ReID tasks in autonomous driving based on the KITTI dataset. Ultimately, our method achieves competitive results on the KITTI MOT leaderboard.

References

1. Shi, S., Wang, X., Li, H.: Pointrcnn: 3d object proposal generation and detection from point cloud. In: Proceedings of the IEEE/CVF Conference on Computer Vision and Pattern Recognition, pp. 770–779 (2019)
2. Sun, J., et al.: DISP R-CNN: stereo 3d object detection via shape prior guided instance disparity estimation. In: Proceedings of the IEEE/CVF Conference on Computer Vision and Pattern Recognition, pp. 10548–10557 (2020)
3. Li, P., Chen, X., Shen, S.: Stereo R-CNN based 3d object detection for autonomous driving. In: Proceedings of the IEEE/CVF Conference on Computer Vision and Pattern Recognition, pp. 7644–7652 (2019)
4. Weng, X., Wang, J., Held, D., Kitani, K.: Ab3dmot: a baseline for 3d multi-object tracking and new evaluation metrics. arXiv preprint arXiv:2008.08063 (2020)
5. Pang, Z., Li, Z., Wang, N.: SimpleTrack: understanding and rethinking 3D multi-object tracking. In: Karlinsky, L., Michaeli, T., Nishino, K. (eds.) Computer - ECCV 2022 Workshops. ECCV 2022, Part I, LNCS, vol. 13801, pp. 680–696. Springer, Cham (2023). https://doi.org/10.1007/978-3-031-25056-9_43
6. Benbarka, N., Schröder, J., Zell, A.: Score refinement for confidence-based 3d multi-object tracking. In: 2021 IEEE/RSJ International Conference on Intelligent Robots and Systems (IROS), pp. 8083–8090. IEEE (2021)

7. Wu, H., Han, W., Wen, C., Li, X., Wang, C.: 3d multi-object tracking in point clouds based on prediction confidence-guided data association. IEEE Trans. Intell. Transp. Syst. **23**(6), 5668–5677 (2021)
8. Wang, X., He, J., Fu, C., Meng, T., Huang, M.: You only need two detectors to achieve multi-modal 3d multi-object tracking. arXiv preprint arXiv:2304.08709 (2023)
9. Huang, K., Hao, Q.: Joint multi-object detection and tracking with camera-lidar fusion for autonomous driving. In: 2021 IEEE/RSJ International Conference on Intelligent Robots and Systems (IROS), pp. 6983–6989. IEEE (2021)
10. Wang, X., Fu, C., He, J., Wang, S., Wang, J.: Strongfusionmot: a multi-object tracking method based on lidar-camera fusion. IEEE Sens. J. **23**, 11241–11252 (2022)
11. Zhang, K., Liu, Y., Mei, F., Jin, J., Wang, Y.: Boost correlation features with 3D-MiIoU-based camera-LiDAR fusion for MODT in autonomous driving. Remote Sens. **15**(4), 874 (2023)
12. Baser, E., Balasubramanian, V., Bhattacharyya, P., Czarnecki, K.: Fantrack: 3d multi-object tracking with feature association network. In: 2019 IEEE Intelligent Vehicles Symposium (IV), pp. 1426–1433. IEEE (2019)
13. Marinello, N., Proesmans, M., Van Gool, L.: Triplettrack: 3d object tracking using triplet embeddings and LSTM. In: Proceedings of the IEEE/CVF Conference on Computer Vision and Pattern Recognition, pp. 4500–4510 (2022)
14. Weng, X., Wang, Y., Man, Y., Kitani, K.M.: Gnn3dmot: Graph neural network for 3d multi-object tracking with 2d–3d multi-feature learning. In: Proceedings of the IEEE/CVF Conference on Computer Vision and Pattern Recognition, pp. 6499–6508 (2020)
15. Kim, A., Ošep, A., Leal-Taixé, L.: Eagermot: 3d multi-object tracking via sensor fusion. In: 2021 IEEE International Conference on Robotics and Automation (ICRA), pp. 11315–11321. IEEE (2021)
16. Wang, X., Fu, C., Li, Z., Lai, Y., He, J.: DeepFusionMOT: a 3d multi-object tracking framework based on camera-lidar fusion with deep association. IEEE Robot. Autom. Lett. **7**(3), 8260–8267 (2022)
17. Kuma, R., Weill, E., Aghdasi, F., Sriram, P.: Vehicle re-identification: an efficient baseline using triplet embedding. In: 2019 International Joint Conference on Neural Networks (IJCNN), pp. 1–9. IEEE (2019)
18. Hao, Y., Wang, N., Li, J., Gao, X.: HSmE: hypersphere manifold embedding for visible thermal person re-identification. In: Proceedings of the AAAI Conference on Artificial Intelligence, vol. 33, pp. 8385–8392 (2019)
19. Li, Y.J., Chen, Y.C., Lin, Y.Y., Du, X., Wang, Y.C.F.: Recover and identify: a generative dual model for cross-resolution person re-identification. In: Proceedings of the IEEE/CVF International Conference on Computer Vision, pp. 8090–8099 (2019)
20. Li, M., Zhu, X., Gong, S.: Unsupervised tracklet person re-identification. IEEE Trans. Pattern Anal. Mach. Intell. **42**(7), 1770–1782 (2019)
21. Luo, H., Gu, Y., Liao, X., Lai, S., Jiang, W.: Bag of tricks and a strong baseline for deep person re-identification. In: Proceedings of the IEEE/CVF Conference on Computer Vision and Pattern Recognition Workshops (2019)
22. Ye, M., Shen, J., Lin, G., Xiang, T., Shao, L., Hoi, S.C.: Deep learning for person re-identification: a survey and outlook. IEEE Trans. Pattern Anal. Mach. Intell. **44**(6), 2872–2893 (2021)
23. He, L., Liao, X., Liu, W., Liu, X., Cheng, P., Mei, T.: Fastreid: a pytorch toolbox for general instance re-identification. arXiv preprint arXiv:2006.02631 (2020)

24. Wu, A., Zheng, W.S., Lai, J.H.: Robust depth-based person re-identification. IEEE Trans. Image Process. **26**(6), 2588–2603 (2017)
25. Karianakis, N., Liu, Z., Chen, Y., Soatto, S.: Person depth REID: robust person re-identification with commodity depth sensors. arXiv preprint arXiv:1705.09882 (2017)
26. Geiger, A., Lenz, P., Urtasun, R.: Are we ready for autonomous driving? the kitti vision benchmark suite. In: 2012 IEEE Conference on Computer Vision and Pattern Recognition, pp. 3354–3361. IEEE (2012)
27. Howard, A., et al.: Searching for mobilenetv3. In: Proceedings of the IEEE/CVF International Conference on Computer Vision, pp. 1314–1324 (2019)
28. Xu, G., Wang, Y., Cheng, J., Tang, J., Yang, X.: Accurate and efficient stereo matching via attention concatenation volume. arXiv preprint arXiv:2209.12699 (2022)
29. Qi, C.R., Su, H., Mo, K., Guibas, L.J.: Pointnet: deep learning on point sets for 3d classification and segmentation. In: Proceedings of the IEEE Conference on Computer Vision and Pattern Recognition, pp. 652–660 (2017)
30. Szegedy, C., Vanhoucke, V., Ioffe, S., Shlens, J., Wojna, Z.: Rethinking the inception architecture for computer vision. In: Proceedings of the IEEE Conference on Computer Vision and Pattern Recognition, pp. 2818–2826 (2016)
31. Shenoi, A., et al.: JRMOT: a real-time 3d multi-object tracker and a new large-scale dataset. In: 2020 IEEE/RSJ International Conference on Intelligent Robots and Systems (IROS), pp. 10335–10342. IEEE (2020)
32. Luiten, J., Fischer, T., Leibe, B.: Track to reconstruct and reconstruct to track. IEEE Robot. Autom. Lett. **5**(2), 1803–1810 (2020)
33. Kim, A., Brasó, G., Ošep, A., Leal-Taixé, L. (2022). PolarMOT: how far can geometric relations take us in 3D multi-object tracking?. In: Avidan, S., Brostow, G., Cissé, M., Farinella, G.M., Hassner, T. (eds.) Computer Vision - ECCV 2022. ECCV 2022. LNCS, Part XXII, vol. 13682, pp. 41–58. Springer, Cham (2022). https://doi.org/10.1007/978-3-031-20047-2_3

Emphasizing Boundary-Positioning and Leveraging Multi-scale Feature Fusion for Camouflaged Object Detection

Songlin Li[1], Xiuhong Li[1(✉)], Zhe Li[2], Boyuan Li[1], Chenyu Zhou[1], Fan Chen[1], Tianchi Qiu[1], and Zeyu Li[3]

[1] School of Information Science and Engineering, Xinjiang University, Xinjiang, China
xjulxh@xju.edu.cn

[2] Department of Electrical and Electronic Engineering, The Hong Kong Polytechnic University, Hong Kong SAR, China

[3] School of Computer Science and Engineering, Dalian Minzu University, Dalian, China

Abstract. Camouflaged object detection (COD) aims to identify objects that blend in with their surroundings and have numerous practical applications. However, COD is a challenging task due to the high similarity between camouflaged objects and their surroundings. To address the problem of identifying camouflaged objects, we investigated how humans observe such objects. We found that humans typically first scan the entire image to obtain an approximate location of the target object. They then observe the differences between the boundary of the target object and its surrounding environment to refine their perception of the object. This continuous refinement process helps humans eventually identify the camouflaged object. Based on this observation, we propose a novel COD method that emphasizes boundary positioning and leverages multi-scale feature fusion. Our model includes two important modules: the Enhanced Feature Module (EFM) and the Boundary and Positioning joint-guided Feature Fusion Module (BPFM). The EFM provides multi-scalc information and obtains aggregated feature representations, resulting in more robust feature representations for the initial positioning of the camouflaged object. In BPFM, we mimic human observation of camouflaged objects by injecting boundary and positioning information into each level of the backbone features, working together to refine the target object in blurred regions progressively. We validated the effectiveness of our model on three benchmark datasets (COD10K, CAMO, CHAMELEON), and the results showed that our proposed method significantly outperforms existing COD models.

Keywords: Camouflaged object detection · boundary · positioning · multi-scale features · contextual information

Supported by the Xinjiang Natural Science Foundation (No. 2020D01C026).

© The Author(s), under exclusive license to Springer Nature Singapore Pte Ltd. 2024
Q. Liu et al. (Eds.): PRCV 2023, LNCS 14436, pp. 508–519, 2024.
https://doi.org/10.1007/978-981-99-8555-5_40

1 Introduction

Camouflage is a common protective mechanism that exists widely in the natural world. Animals or objects blend into their surroundings by utilizing textures, colors, or natural light, making it difficult for them to be detected and achieving the effect of invisibility, thereby evading predation from predators. Thanks to its ability to segment target objects and their highly similar backgrounds, COD is of great value in various fields such as medicine (e.g., polyp segmentation [9] and lung infection segmentation [10]), industry (e.g., equipment safety hazard detection), agriculture (e.g., pest detection), military (e.g., latent target detection), art (e.g., excavation of ancient wall paintings), and scientific research (e.g., the discovery and protection of rare species) and many other fields are of significant value. However, COD is a very challenging task due to the nature of camouflage, that is, the high intrinsic similarities between candidate objects and chaotic backgrounds, which make it difficult to spot camouflaged objects for humans.

Researchers have developed various COD models to address this challenge to improve detection performance. In the early years, several traditional COD methods [1,19] were proposed to segment camouflaged objects by using manually designed features. With the rapid development of deep learning techniques, applying deep learning to COD has become more widespread. Existing deep learning-based COD models have achieved remarkable results. The SINet [8] proposed by Fan *et al.* first uses a search module to locate the camouflaged target roughly and then uses a recognition module to finely segment it. The same research group later developed SINet-v2 [6], which has a better decoder and attention mechanism for detecting camouflaged targets. Although these models have improved camouflaged target detection from a local perspective, they still cannot obtain clear boundaries. In COD, the high similarity between camouflaged objects and their surroundings makes boundary information between the target object and the background particularly important. Therefore, the extraction of boundary information is still a key factor. Zhou *et al.* [29] designed the FAPNet, which adds boundary supervision to COD and uses boundary information to supplement details. Sun *et al.* [22] designed the BGNet, which iteratively aggregates features with boundary information for COD. Although these models have utilized boundary information in COD, they overlook the impact of positioning information on blurry regions.

To this end, we propose a novel Emphasizing Boundary-Positioning and Leveraging Multi-Scale Feature Fusion for COD. Our model imitates human habits of observing imperceptible objects. Firstly, the model scans the global environment to find the rough positioning of the camouflaged object. Then, it observes the differences between the boundary of the target object and its surroundings to refine the target. We propose an Enhanced Feature Module (EFM) to obtain multi-scale features, which uses a series of dilated convolutions to extract and aggregate multi-scale information. It not only enlarges the receptive field but also reduces the number of channels. The features extracted from the Res2Net-50 [11] are input into the EFM to produce stronger and more effective

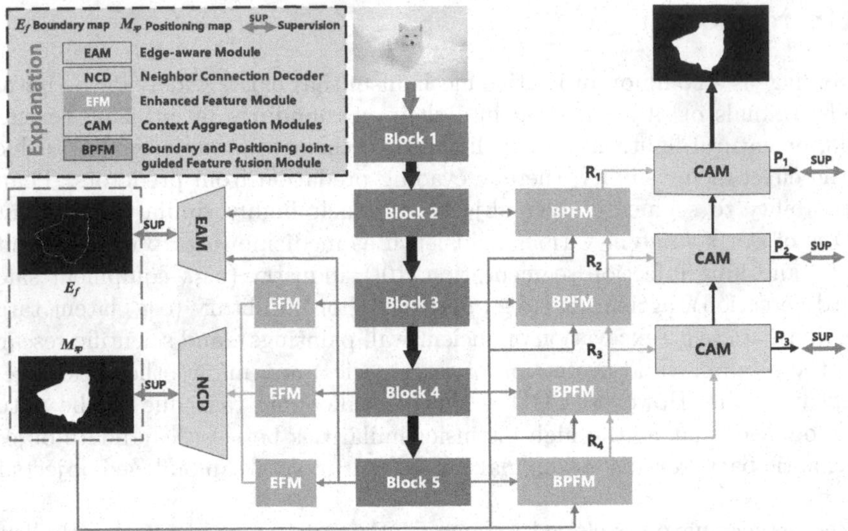

Fig. 1. This paper presents the overall architecture of our model, which utilizes Res2Net-50 as the backbone network and incorporates five key modules: the EFM, NCD [6], EAM [22], BPFM, and CAM [22]. The symbols represent the positioning map, and the symbols represent the boundary map.

feature information. The Boundary and Positioning joint-guided Feature fusion Module (BPFM) is designed to achieve joint-guided learning from the boundary and positioning maps, cross-scale fusion of multi-level features, and detailed boundary delineation between the target and background by combining global contextual information.

The main contributions of this paper are the following:

- We propose a joint-guided COD method using positioning and boundary maps. The method roughly locates the camouflaged target using the positioning map and then gradually refines the boundary using the boundary map. We evaluate the proposed model's performance on three benchmark datasets. Compared with 18 typical deep learning-based models, our model achieves state-of-the-art performance.
- We propose an Enhanced Feature Module (EFM) that utilizes convolution kernels of different sizes and dilation rates, as well as mutual fusion between upper and lower branches, to obtain multi-scale aggregated features.
- We propose a Boundary and Positioning joint-guided Feature fusion Module (BPFM) that can inject semantic information from positioning maps and edge information from boundary maps into the feature representation of the backbone network, fuse multi-level features across scales and refine camouflaged objects by combining rich contextual information.

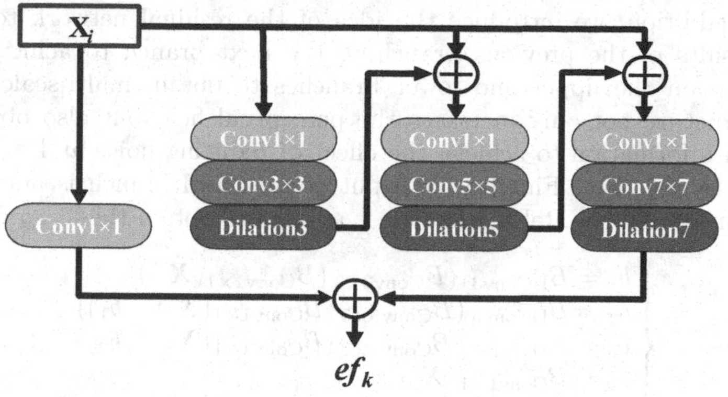

Fig. 2. The detailed architecture of the proposed Enhanced Feature Module (EFM).

2 Our Method

Figure 1 shows the overall architecture of the proposed our model, which consists of two key components: Enhanced Feature Module (EFM) and Boundary and Positioning joint-guided Feature fusion Module (BPFM). Specifically, Res2Net-50 is used as the backbone to extract multi-level features denoted as $\boldsymbol{X}_i(\mathrm{i} = 1, \ldots, 5)$. According to [28], low-level features in shallower convolutional layers contain spatial information that can be used to construct object edges. In comparison, deeper convolutional layers preserve more semantic information related to object positioning. Therefore, the backbone features $\boldsymbol{X}_3$, $\boldsymbol{X}_4$, and $\boldsymbol{X}_5$ are input into the proposed EFM to obtain multi-scale features. The lightweight and robust positioning map M_{sp} is generated by the Neighbor Connection Decoder (NCD) [6] to locate semantic positions. Since $\boldsymbol{X}_1$ contains more noise and has a small receptive field that can affect the determination of the target boundary, we discard the first-level features and use $\boldsymbol{X}_2$ and $\boldsymbol{X}_5$ with the Edge-Aware Module (EAM) [22] to generate a boundary map E_f with edge spatial information. BPFM integrates the positioning and boundary maps with the backbone network features to guide feature learning and enhance positioning and boundary representation. Finally, CAM progressively aggregates multi-level fusion features and performs multi-level supervision training.

2.1 Enhanced Feature Module

The Enhanced Feature Module (EFM) is designed with inspiration from the vacant space convolutional pooling pyramid (ASPP) [18]. As shown in Fig. 2, the EFM has four branches ($b_k, i \in \{1, 2, 3, 4\}$). In each branch, the number of channels is set to 64 using a 1×1 convolutional layer implementation. The first three branches use convolutional kernels with different step sizes and fill rates ($s = 2k + 1$, s denotes the size of the convolutional kernel, and k denotes the branch), after which an inflated convolution with an expansion rate of $(2k + 1)$

is set. In addition, we introduce the idea of the residual network to add the output results of the previous branch to the next branch to achieve mutual fusion between the upper and lower branches to obtain multi-scale features. Such a structure not only increases the perceptual field but also obtains rich contextual information to achieve the effect of reducing noise and highlighting the target positioning. Finally, the input of the third branch is added to the fourth branch, and the ReLU function is applied to obtain the feature ef_k:

$$\begin{cases} b_1 = B_{\mathrm{DConv3}}\left(B_{\mathrm{Conv3\times3}}\left(B_{\mathrm{Conv1\times1}}(\boldsymbol{X}_i)\right)\right) \\ b_2 = B_{\mathrm{DConv5}}\left(B_{\mathrm{Conv5\times5}}\left(B_{\mathrm{Conv1\times1}}(\boldsymbol{X}_i) \oplus b_1\right)\right) \\ b_3 = B_{\mathrm{DConv7}}\left(B_{\mathrm{Conv7\times7}}\left(B_{\mathrm{Conv1\times1}}(\boldsymbol{X}_i) \oplus b_2\right)\right) \\ b_4 = B_{\mathrm{Conv1\times1}}(\boldsymbol{X}_i) \\ ef_k = ReLU\left(b_4 \oplus b_3\right) \end{cases} \tag{1}$$

where $\oplus$ denotes element-wise addition. $B_{\mathrm{Convk\times k}}$ denotes a convolutional kernel of size k, stride of 1, padding of $\frac{k-1}{2}$, and finally batch normalized. B_{DConvd} denotes a convolutional kernel of size 3, stride of 1, padding of d, dilation rate of d, and finally batch normalized. $ReLU(\cdot)$ denotes the output of the Rectified Linear Unit (ReLU) activation function.

2.2 Boundary and Positioning Joint-Guided Feature Fusion Module

To harness boundary information and positional information for detecting camouflaged objects, we propose a novel method that mimics the human visual system, called the Boundary and Position Jointly Guided Feature Fusion Module (BPFM). Specifically, the semantic information of the localization map is infused into the backbone features to enhance the localization expression, emulating human eye positioning. Subsequently, edge information from the boundary map is injected into the backbone features to boost boundary expression, imitating the human process of contrasting camouflaged objects against their background to supplement edge details. Research has shown [28] that different convolutional layers retain various levels of image information. Low-level features are crucial for boundary detection, while high-level features are vital for positional detection. Therefore, the diversity of information scales is essential for detecting camouflaged targets. To ensure a robust feature representation, BPFM implements a cross-scale fusion of multi-level features.

As shown in Fig. 3, the backbone network feature $\boldsymbol{X}_i(\mathrm{i} = 2, \dots, 5)$ and the boundary map (E_f) are given as inputs. $\boldsymbol{X}_i$ and E_f are element-wise multiplied, and the result of their multiplication is added to $\boldsymbol{X}_i$ element-wise. This is followed by 3×3 convolution to obtain the initial fusion feature f_m, which can be denoted as:

$$f_m = \mathrm{Conv}_{3\times3}\left(\boldsymbol{X}_i \otimes D\left(E_f\right) \oplus \boldsymbol{X}_i\right) \tag{2}$$

where D denotes down-sampling and $\mathrm{Conv}_{3\times3}$ is 3×3 convolution. $\otimes$ is element-wise multiplication. To enhance feature representation, inspired by [23], we introduce local attention to explore critical feature channels. Specifically, we aggregate the convolution features (f_m) using a channel-wise global average pooling

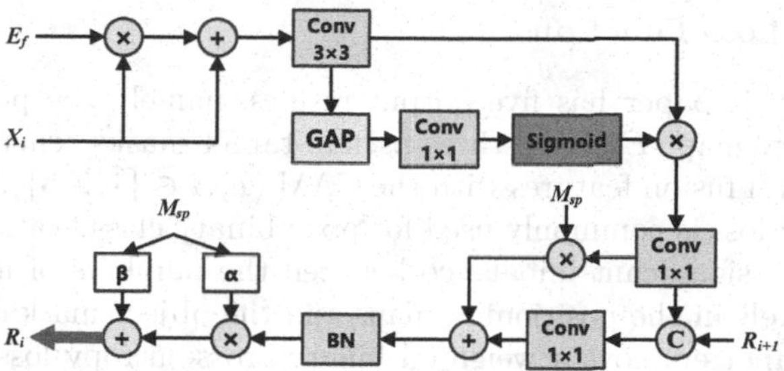

Fig. 3. The detailed architecture of the proposed Boundary and Positioning joint-guided Feature fusion Module (BPFM).

(GAP). Then, we obtain the corresponding channel attention (weight) by the 1D convolutions followed by a Sigmoid function. After that, we multiply the channel attention with the input feature f_m and reduce the channels by 1×1 convolution layer to obtain the final output f_{ei}, *i.e.*,

$$f_{ei} = \mathrm{Conv}_{1\times1}\left(Sig\left(\mathbf{F}_{1D}^{k}\left(GAP\left(f_m\right)\right)\right) \otimes f_m\right) \tag{3}$$

where $\mathbf{F}_{1D}^{k}$ is $1D$ convolution with kernel size k and $Sig(\cdot)$ denotes Sigmoid function. The kernel size k can be set adaptively as $k = \left|(1 + \log_2(C))/2\right|_{odd}$, where $|*|_{odd}$ denotes the nearest odd number and C is the channels of f_{ei}. The kernel size is proportional to the channel dimension.

Next, the feature map f_{ei} is multiplied element-wise with the positioning map M_{sp}. The resulting product is then processed by a 1×1 convolution to fuse the positioning-related information into the feature map of the current stage. This fused output is referred to as E_{mi} and is jointly guided by the boundary and positioning maps. To obtain multi-scale features, the feature map f_{ei} from the i-th stage is concatenated with the output of the subsequent stage's BPFM (denoted as R_{i+1}). This achieves feature fusion between adjacent stages. The concatenated feature map is further processed by a 3×3 convolution, and the resulting output is element-wise added to E_{mi}. The feature map E_{fmi} is then obtained by normalizing E_{mi} using batch normalization. Finally, this paper assigns weights α and β to the foreground and background regions of M_{sp}. This is done to enhance the positioning of the camouflaged target and suppress the interference from the background for detection. The resulting feature map is denoted as R_i, which can be defined as:

$$\begin{cases} E_{mi} = \mathrm{Conv}_{1\times1}\left(D\left(M_{sp}\right) \otimes f_{ei}\right) \\ E_{fmi} = BN\left(Concat\left(f_{ei}, U\left(R_{i+1}\right)\right) \oplus E_{mi}\right) \\ R_i = E_{fmi} \otimes (1 \oplus \alpha)M_{sp} \oplus \beta M_{sp} \end{cases} \tag{4}$$

U denotes upsampling, $Concate(\cdot)$ denotes concatenation operation, BN denotes batch normalization.

2.3 Overall Loss Function

The model in this paper has five output results, namely the positioning map (M_{sp}), boundary map (E_f), and three pseudo-target masks generated by aggregating multi-level fusion features with the CAM ($P_i, i \in \{1, 2, 3\}$). Binary Cross-Entropy (BCE) loss is commonly used for pixel binary classification tasks. However, due to the significant imbalance between the numbers of foreground and background pixels in the positioning maps and the object mask of camouflaged objects, this paper employs a weighted binary cross-entropy loss function [24]. To better constrain the global optimization, a weighted IOU loss function is also utilized. Thus, the loss function for the positioning maps and the object mask of camouflaged objects is defined as follows:

$$L_{em} = L_w BCE + L_w IOU \tag{5}$$

Regarding the boundary maps, this paper utilizes the dice loss (L_{dice}) as described in [25]. Thus, the overall loss function is defined as follows:

$$L_{total} = \sum_{i=1}^{3} L_{em}(P_i, G) + L_{em}(M_{sp}, G) + \lambda L_{dice}(E_f, G_e) \tag{6}$$

where G denotes the ground truth of the camouflaged object, G_e denotes the ground truth of the camouflaged object boundary and set $\lambda = 3$.

3 Experiments and Results

3.1 Implementation Details

In this paper, the model is implemented using PyTorch, with Res2Net-50 [11] pre-trained on ImageNet serving as the backbone. During the training phase, all input images were resized to 416 × 416, and data augmentation was performed by randomly flipping the images horizontally. The model was trained using a batch size 26 and the Adam optimizer [12]. The initial learning rate was set to 1e−4, with a decay period of 40 epochs and a decay rate of 0.1. With the help of an NVIDIA 3090 GPU, the model converged after 50 rounds of training and took approximately 3 h to complete.

3.2 Datasets and Evaluation Metrics

We evaluate our method on three public benchmark datasets: COD10K [20], CAMO [13] and CHAMELEON [8]. Here, the training and test sets are the same as [8]. We utilize four widely used metrics to evaluate our method, *i.e.*, mean absolute error (MAE, $\mathcal{M}$), weighted F-measure (F_β^ω) [15], structure measure (S_α) [3] and mean E-measure (E_ϕ) [5].

Table 1. Four evaluation metrics are employed in this study, namely S_α [4], F_β^ω [15], E_ϕ [7], and $\mathcal{M}$ [17]. The symbols "↑" and "↓" indicate that larger and smaller values are better, respectively. The best results are highlighted in bold.

Method	COD10K-Test				CAMO-Test				CHAMELEON-Test			
	$S_\alpha\uparrow$	$E_\phi\uparrow$	$F_\beta^\omega\uparrow$	$\mathcal{M}\downarrow$	$S_\alpha\uparrow$	$E_\phi\uparrow$	$F_\beta^\omega\uparrow$	$\mathcal{M}\downarrow$	$S_\alpha\uparrow$	$E_\phi\uparrow$	$F_\beta^\omega\uparrow$	$\mathcal{M}\downarrow$
C2FNet21	0.813	0.890	0.686	0.036	0.796	0.864	0.719	0.080	0.888	0.935	0.828	0.032
PFNet	0.800	0.877	0.660	0.040	0.782	0.842	0.695	0.085	0.882	0.931	0.810	0.033
R-MGL	0.814	0.852	0.666	0.035	0.775	0.847	0.673	0.088	0.893	0.923	0.813	0.030
LSR	0.804	0.880	0.673	0.037	0.787	0.854	0.696	0.080	0.893	0.938	0.839	0.033
SINet-v2	0.816	0.888	0.679	0.037	0.818	0.871	0.738	0.071	0.889	0.938	0.815	0.032
C2FNet22	0.808	0.882	0.683	0.037	0.764	0.825	0.677	0.090	0.893	0.950	0.841	0.027
FAPNet	0.820	0.887	0.693	0.036	0.805	0.857	0.724	0.080	0.889	0.938	0.821	0.030
BSANet	0.817	0.887	0.696	0.035	0.804	0.860	0.728	0.079	0.888	0.936	0.826	0.032
PreyNet	0.813	0.881	0.697	0.034	0.790	0.842	0.709	0.077	0.895	0.952	0.844	0.028
BGNet	0.831	**0.901**	0.722	0.033	0.816	0.871	0.751	0.069	0.901	0.944	0.852	0.026
Ours	**0.838**	**0.901**	**0.733**	**0.030**	**0.822**	**0.872**	**0.760**	**0.068**	**0.909**	**0.958**	**0.861**	**0.024**

3.3 Comparison with State-of-the-Art Methods

We compared our proposed method with 18 state-of-the-art models, including C2FNet21 [21], PFNet [16], R-MGL [26], LSR [14], SINet-v2 [6], C2FNet22 [2], FAPNet [29], BSANet [30], PreyNet [27], and BGNet [22]. For C2FNet21, SINet-V2, C2FNet22, FAPNet, BSANet, and BGNet, we retrained these six models using the authors' released code. Previously published papers provided all other results. Additionally, we evaluated all predicted images using the same code.

1. **Quantitative Comparison:** Table 1 presents the quantitative results of different camouflaged object detection methods on three benchmark datasets. The proposed model in this paper outperforms the 18 compared models on each dataset. Besides, BGNet, FAPNet, and BSANet utilize auxiliary edge or boundary information and still fail to locate camouflaged objects, while our model can effectively locate them and achieve the best performance. This is because our model utilizes boundary and positioning information, which jointly operate on ambiguous regions, significantly improving the performance of COD. Specifically, compared with BGNet, on the COD10K dataset, the proposed model in this paper improved S_α by 0.84%, F_β^ω by 1.52%, and decreased $\mathcal{M}$ by 9.09%.
2. **Visual Comparison:** From the visual results in Fig. 4, our model outperforms the comparative models in the visual comparison results of the nine collected test samples. Specifically, in the first and second rows, it can be seen that our model can effectively handle size variations. In the third and fourth rows, our model can effectively handle changes in scene brightness. In the fifth row, the camouflaged objects have similar textures to the background, which poses a severe challenge for identifying them from similar backgrounds. In this case, our model performs better, accurately locating the camouflaged object. In the sixth row, the object has rich edge details, and our method still detects the camouflaged object accurately. Overall, the results demonstrate

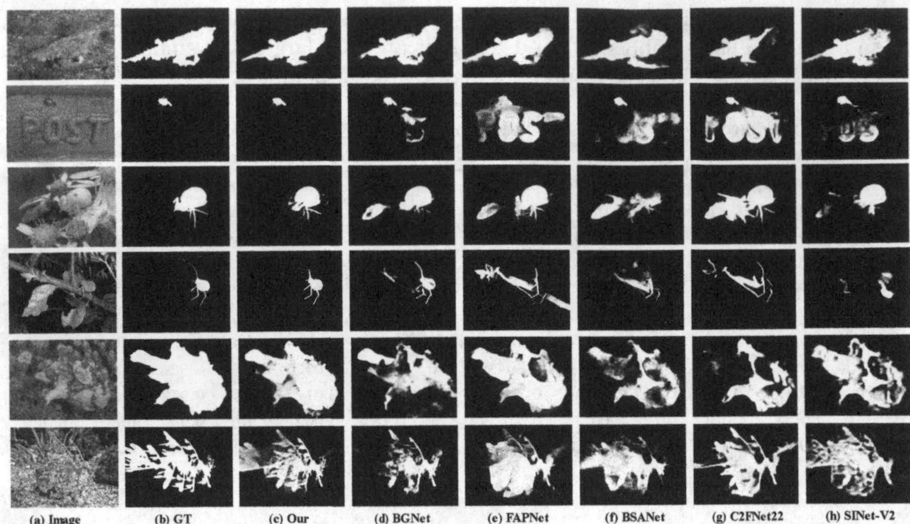

Fig. 4. This section provides a visual comparison between our proposed model and the latest models. Specifically, we show (a) the input image, (b) the ground truth (GT), (c) our method, and (d)–(h) the state-of-the-art models, including BGNet, FAPNet, BSANet, C2FNet22, and SINet-V2 The results demonstrate that our model achieves superior performance to the other compared methods in terms of visual quality.

that our model can perform well in detecting camouflaged objects under different challenging factors.

3.4 Ablation Study

This article conducted several ablation experiments to validate the effectiveness of each module, and the specific results are presented in Table 2. For the baseline model (B), all additional modules (such as EFM, BPFM) were removed, and only the 1×1 convolution in BPFM was retained to reduce the number of channels and along with the initial aggregation operation in the CAM [22] for feature aggregation. FS denotes the positioning map guidance part in the BPFM, BS denotes the boundary map guidance part in the BPFM, and R denotes the cross-level fusion part in the BPFM.

Effectiveness of EFM. According to Table 2, adding the EFM on top of (d) and (f) can improve the model's performance to some extent. The EFM can fuse multi-scale features to enhance the features extracted by the backbone network, resulting in more accurate positioning maps. However, the performance presented in experiment (f) was worse than that of other ablation experiments, mainly due to the blurred positioning map affecting boundary guidance. The negative correlation between the two factors in ambiguous regions further validates the effectiveness of the EFM designed in this article.

Table 2. Four evaluation metrics are employed in this study, namely S_α [4], F_β^ω [15], E_ϕ [7], and $\mathcal{M}$ [17]. The symbols "↑" and "↓" indicate that larger and smaller values are better, respectively. The best results are highlighted in bold.

Method	COD10K-Test				CAMO-Test				CHAMELEON-Test			
	$S_\alpha\uparrow$	$E_\phi\uparrow$	$F_\beta^\omega\uparrow$	$\mathcal{M}\downarrow$	$S_\alpha\uparrow$	$E_\phi\uparrow$	$F_\beta^\omega\uparrow$	$\mathcal{M}\downarrow$	$S_\alpha\uparrow$	$E_\phi\uparrow$	$F_\beta^\omega\uparrow$	$\mathcal{M}\downarrow$
(a) B	0.819	0.887	0.681	0.037	0.814	0.863	0.732	0.075	0.880	0.929	0.802	0.035
(b) B+R	0.834	0.895	0.719	0.032	**0.825**	0.872	0.755	0.073	0.889	0.929	0.825	0.033
(c) B+R+BS	0.837	0.898	0.727	0.031	0.822	0.870	0.755	0.073	0.904	0.938	0.850	0.030
(d) B+R+FS	0.833	0.895	0.717	0.034	0.818	0.863	0.743	0.072	0.896	0.942	0.833	0.031
(e) B+R+FS+EF	0.838	0.900	0.725	0.032	0.821	0.871	0.750	0.072	0.900	0.939	0.842	0.026
(f) B+R+FS+BS	0.828	0.889	0.706	0.035	0.824	**0.872**	0.748	0.072	0.889	0.889	0.822	0.035
(g) Ours	**0.838**	**0.901**	**0.733**	**0.030**	**0.822**	**0.872**	**0.760**	**0.068**	**0.909**	**0.958**	**0.861**	**0.024**

Effectiveness of BPFM. Adding the cross-level fusion part of the BPFM to (a), resulting in (b), leads to a certain degree of decline in the detection capability for camouflaged targets. This is because the indiscriminate fusion of features from different levels leads to noise accumulation, which hinders the detection of camouflaged targets. This also demonstrates the importance of guiding input features. Adding the boundary guidance part of the BPFM to (b), resulting in (c), improves the detection capability. For example, on the COD10K dataset, S_α improved by 0.48%, E_ϕ by 0.34%, F_β^ω by 1.54%, and $\mathcal{M}$ decreased by 6.06%. The effectiveness of the EFM has been demonstrated. To eliminate the adverse effects of rough initial positioning maps on the positioning map guidance part of the BPFM, finely enhanced initial positioning maps from the EFM were directly introduced for experiments to demonstrate the effectiveness of the positioning map guidance part of the BPFM, resulting in (e), which showed significant improvements in all metrics. This demonstrates that the boundary map guidance of the BPFM enhances the details of the detection results, the positioning map guidance enhances the local features of the detection results, and the cross-level fusion part provides multi-scale feature information for boundary and positioning map guidance, all of which prove the effectiveness of the BPFM.

4 Conclusion

We propose a novel approach for Camouflaged Object Detection called Emphasizing Boundary-Positioning and Leveraging Multi-Scale Feature Fusion. This approach is developed based on human observation habits of disguised objects. Our model utilizes the Enhanced Feature Module (EFM) and the Boundary and Positioning joint-guided Feature fusion Module (BPFM) to explore the edge and semantic information of the target and iteratively refines the object structure and boundary to obtain a complete and precise representation. Extensive experiments demonstrate that the proposed model outperforms state-of-the-art models on three benchmark datasets.

References

1. Bhajantri, N.U., Nagabhushan, P.: Camouflage defect identification: a novel approach. In: 9th International Conference on Information Technology (ICIT 2006), pp. 145–148. IEEE (2006)
2. Chen, G., Liu, S.J., Sun, Y.J., Ji, G.P., Wu, Y.F., Zhou, T.: Camouflaged object detection via context-aware cross-level fusion. IEEE Trans. Circuits Syst. Video Technol. **32**(10), 6981–6993 (2022)
3. Fan, D.P., Cheng, M.M., Liu, Y., Li, T., Borji, A.: Structure-measure: a new way to evaluate foreground maps. In: 2017 IEEE International Conference on Computer Vision (ICCV), pp. 4558–4567 (2017). https://doi.org/10.1109/ICCV.2017.487
4. Fan, D.P., Cheng, M.M., Liu, Y., Li, T., Borji, A.: Structure-measure: a new way to evaluate foreground maps. In: Proceedings of the IEEE International Conference on Computer Vision, pp. 4548–4557 (2017)
5. Fan, D.P., Gong, C., Cao, Y., Ren, B., Cheng, M.M., Borji, A.: Enhanced-alignment measure for binary foreground map evaluation. arXiv preprint arXiv:1805.10421 (2018)
6. Fan, D.P., Ji, G.P., Cheng, M.M., Shao, L.: Concealed object detection. IEEE Trans. Pattern Anal. Mach. Intell. **44**(10), 6024–6042 (2022)
7. Fan, D.P., Ji, G.P., Qin, X., Cheng, M.M.: Cognitive vision inspired object segmentation metric and loss function. Scientia Sinica Informationis **6**(6) (2021)
8. Fan, D.P., Ji, G.P., Sun, G., Cheng, M.M., Shen, J., Shao, L.: Camouflaged object detection. In: Proceedings of the IEEE/CVF Conference on Computer Vision and Pattern Recognition, pp. 2777–2787 (2020)
9. Fan, D.P., et al.: PraNet: parallel reverse attention network for polyp segmentation. In: Martel, A.L., et al. (eds.) Medical Image Computing and Computer Assisted Intervention-MICCAI 2020: 23rd International Conference, Lima, Peru, 4–8 October 2020, Proceedings, Part VI 23, pp. 263–273. Springer, Cham (2020). https://doi.org/10.1007/978-3-030-59725-2_26
10. Fan, D.P., et al.: Inf-Net: automatic COVID-19 lung infection segmentation from CT images. IEEE Trans. Med. Imaging **39**(8), 2626–2637 (2020)
11. Gao, S.H., Cheng, M.M., Zhao, K., Zhang, X.Y., Yang, M.H., Torr, P.: Res2Net: a new multi-scale backbone architecture. IEEE Trans. Pattern Anal. Mach. Intell. **43**(2), 652–662 (2019)
12. Kingma, D.P., Ba, J.: Adam: a method for stochastic optimization. arXiv preprint arXiv:1412.6980 (2014)
13. Le, T.N., Nguyen, T.V., Nie, Z., Tran, M.T., Sugimoto, A.: Anabranch network for camouflaged object segmentation. Comput. Vis. Image Underst. **184**, 45–56 (2019)
14. Lv, Y., et al.: Simultaneously localize, segment and rank the camouflaged objects. In: Proceedings of the IEEE/CVF Conference on Computer Vision and Pattern Recognition, pp. 11591–11601 (2021)
15. Margolin, R., Zelnik-Manor, L., Tal, A.: How to evaluate foreground maps? In: Proceedings of the IEEE Conference on Computer Vision and Pattern Recognition, pp. 248–255 (2014)
16. Mei, H., Ji, G.P., Wei, Z., Yang, X., Wei, X., Fan, D.P.: Camouflaged object segmentation with distraction mining. In: Proceedings of the IEEE/CVF Conference on Computer Vision and Pattern Recognition, pp. 8772–8781 (2021)
17. Perazzi, F., Krähenbühl, P., Pritch, Y., Hornung, A.: Saliency filters: contrast based filtering for salient region detection. In: 2012 IEEE Conference on Computer Vision and Pattern Recognition, pp. 733–740. IEEE (2012)

18. Samuels-Lev, Y., et al.: ASPP proteins specifically stimulate the apoptotic function of P53. Mol. Cell **8**(4), 781–794 (2001)
19. Singh, S.K., Dhawale, C.A., Misra, S.: Survey of object detection methods in camouflaged image. IERI Procedia **4**, 351–357 (2013). https://www.sciencedirect.com/science/article/pii/S2212667813000531, 2013 International Conference on Electronic Engineering and Computer Science (EECS 2013), https://doi.org/10.1016/j.ieri.2013.11.050
20. Skurowski, P., Abdulameer, H., Błaszczyk, J., Depta, T., Kornacki, A., Kozieł, P.: Animal camouflage analysis: Chameleon database. Unpublished Manuscript **2**(6), 7 (2018)
21. Sun, Y., Chen, G., Zhou, T., Zhang, Y., Liu, N.: Context-aware cross-level fusion network for camouflaged object detection. arXiv preprint arXiv:2105.12555 (2021)
22. Sun, Y., Wang, S., Chen, C., Xiang, T.Z.: Boundary-guided camouflaged object detection. In: IJCAI, pp. 1335–1341 (2022)
23. Wang, Q., Wu, B., Zhu, P., Li, P., Zuo, W., Hu, Q.: ECA-Net: efficient channel attention for deep convolutional neural networks. In: Proceedings of the IEEE/CVF Conference on Computer Vision and Pattern Recognition, pp. 11534–11542 (2020)
24. Wei, J., Wang, S., Huang, Q.: F^3Net: fusion, feedback and focus for salient object detection. In: Proceedings of the AAAI Conference on Artificial Intelligence, vol. 34, pp. 12321–12328 (2020)
25. Xie, E., Wang, W., Wang, W., Ding, M., Shen, C., Luo, P.: Segmenting transparent objects in the wild. In: Vedaldi, A., Bischof, H., Brox, T., Frahm, J.M. (eds.) Computer Vision-ECCV 2020: 16th European Conference, Glasgow, UK, 23–28 August 2020, Proceedings, Part XIII 16. LNCS, vol. 12358, pp. 696–711. Springer, Cham (2020). https://doi.org/10.1007/978-3-030-58601-0_41
26. Zhai, Q., Li, X., Yang, F., Chen, C., Cheng, H., Fan, D.P.: Mutual graph learning for camouflaged object detection. In: Proceedings of the IEEE/CVF Conference on Computer Vision and Pattern Recognition, pp. 12997–13007 (2021)
27. Zhang, M., Xu, S., Piao, Y., Shi, D., Lin, S., Lu, H.: PreyNet: preying on camouflaged objects. In: Proceedings of the 30th ACM International Conference on Multimedia, pp. 5323–5332 (2022)
28. Zhao, T., Wu, X.: Pyramid feature attention network for saliency detection. In: Proceedings of the IEEE/CVF Conference on Computer Vision and Pattern Recognition, pp. 3085–3094 (2019)
29. Zhou, T., Zhou, Y., Gong, C., Yang, J., Zhang, Y.: Feature aggregation and propagation network for camouflaged object detection. IEEE Trans. Image Process. **31**, 7036–7047 (2022)
30. Zhu, H., et al.: I can find you! Boundary-guided separated attention network for camouflaged object detection. In: Proceedings of the AAAI Conference on Artificial Intelligence, vol. 36, pp. 3608–3616 (2022)

Author Index

© The Editor(s) (if applicable) and The Author(s), under exclusive license
to Springer Nature Singapore Pte Ltd. 2024

Q. Liu et al. (Eds.): PRCV 2023, LNCS 14436, pp. 521–523, 2024.
https://doi.org/10.1007/978-981-99-8555-5

Z

Printed in the United States
by Baker & Taylor Publisher Services